Instructor's Resource Manual
with
Chapter Tests

Instructor's Resource Manual with Chapter Tests

to accompany
College Algebra and Trigonometry, 2/e
College Algebra, 2/e
College Trigonometry, 2/e
Precalculus, 2/e

Richard N. Aufmann
Vernon C. Barker
Richard D. Nation
Palomar College

HOUGHTON MIFFLIN COMPANY BOSTON TORONTO
Geneva, Illinois Palo Alto Princeton, New Jersey

Sponsoring Editor: Maureen O'Connor
Senior Development Editor: Anthony Palermino
Senior Project Editor: Toni Haluga
Editorial Assistant: Tracy Theriault
Marketing Manager: Michael Ginley

Printed in the U.S.A.

ISBN: 0-395-63820-8

123456789-VG-97 96 95 94 93

Table of Contents

PREFACE

The Instructor's Resource Manual for the Aufmann/Barker/Nation Precalculus series contains answers to the Essays and Projects questions, Tips for Using Graphing Calculators, and Chapter Tests.

The answers to the Essay and Projects questions contain the main points that a correct essay should have in their logical order. Occasionally, references are given and examples are supplied. These answers should not be construed to be a complete essay or the only proper answer to the question.

The Tips for Using Graphing Calculators present elementary information for learning the basic functions of the TI-81 and Casio fx-7700G graphing calculators. Some functions include setting the range variables, using the trace key, using the zoom function, creating statistical graphs and operating on matrices. These tips may be used to introduce students to graphing calculators and as starting points for more extended explorations of the power of graphing caculators.

The Chapter Tests contain 3 multiple-choice and 3 free-response tests per chapter, which are designed as 1 hour tests. Users of the *College Algebra*, *College Trigonometry* and *Precalculus* will find chapter tests in separate sections. For chapters that are shared by more than one book, users should refer to the following correlation chart. Answers are provided at the end of the book.

Correlation Chart for College Algebra

College Algebra	Printed Test
Chapter 1	Chapter 1
Chapter 2	Chapter 2
Chapter 3	Chapter 3
Chapter 4	Chapter 4
Chapter 5	Chapter 5
Chapter 6	Chapter 9 (excluding 9.4)
Chapter 7	Chapter 10
Chapter 8	Chapter 11
Chapter 9	Chapter 12

Correlation Chart for College Trigonometry

College Trigonometry	Printed Test
Chapter 2	Chapter 6
Chapter 3	Chapter 7
Chapter 7	Chapter 5

Correlation Chart for Precalculus

Precalculus	Printed Test
Chapter 2	Chapter 3
Chapter 3	Chapter 4
Chapter 4	Chapter 5
Chapter 6	Chapter 7
Chapter 7	Chapter 8
Chapter 9	Chapter 10
Chapter 10	Chapter 11
Chapter 11	Chapter 12

Outlines to Essays and Projects

Chapter 1 Essays and Projects, Page 65

1. Let k and j be integers. Now consider the two odd numbers

$$2k - 1, \text{ and } 2j - 1.$$

The product of the two odd numbers is given by:

$$\begin{aligned}(2k-1)(2j-1) &= 4kj - 2k - 2j + 1 \\ &= 4kj - 2k - 2j + 2 + 1 - 2 \\ &= 2(2kj - k - j + 1) - 1\end{aligned}$$

Let m equal the integer $2kj - k - j + 1$. Then the above product can be written as $2m - 1$. This number is an odd number because it satisfies the definition that it can be written in the form of two times an integer minus one.

2. Let a, b, and c be odd integers and assume that $ax^2 + bx + c$ is factorable over the integers. That is

$$ax^2 + bx + c = (dx + e)(fx + g)$$

with d, e, f, and g are integer such that $df = a$, $eg = c$, and $(ef + dg) = b$. The integers d, e, f, and g must all be odd integers because the product of two integers is odd if and only if both of the integers are odd. Thus the products ef and dg are each odd numbers. Since the sum of any two odd numbers is an even number, we conclude that $(ef + dg) \neq b$.

Our assumption that $ax^2 + bx + c$ is factorable over the integers, has lead us to conclude that $(ef + dg) = b$, and $(ef + dg) \neq b$. This contradiction tells us that if a, b, and c are odd integers, then the trinomial $ax^2 + bx + c$ is *not* factorable over the integers.

3. The next number in the sequence is $2 \cdot 3 \cdot 5 \cdot 7 \cdot 11 + 1 = 2311$, which is a prime number. However, the number $2 \cdot 3 \cdot 5 \cdot 7 \cdot 11 \cdot 13 + 1 = 30{,}031$ is not a prime number, because it has factors of 59, and 509. Therefore the conjecture is false.

4. The *Fundamental Theorem of Arithmetic* states that each composite number can be expressed as a product of prime numbers in exactly one way (disregarding the order of the factors). By using the fundamental theorem of arithmetic and *indirect reasoning*, we can show that if p is a prime number, then $\sqrt{p}$ is not a rational number. The proof is as follows. Assume that there is a rational number $\frac{a}{b}$, such that $\sqrt{p} = \frac{a}{b}$.

Then we can show:

$$\begin{aligned}\sqrt{p} &= \frac{a}{b} \\ p &= \frac{a^2}{b^2} \\ p \cdot b^2 &= a^2\end{aligned}$$

By the fundamental theorem of arithmetic, the positive integers $p \cdot b^2$ and a^2 must have the same prime factorization. Since squares have prime factors that occur in pairs, we know that a^2 has an even number of prime factors. Using the same reasoning, we know that b^2 has an even number of prime factors. Since p is a prime number, $p \cdot b^2$ must have an odd number of prime factors. Now we have a contradiction, since by the fundamental theorem of arithmetic it is impossible for a number to have an odd number of prime factors and an even number of prime factors. Thus we conclude that $\sqrt{p}$ is not a rational number.

5. Any prime triplets other than 3, 5, and 7, must be of the form k, $k+2$, and $k+4$, where k is an integer larger than 3. However, one of the integers k, $k+2$, and $k+4$, must be a multiple of 3 larger than 3, and thus is not a prime number.

To verify that one of the numbers k, $k+2$, or, $k+4$ is a multiple of three consider the following.

Case 1. If k is a multiple of 3, then we are finished.

Case 2. If k is not a multiple of 3, then $\frac{k}{3}$ has a remainder of 1 or 2.

- If $\frac{k}{3}$ has a remainder of 1, then k is 1 larger than a multiple of 3, and therefore $k+2$ will be a multiple of 3.
- If $\frac{k}{3}$ has a remainder of 2, then k is 2 larger than a multiple of 3, and therefore $k+4$ will be a multiple of 3.

6. The number $5! + 2$ is not a prime number because it has a factor of 2.

$$5! + 2 = 5 \cdot 4 \cdot 3 \cdot 2 \cdot 1 + 2 = 2(5 \cdot 4 \cdot 3 \cdot 1 + 1)$$

In a similar manner we can show that:

$5!+3$ is not a prime number because it has a factor of 3.
$5!+4$ is not a prime number because it has a factor of 4.
$5!+5$ is not a prime number because it has a factor of 5.

Assume k is an integer larger than 1.

$k!+2$ is not a prime number because it has a factor of 2.
$k!+3$ is not a prime number because it has a factor of 3.
$k!+4$ is not a prime number because it has a factor of 4.
⋮
$k!+(k-1)$ is not a prime number because it has a factor of $(k-1)$.
$k!+k$ is not a prime number because it has a factor of k.

This procedure shows that there are very large groups of consecutive positive integers none of which is a prime number. For example, consider $k = 1{,}000{,}000{,}001$. Then $k! + 2$, $k! + 3$, ..., $k! + k$, is a sequence of one billion consecutive positive integers none of which is a prime number. This is remarkable, because in Exercise 69, on page 9 we established that there are an infinite amount of prime numbers. Thus even though there are very large regions where no prime numbers appear, we know that if you continue your search, you will discover another prime number, no matter where you start your search.

For more information on this topic, see Prime Desert n-Tuplets, by

George W. Polites, The American Mathematical Monthly, February 1988, page 98.

7. Prime numbers of the form $2^p - 1$, where p is a prime number, are called Mersenne primes, after the french mathematician Marin Mersenne (1588-1648). Mersenne asserted that $2^p - 1$ was a prime if and only if

$$p = 2, 3, 5, 7, 13, 17, 19, 31, 67, 127, \text{ or } 257.$$

It is now known that :

- $2^{67} - 1$, is not a prime number.
- $2^{257} - 1$, is not a prime number.
- $2^{61} - 1$, $2^{89} - 1$, and $2^{107} - 1$, are prime numbers.

Although Mersenne was not completely correct in his assertion, his statement is remarkable because of the size of the numbers involved. The number $2^{257} - 1$, for example has 78 digits.

As of April 1992, only 32 Mersenne prime numbers had been discovered. The largest of these is

$$2^{756839} - 1.$$

Mersenne primes can be used to generate perfect numbers.

Although Mersenne primes offer little direct benefit to society, they do have a curiosity value and the pursuit of ever larger primes has resulted in new techniques that are important in such areas as computer science and cryptography.

For additional information read:

The Little Book of Big Primes, by Paulo Ribenboim, Springer-Verlag, 1991.

Achieving Electronic Privacy, Scientific American, August 1992, page 96-101.

Modern Cryptology: A Tutorial., Gilles Brassard, *Lecture Notes in Computer Science*, Vol. 325, Springer-Verlag, 1988.

8. Georg Canter made use of the following definitions:

> Two sets A and B have the same number of elements, if it is possible establish a one-to-one correspondence between the elements of A and B.

> A set is *denumerable* or *countably infinite* if its elements can be put into a one-to-one correspondence with the set of natural numbers $\{1, 2, 3, 4, \ldots\}$.

Canter was able to show that the set of rational numbers is denumerable. He was also able to show that the set of all real number between 0 and 1, is not denumerable. Since the set of real numbers is the union of the set of rational numbers and the set of irrational numbers, he reasoned that it was the irrational numbers that caused the interval (0, 1) to be non-denumerable. For the details of Cantor's proofs, see Chapter 11, The Non-denumerability of the Continuum, of *Journey Through Genius*, by William Dunham, John Wiley & Sons Inc., 1990.

Chapter 2 Essays and Projects, Page118

1. The conjecture that $x^n + y^n = z^n$ is impossible for all integers $n > 2$, is know as "*Fermat's last theorem*". The actual statement of this conjecture was given by Fermat in 1637. Fermat wrote the conjecture as a paragraph in the margin of the text *The Arithmetic of Diophantus:*

> It is impossible to write a cube as the sum of two cubes, a fourth power as the sum of two fourth powers, and , in general any power beyond the second as the sum of two similar powers. For this I have discovered a truly wonderful proof but the margin of this book is too small to contain it.

Fermat's conjecture remains a conjecture to this day, although mathematicans have been able to prove that there are no positive integral values x, y, and z, that are solutions of $x^n + y^n = z^n$ for $2 < n < 25{,}000$.

Many mathematicians feel that Fermat conjecture is correct. However no mathematician to date has given a valid proof of the "theorem", and no mathematician has come up with a counter example to show that it is false.

If we remove the restriction that n must be larger than 2, and consider the equation $x^2 + y^2 = z^2$, then we can find an infinite number of solutions. For example, we know that $x = 3$, $y = 4$, $z = 5$ are solutions of this equation as are $3k$, $4k$, and $5k$ for any positive integer k. Since $x^2 + y^2 = z^2$ is in the form of the Pythagorean theorem, positive integer solutions of this equation are called Pythagorean triples, and we can generate sets of Pythagorean triples by using the method explained in Exercise100 page 130.

2. a. Let $y = 2x(1 - x)$. Let $x = 0.2$, then $y = 2(0.2)(1 - 0.2) = 0.32$. Now let $x = 0.32$, and compute a new y – value. Continue the process as shown in the following table

$x = 0.2$	$y = 0.32$
$x = 0.32$	$y = 0.4352$
$x = 0.4352$	$y = 0.49160192$
$x = 0.49160192$	$y = 0.4998589445$
$x = 0.4998589445$	$y = 0.4999999602$

b. Let $y = 2x(1 - x)$. Let $x = 0.713$, then

$$y = 2(0.713)(1 - 0.713) = 0.409262.$$

Now let $x = 0.409262$, and compute a new $y -$ value. Continue the process as shown in the following table.

$x = 0.713$	$y = 0.409262$
$x = 0.409262$	$y = 0.4835332307$
$x = 0.4835332307$	$y = 0.499457691$
$x = 0.499457691$	$y = 0.4999994118$
$x = 0.4999994118$	$y \approx 0.5000000000$

Computing a y-value for some initial x-value, and then substituting the y for x, so that a new y-value can be determined is called an iteration process. If we start with the equation $y = 4x$, and x=3, then we can apply iteration to produce the following table.

$x = 3$	$y = 12$
$x = 12$	$y = 48$
$x = 48$	$y = 192$
$x = 192$	$y = 768$
$\vdots$	$\vdots$

Iteration of the equation $y = 4x$, produces larger and larger values that are easy to predict. The *nonlinear equation* $y = nx(1 - x)$ is not so predictable. If $n = 2$, and x is any number between 0 and 1, as in **2a**, and **2b**, then we can show that the iteration process will be produce a sequence of numbers that are closer and closer to 0.5. That is they are attracted to the fixed point 0.5. However, if $n = 3.25$, the iteration process produces a sequence of numbers that oscillates between two numbers, one of which is approximately 0.495, and the other is approximately 0.812. If $n = 4$, and x is any number between 0 and 1, then the iteration process produces sequences of numbers that are not so predictable and are attracted to several fixed points between 0 and 1. In this case the iteration process is said to be *chaotic*. For more information, read Chapter 6, The Dragons of Chaos, in *The Mathematical Tourist*, by Ivars Peterson,1988, W. H. Freeman and Company, New York.

3. a. An expression of the form

$$a_1 + \cfrac{1}{a_2 + \cfrac{1}{a_3 + \cfrac{1}{a_4 + \cfrac{1}{a_5 + \cdots}}}}$$

is called an infinite continued fraction. If there are a finite number of the terms $a_1, a_2, \cdots, a_n$, then the expressions is a finite continued fraction. Every real number can be written as a continued fraction. For example,

$$\frac{51}{14} = 3 + \frac{9}{14}$$

$$= 3 + \cfrac{1}{\cfrac{14}{9}}$$

$$= 3 + \cfrac{1}{1 + \cfrac{5}{9}}$$

Also the irrational number $\sqrt{2}$ can be written as the infinite continued fraction

$$1 + \cfrac{1}{2 + \cfrac{1}{2 + \cfrac{1}{2 + \cfrac{1}{2 + \cdots}}}}$$

An algorithm for converting an irrational number to a continued fraction is given in number theory texts such as *The Higher Arithmetic* by H. Davenport, Harper and Brothers, New York, 1960.

Continued fractions have applications in the areas of number theory, differential equations and physics. In the area of number theory, continued fractions can often be used to provide an exact method of construction for the solution of a problem, where-as other techniques and procedures can only be used to show that a solution exists.

b. To solve the equation substitute x for the portion of the equation shown in the rectangle.

$$x = \cfrac{1}{1 + \boxed{\cfrac{1}{1 + \cfrac{1}{1 + \cfrac{1}{1 + \cdots}}}}}$$

This gives us

$$x = \frac{1}{1+x}$$

Multiplying each side of the equation by $(1 + x)$ gives us:

$$x + x^2 = 1$$
$$x^2 + x - 1 = 0$$

Solving by the quadratic formula:

$$x = \frac{-1 \pm \sqrt{1^2 - 4(1)(-1)}}{2(1)}$$

$$x = \frac{-1 \pm \sqrt{5}}{2}$$

From the original equation we know that $x > 0$, so we have

$$x = \frac{-1 + \sqrt{5}}{2} \approx 0.6180339888$$

4. The following diagram can be used to establish the Pythagorean theorem.

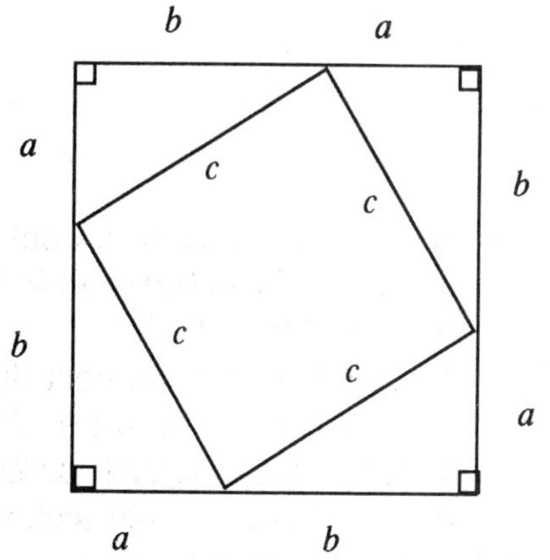

The proof is as follows. The area of the large square is

$$(a + b)^2 = a^2 + 2ab + b^2.$$

The area of the large square can also be determined by adding the areas of the four triangles to the area of the small square (Explain why you know the quadrilateral that measures c units on each side is a square.). This yields:

$$4\left(\frac{1}{2}ab\right)+c^2 = 2ab + c^2$$

Equating these results produces $a^2 + 2ab + b^2 = 2ab + c^2$. Subtracting $2ab$ from each side of this equation gives us the desired result:

$$a^2 + b^2 = c^2.$$

The following diagram was used by President James Garfield to establish the Pythagorean theorem. He developed this proof in 1876, five years before he became the twentieth president of the United States.

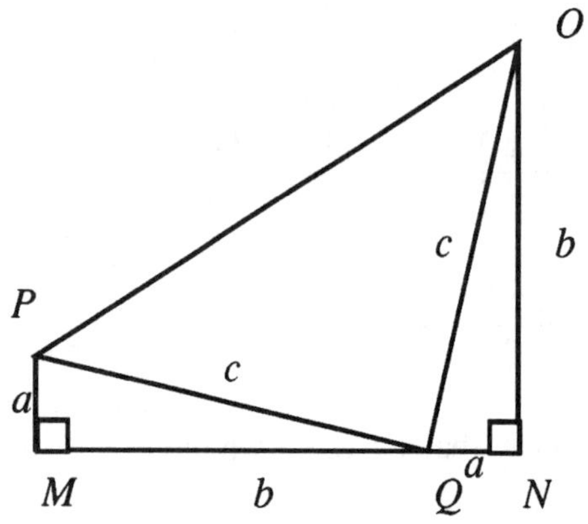

President Garfield's proof is as follows:

The area of trapezoid $MNOP$ is one-half its height times the sum of its bases or:

$$\frac{1}{2}(b+a)(a+b) = \frac{1}{2}(a+b)^2$$

Triangle PQO is a right triangle since:

(1) $m\angle MPQ + m\angle MQP = 90°$

(2) $m\angle MPQ = m\angle NQO$

(3) $\therefore\ m\angle NQO + m\angle MQP = 90°$

(4) $m\angle PQO + (m\angle NQO + m\angle MQP) = 180°$

(5) $\therefore\ m\angle PQO = 90°$

The area of triangle $PMQ = \frac{1}{2}ab$. The area of triangle $QNO = \frac{1}{2}ab$.

The area of triangle $PQO = \frac{1}{2}c^2$.

The area of the three triangles must equal the area of the trapezoid. That is,

$$\frac{1}{2}ab + \frac{1}{2}ab + \frac{1}{2}c^2 = \frac{1}{2}(a+b)^2$$

$$ab + ab + c^2 = (a+b)^2$$

$$2ab + c^2 = a^2 + 2ab + b^2$$

$$c^2 = a^2 + b^2$$

(End of President Garfield's Proof.)

Another version of the proof of the Pythagorean Theorem makes use of the following figure.

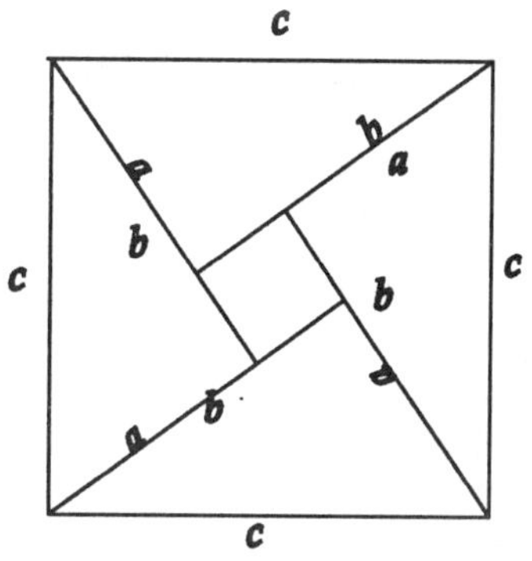

If the length of the hypotenuse of each triangle is c, and the length of the sides of the triangles are a and b, then we know the area of the large outer square is the same as the sum of the areas of the four triangles and the area of the center square. Using our notation we have:

$$c^2 = 4\left(\frac{1}{2}ab\right) + (b-a)^2$$

$$= 2ab + b^2 - 2ab + a^2$$

$$= b^2 + a^2$$

The converse of the theorem of Pythagoras is:

> In a triangle with sides of length a, b, and c, where c is the length of the longest side, if $c^2 = a^2+b^2$, then the triangle is a right triangle.

Carpenters often apply this theorem when they need to make a square corner. For example, if a point A measures 3 feet from the corner C of a deck and a second point B in the same level plane as point A and C, measures 4 feet from the corner, $\angle ACB$ is a right angle if and only if the distance between points A and B is 5.

5. Definition: Let a secant line through point P intersect a circle O at the points A, and B. The number given by $d(P, A) \cdot d(P, B)$ is called the power of point P for circle O.

An amazing theorem from geometry states that for any given circle, and any given point, the power of the point for the circle will be the same reqardless of which secant line is used. Thus for the following drawing

$$d(P, A_1) \cdot d(P, B_1) = d(P, A_2) \cdot d(P, B_2) = d(P, A_3) \cdot d(P, B_3)$$

Note that if $d(P, A_1) = 4$, and $d(P, B_1) = 7$, then the power of the point P for the circle O is $4 \cdot 7 = 28$, and

$$d(P, A_2) \cdot d(P, B_2) = d(P, A_3) \cdot d(P, B_3) = 28.$$

The above result is also valid when P is in the interior of the circle.

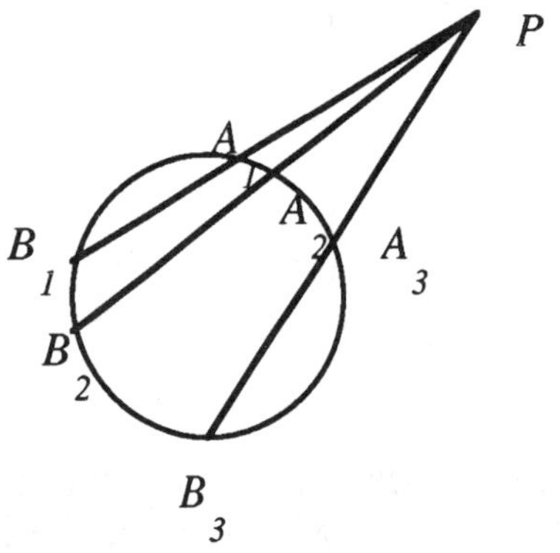

The tangent-secant theorem can be stated as follows:

The square of the length of a segment tangent to a circle from a point is the power of the point for the circle.

Chapter 3 Essays and Projects, Page 196-197

1. Benoit Mandelbrot has often been called the man who reshaped geometry. Newton and Leibnitz had made a detailed study of smooth continuous curves, however, it was Mandelbrot who gave the world a wonderful way to study rough, scabrous curves. It was 1975 when Mandelbrot chose the word *fractal* to represent the fractional and fractured structures that he was able to produce.

One aspect of fractal geometry concerns the Mandelbrot set. A computer drawing of the Mandelbrot set is shown by the white region shown in the following figure.

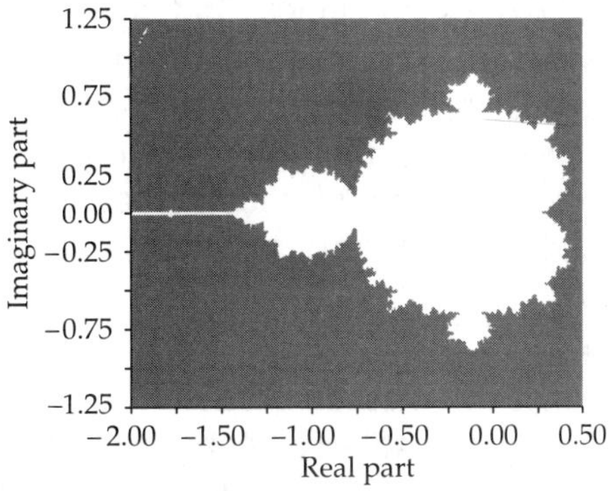

The Mandelbrot set consists of the graph of all complex numbers which when used as seeds in the Mandelbrot replacement procedure (described on pages 62–63) yield sequences, that are bounded. For example, consider the complex numbers $1 + 0i$, $-1 + 0i$, $0.25 + 0i$, and $0 + i$. When $1 + 0i$ is used as a seed in the Mandelbrot replacement procedure, an unbounded sequence is produced. Notice that the graph of $1 + 0i$ (represented by the point (1, 0)) is not part of the Mandelbrot set. However, when $-1 + 0i$, $0.25 + 0i$, and $0 + i$ are used as seeds, the Mandelbrot replacement procedure yields bounded sequences. Thus the points (−1, 0), (0.25, 0), and (0, 1) are part of the Mandelbrot set.

The Mandelbrot set has many interesting properties. (1) It is chaotic in that two points very close together may or may not be an element of the Mandelbrot set. (2) It has the property of self-embeddedness. That is, a close up of some regions of the Mandelbrot set are self similar to the entire set. (3) It is a complicated object that is created with a simple mathematical algorithm. (4) Some regions of the set are similar to elements found in nature. (5) It can be used to study certain events that we once considered random and unexplainable.

For more information on fractals and Benoit Mandelbrot, consult:

The Mathematical Tourist by Ivars Peterson, W. H. Freeman and Company, New York, 1988.

Fractals Everywhere by Michael Barnsley, Academic Press, Inc., San Diego, California, 1988.

Chaos and Fractals, the American Mathematical Society, Providence Rhode Island, 1989.

2. The "cubic formula" is an algorithm for solving by radicals polynomial equations of the form

$$ax^3 + bx^2 + cx + d = 0$$

The development of the cubic formula is a bizarre story. The principal characters consist of Scipione del Ferro (1465–1526), Antonio Fior (ca. 1506–?), Niccolo Fontana (1499–1557), Gerolamo Cardano (1501–1576), and Ludovico Ferrari (1522–1565). The major events are as follows. Scipione del Ferro discovered a method to solve the so-called "depressed cubic" $ax^3 + cx + d = 0$. Before his death, Scipione del Ferro told Antonio Fior about his method. Antonio Fior promptly challenged Niccolo Fontana to a mathematics contest in which Fontana was required to solve several depressed cubics. Fontana, also known as Tartaglia, was able with some time and great effort to determine the method. Gerolamo Cardano pestered Tartaglia for several years until Tartaglia finally told Cardano of the method. He made Cardano promise not to publish or tell anyone of the method. However, Cardano did tell his pupil Ludovico Ferrari. In fact, Cardano learned how to solve the general cubic

$$ax^3 + bx^2 + cx + d = 0.$$

The only problem was that his method consisted of a reduction technique that resulted in the need to solve a depressed cubic, and he had promised not to tell or publish Tartaglia's method. To compound matters, Ludovico Ferrari soon developed a method to solve the general quartic equation

$$ax^4 + bx^3 + cx^2 + dx + e = 0.$$

It also used a reduction technique that eventually required Tartaglia's method of solving a depressed cubic. Cardano and Ferrari had developed some of the most important mathematical work of the century, but they could not publish their work because it would require them to publish Tartaglia's method. Eventually Cardano and Ferrari discovered the written work of Scipione del Ferro. It was at this time that Cardano felt they could publish their work because the method of solving a depressed cubic was originally due to Ferro, and not Tartaglia.

For more of the details of this unusual historical development and the actual method for solving a cubic by radicals, consult the following.

Journey Through Genius by William Dunham, John Wiley and Sons, 1990.

An Introduction to the History of Mathematics by Howard Eves, Saunders College Publishing,.fifth edition, 1983.

3. Let X be a non-empty set. A real valued function d defined on $X \times X$, is called a metric or distance function on x if for all a, b, and c elements of X, the following four properties are true.

(1) $d(a, b) > 0$ if $a \neq b$.

(2) $d(a, a) = 0$.

(3) $d(a, b) = d(b, a)$

(4) $d(a, c) \leq d(a, b) + d(b, c)$

The real number $d(a, b)$ is called the distance from a to b. The metric $d = \sqrt{(x_2 - x_1)^2 + (y_2 - y_1)^2}$ is called the Euclidean metric because it measure the distance between the two points (x_1, y_1), (x_2, y_2) as the length of the line segment that connects the two points just as in Euclidean geometry.

The following functions are metrics:

$$d = (x_2 - x_1)^2 + (y_2 - y_1)^2$$

$$d = (x_2 - x_1)^4 + |y_2 - y_1|$$

$$d = 2|x_2 - x_1| + 3|y_2 - y_1|$$

Not every function is a metric. For example, the function

$$d = |x_2 y_1| + 2|x_1 y_2|$$

is not a metric, because of the following reasons:

The distance d from (0, 0) to any other point such as (5, 1) is 0. This contradicts property (1). Also the distance from (2, 5) to (1, 3) is not equal to the distance from (1, 3) to (2, 5). This contradicts property (3).

The metric $d = |x_2 - x_1| + |y_2 - y_1|$ is called the Manhattan metric because it measure the distance between (x_1, y_1), (x_2, y_2) as the distance from the first point down to the corner point (x_2, y_1), added to the distance from the corner up to the second point. For example, in Manhattan to get from the corner labeled (5, 3) to the corner (7, 8) would require you to travel 2 blocks East followed by 5 blocks North. Thus you would travel

$$|7 - 5| + |8 - 3| = 2 + 5 = 7 \text{ blocks.}$$

Sometimes distance is measured as a function of time. Note that this method of measuring distance is not a metric, since for the points a, and b, it is often not the case that $d(a, b) = d(b, a)$

4. Static symmetry involves the length of line segments and other geometric figures. For example, artists often arrange geometrical figures around an axis so that one side of the figure is a mirror image of the other side about the axis of symmetry. Dynamic symmetry does not involve length, but instead makes use of proportions such as area.

To illustrate dynamic symmetry, consider the following figure. Figure $ABCD$ is a square, with sides of length 1. The diagonal $\overline{AC}$ has a length of $\sqrt{2}$. Now use line segment $\overline{AC}$ to swing an arc that intersects side $\overline{AB}$ extended at point E. The length of line segment $\overline{AE}$ is $\sqrt{2}$. Construct the rectangle $AEFD$ that has $\overline{AE}$, as the base of a rectangle and has a height of $d(A, D)$.

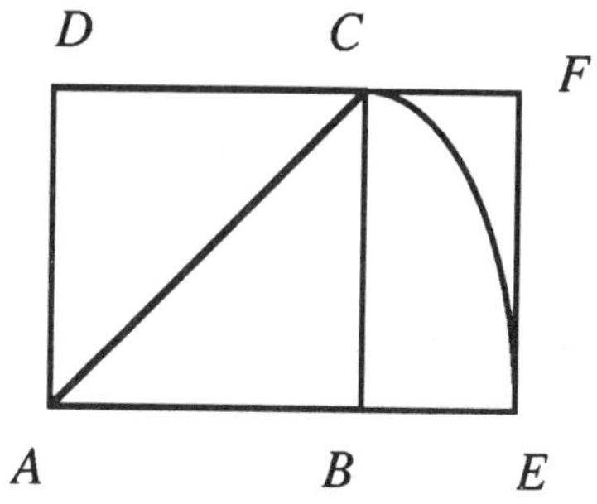

The rectangle $AEFD$ is known as a root-two rectangle. It is called a root-two rectangle because its base can be used to construct larger squares whose area is $\sqrt{2}\sqrt{2} = 2$, and its height can be used to construct squares whose area is $1 \cdot 1 = 1$. The length of each side of the larger square is $\sqrt{2}$. Its area is 2 which is an integral multiple of the area of the original square. Such squares are said to be commensurable. Architects and artists often incorporate in their work, square regions whose areas are commensurable.

For more information on Static and dynamic symmetry, consult: *The Golden Section and Related Curiosa* by Garth E. Runion, Scott, Foresman and Company, 1972.

5.

$\circ$	f	g	h	j	k	l
f	f	g	h	j	k	l
g	g	f	j	h	l	k
h	h	l	f	k	j	g
j	j	k	g	l	h	f
k	k	j	l	g	f	h
l	l	h	k	f	g	j

***Definition of a Group*:**

A nonempty set G on which there is defined a binary operation "•" is called a group provided the following three properties are satisfied.

(i) If a, b, and c are elements of the group, then $(a \bullet b) \bullet c = a \bullet (b \bullet c)$ (associative property)

(ii) There exists an element e of G such that $e \bullet a = a \bullet e = a$ for every element a of G. (existance of an identity element)

(iii) If a is an element of G, there exists an element x of G such that $a \bullet x = x \bullet a = e$. (existance of an inverse element for each element of the set G).

a. The identity element of the composition group is f.

b. The group is not an abelian group, because the operation of composition is not a commutative operation. For example: $h \circ g = l$, whereas $g \circ h = j$.

6. a. Let $m = 10$, $d = 7$, $c = 19$, $y = 41$. Then:

$$z = \left[\!\left[\frac{13m-1}{5}\right]\!\right] + \left[\!\left[\frac{y}{4}\right]\!\right] + \left[\!\left[\frac{c}{4}\right]\!\right] + d + y - 2c$$

$$= \left[\!\left[\frac{13 \cdot 10 - 1}{5}\right]\!\right] + \left[\!\left[\frac{41}{4}\right]\!\right] + \left[\!\left[\frac{19}{4}\right]\!\right] + 7 + 41 - 2 \cdot 19$$

$$= 25 + 10 + 4 + 7 + 41 - 38$$

$$= 49$$

The remainder of 49 divided by 7 is 0. Thus, December 7, 1941 was a Sunday.

b. This one is tricky. Since we are finding a date in the month of January, we must use 11 for the month and we must use the previous year which is 1999. Thus we let $m = 11, d = 1, c = 19, y = 99$. *Then*:

$$z = \left[\!\left[\frac{13m - 1}{5}\right]\!\right] + \left[\!\left[\frac{y}{4}\right]\!\right] + \left[\!\left[\frac{c}{4}\right]\!\right] + d + y - 2c$$

$$= \left[\!\left[\frac{13 \cdot 11 - 1}{5}\right]\!\right] + \left[\!\left[\frac{99}{4}\right]\!\right] + \left[\!\left[\frac{19}{4}\right]\!\right] + 1 + 99 - 2 \cdot 19$$

$$= 28 + 24 + 4 + 1 + 99 - 38$$

$$= 118$$

The remainder of 118 divided by 7 is 6. Thus, January 1, 2000 will be a Saturday.

c. Let m = 5, d = 4, c = 17, y = 76. Then:

$$\mathrm{z} = \left[\!\left[\frac{13m - 1}{5}\right]\!\right] + \left[\!\left[\frac{y}{4}\right]\!\right] + \left[\!\left[\frac{c}{4}\right]\!\right] + \mathrm{d} + \mathrm{y} - 2\mathrm{c}$$

$$= \left[\!\left[\frac{13 \cdot 5 - 1}{5}\right]\!\right] + \left[\!\left[\frac{76}{4}\right]\!\right] + \left[\!\left[\frac{17}{4}\right]\!\right] + 4 + 76 - 2 \cdot 17$$

$$= 12 + 19 + 4 + 4 + 76 - 34$$

$$= 81$$

The remainder of 81divided by 7 is 4. Thus, July 4, 1776 was a Thursday.

Chapter 4 Essays and Projects, Page 248

1. The rational function given by $F(x) = P(x)/Q(x)$, where $P(x)$ and $Q(x)$ have no common factors, will have a parabolic asymptote if the degree of $P(x)$ is exactly two more than the degree of $Q(x)$.

To determine the polynomial that represents the parabolic asymptote, find the quotient of $P(x)/Q(x)$. For example the rational function

$$F(x) = \frac{x^3 + 5x^2 + 6x}{x + 1},$$

has a parabolic asymptote given by the graph of $h(x) = x^2 + 4x + 2$, as shown in the following graph.

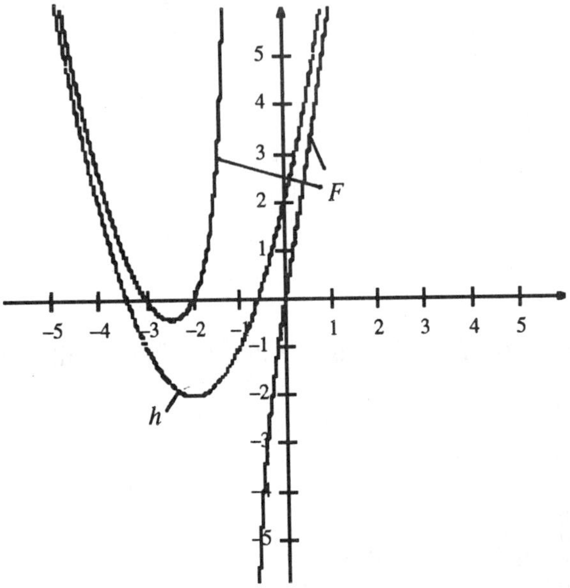

2. The sum of the roots of a polynomial in x written in standard form (of degree one or larger) is the opposite of the coefficient of the $(n - 1)$th degree term. The sum of the product of the roots taken two at a time is the coefficient of the $(n - 2)$th degree term. The sum of the product of the roots taken three at a time is the opposite of the coefficient of the $(n - 3)$th degree term.

⋮

The sum of the product of the n roots taken k at a time, where k is an even integer less than n, is the coefficient of the $(n - k)$th degree term. The sum of the product of the n roots taken j at a time, where j is an odd integer less than n, is the opposite of the coefficient of the $(n - j)$th degree term.

⋮

If the polynomial has even degree, then the product of all the roots is the constant term. If the polynomial has odd degree, then the product of all the roots is the opposite of the constant term.

For example, consider the polynomial

$$x^4 + 5x^3 - 7x^2 - 29x + 30,$$

and its roots: 1, 2, -3, and -5. The sum of the roots is

$$1 + 2 + (-3) + (-5) = -5$$

which is the opposite of the coefficient of the third degree term. The sum of the product of the roots taken two at a time is:

$$1 \cdot 2 + 1(-3) + 1(-5) + 2(-3) + 2(-5) + (-3)(-5) = -7,$$

which is the coefficient of the second degree term. The sum of the product of the roots taken three at a time is:

$$1 \cdot 2(-3) + 1 \cdot 2(-5) + 1(-3)(-5) + 2(-3)(-5) = -29,$$

which is the opposite of the coefficient of the first degree term. The product of all the roots is $(1)(2)(-3)(-5) = 30$, which is the constant term.

3. Carl Friederich Gauss was born on April 30, 1777, in Brunswick Germany. He was the son of poor parents. His father Gerhard Diederich Gauss wanted Carl to become a gardener, or a bricklayer, and was not interested in Carl's education. However his mother Dorothea saw Carl as an outstanding intellectual who amazed everyone who watched his mental development. Some observers went so far as to say "he is not of this earth". Carl Friederich Gauss showed a remarkable talent for mathematics as early as 3 years of age. At a very young age it is reported that Gauss found an error in the weekly payroll for the laborers working under his father. Later in his life he often said that he knew how to reckon before he could talk.

Many mathematicians classify Archimedes, Newton, and Gauss, as the three greatest mathematicians of all time. Gauss made many great mathematical discoveries, but he seldom published his mathematical work. Some people feel that he held back at reporting his work because it was often misunderstood by the mathematicians of his time. Others say that he had so many ideas that he felt that he had little time to waste publishing his results. Although Gauss did not publish most of his work, the mathematicians of the day soon learned to take notice of the work that he did publish.

In 1796, Gauss discovered that a regular 17-gon could be constructed with compass and straightedge. This proof caused a sensation in the mathematical community, because a teenager had been able to prove a result that had been over looked by such greats as Euclid, Archimedes, Newton, and all the other powerful mathematicians that had made a study of the classical problem of constructing regular polygons.

In 1799, Gauss received his doctorate from the University of Helmstadt. It was at this time that he developed the first complete proof of the fundamental theorem of algebra.

The Fundamental Theorem of Algebra
If P is a polynomial of degree $n \geq 1$ with complex coefficients, then P has at least one complex zero.

The following corollary of the fundamental theorem concerns the number of zeros of a polynomial.

If P is a polynomial of degree $n \geq 1$, with complex coefficients, then P has exactly n complex zeros provided each zero is counted according to its multiplicity.

The fundamental theorem of algebra is an important theorem for two reasons. The first is that it was difficult to establish. Many well know mathematicians including Euler, tried to establish the theorem, but fell short of a complete proof. The second reason is the power of the theorem and the above corollary. For example, consider the polynomial $P = x^3 - 8$. According to the corollary of the fundamental theorem, we know that $P = x^3 - 8$ must have exactly 3 zeros. A check will show that

$$2,\ -\frac{1}{2} + \frac{\sqrt{3}}{2}i\text{, and } -\frac{1}{2} - \frac{\sqrt{3}}{2}i$$

are indeed zeros of P. In chapter 4, we will develop methods of finding the zeros of a polynomial. Knowing how many zeros to expect will certainly help in the search.

Much of Gauss's mathematics involved pure mathematics, however in 1807 he was appointed director of the Observatory at Göttingen. It was here that he was required to consider the area of applied mathematics. He excelled in this arena as is shown by his work in determining the orbit of the asteroid Ceres, his work in mapping the earth's magnetic field, and the invention of an electric telegraph some years before the work of Samuel F. B. Morse. Gauss died February 23, 1855 at the age of 78, at the Göttingen Observatory where he was the director for 48 years.

Additional information about Gauss and his work can be found in:
Men of Mathematics by E. T. Bell, Simon and Schuster, Inc. New York, New York, 1965.
Journey Through Genius by William Dunham, John Wiley & Sons, Inc., 1990.

4. René Descartes (1596-1650), was a French philosopher who also became well known for his work in mathematics. René Descartes was born in France at La Haye in 1596. At the age of 8 he attended the Jesuit school at La Flèche. Due to poor health, he was allowed to spend the morning hours in bed while his schoolmates did chores and morning calisthenics. Even in his adult life, Descartes continued this practice of using the morning hours to meditate and to think about philosophy and mathematics. In 1628 he moved to Holland. It was here that he wrote *Le monde,* and *Discours de la méthode pour bien conduire sa raison et chercher la vérité dans les sciences*. The later book a discourse on the method of reasoning had three appendices. The last of these was titled *La géometrie,* and it contained Descartes work in the area of analytic geometry. In the third part of *La géometrie,* Descarte makes use of the theorem that is now know as Descartes's Rule of Signs. Although a complete proof of this theorem is beyond the scope of college algebra, we can prove the following special case in which all of the coefficients of a given polynomial are positive.

Consider the polynomial

$$P(x) = a_n x^n + a_{n-1} x^{n-1} + \cdots + a_2 x^2 + a_1 x + a_0, \qquad a_n \neq 0$$

If all of the coefficients a_n, a_{n-1}, … , a_2, a_1, and a_0 are positive numbers, and x is a positive real number, then all of the terms of the polynomial are positive. The sum of the positive terms is positive. Thus we have shown that if each of the coefficients is positive, then there are no alternation of signs, and there are no positive zeros. All of the zeros of P must be negative numbers or nonreal complex numbers.

Now consider the polynomial

$$P(-x) = a_n(-x)^n + a_{n-1}(-x)^{n-1} + \cdots + a_1(-x) + a_0,$$

All of the coefficients a_n, a_{n-1}, … , a_2, a_1, and a_0 are positive real numbers. Therefore $P(-x)$ has n alternations of signs.

We know that P has n zeros, and we know that the nonreal complex zeros must occur in conjugate pairs. Thus the number of negative zeros of P must be n, or less than n by an even amount. Thus we have established Descartes Rule of Signs for the special case where all of the coefficients are positive numbers.

5. Evariste Galois was born near Paris in 1811. Galois' talent in the area of mathematics went unnoticed by his superiors, even though he showed a great talent for mathematics as a teenager. Galois published two mathematical papers. The first was *Démonstration d'un théorème sur les fractions continues périodiques*. This paper went unnoticed. Galois second paper contained some of the greatest mathematical ideas of the century. The paper contained his work on the solvability of polynomial equations and group theory. It was in this paper that Galois independent of Niels Abel's work established the theorem that the zeros of polynomials of degree 5 and greater cannot be expressed in terms of radicals. It was written when Galois was only seventeen. Galois submitted his work to the French Academy. Galois had hoped that this paper would be read by M. Cauchy. He felt this paper would bring him the recognition that he desired. It is reported that his work was misplaced and not read.

Galois' life was short and filled with many difficulties. For instance, Galois was expelled from school, his father committed suicide, he was imprisoned for a few months, and shortly after his release from prison he was shot and killed in a pistol duel at the age of twenty.

It was not until 1870 and later that Galois work became known by the mathematical community. Today Galois is best recognized for the development of group theory which is a major part of contemporary mathematics.

6. In this case $n = 4$, $m = 6$. Thus Cardano's Formula yields

$$\sqrt[3]{\sqrt{\left(\frac{4}{2}\right)^2 + \left(\frac{6}{3}\right)^3} + \frac{4}{2}} - \sqrt[3]{\sqrt{\left(\frac{4}{2}\right)^2 + \left(\frac{6}{3}\right)^3} - \frac{4}{2}} \approx 0.6258$$

$$\begin{aligned} \text{Check:}\quad (0.6258)^3 + 6(0.6258) &= 4 \\ 0.2450793255 + 3.7548 &= 4 \\ 3.999879326 &\approx 4 \end{aligned}$$

Chapter 5 Essays and Projects Page 291-2

1. John Napier, (1550-1617) was a Scottish mathematician. Napier was born in the town of Merchiston near Edinburgh. He was educated at Saint Salvador's College, in Saint Andrews. He was one of the early Protestant leaders in Scotland, and he was an active public leader for most of his adult life.

He is best known for his invention of logarithms, which was motivated by a desire to simplify the work needed to perform complicated mathematical calculations. He soon realized the practical benefit of logarithms in astronomy and trigonometry, and gave up his work in algebra so that he could produce a set of seven figure logarithmic tables. His complete system of logarithms was published in 1614, under the title *Canonis Descriptio*. To honor Napier, and his work with logarithms, we often refer to natural logarithms as Napierian logarithms, even though Napier did not originate natural logarithms.

Napier also suggested the system that we use for expressing decimals, as well as designing and building several mechanical devices that could perform difficult computations such as extracting square roots, and cube roots. Napier developed many theorems in the area of spherical trigonometry. Two of his better known theorems are called Napier's Analogies, which can be used to solve spherical triangles and Napier's rules of circular parts, which can be used to solve right spherical triangles.

2. Henry Briggs, (1561-1630) was a Professor of Geometry in London. He heard of Napier's work with logarithms when Napier published his *Canonis Descriptio* in 1614. Briggs and Napier soon became good friends, but the friendship was cut short when Napier died in 1617. Briggs admired the work of Napier. It is reported that at one of there first meetings Briggs stated that he was amazed that no one had developed logarithms before Napier, for after reading the work of Napier, it all appeared so easy.

Briggs carried on the work of Napier. He must be given much credit for producing several different logarithmic tables in a very short period of time. Briggs and Napier decided that the logarithm tables that Napier had produced would be more useful if they were changed so that the logarithm of 1 was 0 and the logarithm of 10 would be 1. Today we often call base ten logarithms *Briggsian* or *common logarithms*. Briggs published in 1624 his *Arithmetica logarithmica* which consisted of a fourteen-place table of common logarithms.

3. a. To determine the number of terms n such that $S_n > 1000$, we solve for the smallest integer n that is a solution of the inequality:

$$\ln n > 1000.$$

The above inequality is equivalent to $n > e^{1000}$. Using a powerful calculator yields

$$e^{1000} \approx 1.970071114 \times 10^{434}.$$

Using a computer yields: $e^{1000} \approx$
1970071114017046993888879352243323125316937985323845789952802991385063850782441193474978076563026889930963817987520226935982981730544612899232627836601528252323205351695845667561922715676027880714224668263140068551685086534979416603160453678179380929052997285801328699458564702865343759004565643555891562204223202605188261122886383583722487247252145061504188819374941008712642322484363157605603774399306239597058441895090500470742175682

$$1970071114017046993888879352243323125316937985323845789952802991385063850782441193474978076563026889930963817987520226935982981730544612899232627836601528252323205351695845667561922715676027880714224668263140068551685086534979416603160453678179380929052997285801328699458564702865343759004565643555891562204223202605188261122886383583722487247252145061504188819374941008712642322484363157605603774399306239597058441895090500470742175 68.2$$

Let k equal the digits to the left of the decimal point in the above number. Then the smallest integer n, such that $S_n > 1000$, is given by $n = k + 1$.

b. To determine the number of terms n such that $S_n > 10{,}000$, we solve for the smallest integer n that is a solution of the inequality:

$$\ln n > 10{,}000.$$

The above inequality is equivalent to $n > e^{10{,}000}$. Using a computer yields

$$n \approx e^{10{,}000} \approx 8.8068182256629215 87 \times 10^{4342}.$$

Students that have access to a program such as Mathematica™ can print all 4343 digits of the integer n.

4. Experimenting with various values of a and b (where $a > b > 0$), we notice that the arithmetic mean $\frac{a+b}{2}$, the logarithmic mean $\frac{a-b}{\ln a - \ln b}$, and the geometric mean $\sqrt{ab}$ are related by the following inequality.

For $a > b > 0$, $$\frac{a+b}{2} > \frac{a-b}{\ln a - \ln b} > \sqrt{ab}$$

An elegant proof of this theorem is given in the article, *The Geometric, Logarithmic, and Arithmetic Mean Inequality*, by Frank Burk, The American Mathematical Monthly, June–July, 1987, pages 527–528.

5. Use a graphing calculator or a computer to produce the following graphs. Then use the trace feature to determine the difference in the heights of the two curves at various distances from the origin.

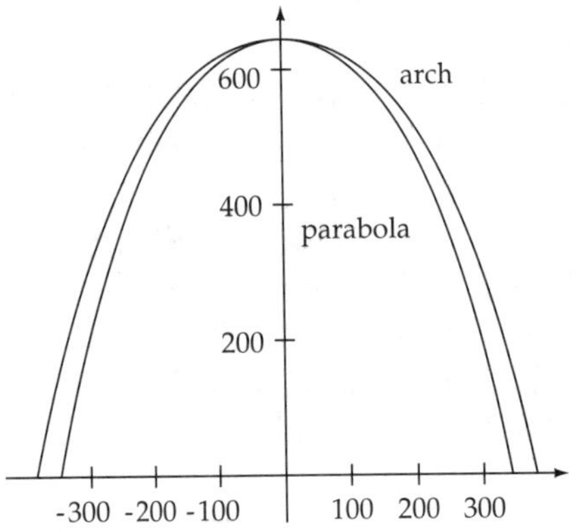

An alternate procedure is to use a graphing calculator or a computer to graph the difference between the two functions. That is graph

$$y = \left(758 - 128\left(\frac{e^{x/128} + e^{-x/128}}{2}\right)\right) - \left(\frac{2}{315}x^2 + 630\right)$$

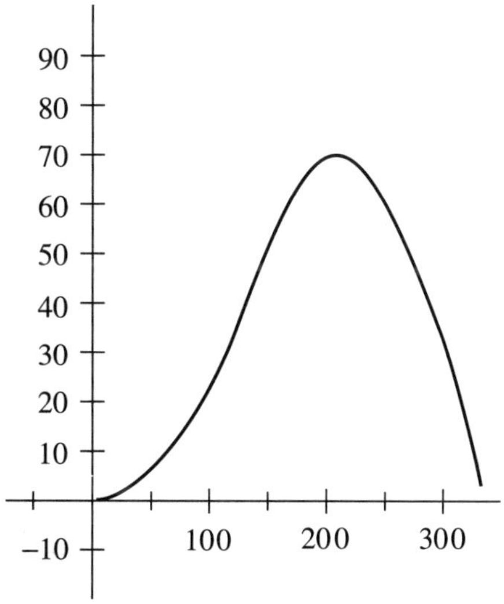

Then use the trace feature to determine the x and the y value of the points (one to the left of the origin and one to the right of the origin) which are the highest on the curve. The above graph shows that the points where the two curves differ in height by the largest amount are

$(x \approx \pm 229.2, y \approx 67.3)$.

The value – 229.2, was determined by making use of symmetry.

6. Semi-log paper is constructed by using a logarithmic scale on the vertical axis, and a uniform scale on the horizontal axis. The following grid uses a logarithmic scale on the vertical axis.

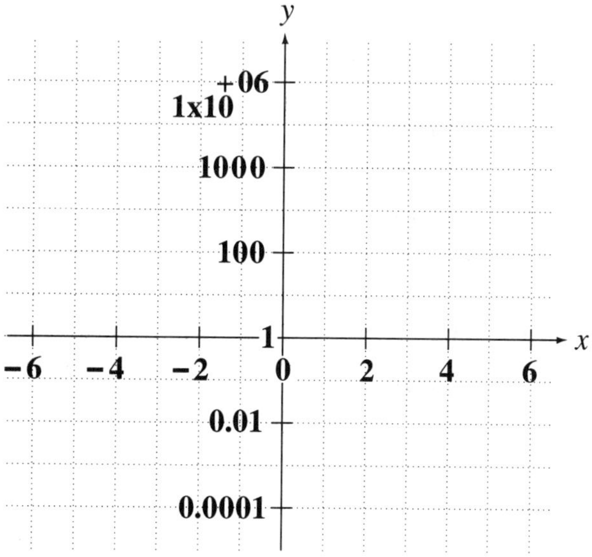

The following graph is the graph of $y = 10^x$ as it appears on semi-log paper. Notice that the exponential function $y = 10^x$ graphs to be a straight line when it is drawn on semi-log paper.

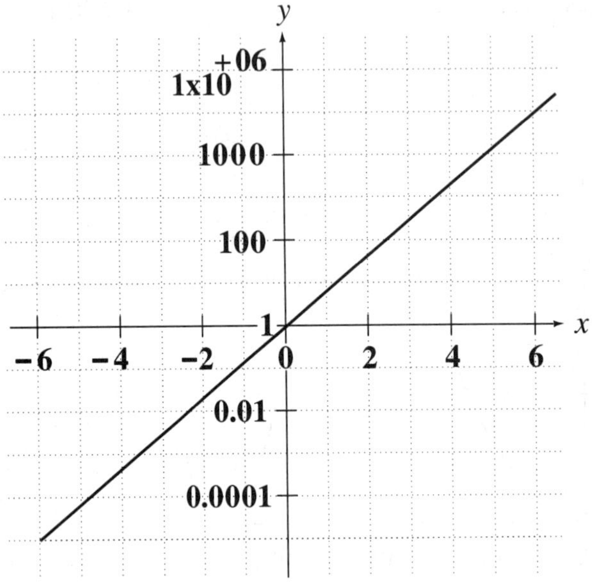

Another advantage of graphing $y = 10^x$ on semi-log paper, is that we can show more of the graph. That is, its graph does not go off the top of our grid as fast as it would using the same linear scale on both the x- and y-axis.

Log-log paper is similar to the semi-log paper shown above, except a logarithmic scale is used on both the horizontal and the vertical axes. With log-log paper a unit on each of the axes are labeled not by their distance from the origin, but rather by the antilogarithm of the actual distances. The graph of $y = x^2$ is a straight line when it is graphed on log-log paper, as shown in the following graph.

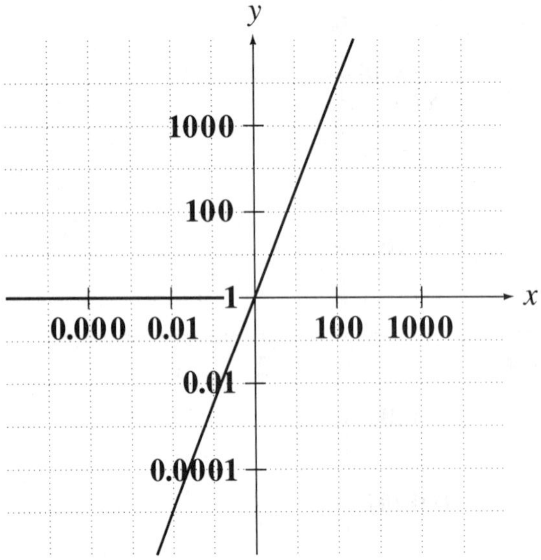

1. Microwave oven uses radiation in the microwave region of the radio spectrum to cook food rapidly. The radiation is generated by a magnetron which is similar to a television picture tube. The radiation is scattered to the walls of the oven where it is reflected. This promotes even cooking of the food. The microwaves cause the water molecules in the food to vibrate, producing heat. Microwaves range from approximately 10^9 hertz to 10^{11} hertz.

   The effect of the watt rating of a microwave can be understood from the equation $E = h\upsilon$, where E is the energy of a wave, h is Planck's constant and υ is the frequency of a wave. Since watt refers to the time rate of change of energy, microwave ovens with a higher number of watts bring the energy of the microwave at a faster rate, thereby cooking the food quicker.

2. Computer operations are based on instruction cycles. Some instructions, such as with registers, require only one cycle. Other instructions, such as floating point arithmetic, may require four cycles before the instruction is processed. In a broad sense, a computer that is rated at 40 megahertz can process 40 million instructions or instruction parts in one second whereas a 25 megahertz computer can process 25 million instructions or instruction parts in one second.

3. This exercise provides a glimpse of the result of integrating sin$x$ along with some practice at numerical approximation. There may be a wide variety of answers to this question. For the partition shown below, the area is approximately 1.41. This is a large error considering the exact value is 2. Using a finer partition, the area can be approximated more accurately. Other approximations to the area may use trapezoids or parabolas.

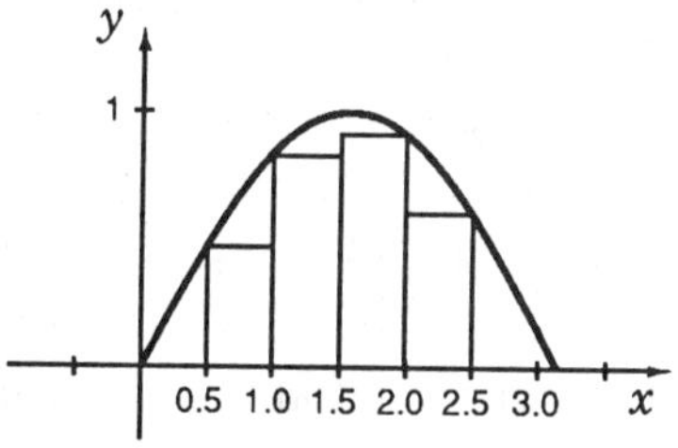

4. Prior to 150 B. C. many special situations between the chord of a circle an its corresponding central angle were studied. These relationships were used by Eratosthenes (c.a 250 B. C.) to measure the circumference of the earth, by Hippocrates (not the physician, but a contemporary from a neighboring island) to measure the relative distances to moon and sun, and by Euclid who effectively stated the law of cosines for acute angles in the language of geometry rather than trigonometry.

   Around 150 B. C., the apparently first trigonometric table was complied by Hipparchus of Nicaea. Because of his work, he is sometimes referred to as the "father of trigonometry."

   The main focus of trigonometry was the relationship between the length of a chord and its central angle. Tables were complied on that basis. For instance, in the figure below, the length of the chord AB was recorded for central angle AOB. In modern terminolgy, $\overline{AB} = 2\sin\frac{1}{2}\theta$, where the radius of the circle is assumed to be 1.

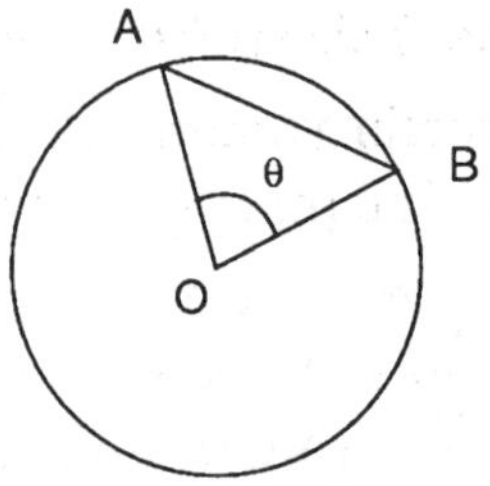

The exact time that the systematic use of the 360^0 circle came into mathematics is not known but Hipparchus is given much of the credit. The use of 360^0 may have been suggested by Babylonian astronomy and the base 60 number system used by Babylonians.

Robert of Chester (circa 1150) is credited with introducing the word 'sine' as a result of his translation of Arabic works in trigonometry. The Hindus had given the name *jiva* for the half chord in trigonometry. The Arabs translated it as *jiba*. However, Robert of Chester came to translate it as the word *jaib* which means 'inlet.' The Latin word for inlet is *sinus*.

5. The assumptions for the solution of this problem are: daylight savings time is not considered; the earliest sunrise is 5:33 AM (5.550 as a decimal); the latest sunrise is 6:02 AM (6.033 as a decimal); sunrise and sunset can be modeled by a sine function.

The amplitude is $\frac{6.033 - 5.550}{2} \approx 0.242$; the period is 365 days; the axis of symmetry is $\frac{6.033 + 5.550}{2} \approx 5.792$.

The equation is $t = -0.242\sin\left(\frac{d+11}{365}\pi\right) + 5.792$

Evaluating this function for various values of d (day of the year) will approximate the sunrise.

6. The graph below that approximates a square wave was generated from the first five terms of the equation $\frac{4k}{\pi}\left(\sin x + \frac{1}{3}\sin 3x + \frac{1}{5}\sin 5x + \ldots\right)$ $(k = 3)$.

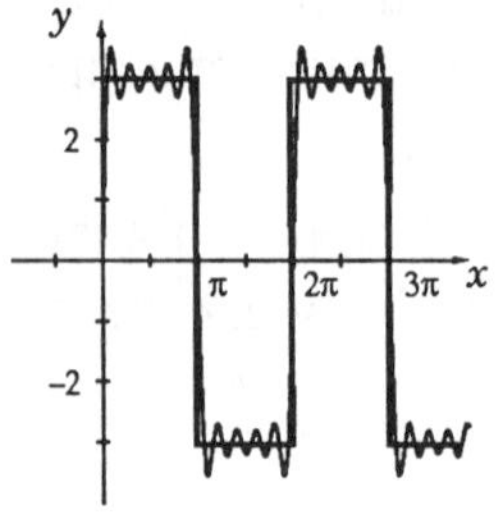

Square waves are used as forcing functions for mechanical systems, as an electromotive force for electrical systems, and in computers.

7. The graphs are shown below. Since the assumption is that the wolves prey on the rabbits, as the rabbit population increases, there is more food for the wolves which in turn allows the wolf population to increase. However, as the wolf population increases, the demand for rabbits increases and a rabbit population will start to decline. This effects a decline in the wolf population. But as the wolf population declines, there is less danger to the rabbits and their population starts to rise. The process then repeats itself.

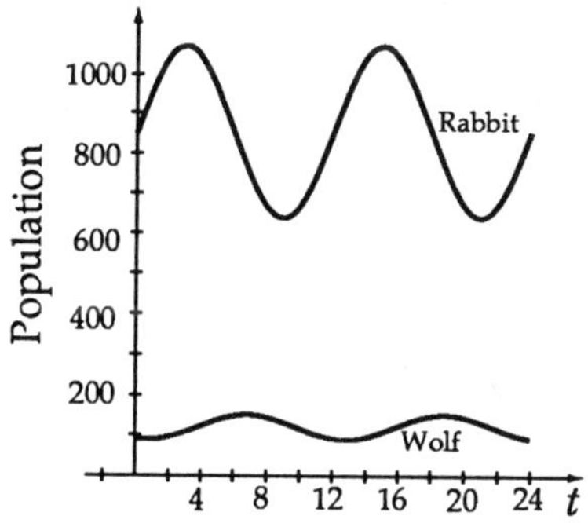

A possible equation for the graph given in the text of the population model for rabbits is $r(t) = \frac{50}{3}t + 200\sin\left(\frac{\pi t}{6}\right) + 400$. This equation is based on the assumption that there is a linear increase in the sine function. Other answers may be given.

Chapter 7 Essays and Projects Page 407-8

1. If a string fixed at both ends is bowed, transverse vibrations travel along the string and are reflected at the fixed ends. This creates a standing wave in the string. If λ is the period of a wave, then the distance between adjacent nodes (that is where the amplitude of the string is zero) is $\lambda/2$. Therefore, in a string of length a, there must be exactly an integral number n of half wavelengths, $\lambda/2$. Thus $\frac{n\lambda}{2} = a$ and $\lambda = \frac{2a}{n}$. Wave velocity $v = \sqrt{\frac{F}{\mu}}$, where F is the force exerted on the string and μ is the mass per unit length of string. Using $\lambda = \frac{2a}{n}$ and $\lambda = \frac{v}{f}$, where f is the frequency of a wave and solving for f, we have

$$f_n = \frac{n}{2a}\sqrt{\frac{F}{\mu}}$$

When $n = 1$, the lowest frequency occurs and is called the fundamental frequency. All other frequencies are overtones. Overtones whose frequencies are integral multiples of the fundamental are said to form a harmonic. The fundamental is the first harmonic, the frequency $2f_1$ is the first overtone or the second harmonic, the frequency $3f_1$ is the second overtone or the third harmonic, and so on.

The first four modes of a vibrating string fixed at both ends is shown below.

2. If bees used square cells one unit on each side for their honeycombs as shown in the following figure, then the area of each cell would be 1 square unit and the perimeter of each cell would be 4 units. However there are other regular polygons that bees could use to construct the cells of their honeycombs. The following table shows the perimeter of each cell that is required to construct a cell of area 1 square unit with the given regular polygon.

regular polygon	**area of each cell**	**perimeter of each cell**
triangular	1	$\frac{6}{\sqrt[4]{3}} \approx 4.56$
square	1	4
hexagon	1	$2\sqrt{2\sqrt{3}} \approx 3.72$

We did not include information for a regular pentagon, because regular pentagons do not tessellate. That is, regular pentagons cannot be used to completely cover a flat surface. Also, circular cells that have an area of 1 square unit have a perimeter of $2\sqrt{p} \approx 3.54$ units, however circles cannot be used to tessellate a flat surface.

The formulas for the volume V, and the surface area S of a regular hexagonal cell whose base is s units on each side, and with a depth of h are:

$$V = \frac{3\sqrt{3}}{2}s^2h, \qquad S = 6sh + \frac{3\sqrt{3}}{2}s^2$$

Additional information can be obtained from:

Honeybees, Hexagons, and the Industrial Workplace, by Samuel F. McInroy, The AMATYC Review, Fall 1992, page 8.

The film *Mathematics of the Honeycomb.* It is available from Moody Institute of Science, 12000 E. Washington Blvd., Whittier, California 90606, telephone: (213) 698-8256.

3. Robert of Chester (circa 1150) is credited with translating, into Latin, much of the Arabic work in trigonometry. When he came to translate *jiba* (meaning half-chord in Arabic) he apparently confused it with the word *jiab*, which meant "bay" or "inlet." Hence he used the word *sinus* which is Latin for bay or inlet. He also used the phrase *sinus versus* or "vertical sine" from which we get "versed sine" which is abbreviated vers θ. Using modern notation, we have the following: vers θ = versed sine θ, verc θ = versed cosine θ, exsecθ = exsecant θ.

$$\text{vers } \theta = 1 - \cos\theta \qquad \text{verc } \theta = 1 - \sin\theta \qquad \text{exsec } \theta = 1 - \sec\theta$$

The following diagram makes use of a unit circle to show the measure of each of these functions The sign of the exsec θ is negative for $0 < \theta < 90°$, and $270° < \theta < 360°$

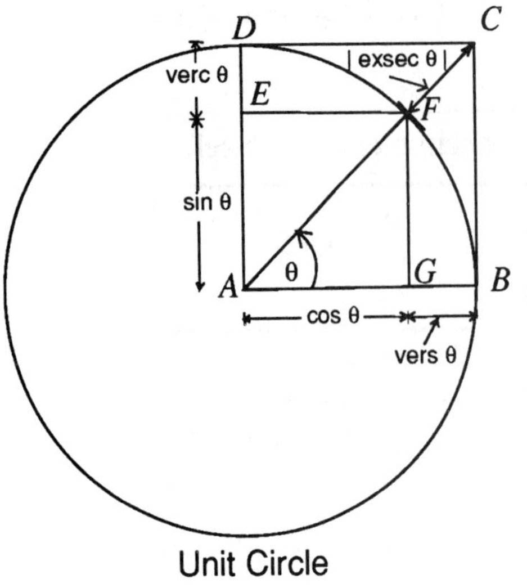

vers θ = $d(G, B)$

verc θ = $d(E, D)$

| exsec θ | = $d(F, C)$

4. The equation $\dfrac{\sin\theta\cos\theta}{\cos\theta} = \sin\theta$ is not an identity because for $\theta = \dfrac{\pi}{2}$, the right side of the equation equals 1, whereas the left side of the equation is undefined.

5. Using $\triangle ADC$ and $\triangle ABC$ we find: $\theta = \alpha - \beta = \cot^{-1}\dfrac{x}{8+a} - \cot^{-1}\dfrac{x}{a}$.

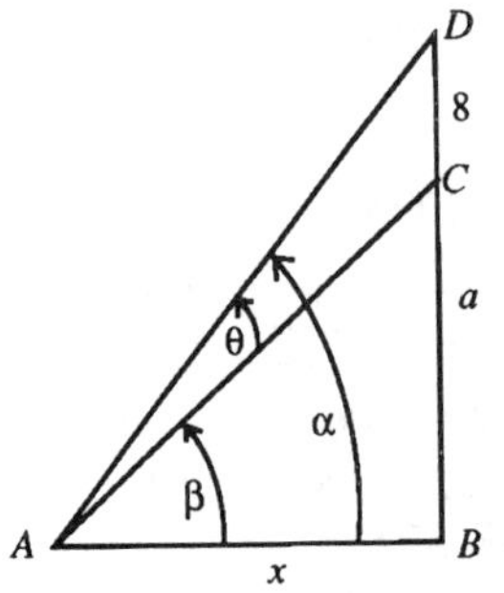

Using a graphing calculator in radian mode we graph θ over the interval [0, 40] to produce:

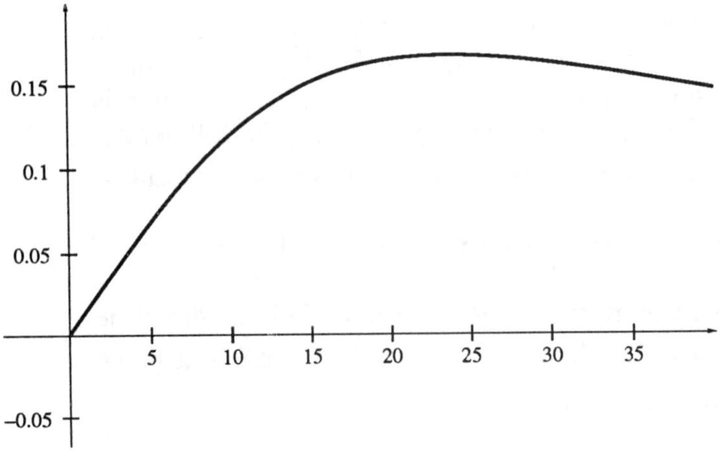

The following graph is a closeup of the previous graph over the domain [20,30] and range [0.165,0.168].

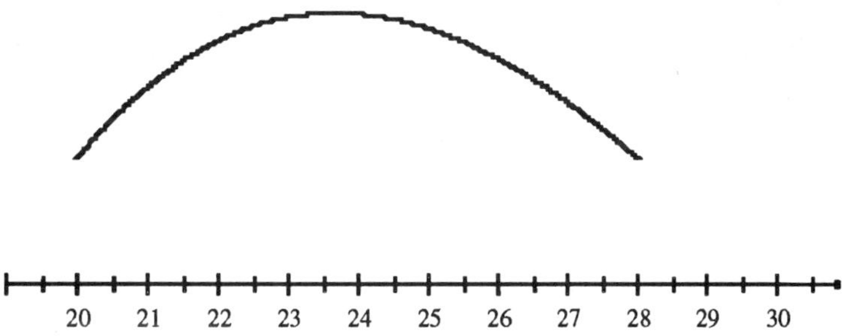

From this graph we estimate that θ is maximized when $x \approx 23.5$ *feet.*

6. In a three dimensional rectangular coordinate system, each point P is represented by an ordered triple (x, y, z) where x is the directed distance from the yz-plane to P, y is the directed distance from the xz -plane to P, and z is the directed distance from the xy - plane to P. The point

$$\left(1, 2, \sqrt{3}\right)$$

is shown in the following rectangular coordinate system.

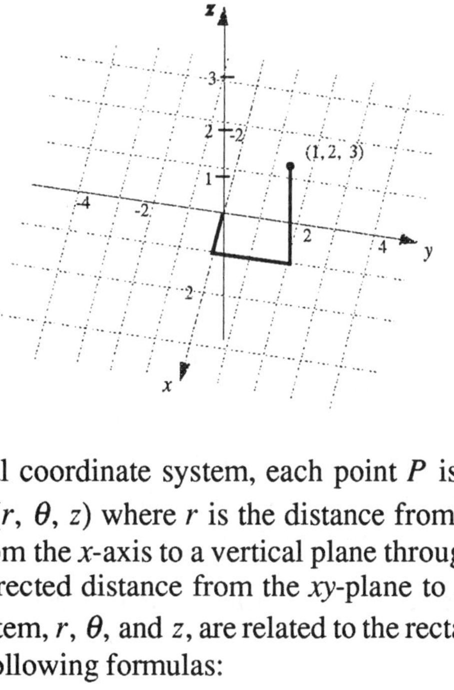

In a cylindrical coordinate system, each point P is represented by an ordered triple (r, θ, z) where r is the distance from P to the origin, θ is the angle from the x-axis to a vertical plane through P and the z-axis, and z is the directed distance from the xy-plane to P. In a cylindrical coordinate system, r, θ, and z, are related to the rectangular coordinates x, y, z by the following formulas:

$$r = \sqrt{x^2 + y^2}, \qquad \theta = \tan^{-1}\frac{y}{x}, \qquad \text{and} \qquad z = z.$$

Thus the point $(1, 2, \sqrt{3})$ in rectangular coordinates is represented by the point $(\sqrt{5}, \tan^{-1} 2, \sqrt{3})$ in a cylindrical coordinate system.

In a spherical coordinate system, a point P is represented by an ordered triple (ρ, θ, ϕ), where ρ is the distance between P and the origin, θ is the same angle as used in the cylindrical coordinate system, and ϕ, is the angle between the positive z-axis and the line through the origin and P $(0 \le \phi \le \pi)$. The relationship between the rectangular coordinates x, y, and z, and the spherical coordinates ρ, θ, and ϕ, are given by

$$\rho = \sqrt{x^2 + y^2 + z^2}, \quad \theta = \tan^{-1}\frac{y}{x}, \text{ and } \phi = \cos^{-1}\left(\frac{z}{\sqrt{x^2 + y^2 + z^2}}\right)$$

The point with rectangular coordinates of (2, 2, 2) is written in spherical coordinates as:

$$\left(2\sqrt{3}, \frac{\pi}{4}, \cos^{-1}\left(\frac{\sqrt{3}}{3}\right)\right) \approx (3.5, 45.0°, 54.7°)$$

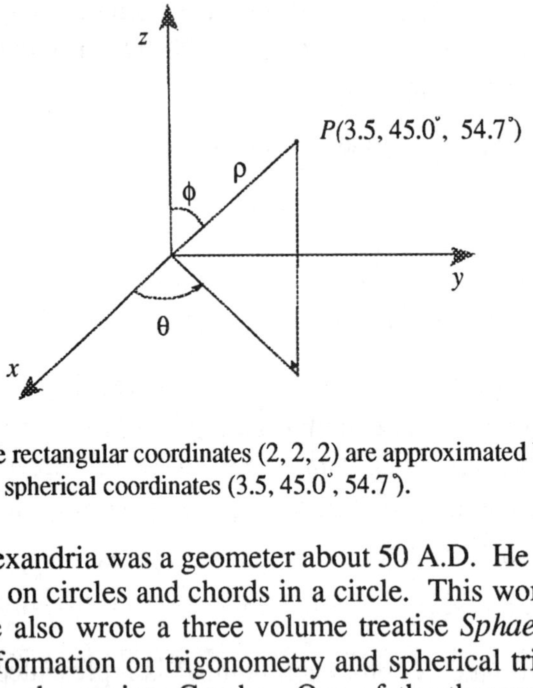

The rectangular coordinates (2, 2, 2) are approximated by the spherical coordinates (3.5, 45.0°, 54.7°).

7. Menelaus of Alexandria was a geometer about 50 A.D. He wrote a six volume treatise on circles and chords in a circle. This work is lost to us, however he also wrote a three volume treatise *Sphaerica* which yields much information on trigonometry and spherical trigonometry as developed by the ancient Greeks. One of the theorems used by Menelaus makes use of the following definitions.

A line that intersects each "side" of triangle ABC (one side is extended) is called a Menelaus line. The points of intersection are called Menelaus points.

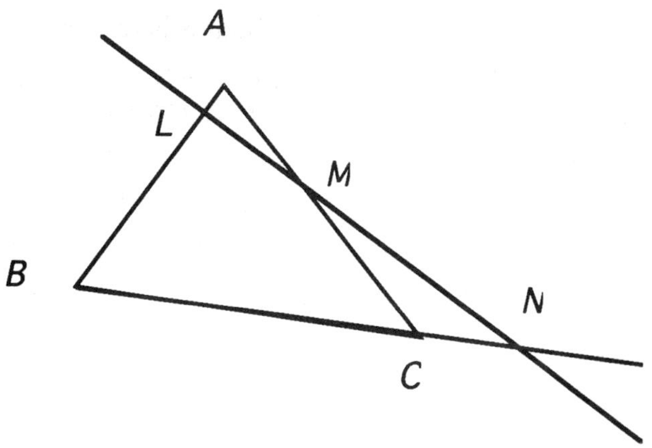

MENELAUS' THEOREM: Given triangle ABC such that a Menelaus line intersects side $\overline{AB}$, at point L, side $\overline{AC}$, at point M, and side $\overline{BC}$, at point N. Then:

$$\frac{AL}{LB} \cdot \frac{BN}{NC} \cdot \frac{CM}{MA} = 1$$

Menelaus theorem has a spherical triangle analogue. Given a spherical triangle ABC such that a great circle intersects side AB, at point L, side AC, at point M, and side BC, at point N. Then:

$$\left(\frac{\sin AL}{\sin LB}\right)\left(\frac{\sin BN}{\sin NC}\right)\left(\frac{\sin CM}{\sin MA}\right) = 1$$

Menelaus used the above theorem to produce many additional theorems concerning spherical trigonometry

8. Some of the meanings of the present names of the trigonometric functions are apparent when we examine a unit circle, with the angle θ placed in standard position. For example in the following diagram, the tan θ is the length of the tangent segment $\overline{PQ}$. The sec θ is the length of the secant segment $\overline{OQ}$. The term cot θ means complements tangent, or the tangent of the complement of θ. In the following figure the cot θ is the length of line segment $\overline{TU}$. The origin of the term *sine* is more difficult to understand. In the following figure the sin θ is given by the length of the half cord $\overline{RS}$. Robert of Chester (circa 1150) is credited with translating, into Latin, much of the Arabic work in trigonometry. When he came to translate *jiba* (meaning half-chord in Arabic) he apparently confused it with the word *jiab*, which meant "bay" or "inlet." Hence he used the word *sinus* which is Latin for bay or inlet.

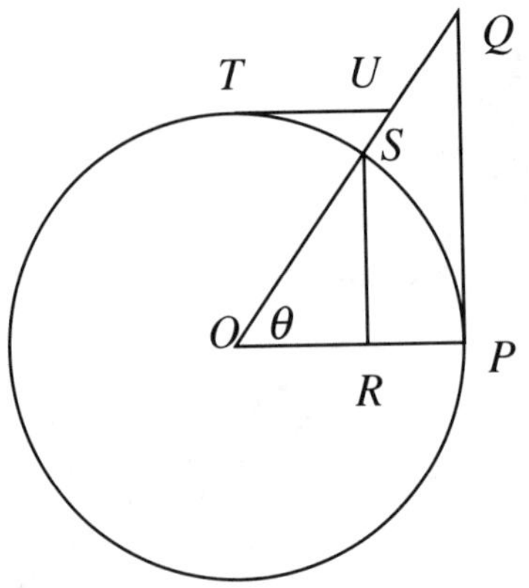

Pappus (ca. 300) was able to determine an extension of the Pythagorean Theorem that we now know as the Law of Cosines. This extension made use of Propositions 12 and 13 of Book II and Proposition 47 of Book I of Euclid's *Elements.* The first proof of the Law of Cosines was first given in 1579 by Francois Viète.

9. A great circle is a circle on a sphere which has its diameter equal to that of the sphere. A great circle is formed by the intersection of a plane and a sphere, where the plane passes through the center of the sphere. A trihedral angle is a polyhedral angle having three faces. The following figure shows the spherical triangle ABC, and the trihedral angle formed by the three sectors ABO, ACO, and BCO.

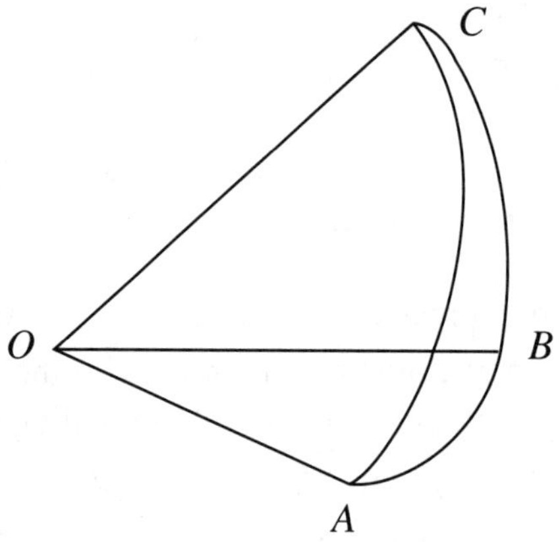

The sum of the measures of the three angles formed by a spherical triangle is always greater than 180°.

There are ten formulas which are useful for solving right spherical triangles. Fortunately, there is no need to memorize all ten of the formulas, because Napier devised two rules (which are now known as Napier's rules of circular parts) that can be used to reconstruct any of the formulas. Consider the following right spherical triangle and the circle to its right that we have divided into five parts. The bars on A, B, and c, mean the complement of the given part. Thus $\bar{A} = 90° - A$. The angles given by a, b, $\bar{A}$, $\bar{c}$, $\bar{B}$ are referred to as circular parts. Every circular part in the circle has two adjacent parts, and two parts that are not adjacent to it. Napier's rules of circular parts are as follows:

- The sine of any part is equal to the product of the cosines of the two nonadjacent parts.
- The sine of any part is equal to the product of the tangents of the two adjacent parts.

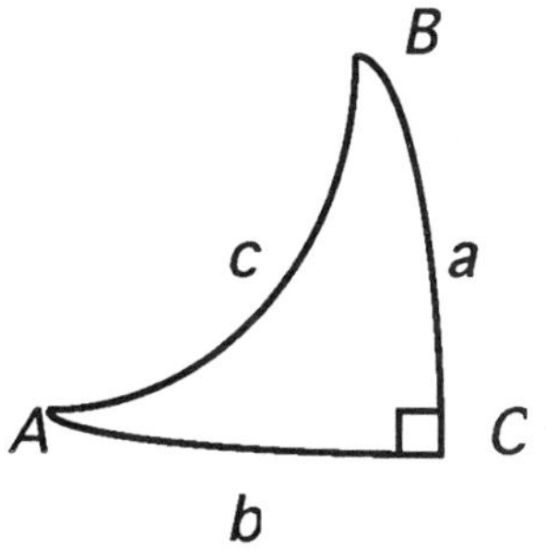

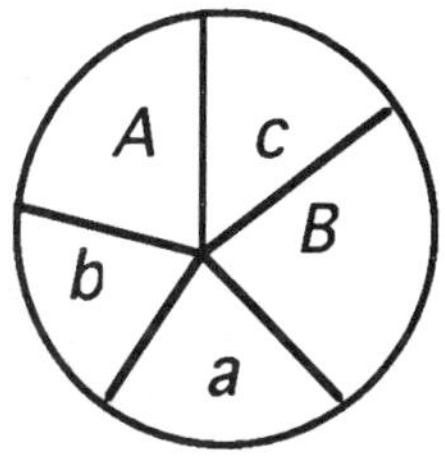

By applying Napier's rules for circular parts, we can construct the following ten formulas.

1. $\sin \bar{A} = \cos \bar{B} \cos a$	2. $\sin \bar{c} = \cos a \cos b$	3. $\sin \bar{B} = \cos \bar{A} \cos b$
4. $\sin a = \cos \bar{A} \cos \bar{c}$	5. $\sin b = \cos \bar{c} \cos \bar{B}$	
6. $\sin \bar{A} = \tan \bar{c} \tan b$	7. $\sin \bar{c} = \tan \bar{A} \tan \bar{B}$	8. $\sin \bar{B} = \tan \bar{c} \tan a$
9. $\sin a = \tan b \tan \bar{B}$	10. $\sin b = \tan a \tan \bar{A}$	

10. The graph of y=ssin t for $0 \le t \le 8$.

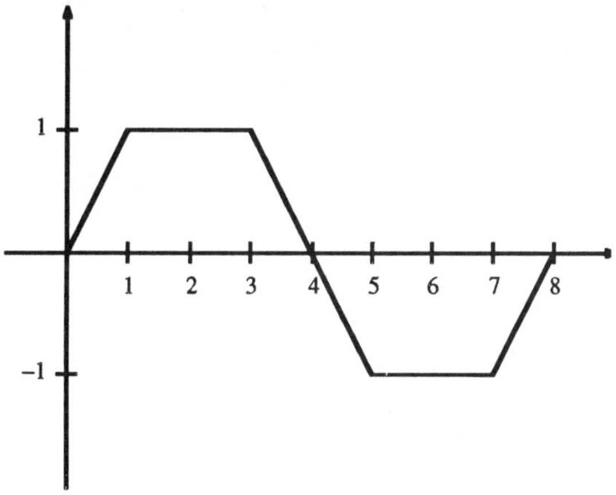

The graph of y = a ssin bt where $a > 0$, and $b > 0$, has the following properties (on the interval $(0, 8/b)$):

- The amplitude is a.
- The period is $8/b$.
- The zeros are 0, $4/b$, and $8/b$.
- The maximum value is a, when $\frac{1}{b} \le t \le \frac{3}{b}$.

  The minimum value is $-a$, when $\frac{5}{b} \le t \le \frac{7}{b}$.
- If $a < 0$, the graph is reflected across the t-axis.

The graph of $y = \text{stan}\, t = \dfrac{\text{ssin}\, t}{\text{scos}\, t}$, for $0 \le t \le 8$:

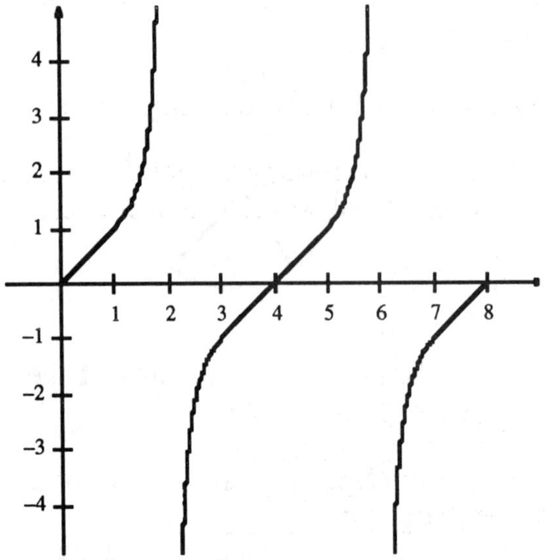

The above graph was achieved by considering the graph of $y = \text{ssin}\, t$, and $y = \text{scos}\, t$ over the intervals

$[0,1]$, $[1, 2) \cup (2, 3]$, $[3,5]$, $[5, 6) \cup (6, 7]$, and $[7,8]$.

For example, $\text{scos}\ t = 1$, and $\text{ssin}\ t = t$, on the interval $[0,1]$. Therefore, the graph of $y = \text{stan}\, t$ is the same as the graph of $y = \dfrac{t}{1}$, on $[0,1]$.

On $[1, 2) \cup (2, 3]$, $\text{scos}\ t = 2 - t$, and $\text{ssin}\ t = 1$. Therefore, the graph of $y = \text{stan}\, t$ is the same as the graph of $y = \dfrac{1}{2-t}$ on

$[1, 2) \cup (2, 3]$.

1. Fermat's principle:

Light traveling from one point to another will follow a path such that, compared with nearby paths, the time required is either a minimum or maximum or will remain unchanged (that is, it will be stationary). Snell's law (which follows) is derived by using Fermat's principle.

Snell's Law: $\dfrac{\sin\theta_1}{\sin\theta_2} = n_{21}$, where n_{21} is a constant called the index of refraction of medium 2 with respect to medium 1. See diagram below. (Note: the index of fraction depends of the wave length. As wavelength increases, the index of refraction decreases.

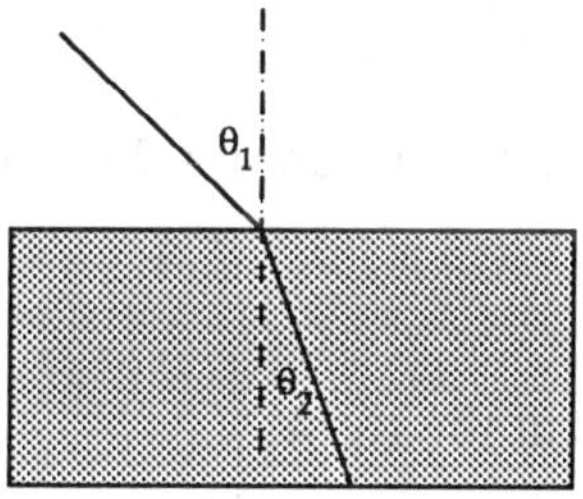

Because a diamond has a higher index of refraction, light entering the diamond will be refracted at a greater angle than the same light entering a piece of glass. The result will be a narrower 'rainbow' as the light leaves the diamond versus the light leaving the glass.

For instance, the index of refraction for certain glass for light at 4000 angstroms is 1.821 and at 5000 angstroms 1.765. For light entering glass at these wave lengths at an angle of 30^0 to the normal, the angle between the refracted beams can be found as follows.

$$\begin{aligned} \sin\theta_1 &= n_{21}\sin\theta_2 \\ \sin 30^0 &= 1.821\sin\theta_2 \\ \theta_2 &= 15.937^0 \end{aligned} \qquad \begin{aligned} \sin\theta_1 &= n_{21}\sin\theta_2 \\ \sin 30^0 &= 1.765\sin\theta_2 \\ \theta_2 &= 16.456^0 \end{aligned}$$

The difference between the waves is 0.519^0

The refractive indices of a certain diamond at the same wavelengths is 2.589 and 2.501. Using these indices, the difference between the waves is 0.397^0.

2. Mollweide's formulas: $\dfrac{a+b}{c} = \dfrac{\cos\frac{1}{2}(A-B)}{\sin\frac{1}{2}C}$ and $\dfrac{a-b}{c} = \dfrac{\sin\frac{1}{2}(A-B)}{\cos\frac{1}{2}C}$

These formulas can be used to verify the solution of a triangle but are not practical for solving a triangle.

Historical note: Mollweide's formulas are due to the German astronomer Karl Mollweide. However, these formulas were known by Thomas Simpson (of Simpson's Rule) in 1748, 26 years prior to the birth of Mollweide.

3. Law of tangents: $\dfrac{a-b}{a+b} = \dfrac{\tan\frac{1}{2}(A-B)}{\tan\frac{1}{2}(A+B)}$

The law of tangents can be applied to a triangle with two known angles and a side opposite one of the angles.

A historical note: The law of tangents is due to Thomas Fincke (circa 1580). He is also the one who introduced the words "tangent" and "secant."

4. Newton's formula: $\dfrac{a+b}{c} = \dfrac{\cos\frac{1}{2}(A-B)}{\sin\frac{1}{2}C}$

This formula is also attributed to Mollweide. See Essays and Projects, question 2.

5. See the diagram at the right.

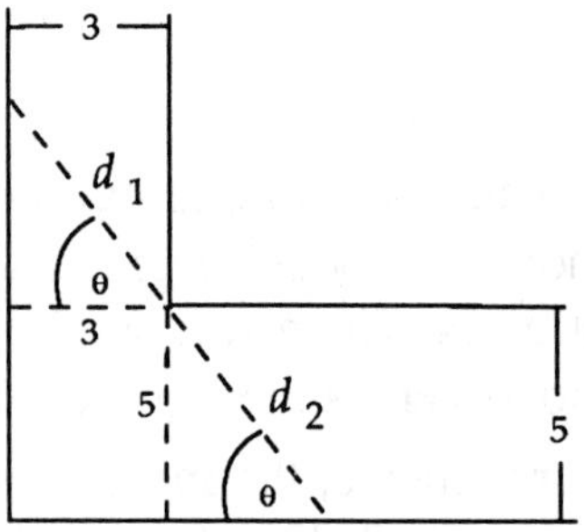

$d_1 = \dfrac{3}{\cos\theta}$ $\qquad d_2 = \dfrac{5}{\sin\theta}$

$L = d_1 + d_2 = \dfrac{3}{\cos\theta} + \dfrac{5}{\sin\theta}$

A graph of L is shown at the right. The minimum value of L is approximately11.194.

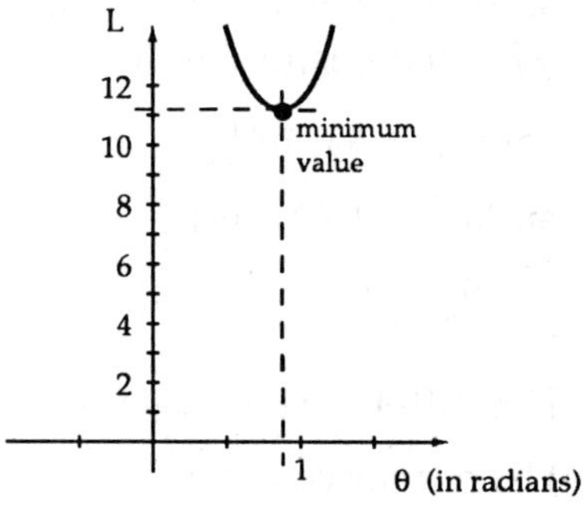

6. a) Graph $y = \dfrac{\sqrt{3}}{3}x - \dfrac{16\left(\frac{4}{3}\right)}{110^2}x^2$ and estimate the maximum value of y.
The maximum value is approximately 47.266 ft when x is 163.733.

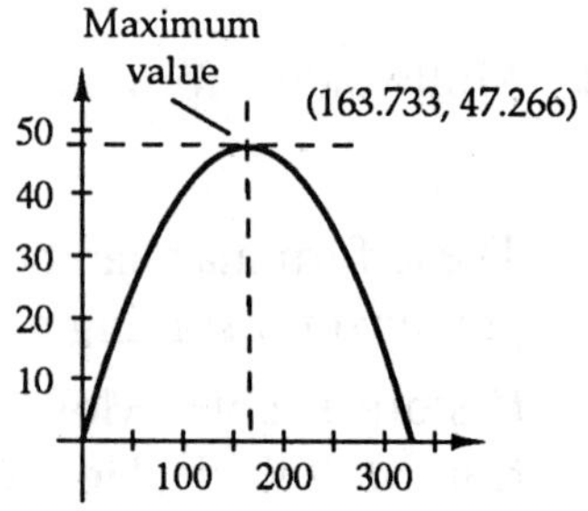

b) Solving $0 = (\tan\theta)x - \frac{16\sec^2\theta}{v_0^2}x^2 \; (x > 0)$,

we have $x = \frac{v_0^2 \sin\theta\cos\theta}{16}$.

Graphing $x = \frac{v_0^2 \sin\theta\cos\theta}{16}$ with x on the vertical axis and θ on the horizontal axis, $v_0 = 110$, the maximum value of x occurs when $\theta = \frac{\pi}{4} \approx 0.785$. At this value, $x = 378.125$ ft

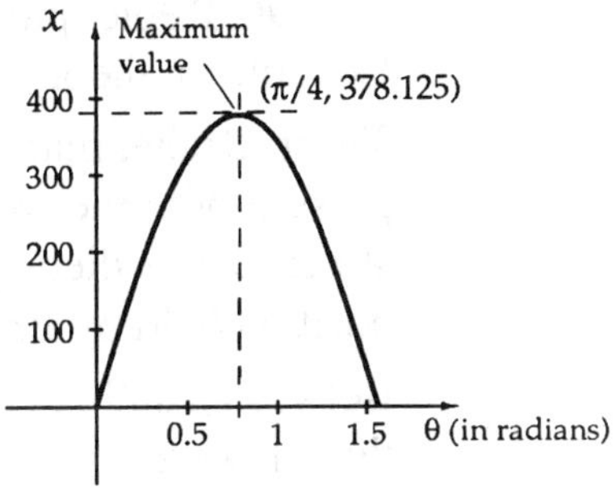

7. The graph is shown at the right. The maximum value is 252 feet when $\theta = \frac{\pi}{3}$

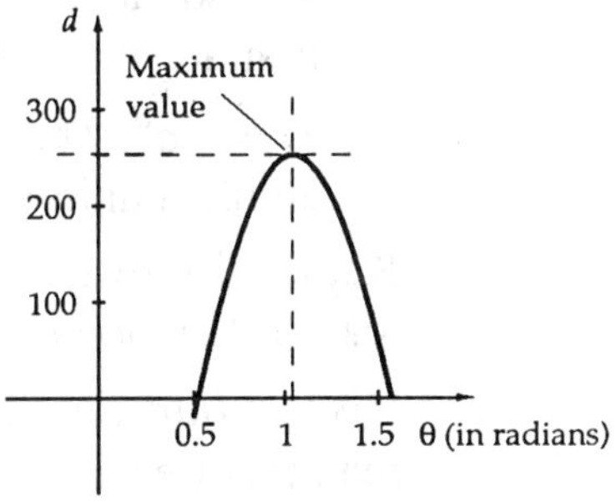

8. Using a transit the angle of elevation from a point a known distance from the base of the building (the distance being found by using the tape measure) to the top of the building can be determined. Using the tangent function, the height of the building can be determined.

One of the problems with this procedure is the practical matter of being able to move far enough from the building to have an accurate measure of the angle of elevation. If the angle of elevation were approximately 70^0 and the distance from the building 100 feet, an error of 1^0 in the measure of the angle would cause an error of approximately 14 to 16 feet in the height.

9. A rope with 12 equally spaced knots could be formed in the shape of a right triangle with legs with 3 and 4 knots and hypotenuse of 5 knots. The right angle could then be used as a template for a corner.

Any length rope that has as many knots as the sum of the three numbers that comprise a Pythagorean triple could be used. For instance, one of 30 knots since 30 is the sum of 5, 12, and 13.

1. Kepler was born Dec. 27, 1571 and died Nov 15, 1630. He was among the first strong supporters of the heliocentric theory.

In 1596, Kepler published Cosmographic Mystery in which he defended the Copernican theory.

Tycho Brahe, mathematician at the court of Emperor Rudolph II, was impressed with the work of Kepler and invited him to Prague as his assistant. When Brahe died the following year (1601), Kepler was appointed to the position held by Brahe.

Between 1609 and 1619, Kepler published his three laws of motion. The three laws are:

1. Each planet moves about the sun in an orbit that is an ellipse, with the sun at one focus of the ellipse.
2. The straight line joining a planet and the sun sweeps out equal areas in space in equal intervals of time.
3. The squares of the sidereal periods of the planets are in direct proportion to the cubes of the semimajor axes of their orbits. That is, $P^2 = ka^3$. The value of k depends on the units of measurement. If astronomical units are used, the $k = 1$.

Kepler also made contributions to optics, telescope lenses, and gave a physical explanation of nova.

His text *Introduction to Copernican Astronomy* was one of the most widely read treatises on astronomy.

A planet's velocity is greatest when at perihelion. This follows from his second law.

The period of Mars is 1.87 years. This follows from the third law.

2. Neptune was discovered as a result of mathematical prediction. The pertubative effects of Jupiter and Saturn on Uranus alone did not allow for discrepancies in Uranus' orbit. Using Newtonian gravitational theory, John Couch Adams produced mathematical evidence that an unknown planet could account for the irregularities in Uranus' orbit.

Adams sent his calculations, along with the region of the sky in which to search for the new planet, to Sir George Airy, the Royal Astronomer, and asked him to begin a search for the planet.

Airy had no faith in Adams' calculations and did not look for Neptune.

In June 1846, Leverrier, a French mathematician, independently reproduced the work of Adams. His calculations also were sent to Airy. This time the Royal Astronomer suggested that Cambridge University begin a search for the new planet.

The work of Challis, director of the Cambridge Observatory, was sloppy and negligent. Consequently, he did not find the planet. In September 1846, Leverrier suggested that Galle of the Berlin Observatory search the *Aquarius* region of the sky for the planet. Galle found the planet during his first observation. The discovery of Neptune was a great achievement for the time and was a major triumph of gravitational theory.

3. Using the eccentricity of Mercury's orbit as 0.206 and the length of the major axis as 7.2×10^7 miles, the apogee and perigee of the orbit can be calculated. Since $a = 3.6\times10^7$

$$\frac{c}{3.6\times10^7} = 0.206$$

$$c = 7.4\times10^6$$

The apogee is $a+c = 3.6\times10^7 + 7.4\times10^6 = 4.3\times10^7$.

The perigee is $a-c = 3.6\times10^7 - 7.4\times10^6 = 2.9\times10^7$.

4. The mirror at the Palomar Observatory is made from Pyrex, is 2 ft. thick at the ends and weighs 14.75 tons. The mirror has been ground to a true paraboloid within 0.0000015 inches. There is a 40 in. diameter hole in the center of the mirror to accommodate other focusing arrangements (Cassegrain, Coude').

An equation of the mirror is $y^2 = 2640x$. This assumes a focal length of 660 inches. A cross section of the mirror is shown below.

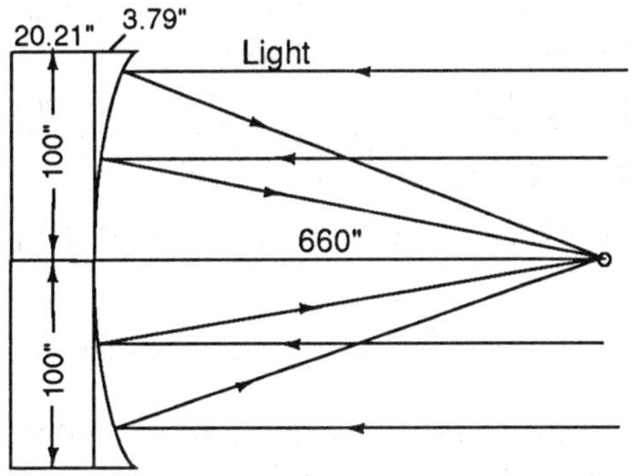

5. Let F be the focus and AB the directrix of a parabola. Fasten a string of length DC at F and C. See the figure below. Move the straightedge so as to keep the string taut. The pencil will trace a parabola.

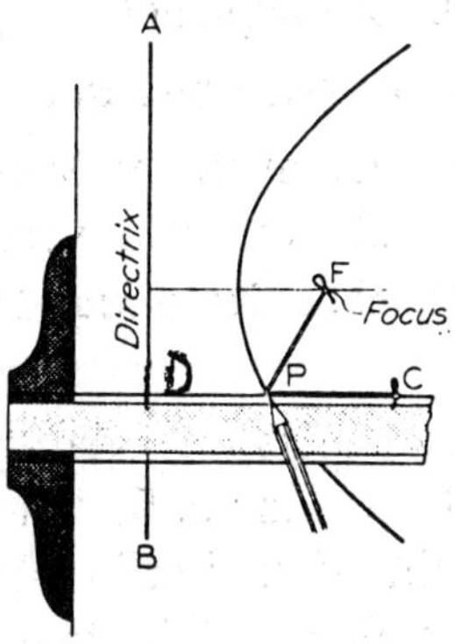

6. Let F and F' be the foci and AB the transverse axis of a hyperbola. See the diagram below. Fasten a string of length $FC - AB$ at F' and C, where C is an arbitrary point on a straight edge. Fasten the straightedge at F. Now revolve the straightedge about F with the pencil point moving against it and with the string taut. The pencil traces one branch of a hyperbola.

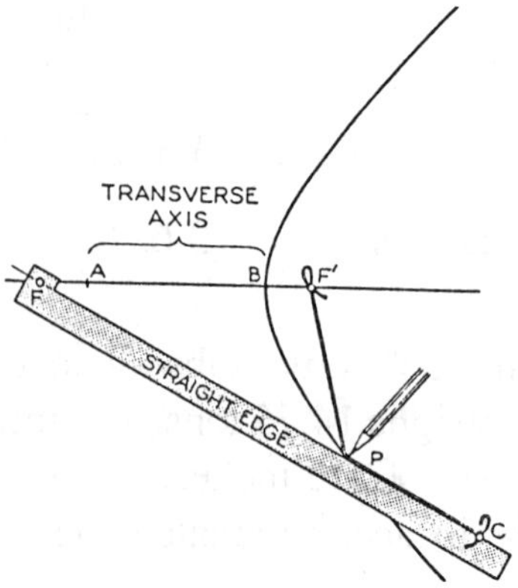

7. A trammel (as used for drawing an ellipse) is a piece of stiff paper or cardboard on which the length of the major and minor axis are drawn. See the diagram below. To draw an ellipse with the trammel, place the trammel with the endpoints on the axes of symmetry. A point on the ellipse is as shown. By moving the trammel to various positions, an ellipse can be drawn.

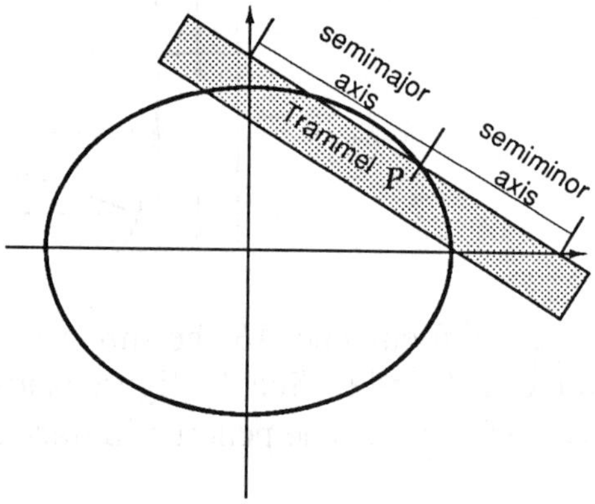

8. The area of an ellipse with semimajor axis a and semiminor axis b is πab. Archimedes, in *Conoids and Spheroids*, states "The areas of ellipses are as the rectangles under their axes." Which is to say, that the area of an ellipse is proportional to *ab*.

Johann Kepler showed that the area of an ellipse can be obtained from a circle of radius a through a transformation under which the ordinate of the circle at each point is shortened according to the ratio $\frac{b}{a}$. Kepler reasoned that the ratio of the areas of an ellipse and circle must be the same ratio as the shortened ordinates. That is, $\frac{\text{Area of ellipse}}{\text{Area of circle}} = \frac{b}{a}$. From this, the equation $A = \pi ab$ is derived.

9. In 1914, Srinivasa Ramanujan developed two formulas for approximating the perimeter of the ellipse given by $\frac{x^2}{a^2} + \frac{y^2}{b^2} = 1$. One of these formulas is $P = \pi(a + b)\left(3 - \frac{\sqrt{(a + 3b)(3a + b)}}{a + b}\right)$. He also stated that the error in this formula was approximately $\pi a \frac{e^{12}}{2^{20}}$. According to Ramanujan's formula perimeter, the perimeter of $\frac{x^2}{4^2} + \frac{y^2}{3^2} = 1$ is given by

$$P = \pi(4 + 3)\left(3 - \frac{\sqrt{(4 + 3 \cdot 3)(3 \cdot 4 + 3)}}{4 + 3}\right) \approx 22.10349179$$

This compared to 22.10349216 obtained by numerically integrating to find the perimeter.

Historical note: Johann Kepler approximated the perimeter of an ellipse using the formula $\pi(a + b)$.

10. The general equation is $\frac{y^2}{b^2} - \frac{x^2}{a^2} = \frac{z}{c}$. The graph for $a = 3$, $b = 4$, and $c = 1$ is shown at the right. The graph is symmetric with respect to the planes $x = 0$ and $y = 0$.

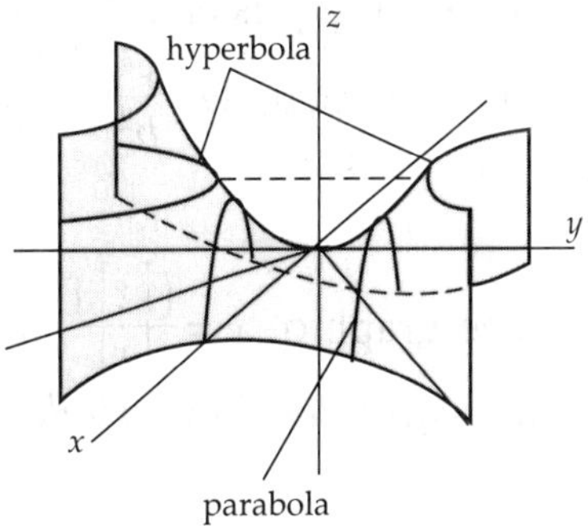

Section in the plane $x = 0$ is $y^2 = \frac{b^2}{c} z$ which is a parabola that opens upward and has vertex at the origin.

Section in the plane $y = 0$ is $x^2 = -\frac{a^2}{c} z$ which is a parabola that opens downward and has vertex at the origin.

If the surface is cut by the plane $z = k > 0$, the section is the hyperbola $\frac{y^2}{b^2} - \frac{x^2}{a^2} = \frac{k}{c}$ whose focal axis is parallel to the y-axis and whose vertices are on the parabola $y^2 = \frac{b^2}{c} z$.

If $k < 0$, the section is the hyperbola $\frac{x^2}{a^2} - \frac{y^2}{b^2} = \frac{|k|}{c}$ whose focal axis is parallel to the x-axis and whose vertices lie on the parabola $x^2 = -\frac{a^2}{c} z$.

11. General equation is $\frac{x^2}{a^2} + \frac{y^2}{b^2} - \frac{z^2}{c^2} = 1$. The graph for $a = 3$, $b = 4$, and $c = 5$ is shown at the right. Graph is symmetric with respect to each of the coordinate planes.

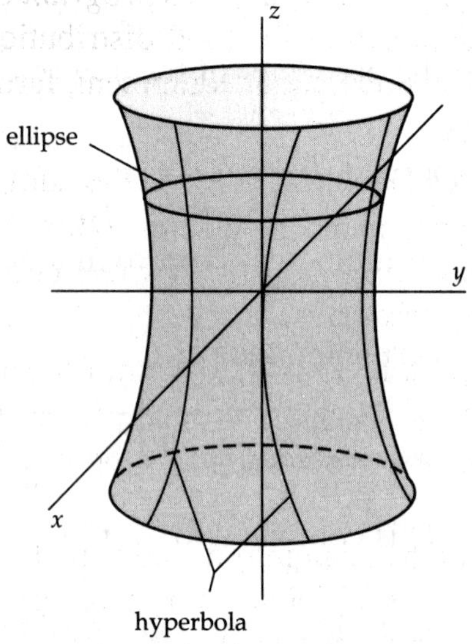

The sections cut by the coordinate planes are

$x = 0: \frac{y^2}{b^2} - \frac{z^2}{c^2} = 1 \qquad y = 0: \frac{x^2}{a^2} - \frac{z^2}{c^2} = 1$

$z = 0: \frac{x^2}{a^2} + \frac{y^2}{b^2} = 1$

The plane $z = k$ cuts the surface in an ellipse with equation $\frac{x^2}{a^2} + \frac{y^2}{b^2} = 1 + \frac{k^2}{c^2}$. The center of the ellipse is on the z-axis and the vertices on the hyperbolas $\frac{y^2}{b^2} - \frac{z^2}{c^2} = 1$ and $\frac{x^2}{a^2} - \frac{z^2}{c^2} = 1$.

12. The graph of $y = \frac{[|x|]}{|x|}$, for $-5 \le x \le 5$ is shown below. The graphs are *equilateral hyperbolas* in the sense that each graph is identical to the other.

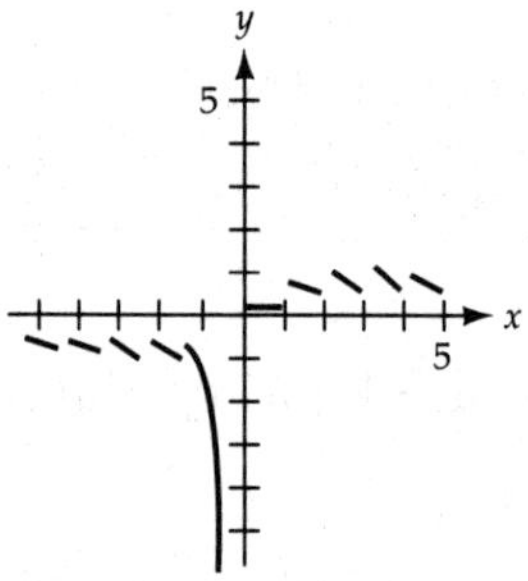

1. Linear programming is a *program* or method of allocating resources. There are applications to food distribution, building rockets, supplying military units with essential equipment, farming, developing economies, plus many others.

   George Danzig developed the 'simplex method' of solving linear programming problems in the late 1940's. The simplex method is basically a matrix method, analogous to row reduction. The result is the best method of allocating resources.

   In the early 1980's, Narendra Karmarkar of AT&T suggested an improvement on the simplex method that greatly reduced the time required for a computer to determine the optimal solution of a linear programming problem.

   L. G. Khachian introduced a method that was supposed to revolutionize the technique of solving linear programming problems. Although his method has some theoretical importance, its practical applications have not been demonstrated.

2. In *Flatland*, people are two-dimensional polygons. A person's station in life is determined by the number of sides of the polygon. People walk by sliding along the plane.

   The Flatlanders visit the people from the world in one dimension. These people live on a line. When the Flatlanders tell the ruler of the one-dimensional people, they could change position with their neighbors, the one dimensional people can not comprehend how such a movement would take place.

   The Flatlanders are dutifully smug about their ability to move in the plane and think they are superior to the one dimensional people.

   Then strange phenomenon begin when a three dimensional person enters the world of the Flatlanders. The Flatlanders cannot understand how the three dimensional people are entering their homes even though all the doors are windows are latched.

   An account of *Flatland* can be found in *World of Mathematics* edited by James Newman.

3. One of the basic postulates of Albert Einstein's special theory of relativity was that all observers should measure the same speed of light. This was contrary to Newtonian physics. In relativity theory there was no absolute time; in Newtonian theory there was absolute time.

   In Newtonian physics, if a pulse of light is sent from one place to another, different observers would agree on the time the journey took but not on the distance of the journey. In relativity theory, different observers would disagree on both the distance and *time* of the journey.

   An excellent account of space-time is *A Brief History of Time* by Stephen Hawking.

4. Some of the best accounts have occurred in *Scientific American*. Descriptions of capabilities of a four-dimensional person range from turning a tennis ball inside out with tearing it to being able to reach into a closed safe and remove the contents.

5. Coordinate lines are drawn parallel to the coordinate axes and form a parallelogram. See the illustration below. A point is located in much the same manner as in a rectangular coordinate system. However, displacement is along the edge of a parallelogram rather than a rectangle.

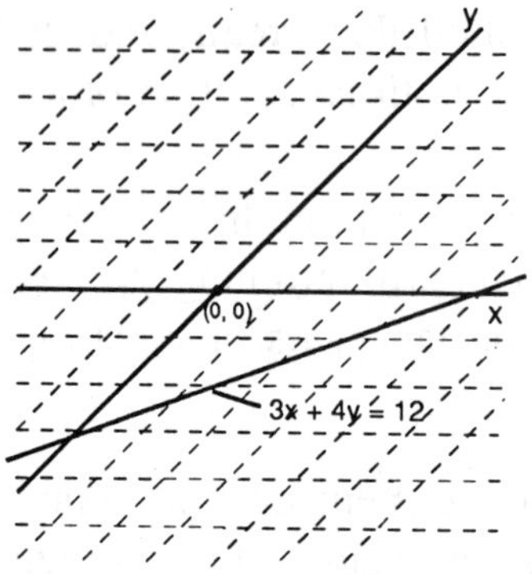

The graph of $3x + 4y = 12$ is a straight line. In fact, all linear equations in two variables have a graph that is a straight line. Here are some other observations a student may include.

Transformation equations between rectangular coordinates and parallelogram coordinates are given by:

$x' = x$, $y' = y - x$, where x' and y' are the coordinates in the parallelogram system.

One way to define the distance d between $P_1(x_1, y_1)$ and $P_2(x_2, y_2)$ is $d = |x_1 - x_2| + |y_1 - y_2|$.

1. Here are just a few applications that students may find:

Physics:	Pauli spin matrices
	Vibrating systems
	Spectroscopy
Probability:	Stochastic processes
Economics	Linear programming
Electrical Engineering	Circuit analysis
Operations Research	Game Theory
Computer Science	Data base, spreadsheets

2. William Rowan Hamilton (1805 - 1865) demonstrated exceptional talents at an early age. By the age of five he read Greek, Hebrew, and Latin. He published a paper in 1833 in which he introduced a formal algebra of ordered pairs that is essentially the operations on complex numbers. For the next ten years he worked to extend his algebra to n-tuples. Then, in 1843, while walking along the Royal Canal, the inspiration for the product of 4-tuples occurred to Hamilton and he engraved the inscription $i^2 = j^2 = k^2 = ijk = -1$ on a stone of Brougham Bridge. The significance of this equation is that it required a noncommutative multiplication and broke the barrier of the so-called "fundamental laws", which up to that time, were supported by the principle of permanence of form. That is, real number axioms were a permanent or fundamental form for all algebra.

   Arthur Cayley (1821 - 1895) was one of the first people to study matrices. Cayley studied transformations and recognized that, given two linear transformations, the order that the transformations were applied was important. Successive applications of transformations was regarded as multiplication of matrices. Cayley also was influential in the development of determinants, which preceded matrices.

   James Sylvester (1814 - 1897) gave the name matrix to any rectangular array from which a determinant could be formed. The idea of using the word matrix was taken from Latin. The word matrix is related to *mater* (mother) and a related word that means womb. A matrix was thought of as a thing from which something else was formed.

   Gabriel Cramer (1704 - 1752) published a theory of determinants in 1750. However, Leibniz had recognized determinants and their basic properties by the late 1600's.

3. There are many textbooks on programming that have the code for writing a program to find the inverse of a matrix. The general procedure is to produce code that will perform elementary row operations. The difficulty with these routines is the large number of multiplications and divisions that are required and the consequent problem of rounding errors. For matrices whose determinant is approximately zero, the rounding errors can be significant.

Approximating the Hilbert matrix $\begin{bmatrix} 1 & \frac{1}{2} & \frac{1}{3} & \frac{1}{4} \\ \frac{1}{2} & \frac{1}{3} & \frac{1}{4} & \frac{1}{5} \\ \frac{1}{3} & \frac{1}{4} & \frac{1}{5} & \frac{1}{6} \\ \frac{1}{4} & \frac{1}{5} & \frac{1}{6} & \frac{1}{7} \end{bmatrix}$ as $\begin{bmatrix} 1 & 0.5 & 0.33 & 0.25 \\ 0.5 & 0.33 & 0.25 & 0.2 \\ 0.33 & 0.25 & 0.2 & 0.17 \\ 0.25 & 0.2 & 0.17 & 0.14 \end{bmatrix}$ and then finding the inverse of the approximation yields $\begin{bmatrix} 10.61 & -34.99 & 4.74 & 25.28 \\ -34.99 & 134.54 & 5.64 & -136.57 \\ 4.74 & 5.64 & -129.80 & 141.08 \\ 25.28 & -136.57 & 141.08 & -14.22 \end{bmatrix}$. This is very different from the actual value of $\begin{bmatrix} 16 & -120 & 240 & -140 \\ -120 & 1200 & -2700 & 1680 \\ 240 & -2700 & 6480 & -4200 \\ -140 & 1680 & -4200 & 2800 \end{bmatrix}$.

The value of the determinant can be found by saving the divisors a_{ii} that were used to make the a_{ii} element a 1. Multiplying these is the value of the determinant. If the matrix is singular, one of the a_{ii} will be zero and the value of the determinant will be zero.

4. Let $A = \begin{bmatrix} 2 & 7 \\ 1 & 4 \end{bmatrix}$. Solving the equation $AX = I$ for X produces $X = A^{-1}$. Multiplying each of the elementary row operations together is effectively the following:

$$\begin{bmatrix} 1 & -\frac{7}{2} \\ 0 & 1 \end{bmatrix}\begin{bmatrix} 1 & 0 \\ 0 & 2 \end{bmatrix}\begin{bmatrix} 1 & 0 \\ -1 & 1 \end{bmatrix}\begin{bmatrix} \frac{1}{2} & 0 \\ 0 & 1 \end{bmatrix}AX = \begin{bmatrix} 1 & -\frac{7}{2} \\ 0 & 1 \end{bmatrix}\begin{bmatrix} 1 & 0 \\ 0 & 2 \end{bmatrix}\begin{bmatrix} 1 & 0 \\ -1 & 1 \end{bmatrix}\begin{bmatrix} \frac{1}{2} & 0 \\ 0 & 1 \end{bmatrix}I$$

$$X = A^{-1}$$

The elementary row operation matrices for a 3 x 3 matrix are:

Multiply a row by a constant: $\begin{bmatrix} k & 0 & 0 \\ 0 & 1 & 0 \\ 0 & 0 & 1 \end{bmatrix}, \begin{bmatrix} 1 & 0 & 0 \\ 0 & k & 0 \\ 0 & 0 & 1 \end{bmatrix}, \begin{bmatrix} 1 & 0 & 0 \\ 0 & 1 & 0 \\ 0 & 0 & k \end{bmatrix}$

Interchange two rows: $\begin{bmatrix} 0 & 1 & 0 \\ 1 & 0 & 0 \\ 0 & 0 & 1 \end{bmatrix}, \begin{bmatrix} 1 & 0 & 0 \\ 0 & 0 & 1 \\ 0 & 1 & 0 \end{bmatrix}, \begin{bmatrix} 0 & 0 & 1 \\ 0 & 1 & 0 \\ 1 & 0 & 0 \end{bmatrix}$

Multiply a row by a constant and add to a second row:

Multiply row 1 by k and add to row 2: $\begin{bmatrix} 1 & 0 & 0 \\ k & 1 & 0 \\ 0 & 0 & 1 \end{bmatrix}$

Multiply row 2 by k and add to row 3: $\begin{bmatrix} 1 & 0 & 0 \\ 0 & 1 & 0 \\ 0 & k & 1 \end{bmatrix}$

These are representative of the third elementary row operation.

To prove that multiplying a row of a matrix by a constant multiplies the determinant of the matrix by a constant, let K be a diagonal n x n matrix with one element c and the remaining elements 1 and A any n x n matrix. Let Det(K) = c and Det(A) = a. The matrix KA is matrix A with one row multiplied by c. Det(KA) = Det(K)Det(A) = ca

1. The early discussion of infinity was primarily theological and related to unlimited power or indefinitely large magnitudes. Galileo considered the infinitely many elements in the collection (*set* was not yet considered) of natural numbers. Galileo's *Dialogue Concerning the Two Chief World Systems* (1632) he makes the observation "There are as many squares as there are numbers because they are just as numerous as their roots." He anticipated Cantor by speculating which of the two sets, natural numbers or squares of natural numbers was larger. No answer was forthcoming in the *Dialogue*. Richard Dedekind was the first to define an infinite set as "A set *M* is infinite if it is equivalent to a proper subset of itself; in the contrary case, *M* is finite."

   Historical note: Dedekind received his Ph.D. under Gauss, yet Gauss was not comfortable about the idea of infinity and said in a letter to a friend "... I must protest most vehemently your use of the infinite as something consummated, as this is never permitted in mathematics."

   Georg Cantor is credited with developing the first coherent set theory in his *Mengenlehre* (Theory of Assemblages). He is also responsible for the concept of *cardinality* in which he tried to measure sets. He stated "Two sets have the cardinal number or have the same power if they are equivalent."

   The cardinality of the natural numbers was the 'smallest' infinite cardinal number and is referred to a *alpeh null.* The cardinality of the integers is *alpeh null.* A set is countably infinite if its cardinality is *alpeh null.*

   The cardinality of the real numbers is *alpeh one* and is frequently designated by *c*. The cardinality of the closed interval [0, 1] is *alpeh one.*

   John Wallis was the first to use the symbol ∞ for infinity (around 1659).

2. Most modern computer languages have evolved to be recursive languages. Even BASIC (Beginners All-purpose Symbolic Instructional Code) is now available in a version that allows for recursive calls to subroutines.

   A recursive function may call itself to do part of its task. This means that one part of the function is suspended while a different part is activated. A recursive function calls a copy of itself to work on a slightly different version of the function.

   Recursive functions make program code more compact and easy to read.

   The disadvantage of recursive functions is that some code may become inefficient as the number of calls increases. Secondly, recursive functions require more memory demands that nonrecursive functions.

3. A Monte Carlo method is a simulation of an experiment using random numbers. The term was coined by John Von Newmann in the 1940's. He and Stanislaw Ulam were studying atomic particles released during atomic reactions. It was their plan to simulate nuclear experiments by using random numbers. Frequently *Monte Carlo* and *simulation* are used synonymously.

4. Here are some applications of geometric series to business finance. Each one of these formulas is derived by adding the terms of a geometric series. In each formula, *PV* is present value, *FV* is future value, *PMT* is payment, n is the number of compound periods, and i is the compounding interest rate per compounding period.

Compound interest: $FV = PV(1+i)^n$

Present value of an annuity: $PV = PMT\left(\dfrac{1-(1+i)^{-n}}{i}\right)$

This formula can be used (by solving for *PMT*) to find the monthly payment for an installment loan such as a mortgage payment or car payment.

Future value of an annuity: $FV = PMT\left(\dfrac{(1+i)^n - 1}{i}\right)$

This formula is used to establish a sinking fund which is a series of payments which are designed to retire a debt.

Present value of a stock: $PV = \dfrac{D}{k_s}$

This formula is the sum of the infinite geometric series given by

$\sum_{t=1}^{\infty} \dfrac{D}{(1+k_s)^t}$, where D is the assumed constant dividend of a stock and k_s is the rate of return required by an investor. This gives the theoretical price of a stock.

Value of a bond: $\sum_{t=1}^{n} I\left(\dfrac{1}{1+k_d}\right)^t + M\left(\dfrac{1}{1+k_d}\right)^n$

where I is the dollars of interest paid per year, M the maturity value of the bond, k_d is the interest rate of the bond and n is the number of years before the bond matures.

Introduction to Graphing Calculators

Tips for Graphing

TEXAS INSTRUMENTS *TI-81*

To produce integer values for rectangular coordinates:

a. Set the range variables as follows: Xmin: –48, Xmax: 47, Xscl: 10, Ymin: –32, Ymax: 31, Yscl: 10, Xres: 1.

b. Graph the function and use the trace key.

To find the coordinates of $(x, f(x))$, $((x + a), f(x + a))$, $(x + 2a, f(x + 2a))$, ..., on the graph of $y = f(x)$:

a. Set the range variables as follows: Xmin: –48, Xmax: 47, Xscl: 10, Ymin: –32, Ymax: 31, Yscl: 10, Xres: 3 (in this example, $a = 3$).

b. Graph the function and use the trace key.

To display the x-coordinate of rectangular coordinates in tenths:

a. Set the range variables as follows: Xmin: –4.8, Xmax: 4.7, Xscl: 1, Ymin: –3.2, Ymax: 3.1, Yscl: 1, Xres: 1.

b. Graph the function and use the trace key.

To produce a pair of rectangular coordinates:

a. Enter the equation; for example, enter $Y_1 = 3X - 4$.

b. Press [2ND] [QUIT].

c. Enter any value for x; for example, enter 3.

d. Press [STO] [X/T] [ENTER].

e. Press [2ND] [Y-VARS] [1] (Y1) [ENTER]. You should get the corresponding y-value for the given x-value; in this example, you should get 5. The point is (3, 5).

f. Repeat steps (c)–(e) to produce as many pairs as desired.

To graph the sum (or difference) of two functions:

a. Enter the first function as Y_1 and the second function as Y_2.

b. Use the Y-VARS menu to enter $Y_1 + Y_2$ (or $Y_1 - Y_2$) into Y_3.

c. Graph Y_3.

CASIO *fx-7700G*

To produce integer values for rectangular coordinates:

a. Set the range variables as follows: Xmin: –47, Xmax: 47, Xscl: 10, Ymin: –31, Ymax: 31, Yscl: 10.

b. Graph the function and use the trace key.

To find the coordinates of the points $(x, f(x))$, $(x + a, f(x + a))$, $(x + 2a, f(x + 2a))$, ..., on the graph of $y = f(x)$:

a. Set the range variables as follows: Xmin: –141, Xmax: 141, Xscl: 10, Ymin: –93, Ymax: 93, Yscl: 10. (In this example, $a = 3$. Note that the Xmin and Xmax values are 3 times the values given above. In general, for $(x, f(x)) \rightarrow (x + a, f(x + a))$, use Xmin: $-47a$, Xmax: $47a$)

b. Graph the function and use the trace key.

To display the x-coordinate of rectangular coordinates in tenths:

a. Set the range variables as follows: Xmin: –4.7, Xmax: 4.7, Xscl: 1, Ymin: –3.1, Ymax: 3.1, Yscl: 1.

b. Graph the function and use the trace key.

To produce a pair of rectangular coordinates:

a. Enter the function; for example enter 3X – 4.

b. Press [SHIFT] [FMEM] [F1] (STO) [1] to store the function in function memory f_1 and press [AC].

c. Enter any value for x; for example, enter 3.

d. Press [→] [X,θ,T] [EXE] to store the value in x.

e. Press [F2] (RCL) [1] [EXE] to find the corresponding y-value for the stored x-value; in this example you should get 5. The point is (3, 5).

f. Repeat steps (c)–(e) to produce as many pairs as desired.

To graph the sum (or difference) of two functions:

a. Enter the first function.

b. Press [SHIFT] [FMEM] [F1] (STO) [1] to store the first function as f_1 and press [AC].

c. Enter the second function.

d. Press [SHIFT] [FMEM] [F1] (STO) [2] to store the second function as f_2 and press [AC].

e. Press [GRAPH] [F3] (fn) [1] [+] [F3] (fn) [2] [EXE]. To graph $f_1 - f_2$, substitute [-] for [+] in the sequence above.

Tips for Using the Zoom Functions

TEXAS INSTRUMENTS *TI-81*

Zoom 1: Box Allows you to define the corners of the new viewing window by positioning the cursor with the arrow keys. After choosing this option, move the cursor to the corner of the desired viewing area and press ENTER. Then use the arrow keys to define the box. Press ENTER again to replot the graph in the new viewing window.

Zoom 2: Zoom In Magnifies the graph from the point where the cursor is positioned.

Zoom 3: Zoom Out Displays a larger portion of the graph from the point where the cursor is positioned.

Zoom 4: Set Factors Allows the magnification or reduction factors used by the other zoom features to be defined.

Zoom 5: Square Replots the graph after squaring the screen.

Zoom 6: Standard Resets the viewing window to $-10 \leq x \leq 10$ and $-10 \leq y \leq 10$ and replots the graph.

Zoom 7: Trig Resets the range values to preset trigonometric values appropriate to either radian or degree measures (according to the mode selection). It then replots the graph.

Zoom 8: Integer Replots the graph after redefining the viewing window so that the x- and y-values are integers. After selecting this option, move the cursor to the center of the desired viewing rectangle and press ENTER. The trace feature will now display integral x-coordinates.

CASIO *fx-7700G*

Box Allows you to define the corners of the new viewing window by positioning the cursor with the arrow keys. After choosing this option, move the cursor to the corner of the desired viewing area and press EXE. Then use the arrow keys to define the box. Press EXE again to replot the graph in the new viewing window.

FCT Allows the magnification or reduction factors used by the other zoom features to be defined. Use this function to set factors which enlarge or reduce a graph.

×f Magnifies the graph from the point where you position the cursor.

×1/f Displays a larger portion of the graph from the point where the cursor is positioned, or replots a graph using the inverse of the factors specified with the FCT function.

ORG Resets the range values to the original values used and replots the graph.

Tips for Statistics

TEXAS INSTRUMENTS *TI-81*

To enter statistical data:

a. Press [2nd] [STAT] and select DATA from the menu. Press [2] (ClrStat) and [ENTER] to clear all old data from the memory.

b. Press [2nd] [STAT] and select DATA again. Press [1] (Edit). Enter each value; press [ENTER] after each entry.

To graph data:

a. Store the data as described above.

b. Press [2nd] [DRAW] and select [1] (ClrDraw) to clear the screen. Press [ENTER].

c. Press [RANGE] and enter values appropriate for the data.

d. Press [2nd] [STAT] and select DRAW. Select [1] (Histogram), [2] (Scatter plot), or [3] (xy Line). Press [ENTER].

To fit an exponential model to a set of data:

a. Store the data as described above.

b. Press [2nd] [STAT] and select CALC to choose a model. Press [4] (ExpReg) [ENTER] to choose an exponential regression curve.

c. Press [VARS] and select LR. Press [4] (RegEQ) to display the regression equation.

CASIO *fx-7700G*

To enter statistical data:

a. Press [MODE] [+] (REG) to use the statistical mode.

b. Press [MODE] [SHIFT] [1] (STO) [F2] (EDIT) [F3] (ERS) [F1] (YES) to select the store mode and to clear all old data.

c. Press MODE SHIFT 4 (NON–) to specify that the data should not be graphed.
d. Enter the x-coordinate of the first data point, then press SHIFT , . Enter the y-coordinate of the first data point, then press F1 (DT). Continue entering data in the same way.
e. Press F2 (EDIT) to review and edit the data.
f. Press PRE to return to the previous menu.

To graph data using an exponential model:

a. Press SHIFT F5 (Cls) to clear the coordinate screen.
b. Press MODE + (REG) and then MODE 6 (EXP) to select the exponential mode.
c. Press MODE SHIFT + (REC) to choose rectangular graphs.
d. Press RANGE and enter values appropriate for the data. To exit the range display, press RANGE twice.
e. Press MODE SHIFT 1 (STO) F2 (EDIT) F3 (ERS) F1 (YES) to select the store mode and clear all old data.
f. Press MODE SHIFT 3 (DRAW) to specify that the data should be graphed.
g. Enter the data as described above. Each point will be displayed on the coordinate screen.

To fit an exponential model to a set of data:

a. Store the data as described above.
b. Press F6 (REG).
c. Press F1 (A) EXE to display the constant term, a.
d. Press F2 (B) EXE to display the regression coefficient, b.
e. Calculate e^b by pressing SHIFT e^x F2 (B) EXE . The exponential model is $y = a \cdot e^b x$.

Tips for Matrices

TEXAS INSTRUMENTS *TI-81*

To enter a matrix:

a. Press MATRX ▶ to select the matrix edit menu.

b. Select either matrix [A], [B], or [C], then press ENTER.

c. Enter the dimensions of the matrix and press ENTER.

d. Enter each matrix element; press ENTER after each entry.

To find the determinant of a matrix:

a. Enter the values of the matrix elements as described above.

b. Press MATRX 5 (det).

c. Enter the name of the matrix by pressing 2nd, followed by [A], [B], or [C]. Press ENTER.

To find the inverse of a matrix:

a. Enter the values of the matrix elements as described above.

b. Press 2nd QUIT to exit the edit menu.

c. Enter the name of the matrix by pressing 2nd, followed by [A], [B], or [C]. Press x^{-1} ENTER.

CASIO *fx-7700G*

To enter a matrix:

a. Press MODE 0 to get to the matrix display.

b. Select matrix A by pressing F1, or matrix B by pressing F2.

c. Press F6 (▼) F1 (DIM) to specify the dimensions of the matrix; press EXE after each entry.

d. Enter each matrix element; press EXE after each entry.

e. Press [PRE] to return to the matrix mode display.

To find the determinant of a matrix:

a. Enter the values of the matrix elements as described above.

b. If the matrix is not displayed, press [F1] (A) or [F2] (B).

c. Press [F3] (|A| or |B|), the determinant key.

To find the inverse of a matrix:

a. Enter the values of the matrix elements and display the appropriate matrix as described above as described above.

b. Press [F4] (A^{-1} or B^{-1}), the determinant key.

Overview of Math Assistant

Overview of the Disk

The *Math Assistant* disk can be used independently by students or by the teacher as a classroom presentation device.

What is the *Math Assistant* software?

The *Math Assistant* software shows students graphic representations of functions, conics, polar graphs, and statistical data, and helps students to understand series, sequences, matrices, and sampling.

What can *Math Assistant* do?

Plotters: These programs allow you to enter equations, decide the scales of the axes, and plot the graphs. The following programs are included on the main menu: **Function Plotter, Equation Solver, Linear Programmer, Inverse Plotter, Conics Plotter,** and **Polar Graph Plotter.**

Quizzes: These programs check understanding by presenting graphs of a given type, and asking the student to determine equations for them. Two quizzes are included on the main menu: **Conics Quiz,** and **Circular Function Quiz.**

Utilities: **Data Analyzer:** a utility program for presenting histograms, scatterplots, and regression lines of spreadsheet data;
Sequence Experiment: a program that generates sequences defined explicitly or recursively, and calculates partial sums;
Matrix Calculator: a utility program that allows students to operate on matrices;
Sampling Experiment: a simulation that chooses samples from a population with specified statistical characteristics.

How can I learn to use the software?

Refer to the User's Manual: Introduction and Worked Examples for detailed instruction on all the options available with the *Math Assistant* disk.

The Worked Examples and Advanced Topics are provided for those who wish to use *Math Assistant* more frequently. They include suggestions of how to use the disk to make further explorations and discoveries.

The Exercises are designed to reinforce specific topics. Students may complete these independently or in small groups. Complete Answers are provided.

User's Manual: Introduction

Follow the instruction sheet that comes with the disk. Once you have booted the disk, you will see the title screen. Press the <Enter> key to get to the main menu, which lists the thirteen options shown at the right.

Use the arrow keys to move up and down this list, then select the highlighted item by pressing <Enter>.

After you select one of the programs from the main menu, it will load and you will see a program menu at the bottom of the screen. The options of a program menu are arranged in the order that you would generally use them. For example, the Function Plotter program has the program menu listed below.

- **Function Plotter**
- **Equation Solver**
- **Data Analyzer**
- **Linear Programmer**
- **Inverse Plotter**
- **Conics Plotter**
- **Conics Quiz**
- **Circular Function Quiz**
- **Polar Graph Plotter**
- **Sequence Experiment**
- **Matrix Calculator**
- **Sampling Experiment**
- **Quit**

HELP EQUATIONS DRAW SCALE ZOOM TABLE OTHER QUIT

Beginners may wish to use the HELP screen information. You must enter your EQUATIONS before you can DRAW the graphs. Likewise, you must DRAW your graphs before you can ZOOM in on them or list a TABLE of values. You may select the SCALE at any time. The OTHER option allows you to specify the grids, labels, and graph styles. Select QUIT to return to the main menu.

Arrow keys

In any menu, one of the menu options will be highlighted. In a vertical list, press the up and down arrow keys to move the highlighting to the option you want. In a horizontal list, press the left and right arrow keys. The arrow keys are also used to move from one set of parentheses to the next when filling in numbers in an equation format. Some sample formats are shown below:

Conics Plotter: ( )(x – ( ))^2 + ( )(y – ( ))^2 = ( )
or ( )x^2 + ( )y^2 + ( )x + ( )y = ( )

Similarly, the arrow keys may be used to enter the elements of a matrix in the Matrix Calculator program, and the elements of a data set in the Data Analyzer.

<Enter>

When making a choice from a menu, press this key to select the currently highlighted option. In the Worked Examples that follow, this key is symbolized by <↵>.

<Esc> (Escape)

Press the <Esc> key to leave an option and return to the previous menu. For example, after entering your EQUATIONS in the Function Plotter program, you may press <Esc> to go back to the Function Plotter program menu. You could then select the DRAW option to graph your equations.

User's Manual: Worked Examples

To familiarize yourself with *Math Assistant,* study the Worked Examples. These are step-by-step introductions to each of the programs on the disk.

In the Worked Examples that follow, the symbol <↵> is used to indicate pressing <Enter>. Also, the term "Select" is used to suggest highlighting a desired function and pressing <Enter>.

Function Plotter

The Function Plotter program allows you to graph up to three functions on the same coordinate plane, including polynomial, absolute value, trigonometric, exponential, and logarithmic functions.

Select Function Plotter from the main menu. Select EQUATIONS. Enter the equations shown:

#1: $y = -2x + 1$

#2: $y = (3/4)x - 3$

Press <Esc>. Select DRAW.

The point of intersection may be estimated by inspecting the graph. A good first approximation would be (1.5, –2.0). To achieve a better approximation, change the boundaries of the graphed region by using the ZOOM feature.

Select ZOOM. Use the arrow keys to move the cross hairs over the intersection. Press <↵>. Use the arrow keys to size the box as small as possible. Press <↵>.

The new approximation might be (1.45, –1.91). If you like, you can zoom in again to get a better approximation. If you wish to see a table of values for both functions, select TABLE. Press <P> to print the table.

The exact algebraic solution is $(1\frac{5}{11}, -1\frac{10}{11})$, or (1.4545. . ., –1.9090. . .).

Select SCALE and highlight "Yes" in the "Standard scale" option. Select EQUATIONS.

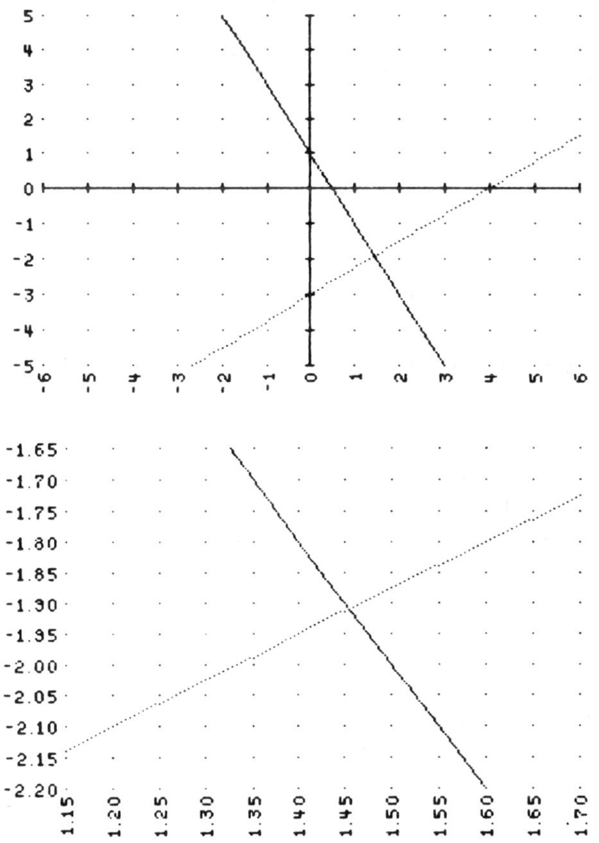

(solid) y = -2x+1
(dotted) y = (3/4)x-3

x	#1: y	#2: y
1.15	-1.3	-2.1375
1.20	-1.4	-2.1
1.25	-1.5	-2.0625
1.30	-1.6	-2.025
1.35	-1.7	-1.9875
1.40	-1.8	-1.95
1.45	-1.9	-1.9125
1.50	-2	-1.875

You can type over the old equations or press <Control-X> to clear each line. Enter the equations shown:

#1: $y = \text{Abs}(x)$

#2: $y = -\text{Abs}(x) + 2$

#3: $y = 3\,\text{Abs}(x - 4) - 5$

Press <Esc>. Select DRAW. The program draws three V-shaped graphs. With the Function Plotter, you can freely select the boundaries of the graph on at all four edges of the screen. To minimize distortion of the grid, you may want to select SCALE and set the following boundaries:

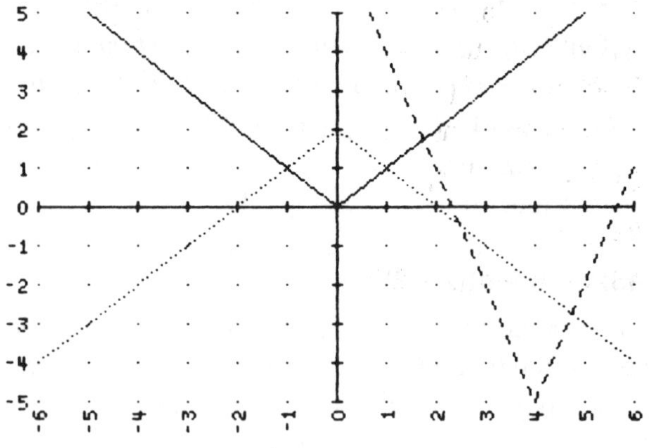

Least x-value: −6
Greatest x-value: 6
Least y-value: −5
Greatest y-value: 2

Press <Esc>. Select DRAW. The appearance of the graph may be improved.

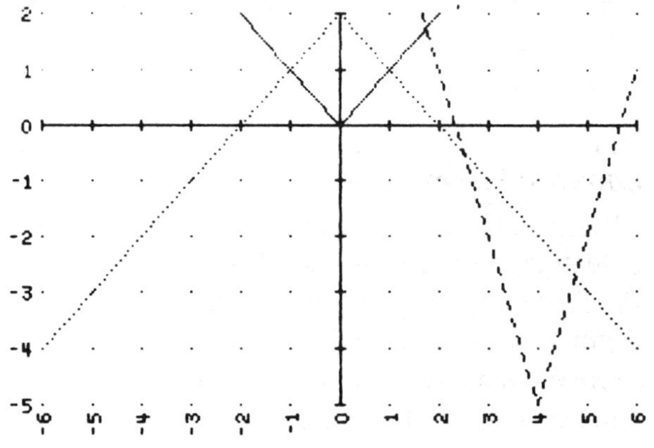

You can print out a copy of the graph by selecting OTHER and highlighting "Yes" in the "Print graph" option. Then press <Esc> and follow the instructions appropriate for your printer. Different printers may distort the grid differently, so you may need to adjust the SCALE until you are pleased with the printed copy.

The Function Plotter program allows you to plot trigonometric functions. It is often convenient to mark the x-axis in multiples of π. To do so, change the scale of the x-axis as follows. Select SCALE. Highlight "Multiples of pi" and use the right-arrow key to indicate "Yes." Press <Esc>.

Select EQUATIONS and enter the following three sinusoidal functions:

#1: $y = 5\sin(x)$

#2: $y = 3\sin(2x)$

#3: $y = \sin(4x)$

Press <Esc>. Select DRAW.

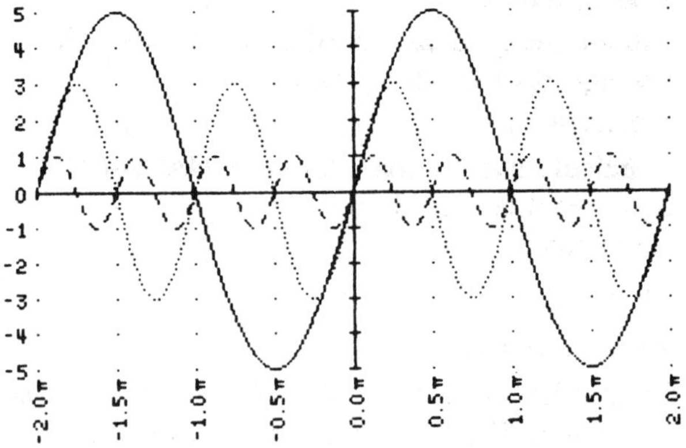

(solid) y = 5Sin(x)
(dotted) y = 3Sin(2x)
(dashed) y = Sin(4x)

Notice that the period of the first function is 2π, the period of the second function is π, and the period of the third function is $\frac{\pi}{2}$. In general, the period of the function $y = a\sin(bx)$ is $\frac{2\pi}{b}$.

You can draw portions of the inverse trigonometric functions. Use the notation: invsin, invcos, invtan, and so on. The irrational number π is programmed into the computer; use the notation PI.

This illustration shows samples of how the Function Plotter program can be used to graph rational functions with asymptotes, exponential functions, and polynomials.

The equations graphed are:

#1: $y = x/(2x + 4)$

#2: $y = e^x$

#3: $y = (x - 1)(x - 2)(x - 4)$

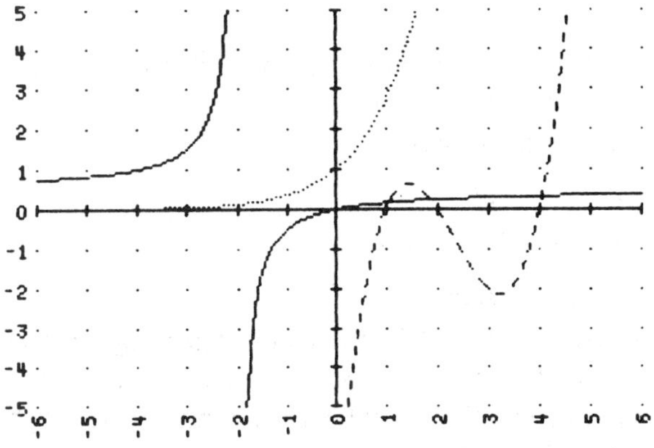

(solid) y = x/(2x+4)
(dotted) y = e^x
(dashed) y = (x-1)(x-2)(x-4)

Equation Solver

The Equation Solver program is like the Function Plotter, but instead of the ZOOM feature, it has a TRACE feature that allows you to move a box (and read the coordinates of points) along the graph of one of the two equations. By moving the center of the box over the intersection of two graphs, you can get a quick approximation of the solution of an equation. This feature is similar to the trace option on graphing calculators.

Problem: A golden rectangle has the property that the ratio of the short side to the long side is equal to the ratio of the long side to the sum of the short and long sides. If the short side of a golden rectangle has length 1, what is the length of the long side?

The problem requires you to solve this equation:

$$\frac{1}{x} = \frac{x}{1+x}$$

One way to solve this is to rewrite it as $x^2 = x + 1$ and use the quadratic formula. Another way is to use Equation Solver to find the two solutions by graphing.

Select Equation Solver from the main menu. Select EQUATIONS. Enter the following equations:

#1: $y = 1/x$

#2: $y = x/(1 + x)$

Press <Esc> and select DRAW. Two hyperbolas are drawn. The first has a vertical asymptote at $x = 0$, and the second has a vertical asymptote at $x = -1$. (The equations are undefined at the these points because division by zero is not defined.)

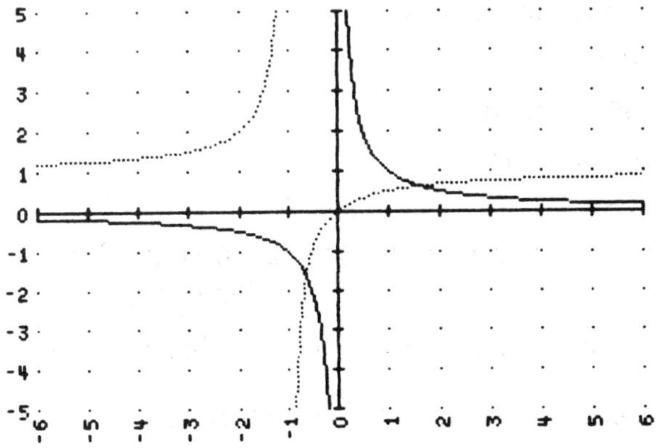

(solid) y = 1/x
(dotted) y = x/(1+x)

By inspecting the graphs, it is clear that there are two solutions to the original equation, one between –1 and 0, and one between 1 and 2.

To get quick approximations of the solutions, you can use the TRACE feature. Select TRACE. Use the arrow keys to move a box along the graph of the first equation. Press the space bar to see the coordinates of the center of the box. When the box contains the solution between 1 and 2, press <↵>.

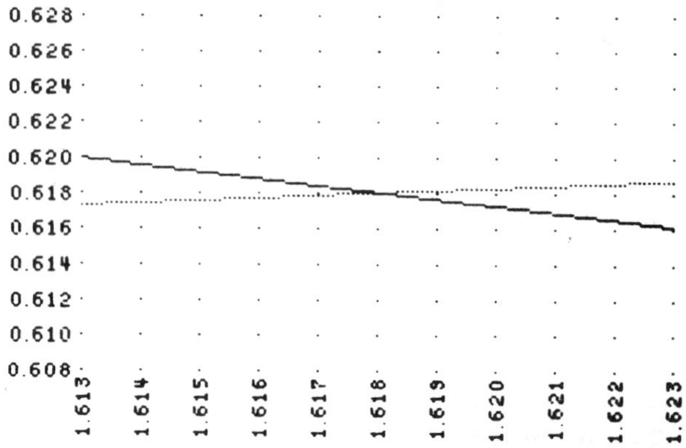

A new graph is drawn showing this intersection in more detail. You may have to use the arrow keys and press the space bar again to position the box directly over the intersection in the new graph. The intersection is approximately (1.618, 0.618). Therefore a rectangle with width 1 and length 1.618 is a golden rectangle, and the ratio of the width to the length is 0.618.

As with the Function Plotter, you can select TABLE to get a table of values of both equations within the bounds of the present screen.

You may get a printout of the graphs at any time by selecting OTHER and highlighting "Yes" in the "Print graph" option.

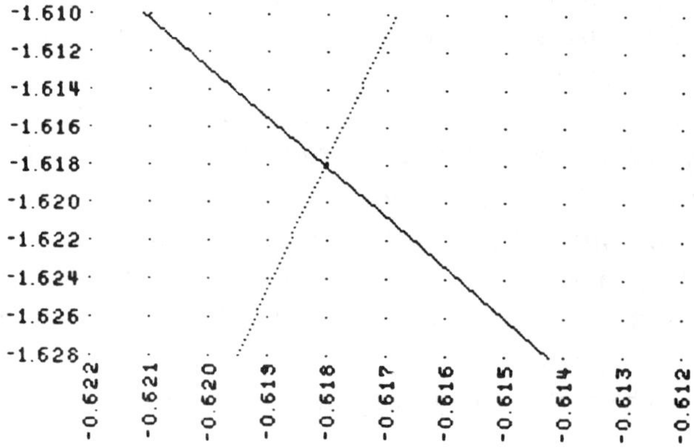

To find the solution that lies between –1 and 0, set the SCALE feature to "Standard scale" again and use TRACE as before.

The solution is approximately (–0.618, –1.618). The value $x = -0.618$ could not be the length of a golden rectangle, but it does satisfy the initial equation.

Data Analyzer

The Data Analyzer program allows you to enter data into a spreadsheet, calculate basic statistical features of the data (such as range, mean, median, and standard deviation), and present one column of data as a histogram or two columns of data as a scatterplot with a line of best fit.

Select Data Analyzer from the main menu. Select ENTER DATA. Press <Control-X> to erase the words "Column A" and type in "Before." Use the tab key to move to "Column B" and press <Control-X> to erase it and type in "After." Use tab and arrow keys to move from cell to cell in the spreadsheet. Enter the data shown on the next page: student names and heartrate measurements before and after exercise. Press <Esc> when all data are entered.

Select CALCULATE. The program will calculate the least and greatest values in each column, as well as the median, mean, and standard deviation of each column. The mean for the *Before* column is 84.1, with values ranging from 69 to 97; the mean for the *After* column is 153.9, with values ranging from 144 to 175.

You can present the data of either column as a histogram. Select GRAPH. Select HISTOGRAM. The first column is automatically highlighted. Press <↵> to accept this column. Its least value is 69. Instruct the computer to group the data in intervals of five.

Least value in first group: 65
Least value in second group: 70

Press <↵>. The histogram shows the frequency of values within each interval of five.

A scatterplot of the two columns can be created. Press <Esc>. Select SCATTER. Each x represents a data pair corresponding to a student's heartrate before and after exercise. [*Note:* If there are more than two columns of data in the table, the program will ask you to select the column to plot along the horizontal axis and then select the column to plot along the vertical axis.]

A line of best fit is drawn. Press the space bar to see the correlation coefficient for the data pairs, and the equation of the line of best fit. In this example, the correlation coefficient is 0.68607549 and the equation of the line of best fit is:
$y = (.80755805)x + (85.984368)$.

Imagine that data for a new student needs to be added to the table. Select EDIT DATA. Choose INSERT and select "Row." Use the arrow key to highlight the row above which you want the new row to appear. Press <↵>. Set the option:

Number of rows: 1

Add the new data: (70, 150).

Select CALCULATE to find new values for mean, median, range, and standard deviation. The new median before exercise is 84 and the new median after exercise is 152. Mean values are slightly reduced as well.

If you wish to save this information, select FILE and use the SAVE option. You will be asked to name the file. Later, you can select FILE and use the LOAD option to retrieve this information.

Item Label	Before	After
1. Walter	69	145
2. Ines	75	153
3. Jean	88	161
4. Oona	97	164
5. Heather	80	144
6. Nat	88	152
7. Si	84	144
8. Tony	78	148
9. Oliver	92	175
10. Naomi	90	153
Number	10	10
Least	69	144
Greatest	97	175
Median	86	152.5
Mean	84.1	153.9
Std Dev	8.1172655	9.5545802

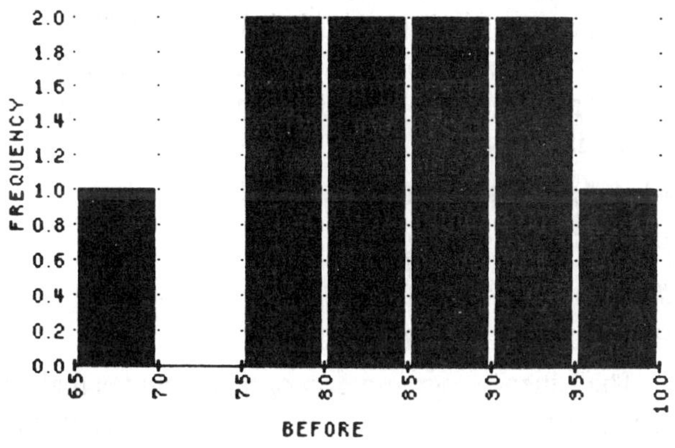

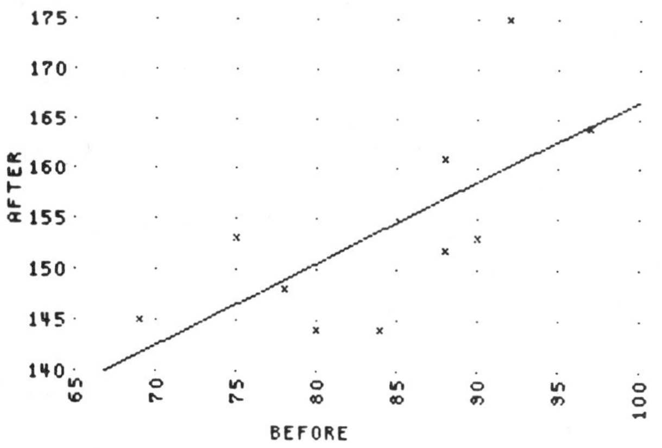

Linear Programmer

The Linear Programmer program presents graphs of three linear inequalities, thus modeling a linear programming problem with three restrictions. You can evaluate an objective function P *at any point inside the critical region defined by the inequalities to find maximum and minimum values.*

Assume that the restrictions that define the feasible region for a linear programming problem are as follows:

$$y \le 1x + 5$$
$$y \le -0.5x + 9.5$$
$$y \le -2x + 20$$

Select Linear Programmer from the main menu. Select DEFINE. Rewrite the inequalities in standard form. Use the right-arrow key to fill in the blank parentheses as shown below. [*Note:* The cursor pauses on the "<" sign, so you can change it to ">" if necessary.]

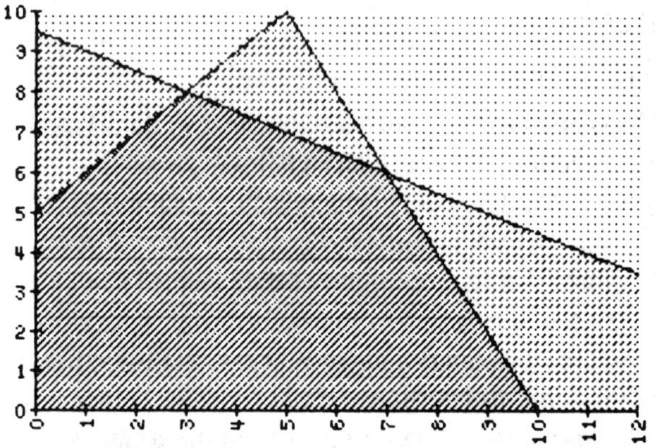

#1: (–1)x + (1)y <= (5)

#2: (0.5)x + (1)y <= (9.5)

#3: (2)x + (1)y <= (20)

```
(-1 )x+(1 )y <= (5 )
(0.5 )x+(1 )y <= (9.5 )
(2 )x+(1 )y <= (20 )
```

Press <Esc>. Select DRAW.

The Linear Programmer program assumes that only positive values of *x* and *y* are realistic. You can use the SCALE feature to change the graph boundaries as you wish. By inspection, the vertices of the most densely shaded area are: (0, 0), (0, 5), (3, 8), (7, 6), and (10, 0). Only those points inside the region defined by these vertices satisfy all three inequalities.

The Linear Programmer allows you to specify a function *P* that depends on *x* and *y* and evaluates it anywhere in the coordinate plane. Select DEFINE. Use the arrow keys to fill in the equation $P = (3)x + (-2)y$. Press <Esc>.

Select EVALUATE. Press the space bar. A box appears, and an information line below the graph lists the coordinates of the center of the box, whether it is inside or outside the critical region, and the value of *P*. Use the arrow keys to move the box across the coordinate plane.

Considering only points inside the critical region, students will find that the maximum and minimum values of *P* will be found at vertices. The maximum value of *P* is 30 at (10, 0) and the minimum value is –10 at (0, 5). (Note: Points at the bottom of the graph may be hidden by the information line. Press the space bar to see the full graph.) Every value of *P* inside the critical region lies somewhere between these two extreme values.

Inverse Plotter

The Inverse Plotter program graphs the inverse of any function you enter. If the inverse is a function, you can guess its equation, input it as a third function, and compare the graph of your guess to the graph of the inverse.

For example, to see what the inverse of the function $y = \frac{1}{2}x^3$ looks like, select EQUATIONS and enter the function shown:

#1: y = (1/2)x^3

Press <Esc> and select DRAW. A dashed line $y = x$ is drawn first. Then a solid and a dotted graph are drawn simultaneously. The solid graph is the entered equation and the dotted curve is its inverse.

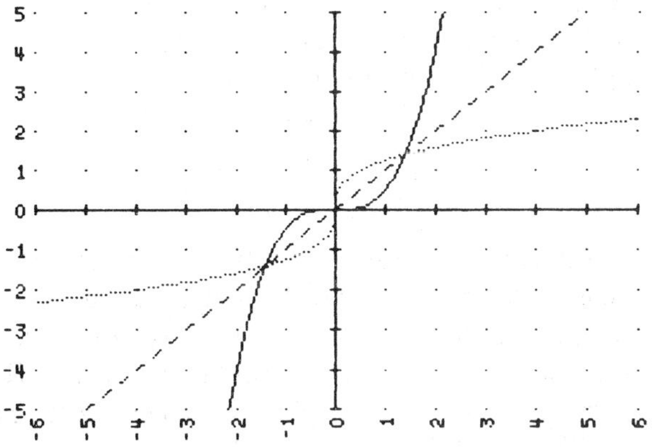

(solid) y = (1/2)x^3

In this case, the inverse equation appears to be a function. Students can try to guess the equation of the inverse and enter this as the optional equation #3. In this case, the inverse of cubing a number and taking half the result is doubling a number and taking the cube root of the result. Select EQUATIONS and enter the function shown:

#3: y = (2x)^(1/3)

[*Note:* You can also enter this as y = (2x)~3, the <~> notation being used to indicate roots.] Press <Esc> and select DRAW. The dashed function (drawn last) should exactly coincide with the dotted inverse function.

You can use the TRACE feature to find the coordinates of the three points where these two functions intersect. First, use SCALE to get closer to one of the solutions.

Least value of x: 1.40
Greatest value of x: 1.42
Least value of y: 1.40
Greatest value of y: 1.42

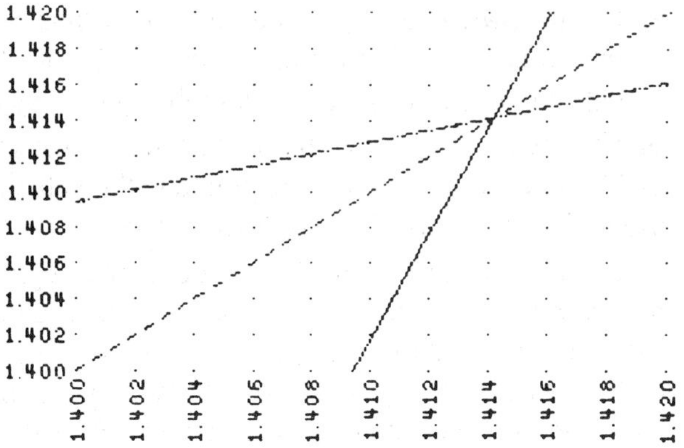

(solid) y = (1/2)x^3
(dashed) y = (2x)^(1/3)

Select TRACE and press the space bar. An open box appears along the graph of the function and a solid box appears along the graph of the inverse function. Use the arrow keys to move these boxes simultaneously. The coordinates of the centers of the boxes are listed below the graph.

The graphs coincide at (–1.414, –1.414), (0, 0), and (1.414, 1.414).

An algebraic solution of the equation

$$\frac{1}{2}x^3 = \sqrt[3]{2x}$$

results in $x = 0$ or $x = \sqrt{2}$ or $x = -\sqrt{2}$, which confirms the graphical result.

Conics Plotter

The Conics Plotter program allows you to investigate parabolas, ellipses, circles, and hyperbolas.

Graph the following system:

$x^2 + y^2 = 16$

$4x^2 + 25y^2 = 100$

Select Conics Plotter from the main menu. Select EQUATIONS. Fill in the first format for each as follows:

#1: (1)(x – (0))^2 + (1)(y – (0))^2 = (16)

#2: (4)(x – (0))^2 + (25)(y – (0))^2 = (100)

Alternatively, you could fill in the second format for each as follows:

#1: (1)x^2 + (1)y^2 + (0)x + (0)y = (16)

#2: (4)x^2 + (25)y^2 + (0)x + (0)y = (100)

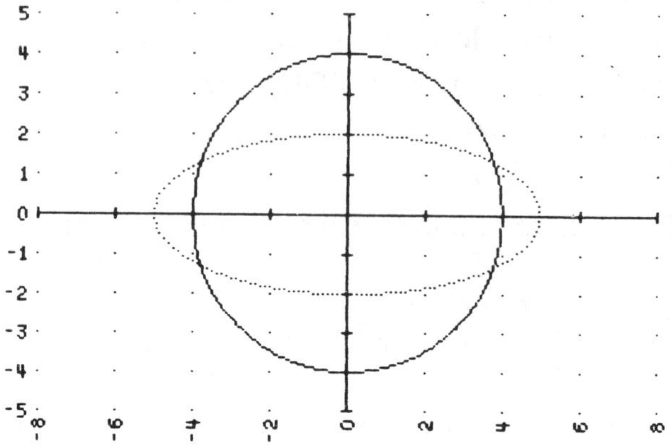

(solid) (1)(x-(0))^2 + (1)(y-(0))^2 = (16)
(dotted) (4)(x-(0))^2 + (25)(y-(0))^2 = (100)

Press <Esc>. Select DRAW.

The first equation represents a circle with radius 4. The second equation represents an ellipse with major axis 10 units and minor axis 4 units. By inspecting the graphs, you can see there are four solutions to the system.

You may find that screen distortion makes circles look like ellipses. You can reduce this distortion by changing the SCALE to display the intervals $-6 \le x \le 6$ and $-4 \le y \le 4$.

Printouts may be be different from the screen display. To achieve an adequate image, you may need to change the scales again. In the example shown, the SCALE was set to display the intervals $-8 \le x \le 8$ and $-5 \le y \le 5$.

The Conics Plotter program can be used to plot parabolas, lines, and hyperbolas as well as ellipses and circles. Using the second format, a parabola will result if the coefficient of x^2 or y^2 is zero. A line will result if both of those coefficients are zero. A hyperbola will result if these coefficients have opposite signs.

Conics Quiz

The Conics Quiz presents various conic graphs and asks you to determine an equation that will generate them. Your guess is graphed along with the original graph so you can visually compare the two and modify your guess if necessary.

Select Conics Quiz from the main menu.

Select OTHER to choose from the following options.

Center/Vertex: At origin Translated
Problem type: Circles Parabolas Ellipses Hyperbolas Mixed

Once the options have been set, press <Esc> to return to the Conics Quiz menu and select PROBLEM.

The computer graphs a conic of the type selected. You must guess the equation of the conic. Select TRY and use the arrow keys to fill in the blanks of the equation format shown below.

( )(x – ( ))^2 + ( )(y – ( ))^2 = ( )

or y – ( ) = ( )(x – ( ))^2 or x – ( ) = ( )(y – ( ))^2

Press <↵> when you are done.

The computer will graph the original conic and your guess so that you can visually compare the two conics. If the conics do not coincide, you may select TRY again. You may obtain the correct equation by selecting ANSWER at any time.

Circular Function Quiz

The Circular Function Quiz presents various trigonometric graphs and asks you to determine an equation that will generate them. Your guess is graphed along with the original graph so you can visually compare the two and modify your guess if necessary. Sometimes there is more than one correct answer.

Select Circular Function Quiz from the main menu.

Select OTHER to choose the difficulty level and type of circular function from the following options.

Difficulty level: Easy Moderate Hard
Problem type: Sin, Cos Tan, Cot Sec, Csc

Once the options have been set, press <Esc> and select PROBLEM.

The computer graphs a circular function of the type selected. You must guess the equation of the function. Select TRY and type in the name of the function. Use parentheses around the argument of a circular function. Type PI for π. An example is shown below.

y = 2SIN(X – PI/2) + 1

Press <↵> when you are done.

The computer will graph the original circular function and your guess so that you can visually compare the two graphs. If the graphs do not coincide, you may select TRY again. You may obtain the correct equation by selecting ANSWER at any time. There may be more than one correct answer.

Polar Graph Plotter

The Polar Graph Plotter program draws graphs of polar equations, in which the radius, r, is a function of the angle from the pole, t (theta). [Note: Most textbooks use the greek letter θ for theta.]

Select Polar Graph Plotter from the main menu. Select EQUATIONS. Enter the equation shown.

#1: r = sin(t/2)

Press <Esc> and select DRAW. The standard scale shows a maximum radius of 3 units. The graph appears to be contained within the unit circle, however. To get a closer view, select SCALE and change the radius to 1.

Greatest value of |r|: 1

The standard scale shows that theta varies from 0 pi to 4 pi. All angle measures in the Polar Graph Plotter program are in radians. Press <Esc> and select DRAW.

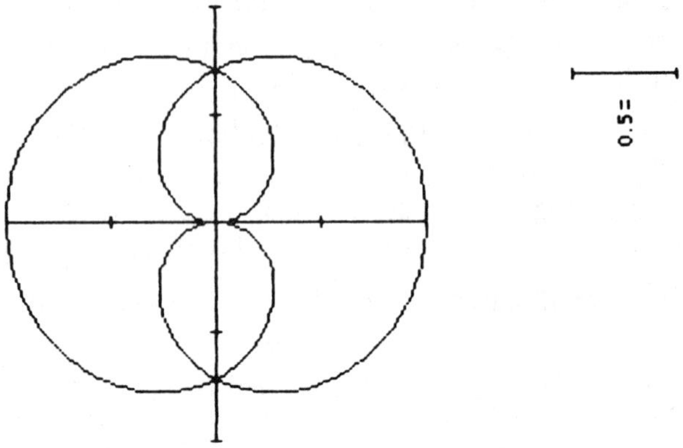

(solid) r = Sin(t/2)

Change the equation as follows.

#1: r = sin(t/4)

In order for the entire graph to be drawn, you will have to set the range of theta from 0 pi to 8 pi. Select SCALE and change the options as shown.

Least value of t (theta): 0 pi
Greatest value of t (theta): 8 pi

Press <Esc> and select DRAW. You can print out a graph at any time by selecting OTHER and highlighting "Yes" in the "Print graph" option.

The number π can be entered using the notation "PI" in any equation.

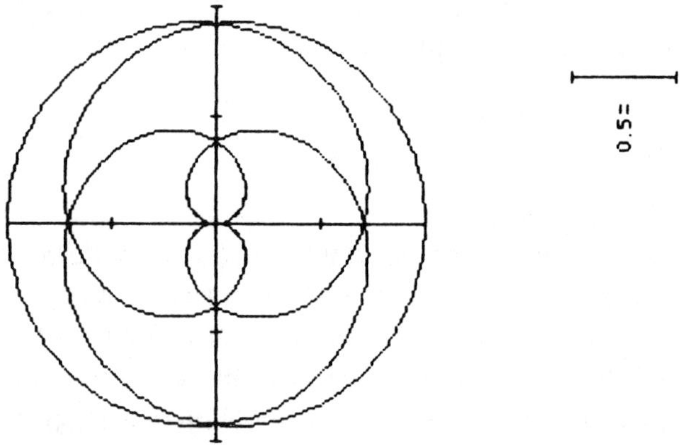

(solid) r = Sin(t/4)

Sequence Experiment

The Sequence Experiment program allows you to specify an explicit or recursive definition for a sequence, and then view the terms of the sequence and the partial sums associated with these terms.

To enter the sequence $\frac{1}{2}, \frac{2}{3}, \frac{3}{4}, \ldots$ using an *explicit* definition, select Sequence Experiment from the main menu. Select DEFINE and enter the equation as shown.

t(n) = n/(n + 1)

The term "t(n)" is used in the program to represent the *n*th term of the sequence. Many textbooks use the notation t_n or a_n for this. The first value of *n* is 1 unless you specify otherwise in the OTHER submenu.

Press <Esc> and select RESULTS. The screen will display the first few terms of the sequence, and the partial sums associated with them. For instance, S(3) = t(1) + t(2) + t(3). To see the next screen of terms, press <S>. To see a continuously scrolling view of the terms, press <C>. You must press the space bar to pause during the continuous scroll. Press <C> again to resume scrolling.

If you wish to print a page of terms and partial sums, select PRINT.

In this case, the sequence of terms appears to have a limit of 1. The sequence of partial sums does not appear to have a limit.

t(n) = n/(n+1)

n	sequence t(n)	partial sums S(n)
1	.5	.5
2	.666666667	1.16666667
3	.75	1.91666667
4	.8	2.71666667
5	.833333333	3.55
6	.857142857	4.40714286
7	.875	5.28214286
8	.888888889	6.17103175
9	.9	7.07103175
10	.909090909	7.98012266

Recursive definitions of sequences define later terms in terms of earlier ones. The Sequence Experiment allows up to four initial terms to be specified. A simple example of a recursive definition that involves two initial terms is the sequence of Fibonacci numbers, whose first two terms are 1 and whose subsequent terms are each the sum of the immediately preceding two terms. Select OTHER and highlight "Recursive" instead of "Explicit." Select DEFINE. Set the following options.

Number of initial values: 2

Initial values: t(1) = 1 t(2) = 1

t(n) = t(n – 2) + t(n – 1)

Press <Esc> and select RESULTS. Press <S> to see the next screen of these Fibonacci numbers. Clearly neither the sequence of terms nor the sequence of partial sums has a limit.

Using a calculator, you can show that the ratio of any two consecutive terms approaches the golden ratio 0.618. Interestingly, the ratio of any two consecutive partial sums also approaches this same number.

t(1) = 1 t(2) = 1
t(n) = t(n-2)+t(n-1)

n	sequence t(n)	partial sums S(n)
1	1	1
2	1	2
3	2	4
4	3	7
5	5	12
6	8	20
7	13	33
8	21	54
9	34	88
10	55	143

Matrix Calculator

The Matrix Calculator program allows you to enter matrices and perform operations on them, including addition, multiplication, inversion, and row reduction.

To enter or edit matrices, or create equations using matrices, use the DEFINE feature. To view the resulting matrices and scalars, use the RESULTS feature.

Select Matrix Calculator from the main menu. Select DEFINE. Choose "New matrix" and enter new variable name *A*. Press <↵>. The default size of a matrix is 3 rows by 3 columns. You can use the arrow keys at this point to change the number of rows or number of columns to any integer from 1 to 6. Press <↵> twice to accept the 3 by 3 dimensions. A 3 by 3 matrix will appear filled with zeros. Use the tab key to move from cell to cell and enter numbers of your own choice to fill in the elements of the matrix *A*.

Press <Esc> when you are done. The letter *A* appears in the command line at the bottom of the screen. If you wish to edit the elements of matrix *A*, highlight this *A* and press <↵> three times until the cursor appears in the matrix.

Select "New equation" and enter new variable name *B*. Define the new matrix *B* with the equation *B* = Inv(*A*). This will place the inverse of matrix *A* into matrix *B*. Press <↵>. [*Note:* You will receive an error message if the matrix is not invertible.] Select "New equation" again and enter new variable name *C*. Define the new matrix *C* with the multiplication *C* = *AB*. [*Note:* No multiplication symbol is needed. Also, to multiply a matrix *M* by itself, type *MM*, not *M*^2.]

Now that matrices *A*, *B*, and *C* have been defined, you can view their contents by returning to the program menu and selecting RESULTS. The letters *A*, *B*, and *C* appear in the command line at the bottom of the screen. Use the arrow keys to select each one in turn. The matrices will appear on the screen. As expected, matrix *C* is an identity matrix, since it is the product of matrix *A* and its inverse. You may find that the elements of matrix *B* are presented as fractions. If you prefer decimals, select DISPLAY MODE from the program menu and highlight "In decimal form always" instead of "In rational form when possible."

Select CLEAR from the program menu and highlight "Yes" to erase all matrices and equations.

You can use the Matrix Calculator to solve systems of equations very quickly. Suppose you have the following system of 3 equations in 3 variables:

$$\begin{aligned} x + y + 2z &= 5 \\ 3x + y - z &= -3 \\ 2x + 2y + 3z &= 6 \end{aligned}$$

Select DEFINE and choose "New matrix" with new variable name *M*. Alter the dimensions of matrix *M* so it has 3 rows and 4 columns. Enter the elements of matrix *M* shown below.

$$M = \begin{bmatrix} 1 & 1 & 2 & 5 \\ 3 & 1 & -1 & -3 \\ 2 & 2 & 3 & 6 \end{bmatrix}$$

Choose "New equation" with new variable name *R*. Define *R* with the equation *R* = Red(*M*). The program will perform row operations automatically to reduce matrix *M* to row-reduced echelon form. Return to the program menu and select RESULTS to view the elements of matrix *R*.

$$R = \begin{bmatrix} 1 & 0 & 0 & 2 \\ 0 & 1 & 0 & -5 \\ 0 & 0 & 1 & 4 \end{bmatrix}$$

Matrix *R* indicates that the solution to the system of three equations is: $x = 2$, $y = -5$, and $z = 4$.

Other available functions are: Det(*M*), which calculates the determinant of matrix *M* (a scalar) and Idn(*n*), which creates an *n* by *n* identity matrix.

Sampling Experiment

The Sampling Experiment program simulates selecting samples from a population with a specified mean and standard deviation. The means of the samples can be plotted on a histogram. Students can discover that when many samples are selected, the results approximate a normal curve.

Select Sampling Experiment from the main menu. Select DEFINE. The computer will create in its memory a "population" of normally distributed numbers having certain default characteristics: a mean of 100 and a standard deviation of 10. Note that the sample size is 20; when the computer samples the population, it will randomly select 20 "individuals" from it.

What do you think the mean of the sample will be? You should not expect the mean to be exactly 100, but you should expect the mean to be near 100. Select SINGLE SAMPLE to confirm or deny this.

Repeat by pressing <↵> several times. A histogram showing the frequency of the sample means obtained from different samples should approximate a normal curve, clustered about the population mean. The MANY SAMPLES program creates such a histogram.

Select MANY SAMPLES. Allow it to run for at least 200 samples. Press the space bar to pause at any time. Press it again to resume sampling.

Experiment with various sample sizes to see how the sample size affects the shape of the histogram of sample means. Select DEFINE. Change the sample size. Select MANY SAMPLES. You may wish to compare printouts of histograms of 200 sample means each for sample sizes of 20, 10, and 5.
[See diagrams on next page.]

What do you notice about the histograms? [The smaller the sample size is, the more dispersed the sample means are.]

What implication does this have for statisticians in the real world? [A sample that is too small may produce a statistic that does not reasonably characterize the population.]

Select DEFINE and change the numbers to:

Population mean:	180	Value to test:	195
Population standard dev.:	8	Least sample mean shown:	150
Sample size:	1	Greatest sample mean shown:	210

What is true about the mean of a sample when the sample size is 1? [The number is the mean of the sample.]

Select MANY SAMPLES, and let the computer create a histogram of about 700 sample means (which also represent 700 individuals from the population, since the sample size is 1). [See diagram on next page.]

What percent of the sample means (and therefore individuals) is less than 195? Select TABLE to find out. [About 97.19%] What percent would therefore be greater than 195? [About 2.81%]

Sample size: 20

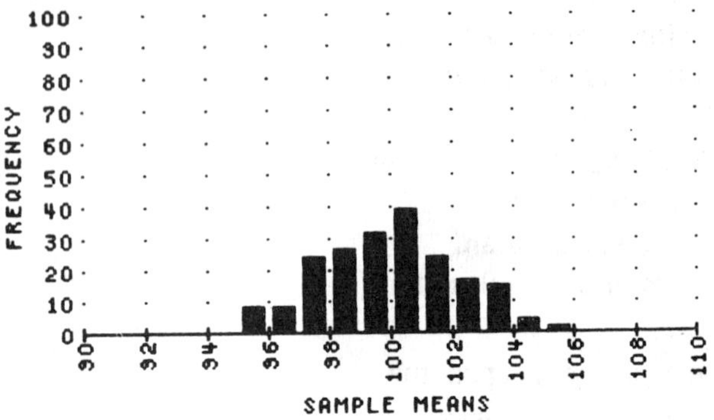

Sample size: 10

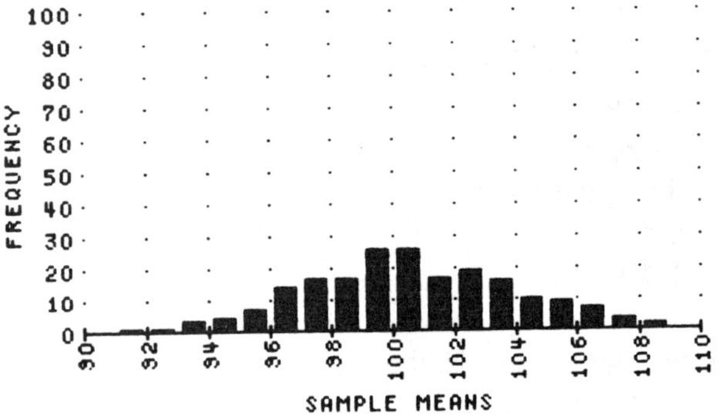

Sample size: 5

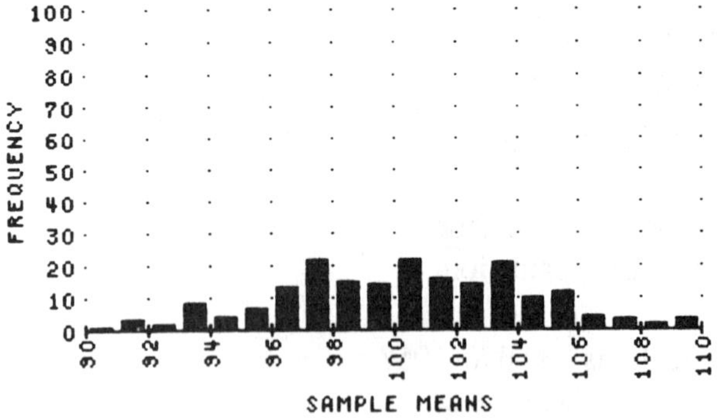

Sample size: 1

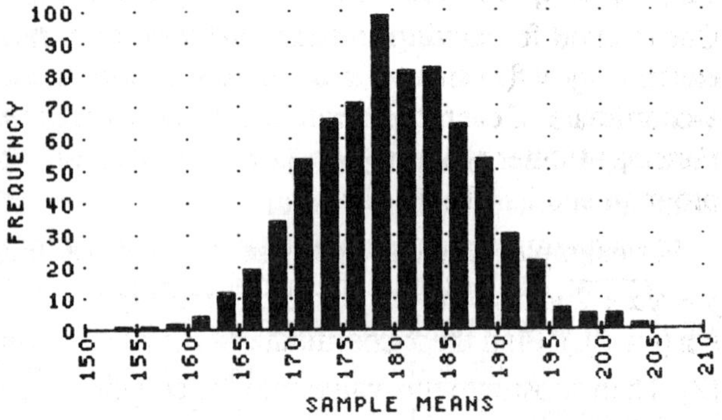

Sample Mean	% of Samples with Smaller Mean	% of Samples with Larger Mean
150.0	0%	100%
152.5	0%	100%
155.0	.14025245%	99.859748%
157.5	.28050491%	99.719495%
160.0	.56100982%	99.43899%
162.5	1.1220196%	98.87798%
165.0	2.8050491%	97.194951%
167.5	5.4698457%	94.530154%
170.0	10.238429%	89.761571%
172.5	17.671809%	82.328191%
175.0	26.928471%	73.071529%
177.5	36.886396%	63.113604%
180.0	50.631136%	49.368864%
182.5	61.991585%	38.008415%
185.0	73.492286%	26.507714%
187.5	82.468443%	17.531557%
190.0	89.901823%	10.098177%
192.5	94.109397%	5.8906031%
195.0	97.194951%	2.8050491%
197.5	98.176718%	1.8232819%
200.0	98.87798%	1.1220196%
202.5	99.579243%	.42075735%
205.0	99.859748%	.14025244%
207.5	99.859748%	.14025244%
210.0	99.859748%	.14025244%

Percent less than 195: 97.194951%
Sample size: 1 Population mean: 180
Mean of sample means: 180.04758

Worked Examples and Advanced Topics

Solving Simple Radical Equations by Graphing

One method for solving an equation of the form $f(x) = g(x)$ involves graphing the equations $y = f(x)$ and $y = g(x)$ on the same set of axes and then determining the x-coordinate of each point where the two curves intersect. You can use the Function Plotter program and its ZOOM feature to do this, or the Equation Solver program and its TRACE feature.

For example, you can solve the radical equation $\sqrt{x+2} = 4 - x$ by graphing $y = \sqrt{x+2}$ and $y = 4 - x$. (Note that the expression $\sqrt{x+2}$ is typically entered as sqr(x + 2).) Find the x-coordinate of the point of intersection of the two curves. [2] Then substitute this value into the equation $\sqrt{x+2} = 4 - x$ to confirm that it is a solution.

Now, use the same technique to solve $\sqrt{x+2} = 5 - x$ by graphing. In this case the x-coordinate of the point of intersection is irrational $\left(\frac{11 - \sqrt{29}}{2}\right)$, so you must ZOOM or TRACE to get a reasonable approximation. [2.8]

An alternate approach to solving an equation of the form $f(x) = g(x)$ involves rewriting the equation as $f(x) - g(x) = 0$. When $y = f(x) - g(x)$ is graphed, any x-intercepts will be solutions of the equation. For example, you can solve $\sqrt{x+2} = 4 - x$ by graphing $y = \sqrt{x+2} - (4 - x)$ and determining the x-coordinates of the point(s) where the graph crosses the x-axis.

Graphing to Find *n*th Roots of Real Numbers

You can approximate the nth root of a number a by using the Equation Solver program to graph $y = x^n - a$. For example, use the Equation Solver program to graph $y = x^2 - 2$. By using TRACE, you can find the x-coordinates of the points where the parabola crosses the x-axis, which are $\pm\sqrt{2}$.

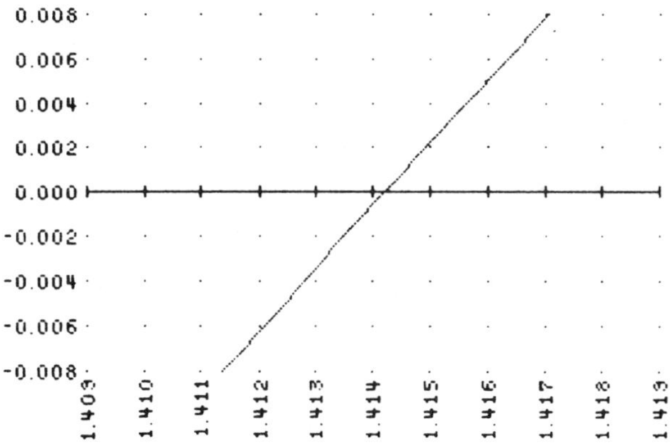

Recall that you may have to press <↵> and then move the trace box again several times to get a good approximation of the roots. Roughly one decimal place of accuracy is added for each new approximation. For example, a typical series of approximations of $\sqrt{2}$ using this method is: first approximation: 1.42857143; second approximation: 1.41285714; third approximation: 1.41420833.

You might ask yourself to explain why this method works. [On the x-axis, $y = 0$, so the equation becomes $0 = x^2 - a$. The roots of this equation are $\pm\sqrt{a}$.]

Likewise, you can approximate $\sqrt[3]{-4}$ by using TRACE to find where the graph of $y = x^3 + 4$ crosses the x-axis. A typical series of approximations is: first approximation: –1.61904762; second approximation: –1.58857143; third approximation: –1.58741667.

Note that when n is even, the graph of $y = x^n - a$ (where a is a positive real number) crosses the x-axis at two points (producing a positive and a negative nth root), and when n is odd, the graph of $y = x^n - a$ (where a is any real number) crosses the x-axis at only one point (producing a single nth root).

You may prefer to use the ZOOM feature of the Function Plotter program rather than the TRACE feature of the Equation Solver program. Or, you may prefer to use the TABLE feature to determine two values between which the root lies, and then use the SCALE feature to enter these two values as the new x-axis boundaries before viewing the TABLE again.

Determining the Domain and Range of Functions by Graphing

You can use the Function Plotter program to determine the domain and range of each function given. You will need to mentally project all graphed points onto the x-axis to obtain the domain and onto the y-axis to obtain the range. If a graph extends beyond the limits of the computer's screen, you should use SCALE and redefine the boundaries until they are sure of the extent of the graph.

1. $y = \sqrt{9 - x^2}$

   $[D = \{x\colon -3 \leq x \leq 3\}, R = \{y\colon 0 \leq y \leq 3\}]$

2. $y = x^2 + 1$

   $[D = \{\text{real numbers}\}, R = \{y\colon y \geq 1\}]$

3. $y = \dfrac{2x}{x^2 + 1}$

   $[D = \{\text{real numbers}\}, R = \{y\colon -1 \leq y \leq 1\}]$

4. $y = \dfrac{x}{|x|}$ — Remember that it may not be clear from the graph that the function is undefined at zero. Use TABLE to see a dashed line (rather than a function value) indicating this.

   $[D = \{x\colon x \neq 0\}, R = \{1, -1\}]$

Functions with Inverses on a Restricted Domain

To familiarize yourself with functions that do not have inverse functions, use the Function Plotter program to graph each of the functions given below. [When entering functions, you may use x^2 for x^2, abs(x) for $|x|$, and sqr(x) for $\sqrt{x}$.]

Although none of the graphs satisfies the horizontal-line test, some *portion* of each graph does. For each graph, then, you can identify a portion that satisfies the horizontal-line test by stating the interval on the x-axis over which the portion lies. In other words, you can determine a *restricted* domain on which each function has an inverse function. [Answers may vary.]

1. $y = (x + 1)^2$ $[\{x\colon x \leq -1\}$ or $\{x\colon x \geq -1\}]$
2. $y = |2x - 1|$ $[\{x\colon x \leq \frac{1}{2}\}$ or $\{x\colon x \geq \frac{1}{2}\}]$
3. $y = \sqrt{x^2 + 4}$ $[\{x\colon x \leq 0\}$ or $\{x\colon x \geq 0\}]$
4. $y = \dfrac{4}{x^2 + 1}$ $[\{x\colon x \leq 0\}$ or $\{x\colon x \geq 0\}]$

Piecewise Defined Functions

The Function Plotter program allows you to define a function in pieces over different intervals by using "comparison expressions" such as $(x < 2)$, $(x <= -3)$, $(x = 7)$, and $(x <> 5)$ (that is, $x \neq 5$). As is common in computer programming, each such expression is given the value 1 by the program if it is true, and the value 0 if it is false. For example:

$$y = (x < 2) = \begin{cases} 1 & \text{if } x < 2 \\ 0 & \text{if } x \geq 2 \end{cases}$$

Comparison expressions can be combined with each other and with ordinary expressions using any ordinary mathematical means. Thus:

$$y = -3(x < 1) + 2(x >= 1) = \begin{cases} -3 & \text{if } x < 1 \\ 2 & \text{if } x \geq 1 \end{cases}$$

This function is pictured at the left below. A more complicated example, pictured at the right below, is:

$$y = (2x+8)(x <= -2) + x^2(-2 < x)(x < 0) + \sin(x)(x >= 0) = \begin{cases} 2x+8 & \text{if } x \leq -2 \\ x^2 & \text{if } -2 < x < 0 \\ \sin(x) & x \geq 0 \end{cases}$$

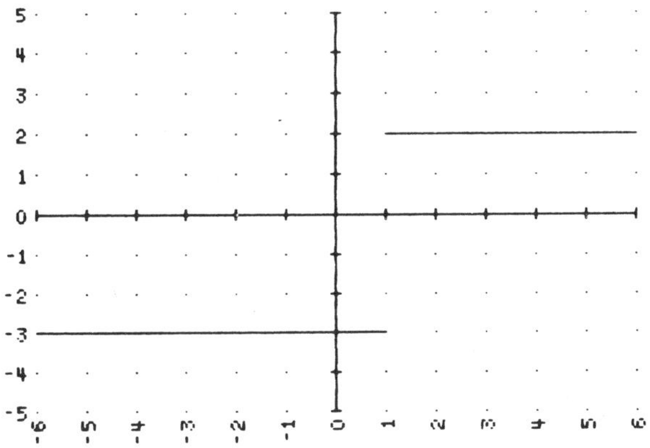

(solid) y = -3(x<1)+2(x>=1)

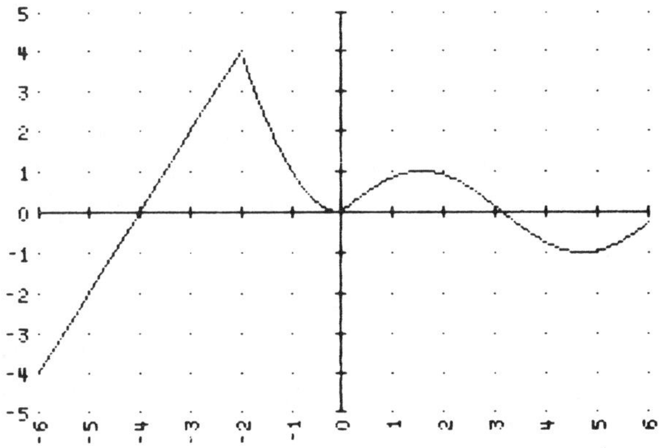

(solid) y = (2x+8)(x<=-2)+x^2(-2<x)(x<0)+Sin(x)(x>=0)

By graphing a sequence of functions of increasing complexity, you can discover the meaning of the comparison expression notation. You may then use the program to investigate the continuity of piecewise-defined functions, as well as left and right limits of functions at possible points of discontinuity.

A Statistics Project

You can gather information from friends and families, and analyze it using the Data Analyzer program. When the data is typed in, the file can be saved to a data disk formatted in DOS .

Working as a class, individually, or in small groups, you can produce and print histograms for each column of data. In addition, the SCATTER option of the GRAPH menu will draw a scatter plot using any two columns of data, and will plot the line of best fit as well as calculate the correlation coefficient.

If you wish to make use of the correlation capabilities of the program, choose categories of data that are likely to be correlated strongly, as well as some categories that are not likely to be correlated.

Iterated Functions

Students can graph $\cos(x)$ and $y = x$ on the same axes using Function Plotter. Where do the graphs intersect? [0.739]

Set your scientific calculator to radian mode and enter any number. Press the cosine key repeatedly, to generate the series $\cos(x)$, $\cos(\cos(x))$, $\cos(\cos(\cos(x)))$, … . What number does this iteration tend towards? Does it matter what initial value of x was chosen?

Fibonacci Numbers

If you change the initial values of the two first terms of a Fibonacci sequence, the numbers generated will be different. For instance, set the following values.

Initial values: t(1) = 3 t(2) = 4

t(n) = t(n – 2) + t(n – 1)

The numbers generated are quite different from the typical Fibonacci sequence generated when both initial values are 1. However, using a calculator, you can show that the ratio of any two consecutive terms or partial sums still approaches the golden ratio, 0.618. Is this the case with any two nonzero initial terms?

```
t( 1) = 3        t( 2) = 4
t(n) = t(n-2)+t(n-1)
```

n	sequence t(n)	partial sums S(n)
1	3	3
2	4	7
3	7	14
4	11	25
5	18	43
6	29	72
7	47	119
8	76	195
9	123	318
10	199	517

Derivative of the Sine Function

Use the Function Plotter program to DRAW $y = \sin(x)$ for $-1 \le x \le 5$, and $-1.5 \le y \le 1.5$. (Estimating slopes will be easier if you do *not* set the SCALE to show multiples of π.)

Since $y = \sin(x)$ is a smooth function, if any region of the graph is magnified sufficiently, it will appear straight; that is, it will very nearly coincide with its tangent line. Estimating the derivative is therefore easy; simply calculate the slope of this almost-straight section of the graph.

With a piece of graph paper and a calculator you may choose different points on the graph to ZOOM in on. [At the very least, consider (0, 0), the maximum point of the function, and several points between.] Alternatively, a group of students can work on non-overlapping sets of points; their results can be combined on a blackboard or on a large piece of graph paper.

At each point chosen, you can ZOOM several times until the graph appears straight. If you then write down the coordinates of two points on the graph (the TABLE option is useful for this), you can calculate the slope, and hence the derivative. Once a slope $f'(x)$ is calculated, the point $(x, f'(x))$ should be plotted on the graph paper. After a number of points are generated in this way, the cosine curve will emerge.

College Algebra and Trigonometry
Chapter Tests and Answers

1. Identify the property that is illustrated by $(a + b) + c = (b + a) + c$.

2. Given $A = \{2,3,4,8,9\}$ and $B = \{2,4,5,7,9\}$, find $A \cap B$

3. Write $-|4| + |-6| + 2$ without absolute value symbols.

4. Find the distance between 4 and -9 on the number line.

5. Simplify: $(-3x^{-3}y^2)^{-3}(-2x^2y^0)^{-2}$

6. Simplify: $\dfrac{(4x^{-2}y^3)^{-2}(-2x^4y^{-1})^3}{-6x^{-2}y^2}$

7. Write 0.000000129 in scientific notation.

8. Simplify: $(2x - 5)(x^3 + 3x - 4)$

9. Evaluate the polynomial $x^4 - 3x^2 + 4x - 7$ for $x = -4$.

10. Factor: $7x^2 - 23x + 6$

11. Factor: $3ax - 2ay + 3bx - 2by$

12. Factor: $3x^4 - 3xy^3$

13. Factor: $x^5 + x^2y^3 - y^2x^3 - y^5$

14. Simplify: $\dfrac{x^4 - 2x^3 - 8x + 16}{3x^3 + 6x^2 + 12x}$

15. Simplify: $\dfrac{3x}{x^2 + 2x - 8} - \dfrac{2}{x^2 - 4}$

16. Simplify: $\dfrac{x^2 + 2x - 8}{x^3 - 1} \div \dfrac{x^2 + 6x + 8}{x^3 + x^2 + x}$

17. Simplify: $\dfrac{a}{x - y} \cdot \dfrac{x^2 - y^2}{ax + ay} - \dfrac{1}{x - y}$

18. Simplify: $\dfrac{\dfrac{y}{y-2}}{\dfrac{1}{y} + \dfrac{2}{y-2}}$

19. Simplify: $\dfrac{a^{2/3}b^{-3/2}}{a^{-1/3}b^{3/4}}$

20. Simplify: $3\sqrt[3]{5x^3} - x\sqrt[3]{625}$

21. Simplify: $\dfrac{4a}{\sqrt[4]{16a^5}}$

22. Simplify: $\dfrac{y}{\sqrt{y} - \sqrt{2}}$

23. Simplify 2(3 - 4i) - 5(6 + 2i) and write the complex number in standard form.

24. Simplify (2 - 3i)(4 - 7i) and write the complex number in standard form.

25. Write $\dfrac{3i}{2 - 3i}$ as a complex number in standard form.

1. Which of the numbers -8, 0.17, 0, $0.7\overline{4}$, 0.141141114..., $\sqrt{11}$ are rational numbers?

2. Identify the property that is illustrated by (a + b) + c = c + (a + b).

3. Write $-|12-23|-|17|$ without absolute value symbols.

4. Find the distance between the points 4 and -14 on the number line.

5. Write the inequality $-4 \leq x < -1$ using interval notation.

6. Simplify: $(-6a^{-2}b^{3})(2ab^{-4})^{2}$

7. Simplify: $\dfrac{(3x^{-2}y^{4})^{-2}}{(-2x^{0}y^{-3})(3x^{-4}y^{2})^{3}}$

8. Write 2,340,000,000 in scientific notation.

9. Simplify: $(x-y+2)(2x+y-4)$

10. Simplify: $(2a-b)^{2}-(3a+2b)^{2}$

11. The number of committees consisting exactly of 3 people that can be formed from a group on n people is given by the polynomial $\frac{1}{6}n^{3}-\frac{1}{2}n^{2}+\frac{1}{3}n$. Find the number of committees consisting of exactly 3 people that can be formed from a group of 36 people.

12. Factor $7x^2 - 28xy + 21y^2$.

13. Factor $2ax + ay - 4bx - 2by$.

14. Factor $2ax^2 - 3bx^2 - 2ay^2 + 3by^2$.

15. Simplify: $\dfrac{8x^3 - 27y^3}{4x^2 - 12xy + 9y^2}$

16. Simplify: $\dfrac{x}{x^2 + 5x + 6} - \dfrac{2}{x^2 + 2x - 3}$

17. Simplify: $\dfrac{x^4 + 2x^3 - x - 2}{3x^4 + 3x^3 + 3x^2} \cdot \dfrac{4x^2 - 12x + 9}{2x^2 + x - 6}$

18. Simplify: $a - \dfrac{a}{1 + \dfrac{1}{a}}$

19. Simplify; $\dfrac{a^{3/4}b^{-1/3}}{a^{-1/2}b^{-3/2}}$

20. Simplify: $3\sqrt[3]{16a^5b^7} - ab\sqrt[3]{54a^2b^4}$

21. Simplify: $\left(\sqrt{a+2} - 3\right)^2$

22. Simplify $\dfrac{\sqrt{2}}{\sqrt{x}-\sqrt{2}}$

23. Simplify (3 - 2i)(4 + 8i) and write the complex number in standard form.

24. Write $\dfrac{3}{-3-i}$ as a complex number in standard form.

25. Simplify: i^{457}

1. Which of the numbers -5, $3.\overline{4}$, $\sqrt{19}$, $\frac{7}{11}$, π, and 0.565565556... are irrational numbers?

2. Identify the property that is illustrated by the following: 6(a + b) = (a + b)6.

3. Write $-|-7+3|-|10-4|$ without absolute value symbols.

4. Graph the inequality $x < -3$ or $x \geq 0$.

5. Simplify: $\left(-2a^{-2}b^{3}\right)^{-1}\left(-3a^{0}b^{2}\right)^{2}$

6. Simplify: $\dfrac{\left(-2a^{-1}b\right)^{-3}\left(-3a^{-2}b^{2}\right)^{4}}{\left(6a^{2}b^{-3}\right)^{2}}$

7. Write 6.1×10^{-7} in decimal form.

8. Simplify: $\left(x^{2}+2x-3\right)\left(x^{2}-2x+3\right)$

9. On an expressway, the recommended safe distance between cars in feet is given by $0.015v^{2}+v+10$, where v is the velocity of the car in miles per hour. Find the safe distance when v = 55 mph.

10. Factor: $7x^{2}+10x-8$

11. Factor: $2ax^{2}-2ax-bx^{2}+bx$

12. Factor: $8x^4y - 27xy^7$

13. Factor: $x^2y^3 + x^2 - y^3 - 1$

14. Simplify: $\dfrac{6}{x^2 - 16} - \dfrac{5}{x^2 + x - 12}$

15. Simplify: $\dfrac{6x^2 + 42x}{6x^2 - 54} \div \dfrac{x^2 + 2x - 35}{x^2 - 2x - 15}$

16. Simplify: $\sqrt[3]{96x^4y^8z^{11}}$

17. Simplify: $\dfrac{\dfrac{x}{x+5}}{\dfrac{1}{x} + \dfrac{x-6}{x+5}}$

18. Simplify: $\dfrac{a^{-3/2}b^{5/3}}{a^{1/2}b^{-3/2}}$

19. Simplify: $\left(3\sqrt{x} + 2\sqrt{y}\right)^2$

20. Simplify: $\dfrac{y}{\sqrt[4]{3^3y}}$

21. Simplify: $xy\sqrt{18x^2y} - \sqrt{32x^4y^3}$

22. Simplify: $\dfrac{2x}{\sqrt{2}-x}$

23. Simplify $\left(2-\sqrt{-1}\right)\left(3+\sqrt{-9}\right)$ and write the complex number in standard form.

24. Write $\dfrac{-3i}{-1+4i}$ as a complex number in standard form.

25. Simplify: i^{766}

1. Which of the numbers $-4, 0.12, 3.5\overline{9}, 11, \frac{2}{9}, \sqrt{41}, 2.131331333\ldots$ are irrational numbers?

   a. $\frac{2}{9}, \sqrt{41}, 2.131331333\ldots$
   b. $3.5\overline{9}, \sqrt{41}, \frac{2}{9}$
   c. $\sqrt{41}, 2.131331333\ldots$
   d. $3.5\overline{9}, \sqrt{41}, 2.131331333\ldots$

2. Given $A = \{2, 4, 6, 7, 8, 9\}$ and $B = \{1, 2, 3, 5, 6, 9\}$, find $A \cup B$.

   a. $\{2,6.9\}$
   b. $\{1, 2, 3, 4, 5, 6, 7, 8, 9\}$
   c. $\{4,7,8\}$
   d. $\{1,3,5\}$

3. Write $-2|5-8| - |8-5|$ without absolute value symbols.

   a. -3
   b. 9
   c. 3
   d. -9

4. Find the distance between the points -4 and -21 on the number line.

   a. 17
   b. -25
   c. 25
   d. -17

5. Simplify: $(2x^{-2}y^3)^{-4}(-3x^{-7}y^2)^3$

   a. $432x^6y^6$
   b. $-\dfrac{27}{16x^{13}y^6}$
   c. $-\dfrac{27x^{10}}{16y^{18}}$
   d. $\dfrac{16x}{27y^5}$

6. Simplify: $\dfrac{(2xy^{-3})^{-2}(-2x^{-2}y)^2}{4x^2y^2}$

   a. $\dfrac{16y^6}{4x^6}$
   b. $\dfrac{16y^4}{x^8}$
   c. $\dfrac{y^8}{x^8}$
   d. $\dfrac{y^6}{4x^8}$

7. Write 7.2×10^{-9} in decimal form.

   a. 720,000,000
   b. 7,200,000,000
   c. 0.000000072
   d. 0.0000000072

8. Simplify: $(x^3 - 7x^2 + 7) - (3x^3 - 7x + 12)$

   a. $-2x^3 - 7x^2 + 7x - 5$ b. $-4x^3 - 7x^2 + 7x - 5$

   c. $-2x^3 + 19$ d. $-2x^3 - 7x^2 - 7x + 19$

9. Simplify: $(x^2 - 2x + 3)(x^2 - 3)$

   a. $x^4 - 2x^3 - 6x^2 + 6x - 9$ b. $x^4 - 2x^3 + 6x - 9$

   c. $x^4 - 5x^3 + 6x + 9$ d. $x^4 - 2x^3 + 6x^2 + 6x + 9$

10. Factor: $8x^2 + 9xy - 14y^2$

    a. $(4x - 7y)(2x + 2y)$ b. $(8x + 7y)(x - 2y)$

    c. $(8x - 7y)(x + 2y)$ d. $(8x + 7y)(x + 2y)$

11. Factor: $2ax - 4bx - ay + 2by$

    a. $(a - 2b)(2x - y)$ b. $(a - 2x)(2b - y)$

    c. $(a + 2b)(2x - y)$ d. $(2x - b)(y - a)$

12. Factor: $x^4yz + 27xy^4z^4$

    a. $xyz(x + 3yz)(x^2 + 6xyz + 9y^2z^2)$ b. $xyz(x + 3yz)(x^2 - 3xyz + 9y^2z^2)$

    c. $x^4y^4z^4(yz + 27x)$ d. $xyz(x - 9yz)(x^2 + 9xyz + 9y^2z^2)$

13. Factor: $(x - 9)^2 - 100$

    a. $(x - 9)(x + 10)$ b. $(x - 13)(x + 10)$

    c. $(x - 19)(x + 1)$ d. $(x^2 - 10)(x^2 + 10)$

14. Simplify: $\dfrac{5}{x^2 - 25} - \dfrac{4}{x^2 - x - 20}$

    a. $\dfrac{9x - 40}{(x - 5)(x + 5)(x + 4)}$ b. $\dfrac{x + 40}{(x - 5)(x + 5)(x - 4)}$

    c. $\dfrac{1}{(x - 5)(x + 5)(x + 4)}$ d. $\dfrac{x}{(x - 5)(x + 5)(x + 4)}$

15. Simplify: $\dfrac{3x^2 - 12x}{3x^2 + 12x + 12} \bullet \dfrac{x^2 - 3x - 10}{x^2 - 9x + 20}$

a. $\dfrac{x}{(x+2)}$

b. $\dfrac{3(x-5)}{(x+2)(x-4)}$

c. $\dfrac{x(x-4)}{(x+2)(x+4)}$

d. $\dfrac{3(x+4)}{(x+2)(x-4)}$

16. Simplify: $\dfrac{3}{x-2} \bullet \dfrac{2x^2 - 7x + 6}{12x - 18} + \dfrac{1}{x-2}$

a. $\dfrac{1}{2}$

b. $\dfrac{x-2}{2(x+2)}$

c. $\dfrac{x-2}{x}$

d. $\dfrac{x}{2(x-2)}$

17. Simplify: $\dfrac{\dfrac{5}{x+5}}{\dfrac{x-7}{x+5} + \dfrac{1}{x}}$

a. $\dfrac{5x}{(x+5)(x+1)}$

b. $\dfrac{5x}{2x-2}$

c. $\dfrac{5x}{(x-5)(x-1)}$

d. $\dfrac{5}{x(x+5)}$

18. Simplify: $\dfrac{x^{3/5}y^{-2/3}}{x^{1/2}y^{1/3}}$

a. $\dfrac{x^{1/10}}{y}$

b. $x^{1/10}y$

c. $\dfrac{x^{11/10}}{y}$

d. $\dfrac{x^{3/10}}{y^{2/9}}$

19. Simplify: $\sqrt[5]{-128x^{17}y^{12}}$

a. $-8x^8y^6\sqrt[5]{2x}$

b. $-2x^3y^2\sqrt[5]{4x^2y^2}$

c. $-4x^2y^2\sqrt[5]{4x^7y^2}$

d. $-2x^3y^2\sqrt[5]{2x^2y^2}$

20. Simplify: $-a\sqrt{18a^3b^3} + 5\sqrt{50a^5b^3}$

a. $2a^2b\sqrt{4ab}$ b. $-3ab\sqrt{2ab}$

c. $28a^2b\sqrt{4ab}$ d. $22a^2b\sqrt{2ab}$

21. Simplify: $(2\sqrt{x} - 3)(2\sqrt{x} + 3)$

a. $4x - 9$ b. $4x - 6\sqrt{x} + 9$

c. $4x - 12\sqrt{x} + 9$ d. $2\sqrt{x} - 9$

22. Simplify: $\dfrac{x}{\sqrt{2} - \sqrt{x}}$

a. $\dfrac{\sqrt{2x} - x}{2 - x}$ b. $\dfrac{x\sqrt{2} + x\sqrt{x}}{2 - x}$

c. $\dfrac{x\sqrt{2} + x}{4 - x^2}$ d. $\dfrac{x\sqrt{2} + x\sqrt{x}}{4 - x}$

23. Simplify $(3 - 2i)(-2 + 4i)$ and write the complex number in standard form.

a. $-14 + 16i$ b. $-14 + 8i$

c. $2 + 16i$ d. $-2 + 16i$

24. Write $\dfrac{-2}{3 + 2i}$ as a complex number in standard form.

a. $-\dfrac{6}{13} + \dfrac{4}{13}i$ b. $-\dfrac{3}{5} + \dfrac{4}{5}i$

c. $\dfrac{3}{5} - \dfrac{2}{5}i$ d. $-\dfrac{6}{13} - \dfrac{2}{13}i$

25. Simplify: i^{1927}

a. $-i$ b. -1

c. 1 d. i

1. Which of the numbers $-4.5, 11, 7.\overline{92}, \sqrt{17}, \frac{13}{17}, 1.131131113\ldots$ are irrational numbers?

a. $\sqrt{17}, 1.131131113\ldots$
b. $7.\overline{92}, \sqrt{17}, \frac{13}{17}, 1.131131113\ldots$
c. $7.\overline{92}, \sqrt{17}, 1.131131113\ldots$
d. $7.\overline{92}, \sqrt{17}, \frac{13}{17}$

2. Identify the property that is illustrated by the following: If $p + 2s = t$ and $t = 7 - r$, then $p + 2s = 7 - r$

a. Distributive Property
b. Reflexive
c. Symmetric
d. Transitive

3. Write $|1.5 - \pi|$ without absolute value symbols.

a. $\pi - 1.5$
b. $-\pi - 1.5$
c. $1.5 - \pi$
d. $1.5 + \pi$

4. Given that $A = \{1, 2, 5, 9, 14, 17\}$ and $P = \{x \mid x \text{ is a prime number less than } 20\}$, find $A \cap P$.

a. $\{1, 2, 5, 7, 9, 11, 13, 14, 17, 19\}$
b. $\{1, 2, 5, 17\}$
c. $\{2, 5, 17\}$
d. $\{1, 2, 5, 9, 17\}$

5. Write the inequality $x \le -2$ or $x > 4$ using interval notation.

a. $[-\infty, -2] \cup (4, \infty)$
b. $(-\infty, -2] \cup [4, \infty)$
c. $[-\infty, -2] \cup [4, \infty)$
d. $(-\infty, -2] \cup (4, \infty)$

6. Simplify: $(-a^3b^{-4})^{-1}(3a^{-2}b^3)^2$

a. $\frac{b^{10}}{9a^7}$
b. $\frac{-9b^{10}}{a^7}$
c. $\frac{-9a}{b^2}$
d. $-9a^3b^{10}$

7. Simplify: $\frac{(-5x^{-4}y^3)^{-4}}{(-2x^0y)^5(4x^3y^{-1})^{-3}}$

a. $-\frac{2x^{25}}{625y^{20}}$
b. $-1250x^{25}y^{20}$
c. $\frac{2048x^7}{625y^{10}}$
d. $-\frac{64x^7}{625y^{10}}$

8. Write 0.000000000000239 in scientific notation.

a. 239×10^{-15}
b. 2.39×10^{-13}
c. 2.39×10^{13}
d. 2.39×10^{15}

9. Simplify: $(2x - y^2)(4x^2 + 2xy^2 + y^4)$

   a. $8x^3 - 8x^2y^2 + 4xy^4 - y^6$ b. $8x^3 + 4xy^4 + y^6$

   c. $8x^3 - y^6$ d. $8x^3 - 8x^2y^2 - y^6$

10. Simplify: $(\sqrt{2x - 3} - 4)^2$

    a. $2x - 8\sqrt{2x - 13} + 13$ b. $4x^2 - 9x + 16$ c. $2x - 7$ d. $2x - 4\sqrt{2x - 3} + 19$

11. On an expressway, the recommended safe distance between cars in feet is given by $0.015v^2 + v + 10$, where v is the velocity of the car in miles per hour. Find the safe distance when v is 40 mph.

    a. 56 ft b. 92 ft c. 64 ft d. 74 ft

12. Factor: $12x^2 - 25x + 12$

    a. $(4x - 3)(3x - 4)$ b. $(4x - 3)(3x + 4)$

    c. $(12x - 1)(x - 12)$ d. $(6x - 4)(2x - 3)$

13. Factor: $14ax + 21bx - 6ay - 9by$

    a. $(2a - 3y)(3b + 7x)$ b. $(3a - 2b)(3y + 7x)$

    c. $(2a - 3b)(7x + 3y)$ d. $(2a + 3b)(7x - 3y)$

14. Factor: $81x^3 - 3y^3$

    a. $(9x - y)(81x^2 + 27xy + y^2)$ b. $3(3x - y)(3x^2 + xy + y^2)$

    c. $3(3x - y)(9x^2 + 3xy + y^2)$ d. $3(3x + y)(8x^2 - 3xy + y^2)$

15. Factor: $x^4 + 2x^3 - x - 2$

    a. $(x - 2)(x + 1)(x^2 - x + 1)$ b. $(x + 2)(x - 1)(x^2 + x + 1)$

    c. $(x + 2)(x + 1)(x^2 - x + 1)$ d. $x(x + 2)(x + 1)(x - 2)$

16. Simplify: $\dfrac{4x}{x^2 + 6x + 8} - \dfrac{2}{3x + 6}$

    a. $\dfrac{2(5x - 4)}{3(x + 4)(x + 2)}$ b. $\dfrac{2}{3(x + 2)}$ c. $\dfrac{12x^2 - 2x - 8}{3(x + 4)(x + 2)}$ d. $\dfrac{8x - 8}{3(x + 4)(x + 2)}$

17. Simplify: $\dfrac{x^3 + 125}{2x^2 + 4x - 30} \div \dfrac{6x^2 + 16x - 70}{4x^2 - 100}$

    a. $\dfrac{(x - 5)(x^2 - 5x + 25)}{(x - 3)(3x - 7)}$ b. $\dfrac{x^2 - 5x + 25}{3x - 7}$ c. $\dfrac{(x - 5)(x^2 - 10x + 25)}{(x - 3)(3x - 7)}$ d. $\dfrac{4(x - 5)(x - 5)}{(x - 3)(3x - 7)}$

18. Simplify: $\dfrac{\dfrac{x-2}{x+5}}{\dfrac{2}{x}+\dfrac{x-2}{x+5}}$

a. $\dfrac{x}{x^2+4x+10}$ b. $\dfrac{x(x-2)}{x^2+10}$ c. $\dfrac{x(x-2)}{x+5}$ d. $\dfrac{x-2}{x^2+10}$

19. Simplify: $\dfrac{a^{-3/2}b^{2/3}}{a^{2/3}b^{-5/3}}$

a. $a^{5/6}b^{7/3}$ b. $\dfrac{b}{a^{13/6}}$ c. $a^{5/6}b^{2/3}$ d. $\dfrac{b^{7/3}}{a^{13/6}}$

20. Simplify: $\sqrt[4]{243a^7b^5}$

a. $3a^2b\sqrt[4]{9ab}$ b. $9\sqrt[4]{3a^3b}$ c. $3ab\sqrt[4]{3a^3b}$ d. $3ab\sqrt[4]{3ab}$

21. Simplify: $3a\sqrt[3]{-27ab^5}+b\sqrt[3]{125a^4b^2}$

a. $-4ab\sqrt[3]{ab^2}$ b. $-13ab\sqrt[3]{ab}$ c. $4ab\sqrt[3]{ab}$ d. $-4a^2b\sqrt[3]{ab}$

22. Simplify: $\dfrac{2-\sqrt{3}}{\sqrt{x}+\sqrt{3}}$

a. $\dfrac{2\sqrt{x}-2\sqrt{3}+\sqrt{3x}+9}{x-9}$ b. $\dfrac{2\sqrt{x}-2\sqrt{3}-\sqrt{3x}+3}{x-3}$

c. $\dfrac{2\sqrt{x}-2\sqrt{3}+\sqrt{3x}+9}{x-9}$ d. $\dfrac{\sqrt{2x}-2\sqrt{3}+\sqrt{3x}+3}{x-9}$

23. Simplify $(3-4i)(6+2i)$ and write the complex number in standard form.

a. $10-18i$ b. $10+30i$ c. $26-18i$ d. $18+30i$

24. Simplify $\dfrac{3i}{1-3i}$ and write the complex number in standard form.

a. $-\dfrac{9}{8}+\dfrac{3}{8}i$ b. $\dfrac{3}{10}-\dfrac{9}{10}i$ c. $-\dfrac{9}{10}+\dfrac{3}{10}i$ d. $-\dfrac{9}{4}+\dfrac{3}{4}i$

25. Simplify: i^{1002}

a. 1 b. i c. -1 d. $-i$

1. Which of the numbers $-11, 1, 0.11, 0.\overline{325}, \frac{5}{13}, \sqrt{19}, 0.121221222\ldots$ are irrational numbers?

a. $0.\overline{325}, \sqrt{19}, 0.121221222\ldots$ b. $0.11, 0.\overline{325}, 0.121221222\ldots$

c. $-11, \frac{5}{13}, \sqrt{19}$ d. $\sqrt{19}, 0.121221222\ldots$

2. Identify the property that identifies the following: $25(a + b) = 25(b + a)$.

a. Distributive Property b. Commutative Property of Addition

c. The Symmetric Property d. Associative Property of Addition

3. Write $-|\pi - 2|$ without absolute value symbols.

a. $\pi - 2$ b. $-\pi - 2$ c. $2 - \pi$ d. $\pi + 2$

4. Find the distance between the points -23 and 12 on the number line.

a. 35 b. -35 c. 11 d. -11

5. Write the interval $-12 \le x \le 3$ using interval notation.

a. $(12, 3)$ b. $(-\infty, -12] \cap [3, \infty)$ c. $(-\infty, -12] \cup [3, \infty)$ d. $[-12, 3]$

6. Simplify: $(-3a^{-2}b^{-3})(2a^{-5}b^{3})^{-4}$

a. $-48a^{22}b^{9}$ b. $\frac{-48a^{18}}{b^{9}}$ c. $-\frac{3a^{18}}{16b^{15}}$ d. $-\frac{3a^{22}}{16b^{9}}$

7. Simplify: $\frac{(2a^{2}b^{-3})^{-3}}{(6^{-1}a^{4}b^{-1})^{2}}$

a. $\frac{9b^{11}}{2a^{14}}$ b. $-\frac{8b^{7}}{9a^{2}}$ c. $\frac{9b^{7}}{2a^{2}}$ d. $\frac{9b^{7}}{8a^{14}}$

8. Write 2.75×10^{-9} in decimal form.

a. 2,750,000,000 b. 0.000000275

c. 0.00000000275 d. 275,000,000

9. Simplify: $(a + b - 3)(a + b + 3)$

a. $a^2 + 2ab + b^2 - 9$ b. $a^2 - 3a + b^2 - 3b + 9$

c. $a^2 + 2ab + 6a + b^2 + 6b - 9$ d. $a^2 + 2ab + b^2 + 6b - 9$

10. Simplify: $(2x - 3y)^2 - (3x - y)^2$

a. $13x^2 - 18xy + 10y^2$ b. $-5x^2 - 6xy + 8y^2$

c. $13x^2 - 6xy + 10y^2$ d. $5x^2 - 18xy + 8y^2$

11. The average time in seconds that it takes a particular computer to determine whether an n-digit number is prime or composite is given by $0.002n^2 + 0.002n + 0.009$, $20 \le n \le 40$. Estimate the average time it takes the computer to determine whether a 25-digit number is a prime or a composite.

    a. 1.309 s b. 12.309 s c. 2.209 s d. 0.109 s

12. Factor: $28x^2 - 53x + 7$

    a. $(14x - 7)(2x - 1)$ b. $(4x - 1)(7x - 7)$

    c. $(4x - 7)(7x - 1)$ d. $(4x + 7)(7x - 1)$

13. Factor: $ax - ay + bx - by$

    a. $(x - y)(a - b)$ b. $(x + y)(a - b)$

    c. $(x - a)(y + b)$ d. $(x - y)(a + b)$

14. Factor: $2a^4b - 16ab^4$

    a. $2ab(a - 2b)(a^2 + 2ab + 4b^2)$ b. $2ab(a - 4b)(a - 2b)$

    c. $2ab(a - 2b)(a^2 + 6ab + 9b^2)$ d. $2ab(a + 2b)(a^2 - 2ab + 4b^2)$

15. Factor: $3x^3 - 4x^2y - 3xy^2 + 4y^3$

    a. $(x^2 + y^2)(3x - 4y)$ b. $(x - y)(x + y)(3x - 4y)$

    c. $(x - y)(x + y)(3x + 4y)$ d. $xy(3x - 4y)(x - y)(x + y)$

16. Simplify: $\dfrac{4x}{x^2 + 7x + 12} - \dfrac{3}{5x + 20}$

    a. $\dfrac{x - 9}{(x + 3)(x + 4)}$ b. $\dfrac{17x - 9}{5(x + 3)}$ c. $\dfrac{17x - 9}{5(x + 3)(x + 4)}$ d. $\dfrac{23x + 9}{5(x + 4)}$

17. Simplify: $\dfrac{80x^2 + 20x + 5}{25x^2 - 25} \bullet \dfrac{20x^2 - 25x + 5}{64x^3 - 1}$

    a. $\dfrac{16x^2 + 4x + 1}{(x + 1)(4x - 1)}$ b. $\dfrac{1}{x + 1}$ c. $\dfrac{4x - 1}{5(x + 1)}$ d. $\dfrac{5(16x^2 + 4x + 1)}{(x - 1)(x + 1)(4x - 1)}$

18. Simplify: $\dfrac{\dfrac{1}{x} - \dfrac{x - 2}{x + 1}}{\dfrac{x}{x + 1}}$

    a. $1 + 3x$ b. $\dfrac{1 - x^2}{x^2}$ c. $\dfrac{1 + 3x - x^2}{x^2}$ d. $\dfrac{x^2 - x - 2}{x(x + 1)}$

19. Simplify: $\dfrac{x^{-2/3}y^{3/4}}{x^{1/3}y^{-1/2}}$

a. $\dfrac{y^{5/4}}{x^{1/3}}$ b. $\dfrac{y^{1/4}}{x^{1/3}}$ c. $xy^{1/4}$ d. $\dfrac{y^{5/4}}{x}$

20. Simplify: $\sqrt[3]{-250a^{10}b^6}$

a. $-5a^3b^2\sqrt[3]{2a}$ b. $-5a^3b^2\sqrt[3]{2a^2b}$ c. $-5a^2b\sqrt[3]{2a^2b}$ d. $-25a^2b\sqrt[3]{2a}$

21. Simplify: $(3\sqrt{2} - 2\sqrt{5})^2$

a. $-2 - 12\sqrt{10}$ b. $38 - 2\sqrt{5} + 3\sqrt{2}$

c. $38 - 12\sqrt{10}$ d. $18 - 6\sqrt{10}$

22. Simplify: $\dfrac{3}{\sqrt{a} - \sqrt{b}}$

a. $\dfrac{\sqrt{3a} + \sqrt{3b}}{a - b}$ b. $\dfrac{3\sqrt{a} + 3\sqrt{b}}{a - b}$ c. $\dfrac{3\sqrt{a} + 3\sqrt{b}}{a^2 - b^2}$ d. $\dfrac{\sqrt{a} + \sqrt{b}}{a + b}$

23. Simplify: $(2 - \sqrt{-4})(3 + \sqrt{-9})$

a. $12 - 12i$ b. $-12i$ c. 12 d. -12

24. Write $\dfrac{3i}{2 + 3i}$ as a complex number in standard form.

a. $-\dfrac{9}{5} + \dfrac{6}{5}i$ b. $\dfrac{9}{13} + \dfrac{6}{13}i$ c. $\dfrac{6}{13} - \dfrac{9}{13}i$ d. $-\dfrac{9}{13} - \dfrac{6}{13}i$

25. Simplify: i^{844}

a. -1 b. $-i$ c. 1 d. i

1. Solve: $2-13x=-4(4x-5)$.

2. Solve: $\frac{3}{x+2}+\frac{7}{x+2}=\frac{5}{x-2}$.

3. Solve $\frac{1}{x}=\frac{1}{a}+\frac{a}{b}$ for x.

4. Solve: $x^2-8x-26=0$ by completing the square.

5. Solve: $4x^2-7x+14=0$.

6. Solve: $x^4-x^3+8x-8=0$.

7. Solve: $\sqrt{x+1}-3=\sqrt{x+4}$.

8. Solve: $2x^{1/2}-x^{1/4}-3=0$

9. Solve: $(2x-5)^{1/3}=3$

10. Solve: $-2(2x+3)\leq 3-5x$

11. Solve: $\frac{(x-1)(x+4)}{x+1}\leq 0$

12. Solve: $-2\leq 2x+5\leq 7$

13 Solve: $|3x-11|=8$

14. Solve: $|2x+3|<5$

15. Solve: $|5x+7|\geq 17$

16. The length of a rectangle is 3 feet more than twice the width of the rectangle. If the perimeter of the rectangle is 90 feet, find the length of the rectangle.

17. An accountant invests a bonus of \$8000 into two simple interest accounts. On one account the annual simple interest rate is 7.5%, and on the second account the annual simple interest rate is 5.2%. The amount of interest earned for 1 year was \$542.50. How much was invested in each account?

18. The first leg of a trip to the country was 150 miles of highway. The second leg, 180 miles of backroads, took 1.5 hours longer than the first part. If the driver was able to travel 10 mph faster on the highway than on the backroads, find the average rate on the backroads.

19. A mason can build a wall in one-third the time it takes an apprentice. Together they build the wall in 9 hours. How long would it take the apprentice working alone to build the wall?

20. How many grams of pure silver must a silversmith mix with a 30 percent alloy to produce 840 grams of a 50 percent alloy?

1. Solve: $\frac{x}{3} - 4 = \frac{3x}{2}$

2. Solve: $\frac{3}{x-3} + \frac{5}{x+3} = \frac{1}{x-3}$

3. Solve $12 - 3bx = 2(x - 3b)$ for x.

4. Solve $x^2 - 3x + 14 = 0$ by completing the square.

5. Solve: $3x^2 - 5x + 1 = 0$

6. Solve: $x^4 - 3x^3 + 8x - 24$

7. Solve: $\sqrt{2x+3} - 1 = \sqrt{3x-5}$

8. Solve: $12(x-2)^2 + 11(x-2) - 5 = 0$

9. Solve: $(2x-3)^{3/5} = 8$

10. Solve: $3 - 5x \le -2(2 - 5x)$

11. Solve: $\frac{x^2 - 7x - 8}{x+6} \ge 0$

12. Solve: $-2 < 3x - 5 \le 10$

13. Solve: $|2-5x|=3$

14. Solve: $|5x-1|\leq 9$

15. Solve: $|4-3x|>7$

16. A biathlon included 12 miles of running and 30 miles of cycling. One contender cycled for 30 minutes longer than he ran, even though his rate of cycling was 7 mph faster than his rate of running. Find his rate of cycling.

17. A chemist mixes a 10 percent acid solution with a 15 percent acid solution. How many milliliters of the 10 percent solution should the chemist use to produce a 1500-milliliter solution that is 12 percent acid?

18. An experienced word processor can input a report in half the time that a trainee can do the job. Together they can complete the task in 20 minutes. How long would it take the trainee to input a report working alone?

19. $10,000 is invested, part at 8 percent simple interest and the remainder at 6.5 percent simple interest. If the two investments earn a total of $707 in interest each year, how much was invested at 8 percent?

20. You can rent a car for the day from Company A for $20 plus $0.09 a mile. Company B charges $13 plus $0.15 a mile. At what point, in terms of miles driven per day, is it cheaper to rent from Company A?

1. Solve: $3x-(3-4x)=x+2(3-x)$

2. Solve: $\frac{3}{x}+\frac{2}{2x-3}=\frac{5}{2x-3}$

3. Solve: $\frac{1}{x}=\frac{b}{a}-\frac{1}{b}$ for x.

4. Solve: $x^2-9x+4=0$ by completing the square.

5. Solve: $(2x-3)(x+2)-x(x+4)=0$

6. Solve: $2x^4-3x^3-16x+24=0$

7. Solve: $\sqrt{2x-1}-1=\sqrt{x+2}$

8. Solve: $8x^{2/3}-41x^{1/3}+5=0$

9. Solve: $x^4-4x^2-32=0$

10. Solve: $3(2x-2)-3x\geq 4x-7$

11. Solve: $\frac{x-2}{x+1}\geq 2$

12. Solve: $-4<5x+3\leq 13$

13. Solve: $|4 - x| = 5$

14. Solve: $|2x - 5| \leq 2$

15. Solve: $|5 - 3x| > 8$

16. After traveling 110 miles on a 200-mile trip, a motorist ran over a rock, which caused the car to vibrate at higher speeds. The motorist decreased his speed by 20 mph for the final 90 miles of the trip. The last leg took 4/5 hours longer than the first leg. Find the motorist's rate after he ran over the rock.

17. Pump A can fill a pool in 6 hours and pump B can fill the same pool in 14 hours. How long will it take to fill the pool if both pumps are used?

18. The length of a rectangle is 3 feet more than three times the width of the rectangle. If the area of the rectangle is 90 square feet, find the width of the rectangle.

19. A radiator contains 5 liters of a 40 percent antifreeze solution. How much should be drained and replaced with pure antifreeze to produce a 60% antifreeze solution?

20. A concert performance brought in $130,800 on the sale of 12,000 tickets. If the tickets sold for $9 and $15, how many of each were sold?

1. Solve: $5x + 12 = 3x - 2(x + 2)$.

a. 4 b. 2 c. -4 d. $-\frac{8}{5}$

2. Solve: $\frac{4}{x+5} + \frac{2}{x+5} = \frac{9}{x-5}$

a. $-\frac{5}{2}$ b. 5 c. $\frac{5}{2}$ d. -25

3. Solve: 13 - 8mx = 5(x + 2m) for x.

a. $\frac{13-10m}{5+8m}$ b. $\frac{\sqrt{5}-10m}{5}$ c. $\frac{13+10m}{5-8m}$ d. $\frac{13-10m}{5-2m}$

4. Solve $x^2 - 7x - 15 = 0$ by completing the square.

a. $\frac{7 \pm \sqrt{49}}{2}$ b. $\frac{7 \pm \sqrt{109}}{2}$ c. $\frac{7}{2}$ or 0 d. $\frac{9}{2}$ or -1

5. Solve: $3x^2 - x - 1 = 0$

a. $\frac{1 \pm \sqrt{13}}{6}$ b. $\frac{-1 \pm \sqrt{13}}{2}$ c. $\frac{-1 \pm \sqrt{11}}{2}$ d. $-\frac{5}{2}$ or 2

6. Solve: $8x^4 + 32x^3 - x - 4 = 0$

a. $-\frac{1}{2}, 4, \pm\frac{1}{4}$ b. $-1 \pm i\sqrt{3}$

c. $-\frac{1}{4}, 4, -1 \pm i\sqrt{3}$ d. $-4, \frac{1}{2}, -\frac{1}{4} \pm \frac{\sqrt{3}}{4}i$

7. Solve: $\sqrt{4x+1} = x - 1$

a. -1, 1 b. $\frac{3 \pm \sqrt{2}}{2}$ c. 6 d. $\frac{3 \pm 2\sqrt{11}}{2}$

8, Solve: $2x^{2/5} + 7x^{1/5} - 4 = 0$

a. $-\frac{1}{2}, 4$ b. $-1024, \frac{1}{32}$ c. 4 d. $-\frac{1}{32}, 1024$

9. Solve: $x^4 - 37x^2 + 36 = 0$

a. $\pm i, \pm 6$ b. $\pm 1, \pm 6$ c. $\pm 1, \pm 6i$ d. $\pm i, \pm 6i$

10. Solve: $-3(x-5)+2x \geq 7x-14$

a. $x \geq \frac{29}{6}$ b. $x \geq \frac{1}{8}$ c. $x \leq \frac{29}{8}$ d. $x \leq -\frac{1}{8}$

11. Solve: $\frac{(x+1)(x-8)}{x+6} \geq 0$

a. $(-6,-1] \cup [8,\infty)$ b. $(-\infty,1) \cup [-1,8]$

c. $[-6,-4] \cup [8,\infty)$ d. $[-8,1] \cup [6,\infty)$

12. Solve: $-8 \leq 3-2x < 5$

a. $-\frac{11}{2} \leq x < 1$ b. $-1 < x \leq \frac{11}{2}$ c. $2 < x \leq \frac{5}{2}$ d. $-\frac{5}{2} < x \leq 1$

13. Solve: $|7-5x| = 12$

a. -1 b. $-1, \frac{19}{5}$ c. $\frac{19}{5}$ d. $-\frac{19}{5}, 1$

14. Solve: $|2x-5| < 5$

a. $0 < x < 5$ b. $-5 < x < 5$ c. $5 < x < 7$ d. $-\frac{5}{2} < x < \frac{5}{2}$

15. Solve: $|7-3x| \geq 5$

a. $\left(-\infty, \frac{2}{3}\right] \cup [4,\infty)$ b. $\left(-\infty, \frac{2}{3}\right] \cup [4,\infty)$

c. $\left[-\infty, \frac{2}{3}\right] \cup [4,8]$ d. $\frac{2}{3} \leq x \leq 4$

16. A jogger started a course at an average speed of 6 mph. One-half hour later, a cyclist started the same course at an average speed of 10 mph. How long after the jogger started did the cyclist overtake the jogger?

a. 0.75 h b. 1.25 h c. 2 h d. 1.5 h

17. How many liters of water should be evaporated from 80 liters of a 10% saline solution so that the solution that remains is a 15% solution?

a. 23.3 L b. 26.7 L c. 33.3 L d. 20 L

18. An investment of $4500 is made at an annual simple interest rate of 8 percent. How much additional money must be invested at an annual simple interest rate of 6 percent so that the total interest earned is $600?

a. $3000 b. $4000 c. $3500 d. $4500

19. A mason can build a wall in one-fourth the time it takes an apprentice. Together they build the wall in 8 hours. How long would it take the apprentice working alone to build the wall?

a. 10 h b. 20 h c. 40 h d. 30 h

20. One side of a triangle is 2 feet longer than the second side. The third side of the triangle is 1 foot more than twice the second side. The perimeter of the triangle is 15 feet. Find the measures of the three sides of the triangle.

a. 4 ft, 5 ft, 6 ft b. 3 ft, 5 ft, 7 ft

c. 2 ft, 4 ft, 9 ft d. 2 ft, 5 ft, 8 ft

1. Solve: $\frac{x}{4} - 3 = \frac{3}{5}$

   a. $\frac{72}{5}$ b. $\frac{48}{5}$ c. $-\frac{48}{5}$ d. 3

2. Solve: $\frac{7}{x-6} + \frac{3}{x-6} = \frac{11}{x+6}$

   a. $\frac{2}{7}$ b. -88 c. $-\frac{2}{7}$ d. 126

3. Solve: $3bx + 9 = 4(3x - b) + 2$ for x.

   a. $-\frac{4b-7}{3b-12}$ b. $-\frac{4b+7}{3b+12}$ c. $-\frac{4b-7}{3b+12}$ d. $\frac{7+4b}{12-3b}$

4. Solve: $x^2 + 5x - 1 = 0$ by completing the square.

   a. $-\frac{5}{2} \pm \frac{\sqrt{21}}{2}i$ b. $-\frac{5}{2} \pm \frac{\sqrt{29}}{2}$ c. $\frac{5}{2} \pm \frac{\sqrt{29}}{2}i$ d. $-\frac{5}{4} \pm \frac{\sqrt{29}}{4}i$

5. Solve: $3x^2 - 2x + 4 = 0$

   a. $\frac{-1 \pm \sqrt{11}}{3}$ b. $-\frac{2}{3} \pm \frac{\sqrt{11}}{3}i$ c. $\frac{2 \pm \sqrt{11}}{3}$ d. $\frac{1}{3} \pm \frac{\sqrt{11}}{3}i$

6. Solve: $x^5 + x^3 + x^2 + 1 = 0$

   a. $-1, 1, i, -i$ b. $-1, 1, i, -i, \frac{1}{2} \pm \frac{\sqrt{3}}{2}i$ c. $-1, -i, i, \frac{1}{2} \pm \frac{\sqrt{3}}{2}i$ d. $-1, -i, i, \frac{1 \pm \sqrt{3}}{2}$

7. Solve: $\sqrt{6x + 19} = x - 6$

   a. 17   b. $3 \pm 2\sqrt{2}\,i$ c. $3 \pm 2\sqrt{2}$ d. 4, 9

8. Solve: $2(x + 3)^{1/2} + 3(x + 3)^{1/4} - 2 = 0$

   a. $1, \frac{49}{16}$ b. $\frac{23}{8}$ c. $-2, \frac{1}{2}$ d. $-\frac{47}{16}$

9. Solve: $(x + 2)^{3/5} = 27$

   a. 245 b. 241 c. –1 d. 3

10. Solve: $x^2 + 12x \geq 45$

    a. $(-\infty, -15] \cup [3, \infty)$ b. $(-\infty, 3) \cup (15, \infty)$
    c. $(-\infty, 3] \cup [15, \infty)$ d. $[15, 3]$

11. Solve: $\dfrac{x^2 + 7x - 78}{x + 2} < 0$

    a. $(-13, -2) \cup (6, \infty)$ b. $(-\infty, -2) \cup (6, \infty)$
    c. $(-\infty, -13) \cap (-2, 6)$ d. $(-\infty, -13) \cup (-2, 6)$

12. Solve: $9 < 2x + 1 \leq 33$

    a. $5 < x \leq 17$ b. $4 < x \leq 16$ c. $-17 \leq x < -5$ d. $8 < x \leq 32$

13. Solve: $|2x - 1| = 9$

    a. 4, –5 b. –5, –4 c. –4, 5 d. –4

14. Solve: $|4x + 3| < 15$

    a. $-6 < x < 3$ b. $-\frac{9}{2} < x < 3$ c. $-6 < x < \frac{9}{2}$ d. $x < 3$

15. Solve: $|3x - 2| \geq 7$

    a. $\left(-\infty, -\frac{5}{3}\right] \cup [3, \infty)$ b. $-3 \leq x \leq \frac{5}{3}$ c. $\left(-\infty, \frac{5}{3}\right] \cup (3, \infty)$ d. $-\frac{5}{3} \leq x \leq 3$

16. An experienced painter can complete the exterior of one unit in a condominium development in one-fifth the time that it takes the trainee. If working together the two painters can complete the job in 2.5 days. How long would it take the trainee, working alone, to complete the job?

    a. $3\frac{3}{4}$ days b. 12 days c. 3 days d. 15 days

17. A jogger started a course at an average speed of 4.5 mph. A cyclist started the same course 1 hour later at an average speed of 14 mph. How long after the jogger started did the cyclist overtake the jogger? Round to the nearest tenth of an hour.

a. 1.5 h b. 1.2 h c. 0.8 h d. 1.8 h

18. The price of a magazine subscription rose 8% this year. If the subscription now costs $29.16, how much did the subscription cost last year?

a. $26.83 b. $27.00 c. $28.00 d. $27.23

19. A total of $25,000 is deposited into two simple interest accounts. On one account the annual simple interest rate is 7.4%, and on the second account the annual simple interest rate is 5.5%. The amount of interest earned for one year was $1527. How much is invested in the 5.5% account?

a. $17,000 b. $8000 c. $11,800 d. $15,000

20. A coffee shop decides to blend a coffee that sells for $10 per pound with a coffee that sells for $6 per pound to produce a blend that will sell for $7.50 per pound. How much of the $6 coffee should be used to yield 50 pounds of the new blend:

a. 22 lb b. 35 lb c. 18.75 lb d. 31.25 lb

1. Solve: $2x - 3(3 - 4x) = -4(2x - 8)$

   a. $\frac{7}{2}$ b. $-\frac{7}{2}$ c. $\frac{41}{22}$ d. $\frac{41}{6}$

2. Solve: $\frac{2x - 3}{x + 1} - 2 = \frac{3}{x}$

   a. $-\frac{3}{8}$ b. $\frac{4 \pm \sqrt{6}}{2}$ c. $-\frac{2}{3}$ d. $-\frac{1}{8}$

3. Solve $\frac{1}{f} = \frac{1}{a} + \frac{1}{b} + \frac{1}{c}$ for f.

   a. $\frac{a + b + c}{ac + ab + bc}$ b. $\frac{abc}{ac + ab + bc}$ c. $a + b + c$ d. $\frac{abc}{a + b + c}$

4. Solve $x^2 - 11x + 33 = 0$ by completing the square.

   a. $\frac{11 \pm \sqrt{253}}{4}$ b. $\frac{11 \pm \sqrt{11}}{2}$ c. $-\frac{11}{2} \pm \frac{\sqrt{11}}{2} i$ d. $\frac{11}{2} \pm \frac{\sqrt{11}}{2} i$

5. Solve: $2x^2 - x + 8 = 0$

   a. $\frac{1}{4} \pm \frac{3\sqrt{7}}{4} i$ b. $\frac{-1 \pm \sqrt{65}}{2}$ c. $-\frac{1}{2} \pm \frac{3\sqrt{7}}{2} i$ d. $\frac{1}{2} \pm \frac{3\sqrt{7}}{2} i$

6. Solve: $x^4 - x^3 + 8x - 8 = 0$

   a. $-2, 2, \frac{1 \pm \sqrt{3}}{2}$ b. $-2, 1, 1 \pm i\sqrt{3}$

   c. $-1, 2, 1 \pm i\sqrt{3}$ d. $1, 8, 1 \pm i\sqrt{3}$

7. Solve: $\sqrt{x - 3} + 1 = \sqrt{x + 1}$

   a. $\frac{3}{4}$ b. $-\frac{3}{4}$ c. $\frac{21}{4}$ d. $\frac{15}{4}$

8. Solve: $4x^{2/3} + 19x^{1/3} - 5 = 0$

   a. $-5, \frac{1}{4}$ b. $-125, \frac{1}{64}$ c. $-125, 64$ d. $\frac{1}{16}, 25$

9. Solve: $(x - 8)^{4/5} = 16$

a. 24 b. 32 c. 40 d. 64

10. Solve: $-4(x - 5) - 7 \geq 2x + 6$

a. $x \geq -\frac{7}{6}$ b. $x \leq \frac{7}{6}$ c. $x \leq \frac{19}{6}$ d. $x \leq -\frac{19}{2}$

11. Solve: $\frac{x - 3}{2x - 3} \leq 2$

a. $(-\infty, 1] \cup \left(\frac{3}{2}, \infty\right)$ b. $(-\infty, -1) \cup \left[\frac{3}{2}, \infty\right)$

c. $(-\infty, -1) \cup \left(\frac{3}{2}, \infty\right)$ d. $(-\infty, 1) \cup [3, \infty)$

12. Solve: $-12 \leq 3 - 5x < 8$

a. $[-3, 1)$ b. $\left[-\frac{9}{5}, \frac{11}{5}\right]$ c. $(1, 3)$ d. $(-1, 3]$

13. Solve: $|4x - 7| = 9$

a. $-\frac{1}{2}, 4$ b. 4 c. $-4, \frac{1}{2}$ d. $\frac{9}{2}, \frac{1}{2}$

14. Solve: $|2x + 5| < 7$

a. $(-1, 1)$ b. $(-6, 6)$ c. $(-6, 1)$ d. $(-1, 6]$

15. Solve: $|4x - 3| \geq 7$

a. $\left[-1, \frac{5}{2}\right]$ b. $(-\infty, -1] \cup \left[\frac{5}{3}, \infty\right)$

c. $\left(-\infty, -\frac{5}{2}\right) \cup [1, \infty)$ d. $(-\infty, -1] \cup [1, \infty)$

16. A biathlon consisted of 20 miles of running and 40 miles of cycling. One contestant ran 2/5 hour longer than he cycled. His cycling rate was 5 mph more than twice that of his running rate. Find his running rate.

a. 12.5 mph b. 7.5 mph c. 9 mph d. 10 mph

17. A radiator contains 6 liters of a 25% antifreeze solution. How much should be drained and replaced with pure antifreeze to produce a 75% antifreeze solution?

a. 2.5 L b. 4 L c. 3 L d. 4.5 L

18. The perimeter of a rectangle is 38 inches and the area is 84 square feet. Find the length of the rectangle.

a. 7 in. b. 12 in. c. 9 in. d. 14 in.

19. Printer A can print a report in 24 minutes and printer B can print the report in 40 minutes. How long would it take both printers working together to print the report?

a. 15 min b. 18 min c. 12 min d. 9 min

20. You can rent a car for the day from Company A for $20 plus $0.08 a mile. Company B charges $8 a day plus $0.16 a mile. At what point, in terms of miles driven per day, is it cheaper to rent from Company A?

a. Less than 150 miles per day b. Greater than 117 miles per day

c. Greater than 150 miles per day d. Greater than 175 miles per day

1. Find the midpoint of the line segment with endpoints (-4, 5) and (3, -2).

2. Determine the x- and y-intercepts, and then graph the equation $y = 2x^2 - 4$.

3. Graph the equation $y = |x - 2| - 1$.

4. Find the center and radius of the circle that has the general form $x^2 + y^2 + 10x - 6y + 18 = 0$.

5. Determine the domain of the function $f(x) = \sqrt{9 - x^2}$.

6. A multi-media system was purchased for $16,000. Assuming that the system depreciates at a rate of $3000 per year (straight line depreciation) for the first four years, write the value v of the system as a function of the time t (measured in years) for $0 \le t \le 4$.

7. Graph $f(x) = -3|x + 1| - 1$. Identify the intervals over which the function is a. increasing, b. constant, or c. decreasing.

8. Graph the function $f(x) = -x^2 - 2$. From the graph, find the domain and range of the function.

9. Sketch the graph of the function $f(x) = (x + 3)^2 - 2$.

10. Use the graph of $f(x) = x^2$ to graph $y = -f(x - 1) + 1$.

11. Which of the following are even functions?
a. $f(x) = x^2 - 5x$ b. $g(x) = x^8$ c. $h(x) = x^5 - x$

12. Find the slope-intercept form of the equation of the line that passes through (-2, 5) and is parallel to the graph of $2x + 5y = 7$.

13. Find the maximum or minimum value of the function $f(x) = 2x^2 - 3x + 4$. State whether this value is a maximum or a minimum value.

14. Let $f(x) = 4x - 9$ and $g(x) = x^2 + 3$. Find $(f + g)$ and $\left(\frac{f}{g}\right)$.

15. Find the difference quotient of the function $f(x) = x^2 + 8x + 3$.

16. Evaluate $(f \circ g)$, where $f(x) = x^2 + 2x$ and $g(x) = 2x + 5$.

17. Find the inverse of $f(x) = x^2 + 4, x \geq 0$. State the domain and range of both f and f^{-1}.

18. Find the inverse of the function given by the equation $f(x) = \frac{1}{2}x + 2$. Graph f and f^{-1} on the same coordinate axis.

19. t varies directly as the square of r and inversely as s. When $r = 6$ and $s = 4$, $t = 27$. Find the variation constant.

20. The loudness, measured in decibels, of a stereo speaker is inversely proportional to the square of the distance of the listener from the speaker. The loudness is 8 decibels at a distance of 9 feet. What is the loudness when the listener is 5 feet from the speaker?

1. Find the midpoint of the line segment with endpoints (-5, 11) and 3, 6).

2. Determine the x- and y-intercepts, and then graph the equation $y = 3(x-1)^2$

3. Graph the equation $x = |y-2| - 2$.

4. Find the center and radius of the circle that has the general form $x^2 + y^2 - 2x + 18y + 33 = 0$.

5. Determine the domain of the function $f(x) = -\sqrt{x^2 - 25}$.

6. A manufacturer produces a product at a cost of $50 per unit. The manufacturer has a fixed cost of $800 per day. Each unit retails for $87.50. Let x represent the number of units produced in a 5-day period.
   a. Write the total cost C as a function of x.
   b. Write the revenue R as a function of x.
   c. Write the profit P as a function of x.

7. Graph $f(x) = |x-2| - |x|$. Identify the intervals over which the function is a. increasing, b, constant, or c, decreasing.

8. Graph the function $f(x) = x^3 - 2$. From the graph, find the domain and range of the function.

9. Which of the following define y as a function of x?
   a. $x^2 + 5y^2 = 10$ b. $x^2 + 11y = 3$ c. $2x + 9y = 7$

10. Use the graph of $f(x) = x^2$ to graph $y = -(x-2)^2 + 4$.

11. Which of the following are odd functions?
a. $f(x)=x^7-x$ b. $g(x)=x^4-3x^3$ c. $h(x)=x^{10}$

12. Find the slope-intercept form of the equation of the line that passes through $(-3, 4)$ and is perpendicular to the graph of $2x-5y=-4$.

13. Find the maximum or minimum value of the function $f(x)=x^2-14x+19$. State whether this value is a maximum or a minimum value.

14. Let $f(x)=x^2-2x+1$ and $g(x)=x+3$. Find $(g-f)$. and (fg).

15. Find the difference quotient of the function $f(x)=x^2+15x-7$.

16. Evaluate $(f \circ g)(4)$, where $f(x)=-3x+1$ and $g(x)=x^2+4$.

17. Find the inverse of $f(x)=x^2-4, x \geq 0$. State the domain and range of both f and f^{-1},

18. Find the inverse of the function given by the equation $f(x)=3x+1$. Graph f and f^{-1} on the same coordinate axis.

19. Hooke's Law states that the distance a spring stretches varies directly as the weight on the spring. A weight of 4 pounds stretches a spring 0.30 inches. How far will a weight of 50 pounds stretch the spring?

20. The sum of the length L and the width W of a rectangular area is 200 meters. Write the area A as a function of L. Find the dimensions that produce the greatest area.

1. Find the distance between the points whose coordinates are (3,-4) and (-2, 7).

2. Which of the following equations defines y as a function of x?
a. $x^2 - 4y^2 = 12$ b. $x^2 + 6y = 15$ c. $8x + 2y = 11$

3. Graph the equation $y = x^2 - 2x + 1$.

4. Find the center and radius of the circle that has the general form $x^2 + y^2 - 4x + 2y - 3 = 0$.

5. Determine the domain of the function $f(x) = \sqrt{6 + x}$.

6. A motel was purchased for \$450,000. Assuming the motel depreciates at a rate of \$15,000 per year (straight line depreciation) for 30 years, write the value v of the motel as a function of the time *t* (measured in years) for $0 \le t \le 30$.

7. Graph $f(x) = \begin{cases} 2, \text{ if } x \le -1 \\ x^2 - 1, \text{ if } -1 < x \le 1 \\ x - 1, \text{ if } x > 1 \end{cases}$

8. Graph the function $f(x) = |x - 2| + 2$. From the graph, find the domain and range of the function.

9. Sketch the graph of the function $f(x) = x^3 - 1$.

10. Use the graph of $f(x) = x^2$ to graph $y = -(x + 2)^2 + 3$.

11. Which of the following are odd functions?
a. $f(x) = x^5 - x^2$ b. $g(x) = 4x^6 + 2x^2$ c. $h(x) = x^{13}$.

12. Find the slope-intercept form of the equation of the line that passes through (2, -3) and is parallel to the graph of 2x - 5y = 23.

13. Find the maximum or minimum value of the function $f(x) = 2x^2 - 3x + 4$. State whether this value is a maximum or a minimum value.

14. Let $f(x) = 2x^2 - x + 4$ and $g(x) = x^2 + 5$. Find (f - g) and (fg).

15. Find the difference quotient of the function $f(x) = -6x + 3$.

16. Evaluate $(f \circ g)$, where $f(x) = 8x + 3$ and $g(x) = -3x^2 + 1$.

17. Find the inverse of $f(x) = \dfrac{7x}{x+5}$, $x \neq -5$. State the domain and range of both f and f^{-1}.

18. Find the inverse of the function given by the equation $f(x) = 3x - 4$. Graph f and f^{-1} on the same coordinate axis.

19. The current I in a wire varies directly as the voltage v and inversely as the resistance r. The current is 4 amperes when r is 6 ohms and v is 24 volts. Find I when v is 66 volts and r is 12 ohms.

20. A lighthouse is 9 miles north of a port. A ship leaves port sailing East at 12 mph. If t is the number of hours the ship has been sailing, express the distance d between the ship and the lighthouse as a function of time.

20. A lighthouse is 9 miles north of a port. A ship leaves port sailing at 12 mph. If t is the number of hours the ship has been sailing, express the distance d between the ship and the lighthouse as a function of time.

1. Find the midpoint of the line segment with endpoints (4, –2) and (–6, –7).

   a. $\left(5, -\frac{9}{2}\right)$ b. $\left(-1, -\frac{9}{2}\right)$ c. $\left(-1, -\frac{5}{2}\right)$ d. $\left(-1, \frac{9}{2}\right)$

2. Graph the equation $(x-1)^2 + (y+1)^2 = 16$.

   a.
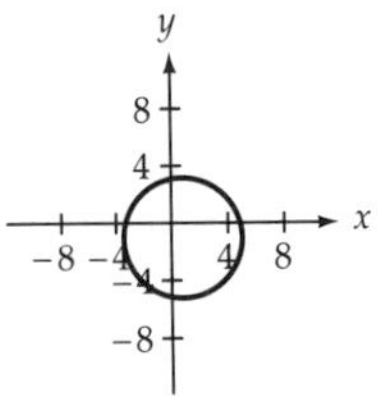

   b.
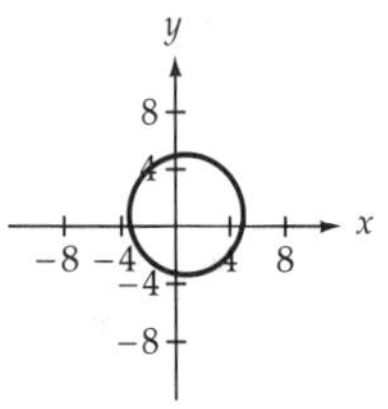

   c.
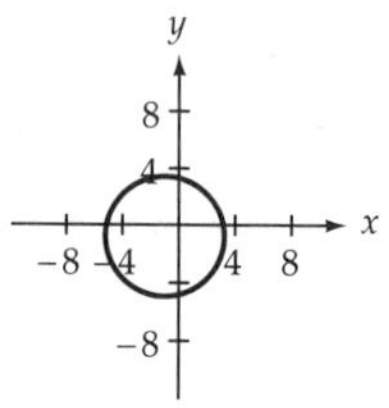

   d.
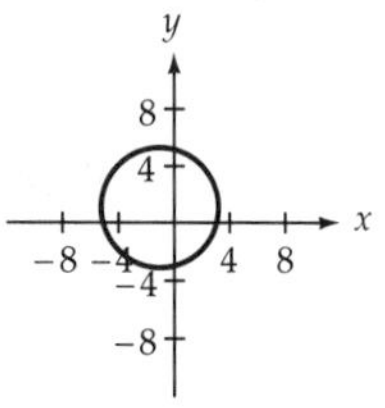

3. Graph the equation $y = |x-2| - 2$.

   a.
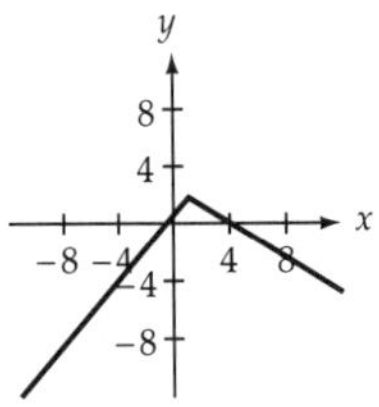

   b.
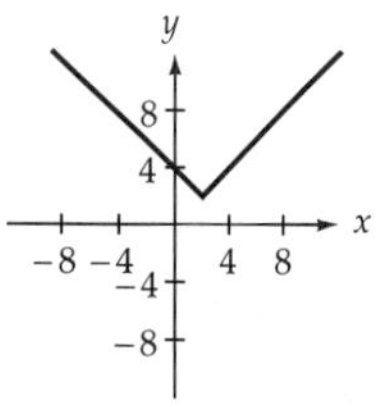

   c.
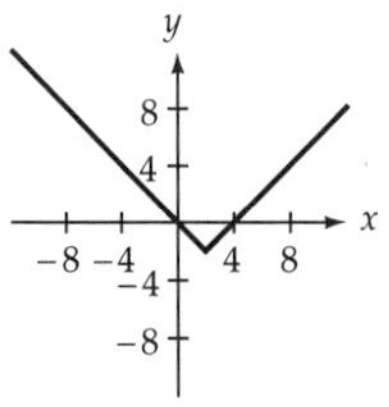

   d.
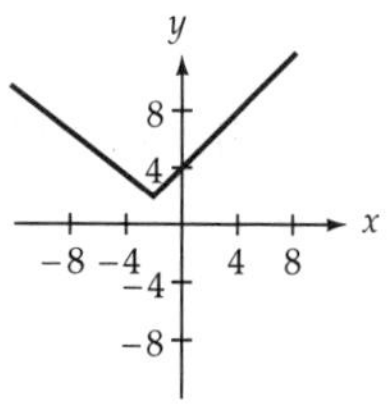

4. Find the center and radius of the circle that has the general form $x^2 + y^2 - 2x + 6y + \frac{39}{4} = 0$.

   a. center: (–1, 3), radius: $\frac{1}{2}$

   b. center: (1, 9), radius: $\frac{\sqrt{79}}{2}$

   c. center: (1, –3), radius: $\frac{1}{2}$

   d. center: (1, –9), radius: $\frac{1}{4}$

5. Determine the domain of the function $f(x) = \sqrt{1 - x^2}$.

   a. $[0, \infty)$ b. $(-\infty, -1] \cup [1, \infty)$ c. $[-1, 1]$ d. $(-\infty, 0]$

6. A business purchases a computer system for \$11,400. After 6 years the system will be obsolete and have no value. Find a linear function that expresses the value V of the computer in terms of the number of years n that it has been used, where $0 \le n \le 6$.

a. $V = 1900n - 11{,}400$ b. $V = -1900n + 11{,}400$

c. $V = 11{,}400n - 1900$ d. $V = \frac{1}{1900}n + 11{,}400$

7. Graph $f(x) = \begin{cases} -7, & x \ge -4 \\ |x|, & x > -4 \end{cases}$

a.
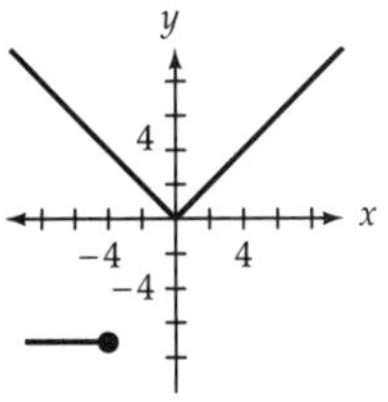
b.
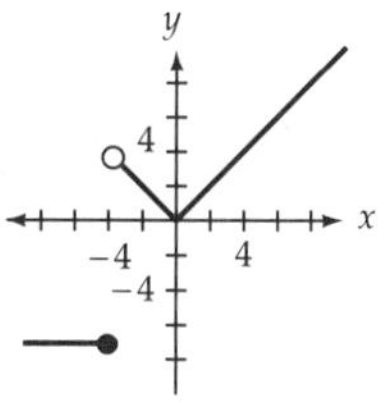
c.
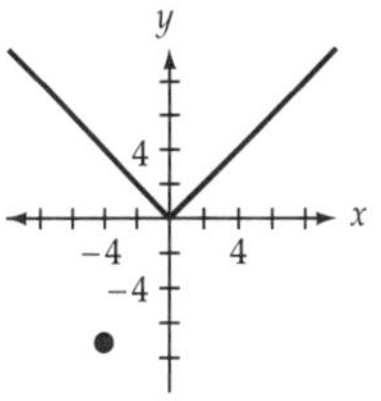
d.
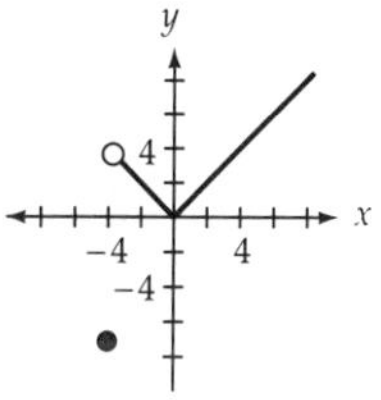

8. Graph the function $f(x) = x^2 + 4x + 4$.

a.
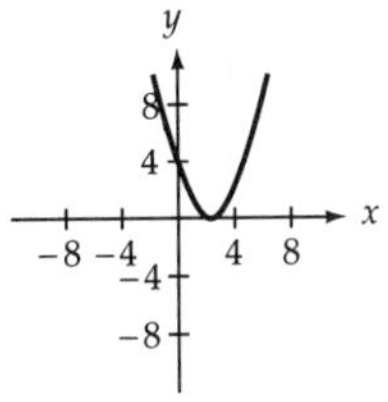
b.
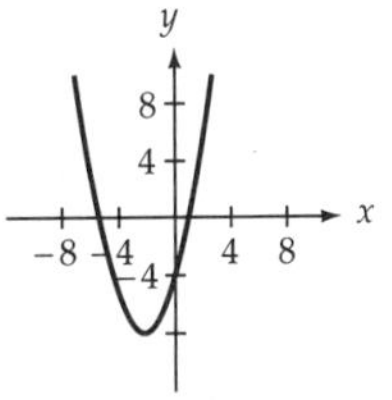
c.
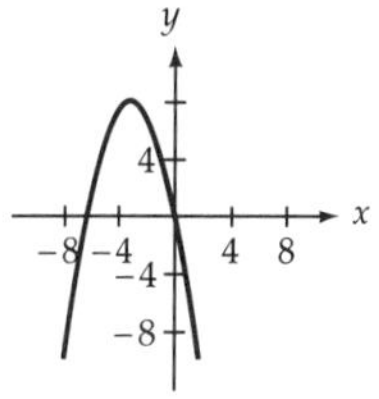
d.
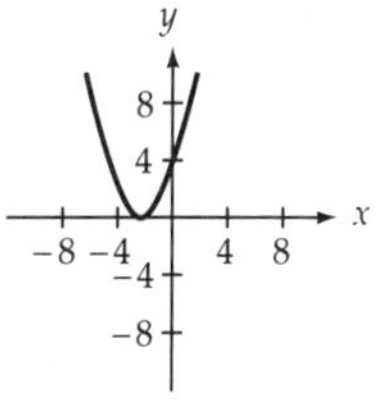

9. Which of the following are odd functions: i. $f(x) = -5|x|$ ii. $g(x) = x^3 + 7$ iii. $h(x) = x^5$

a. i b. ii c. iii d. none

10. Use the graph of $f(x) = |x|$ to graph $y = -f(x-2) + 1$.

a.
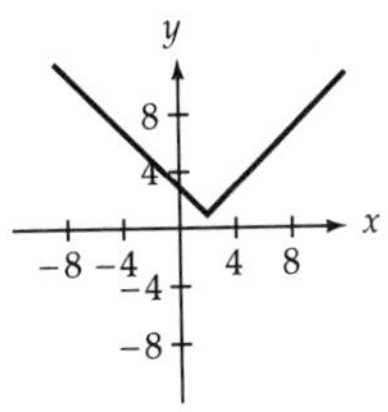
b.
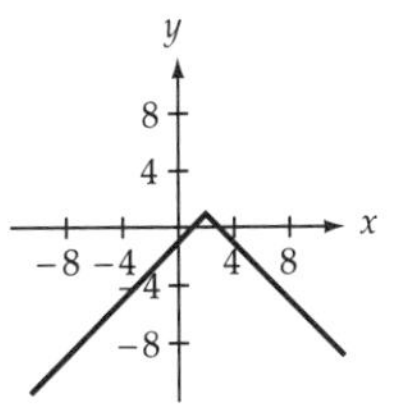
c.
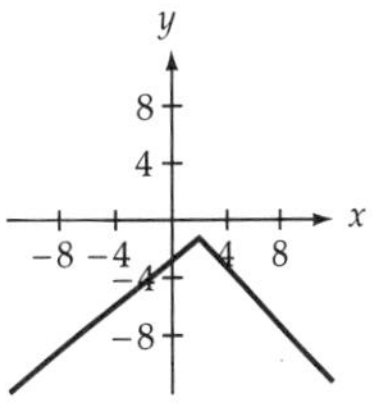
d.
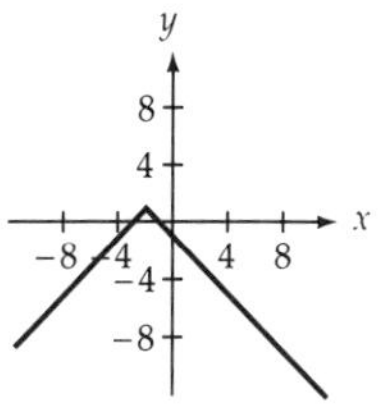

11. Which of the following define y as a function of x?
    i. $2x + 15y = 18$ ii. $x^2 + 9y^2 + 12$ iii. $x^2 + 5y = 7$

    a. i, ii b. ii, iii c. i, iii d. i

12. Find the slope-intercept form of the equation of the line that passes through (2, –4) and is perpendicular to the graph of $4x + 3y = 16$.

    a. $y = -\frac{4}{3}x - \frac{4}{3}$ b. $y = \frac{3}{4}x + \frac{5}{2}$

    c. $y = \frac{3}{4}x - \frac{11}{2}$ d. $y = -\frac{4}{3}x + \frac{20}{3}$

13. Find the maximum or minimum value of the function $f(x) = -2x^2 - 4x + 7$. State whether this value is a maximum or a minimum value.

    a. 1, maximum b. 1, minimum c. 9, minimum d. 9, maximum

14. Let $f(x) = x^2 + 2x + 3$ and $g(x) = 4x^2 - 9$. Find $(f - g)$.

    a. $5x^2 - 11x + 3$ b. $-3x^2 - 2x - 12$

    c. $5x^2 + 2x - 6$ d. $-3x^2 + 2x + 12$

15. Find the difference quotient of the function $f(x) = x^2 + 3x + 2$.

    a. $2x + h + 3$ b. $2x + 3$ c. $2xh + h^2 + 3$ d. $\frac{2x + 3}{h}$

16. Evaluate $(f \circ g)$, where $f(x) = -2x^2 + 3$ and $g(x) = 3x - 4$.

    a. $-6x^2 + 5$ b. $-6x^2 + 9x - 4$

    c. $-18x^2 + 48x - 29$ d. $-18x^2 + 48x - 32$

17. Find the inverse of $f(x) = \sqrt{4 - x}, x \leq 4$.

    a. $y = \sqrt{x - 4}$ b. $x = \sqrt{4 - y}$ c. $y = 4 - x^2$ d. $y = x^2 - 4$

18. Graph $f(x) = 3x - 2$ and f^{-1} on the same coordinate axis.

a. 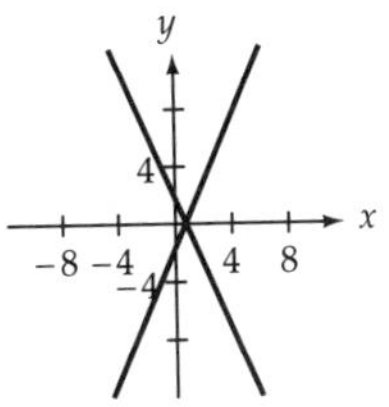

b. 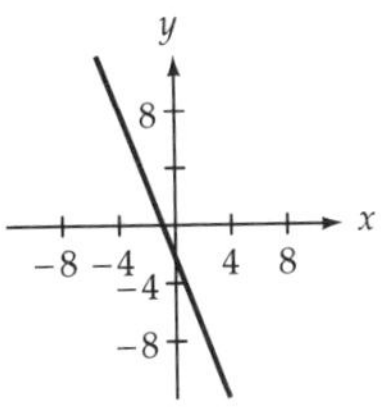

c. 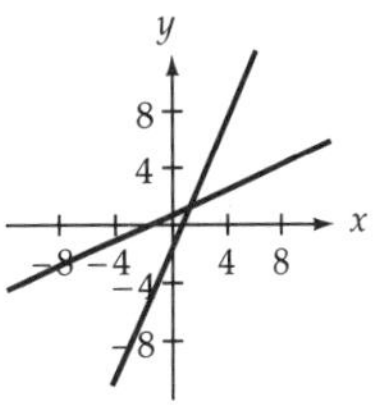

d. 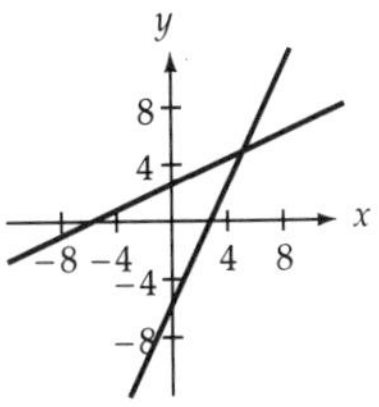

19. u varies directly as v and inversely as the square root of w and u = 10 when v = 80 and w = 16. Find u when v = 120 and w = 4.

a. 60 b. 25 c. 30 d. 40

20. The distance that an object falls from rest (neglecting air resistance) varies directly as the square of the time t that it has been falling. If an object falls 144 feet in 3 seconds, how far will it fall in 7.5 seconds?

a. 900 ft b. 1200 ft c. 120 ft d. 600 ft

1. Find the midpoint of the line segment with endpoints $(-7, 4)$ and $(3, 8)$.

a. $(-5, -2)$ b. $(-2, 1)$ c. $(-2, 6)$ d. $(-2, 2)$

2. Graph the equation $y = -3x^2 + 4$.

a.
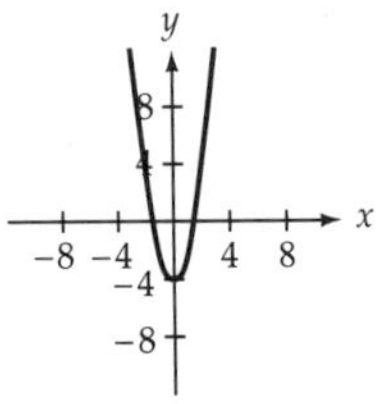

b.
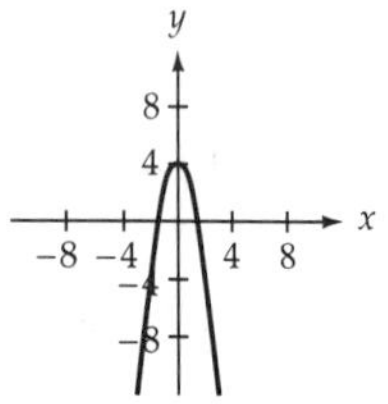

c.
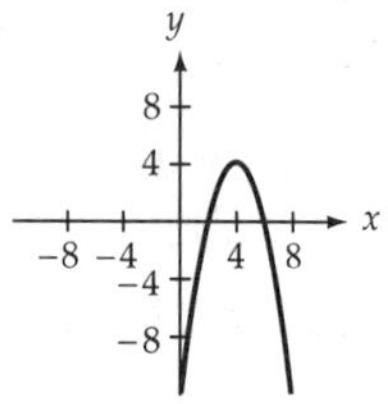

d.
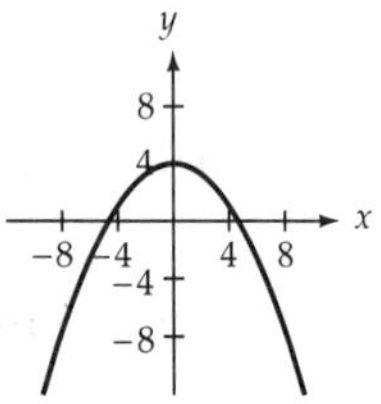

3. Graph the equation $y = |x + 6| - 3$.

a.
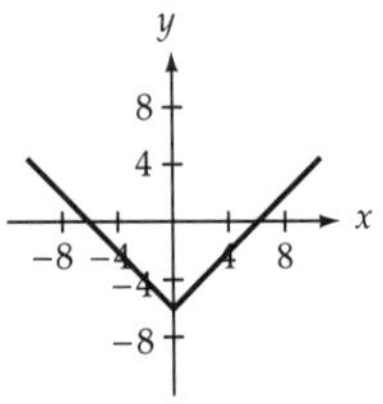

b.
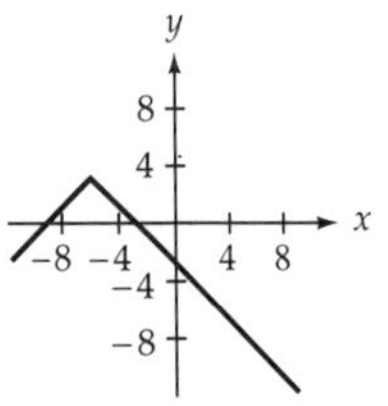

c.
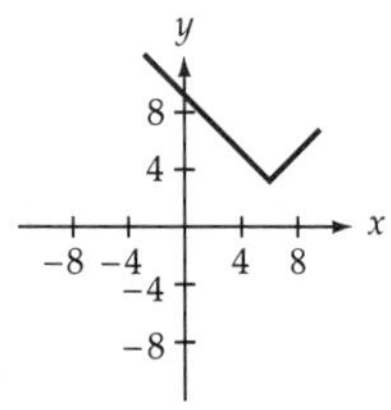

d.
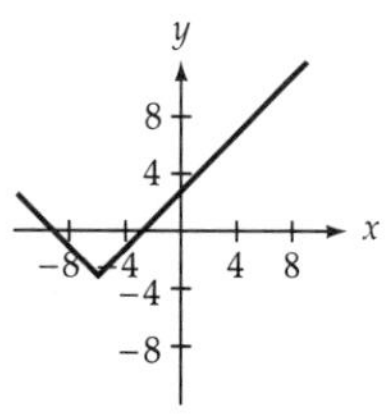

4. Find the center and radius of the circle that has the general form $x^2 + y^2 + 4x - 12y - 24 = 0$.

a. center: $(-2, 6)$, radius: 8 b. center: $(2, -4)$, radius: $2\sqrt{6}$

c. center: $(2, -6)$, radius: 64 d. center: $(2, -6)$ radius: 4

5. Determine the domain of $f(x) = \sqrt{x^2 - 9}$.

a. $[-3, 3]$ b. $(-\infty, -3] \cup [3, \infty]$

c. $[-\infty, -3] \cup [3, \infty]$ d. $(-\infty, -3) \cup (3, \infty)$

6. The profit function P for a company selling x items is given by $P(x) = -2x^2 + 320x - 2500$. What value of x will maximize the profit? What is the maximum profit?

a. 80, \$25,600 b. 40, \$7100 c. 40, \$12,800 d. 80, \$10,300

7. Graph $f(x) = \begin{cases} -5, & \text{if } x \le -2 \\ x^2 - 4, & \text{if } -2 < x \le 2 \\ x - 2, & \text{if } x > 2. \end{cases}$

a. 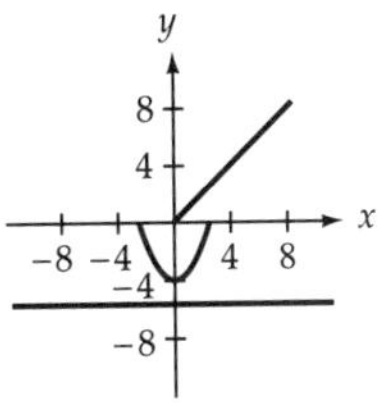

b. 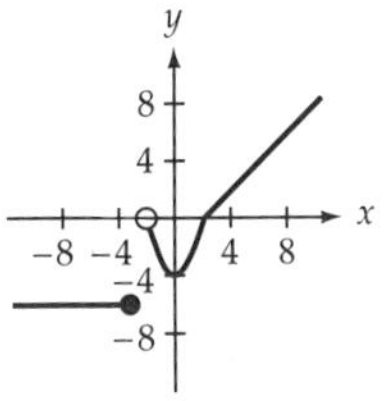

c. 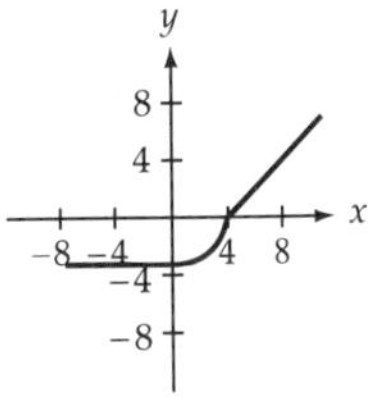

d. 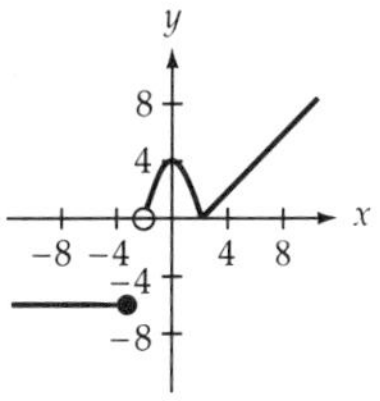

8. Find the domain and the range of $y = -\sqrt{16 - x^2}$.

a. domain: $(-\infty, 4) \cup (4, \infty)$, range: $(4, \infty)$

b. domain: $(-4, 4)$, range: $(-4, 0)$

c. domain: $[-4, 4]$, range: $[-4, 0]$

d. domain: $(-\infty, 4) \cup (4, \infty)$, range: $(-\infty, 0]$

9. Which of the following define y as a function of x?

i. $\frac{x^2}{4} + \frac{y^2}{16} = 1$ ii. $x^3 - 3y = 4$ iii. $x = |y|$

a. i b. ii c. iii d. ii and iii

10. Use the graph of $y = x^2$ to graph $y = -(x - 3)^2$.

a. 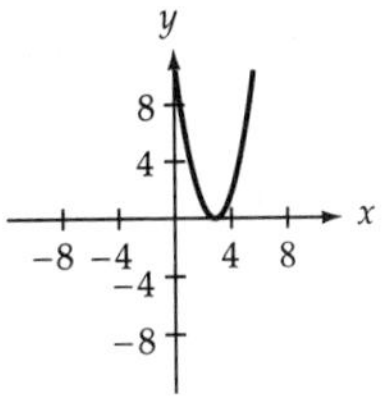

b. 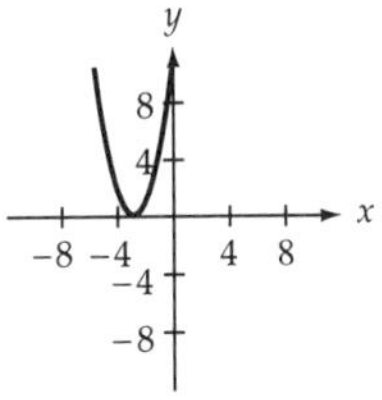

c. 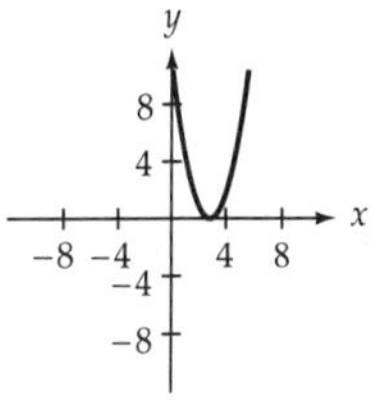

d. 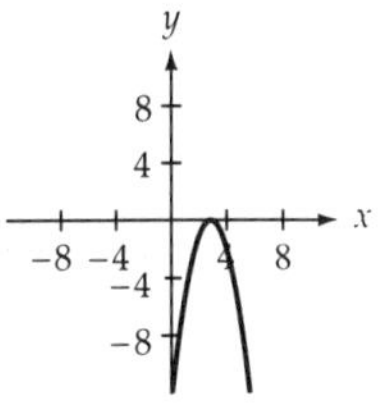

11. Which of the following are even functions?

i. $f(x) = |x| - 2$ ii. $g(x) = x^2 - x$ iii. $x^2 + y = 4$

a. i, ii b. i, iii c. ii, iii d. iii

12. Find the slope-intercept form of the equation of the line that passes through (2, 0) and is perpendicular to the graph of $3x - 7y = 33$.

a. $y = \frac{3}{7}x - \frac{6}{7}$
b. $y = \frac{7}{3}x - \frac{6}{7}$
c. $y = -\frac{7}{3}x + \frac{14}{3}$
d. $y = -\frac{3}{7}x + \frac{14}{3}$

13. Find the maximum or minimum value of the function $f(x) = 2x^2 - 4x - 1$. State whether this value is a maximum or a minimum value.

a. -1, minimum
b. -3, minimum
c. 1, maximum
d. 3, maximum

14. Let $f(x) = x^2 - 5x + 6$ and $g(x) = 3x + 4$. Find (fg).

a. $3x^3 - 19x^2 - 4x + 24$
b. $3x^3 - 5x^2 + 2x + 6$
c. $3x^3 + 11x^2 - 4x + 6$
d. $3x^3 - 11x^2 - 2x + 24$

15. Find the difference quotient of the function $f(x) = 2x^2 - 3x + 1$.

a. $4x + 2h - 3$
b. $x^2 + 2xh - 3h$
c. $4x - 3$
d. $2h - 3$

16. Evaluate $(f \circ g)$, where $f(x) = 5x^2 - 4$ and $g(x) = \sqrt{2x - 4}$.

a. x
b. $10x - 24$
c. $\sqrt{10x^2 - 12}$
d. $5\sqrt{2x - 4} - 4$

17. Find the inverse of $f(x) = \frac{9x}{x + 2}$.

a. $f^{-1}(x) = \frac{x + 2}{9x}$
b. $f^{-1}(x) = \frac{9x}{x - 2}$
c. $f^{-1}(x) \frac{x - 9}{2}$
d. $f^{-1}(x) = \frac{2x}{x - 9}$

18. Graph the function given by the equation $f(x) = 4x - 3$ and its inverse f^{-1} on the same coordinate axis.

a. 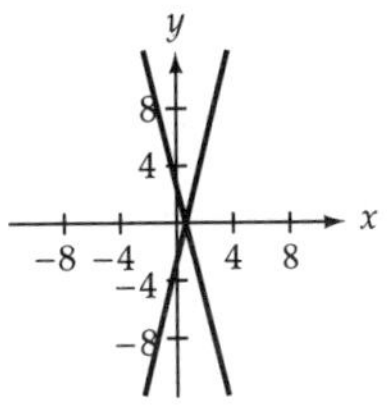

b. 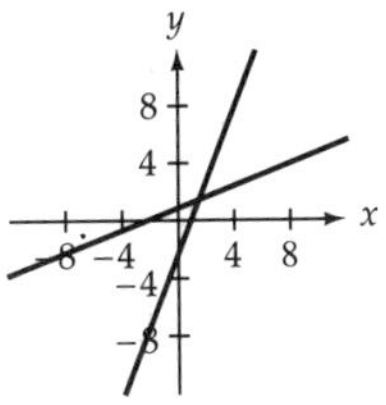

c. 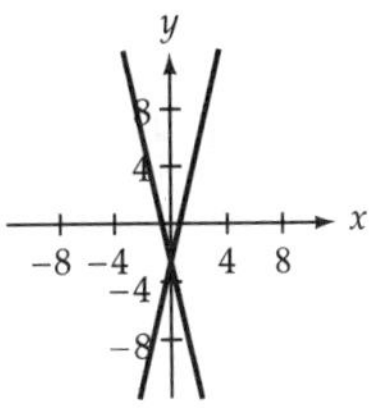

d. 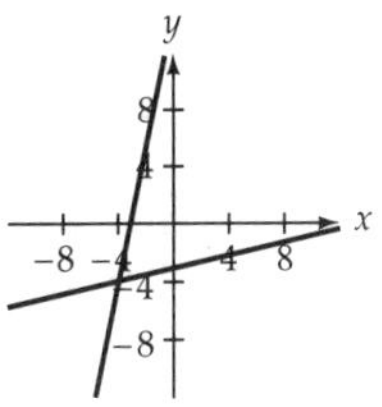

19. Write an equation that represents the relationship "p varies jointly as m^2 and n." Use k as the variation constant.

a. $p = km^2n$

b. $p = \dfrac{kn}{m^2}$

c. $p = \dfrac{km^2}{n}$

d. $p = \dfrac{k}{m^2n}$

20. The illumination that a source of light provides is inversely proportional to the square of the distance from the source. If the illumination at a distance of 5 feet is 12 lumens, what is the illumination at a distance of 20 feet?

a. 15 lumens

b. 1.25 lumens

c. 0.75 lumens

d. 3 lumens

1. Find the midpoint to the line segment with endpoints (–6, 1) and (4, –3).

   a. (5, 2) b. (–1, –1) c. (–5, 1) d. (1, –2)

2. Graph $y = -2x^2 + 3$.

   a. 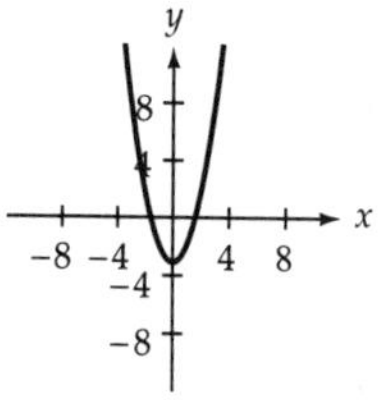

   b. 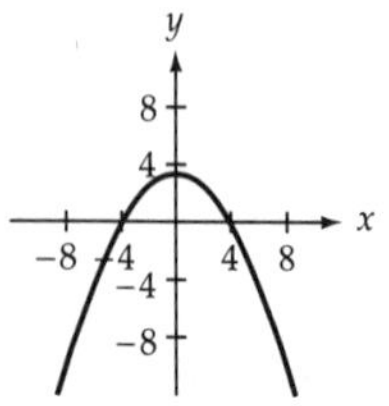

   c. 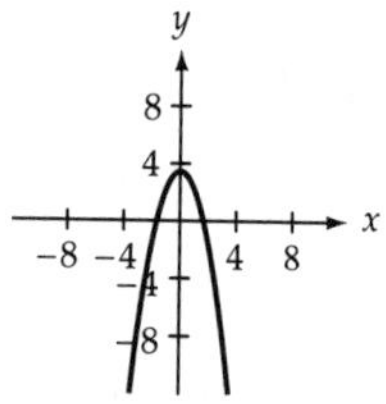

   d. 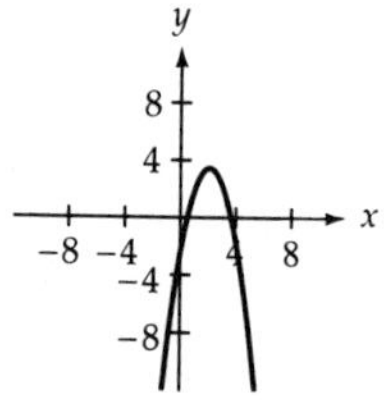

3. Graph $y + \frac{1}{2}|x + 2| - 2$.

   a. 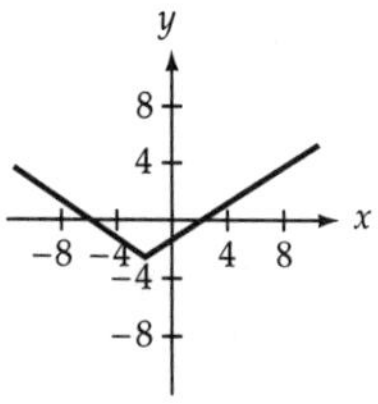

   b. 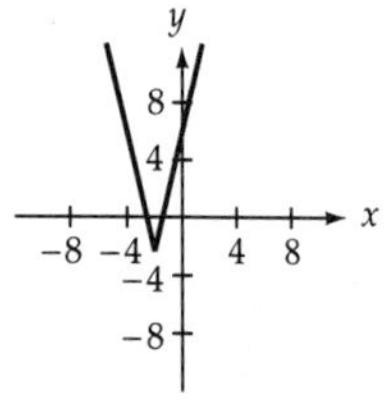

   c. 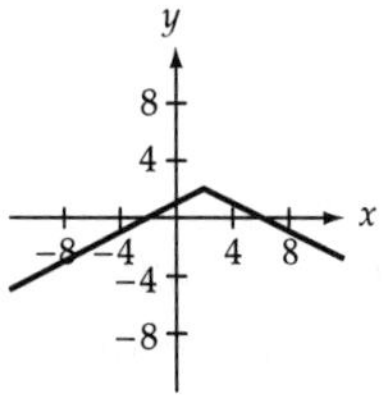

   d. 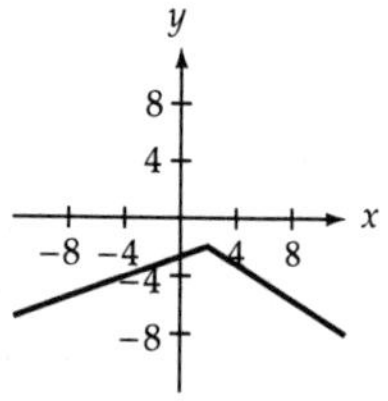

4. Find the center and radius of the circle that has the general form $x^2 + y^2 + 18x - 8y + 93 = 0$.

   a. center: (9, –4), radius: 4 b. center: (18, –8), radius: $\sqrt{93}$

   c. center: (–9, 4), radius: 4 d. center: (–9, 4), radius: 2

5. Determine the domain of the function $f(x) = -\sqrt{4 - x^2}$.

   a. [–2, 2] b. (–4, 4)

   c. $(-\infty, -2] \cup [2, \infty)$ d. [0, 2]

6. The sum of the length L and the width W of a rectangular area is 200 feet. Write the area A as a function of L.

   a. $A = L^2 - 200L$ b. $A = 200L - L^2$

   c. $A = L(400 - L)$ d. $A = L^2 - 400L$

7. Graph $f(x) \begin{cases} -5, \text{ if } x \leq -1 \\ 3x - 2, \text{ if } -1 < x \leq 2 \\ 4, \text{ if } x > 2 \end{cases}$

a.
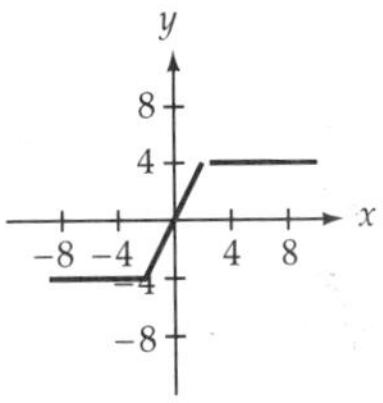

b.
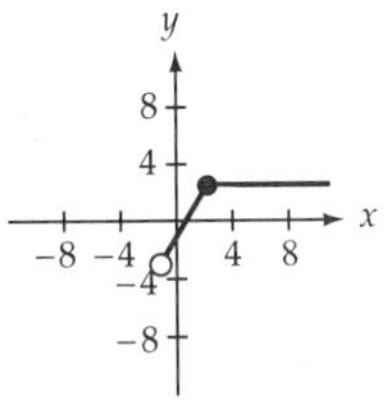

c.
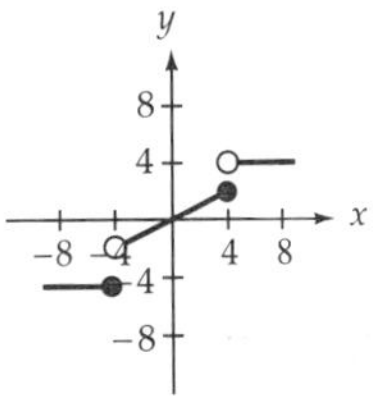

d.
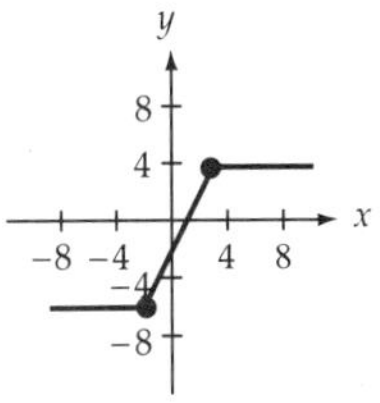

8. Find the domain and range given by the equation $y = -2x^2 + 3$.

a. domain: $(-\infty, \infty)$, range: $(-\infty, 3]$

b. domain: $[-\infty, \infty]$, range: $(-\infty, 3]$

c. domain: $\left[-\frac{3}{2}, \frac{3}{2}\right]$, range: $(-2, 3)$

d. domain: $(-\infty, 0]$, range: $[3, \infty)$

9. Which of the following equations define y as a function of x?

i. $5x - 7y = 9$ ii. $x^2 - 3y^2 = 11$ iii. $x^2 + 8y = 2$

a. i, ii b. ii, iii c. i, iii d. i

10. Use the graph of $f(x) = x^2$ to graph $y = 2(x + 3)^2 - 2$.

a.
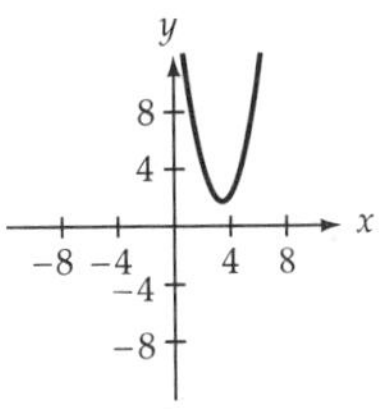

b.
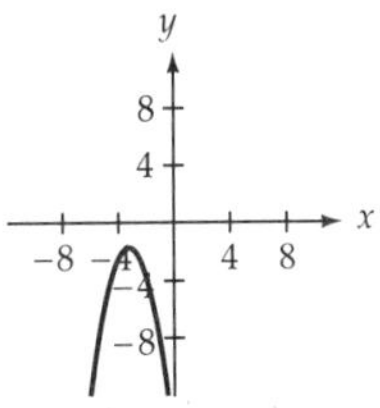

c.
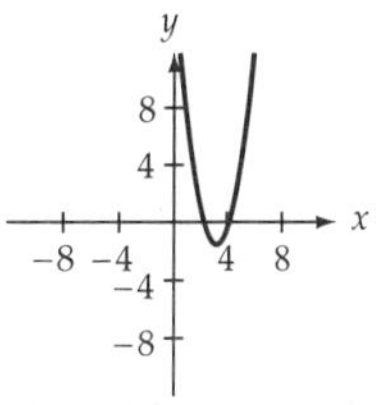

d.
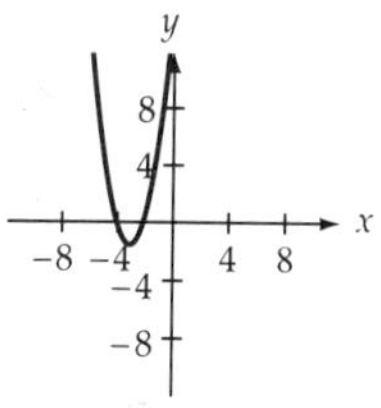

11. Which of the following are odd functions?

i. $f(x) = x^5 - x^3 - 1$ ii. $g(x) = x^9 - x^3$ iii. $h(x) = x^4 - x^2 - 1$

a. i b. ii c. iii d. i, ii

12. Find the slope-intercept form of the equation of the line that passes through (–4, 5) and is parallel to the graph of $5x + 3y = -12$.

a. $y = -\frac{5}{3}x - \frac{5}{3}$

b. $y = -\frac{5}{3}x - \frac{35}{3}$

c. $y = \frac{3}{5}x + \frac{37}{5}$

d. $y = -\frac{3}{5}x - \frac{13}{5}$

13. Find the maximum or minimum value of the function $f(x) = -3x^2 - 2x + 7$. State whether this value is a maximum or a minimum value.

a. 7, minimum b. 2, maximum c. 8, maximum d. $\frac{22}{3}$, maximum

14. Let $f(x) = 3x^2 - 2x - 2$ and $g(x) = x^2 - x - 1$. Find (fg).

a. $3x^4 - x^3 - 7x^2 + 2$

b. $3x^4 - 5x^3 - 3x^2 + 4x + 2$

c. $3x^4 + 5x^3 - 2x^2 + 2x + 2$

d. $3x^4 - 5x^3 + 3x^2 - 4x + 2$

15. Find the difference quotient of the function $f(x) = x^2 + x + 1$.

a. $\frac{x^2 + 2xh + h^2 + x + h + 1}{h}$

b. $2x + 1$

c. $2x + h$

d. $2x + h + 1$

16. Evaluate $(g \circ f)$, where $f(x) = 8x + 2$ and $g(x) = x^2 - 6$.

a. $64x^2 + 32x - 2$

b. $8x^2 - 46$

c. $64x^2 + 16x + 4$

d. $8x^2 + 16x - 46$

17. Find the inverse of $f(x) = x^2 + 8x, x \geq -4$.

a. $f^{-1}(x) = \sqrt{8 - x}$

b. $f^{-1}(x) = -4 + \sqrt{16 + x}$

c. $f^{-1}(x) = -4 - \sqrt{16 + x}$

d. $f^{-1}(x) = -4 \pm \sqrt{x^2 - 4}$

18. Graph the function given by the equation $f(x) = 3x - 5$ and its inverse f^{-1} on the same set of axes.

a.
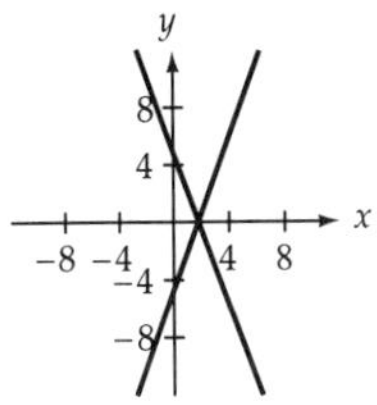

b.
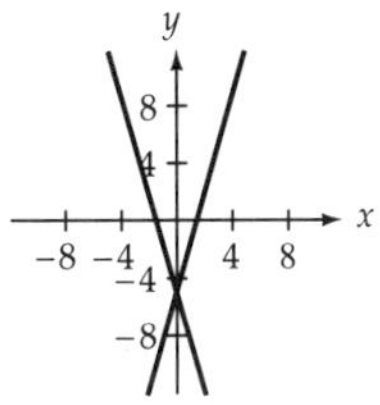

c.
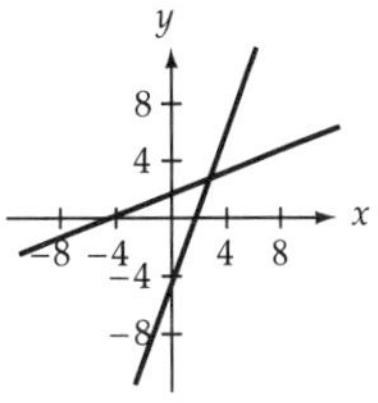

d.
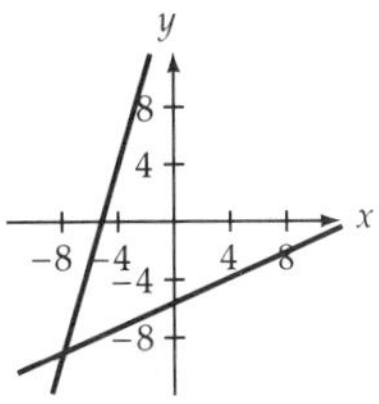

19. A lighthouse is 7 miles north of a port. A ship leaves port sailing East at 15 mph. If t is the number of hours the ship has been sailing, express the distance d between the ship and the lighthouse as a function of time.

a. $d = 15$

b. $d = \sqrt{49 + 225t^2}$

c. $d = 7 + 15t$

d. $d = \sqrt{225t^2 - 49}$

20. The cost of insulating the ceiling of a house varies jointly with the thickness of the insulation and the area of the ceiling. It costs \$550 to insulate a 1600-square-foot ceiling with 3-inch-thick insulation. Find the cost of insulating a 2400-square-foot ceiling with insulation that is 5 inches thick.

a. \$1095 b. \$1500 c. \$1375 d. \$900

1. Use synthetic division to divide: $(4x^3 - 12x^2 + 4x - 7) \div (x - 2)$.

2. Use the Remainder Theorem to find P(-3) if $P(x) = -2x^3 + 5x^2 - x + 5$.

3. Show that x + 2 is a factor of $x^4 - x^3 - 3x^2 + 3x - 6$.

4. Determine the far left and far right behaviour of the function given by the equation $P(x) = -x^6 - 3x^5 + 7x + 11$.

5. Find the real zeros of the polynomial function given by the equation $x^3 - 5x^2 + 6x = 0$.

6. Use the Zero Location Theorem to verify that $P(x) = 3x^3 - x^2 + x - 5$ has a zero between 1 and 2.

7. Find the zeros of the function given by the equation $P(x) = x^3(x^2 - 9)^2(3x - 4)$ and state the multiplicity of each.

8. Use the Rational Zero Theorem to list all possible rational zeros of $P(x) = 5x^3 - 11x^2 + x - 21$.

9. Find the smallest positive integer and the largest negative integer that are upper and lower bounds for the polynomial given by the equation $P(x) = 2x^4 - x^3 - 14x^2 - 5x + 6$.

10. Use Descartes' Rule of Signs to state the number of possible positive and negative real zeros of $P(x) = -11x^5 + 6x^3 - x^2 + x + 15$.

11. Find the zeros of the polynomial given by the equation

$P(x) = 2x^3 - 7x^2 - 5x + 4.$

12. Given that $-1+4i$ is a zero of $P(x) = x^4 + 2x^3 + 18x^2 + 2x + 17$, find the remaining zeros.

13. Find all the zeros of the polynomial $P(x) = x^5 - 5x^4 + 7x^3 - 5x^2 + 6x$.

14. Find a polynomial of lowest degree that has real coefficients and zeros $1-i$, 2, and -2.

15. Find all vertical asymptotes of the rational function given by the equation $f(x) = \dfrac{x^2+6}{x^2-3x-18}$.

16. Find all horizontal asymptotes of the rational function given by the equation $f(x) = \dfrac{4x^2-2x+1}{3x^2-4}$.

17. Sketch $f(x) = \dfrac{x^2+5}{x^2-3x-10}$.

18. Sketch $f(x) = \dfrac{-2x^2}{x^2+4}$.

19. Sketch $f(x) = \dfrac{x^2-2x+1}{x-2}$.

20. Approximate the real zero in the interval $0 < x < 1$ of the polynomial $P(x) = x^3 - 5x + 3$.

1. Use synthetic division to divide: $(2x^3 - 4x^2 + 2x - 3) \div (x + 2)$.

2. Use the Remainder Theorem to find P(-3) if $P(x) = -2x^4 - 2x^3 + 4x^2 - 2x + 1$.

3. Show that x - 4 is a factor of $x^4 - 8x^3 + 15x^2 + 8x - 16$.

4. Examine the leading term of the function given by the equation $P(x) = -6x^4 - 3x^2 + 9x - 8$ and determine the far-left and far-right behavior of the graph of the polynomial function.

5. Find the real zeros of the polynomial function given by the equation $x^3 + 6x^2 + 11x + 6 = 0$.

6. Use the Zero Location Theorem to verify that $P(x) = x^4 - 3x^2 + 2x - 1$ has a zero between 1 and 2.

7. Find the zeros of the function given by the equation $P(x) = (3x + 4)(x^2 - 9)^2(x + 4)^3$ and state the multiplicity of each.

8. Use the Rational Zero Theorem to list the possible rational zeros for the function given by the equation $P(x) = 6x^3 - 12x^2 + 34x - 5$.

9. Find the smallest positive integer and the largest negative integer that are upper and lower bounds for the polynomial given by the equation $P(x) = 3x^4 - x^2 + x - 9$.

10. Use Descartes' Rule of Signs to state the number of possible positive and negative real zeros of $P(x) = -9x^5 + 3x^3 + 2x^2 - 1$.

11. Find the zeros of the polynomial given by the equation $P(x) = 6x^3 - 17x^2 - 7x + 12$.

12. Given that -2+i is a zero of $P(x) = x^4 + 4x^3 + 14x^2 + 36x + 45$, find the remaining zeros.

13. Find all the zeros of the polynomial $P(x) = 5x^3 + 27x^2 - 23x + 3$.

14. Find a polynomial of lowest degree that has real coefficients and zeros i, −2, 0, and 1/2.

15. Find all vertical asymptotes of the rational function given by the equation $f(x) = \dfrac{5x^2 - 3x + 7}{3x^3 + 2x^2 - 8x}$.

16. Find all horizontal asymptotes of the rational function given by the equation $f(x) = \dfrac{x^2 + 1}{3x^3 + 2x^2 - x - 1}$.

17. Sketch $f(x) = \dfrac{x^2 - 5}{x^2 - x - 6}$.

18. Sketch $f(x) = \dfrac{3x^2 - 5x - 6}{2x - 3}$.

19. Sketch $f(x) = \dfrac{4x^2}{x^2 + 4}$.

20. Approximate the zero in the interval $2 < x < 3$ of the polynomial $P(x) = x^3 - 5x + 3$ to within one-tenth of a unit.

1. Use synthetic division to divide: $(2x^3 + x^2 - 9) \div (x - 5)$.

2. Use the remainder Theorem to find P(-2) if $P(x) = 5x^4 - 2x^2 + x - 9$.

3. Show that x + 1 is a factor of $x^5 + x^4 - 2x^3 - 8x^2 - 8x - 2$.

4. Examine the leading term of the function given by the equation $P(x) = 11x^9 + 3x^7 - 5x^4 - 3x^2 + 2$ and determine the far-left and far-right behavior of the graph of the polynomial function.

5. Find the real zeros of the polynomial function given by the equation $x^4 - x^3 - 7x^2 + x + 6 = 0$.

6. Use the Zero Location Theorem to verify that $P(x) = 3x^3 - 2x^2 - 3x - 4$ has a zero between 1 and 2.

7. Find the zeros of the function given by the equation $P(x) = x(x^2 - 1)^2 (x + 9)^4$ and state the multiplicity of each.

8. Use the Rational Zero Theorem to list the possible rational zeros for the function given by the equation $P(x) = 7x^3 + 4x^2 - x + 15$.

9. Find the smallest positive integer and the largest negative integer that are upper and lower bounds for the polynomial given by the equation. $P(x) = 2x^4 - x^3 + 8x - 10$..

10. Use Descartes' Rule of Signs to state the number of possible positive and negative real zeros of $P(x) = -6x^5 + 8x^3 + x^2 - 11$.

11. Find the zeros of the polynomial given by the equation $P(x) = 7x^3 + 43x^2 + 27x + 3$.

12. Given that 1 + i is a zero of $P(x) = x^4 - 2x^3 + 9x^2 - 14x + 14$, find the remaining zeros.

13. Find all the zeros of the polynomial $P(x) = 5x^3 + 41x^2 + 13x + 1$.

14. Find a polynomial of lowest degree that has real coefficients and zeros of $2 - i, -3, 2$ and 0.

15. Find all vertical asymptotes of the rational function given by the equation $f(x) = \dfrac{x^3 + 1}{3x^3 + x^2 - 4x}$.

16. Find all horizontal asymptotes of the rational function given by the equation $f(x) = \dfrac{7x^2 + 3x - 8}{14x^2 + 8x}$.

17. Sketch $f(x) = \dfrac{x^2 + 1}{x^2 - 2x - 8}$.

18. Sketch $f(x) = \dfrac{x^2 + x + 3}{x + 2}$.

19. Sketch $f(x) = \dfrac{-7x^2}{x^2 + 1}$.

20. Approximate the zero in the interval $3 < x < 4$ of the polynomial $P(x) = x^3 - 7x - 9$ to within one-tenth of a unit.

1. Use synthetic division to divide: $(x^4+6x^3-8x^2-6x+7)\div(x-1)$.

   a. $x^3+5x^2-13x+7$ b. x^3+7x^2-x-7

   c. $x^3+5x^2-3x-9+\dfrac{2}{x-1}$ d. $x^3+5x^2-13x-9+\dfrac{2}{x-1}$

2. Use the Remainder Theorem to find P(-3) if $P(x)=-3x^4+7x^3+12x^2+21x+9$.

   a. 108 b. 0 c. -378 d. -216

3. Use synthetic division and the factor theorem to determine which of the following is a factor of $x^4+4x^3+3x^2-4x-4$.

   a. x - 4 b. x + 4 c. x - 2 d. x + 2

4. Examine the leading term of the function given by the equation $P(x)=3x^5-6x^4+2x-8$ and determine the far-left and far-right behavior of the graph of the polynomial function.

   a. Up to the far left and up to the far right
   b. Up to the far left and down to the far right
   c. Down to the far left and up to the far right
   d. Down to the far left and down to the far right

5. Find the real zeros of the polynomial function given by the equation $3x^3-25x^2+28x=0$.

   a. 9, 2, 14 b. 0, $\frac{4}{3}$, 7 c. 0, -7, -4 d. 0, $\frac{3}{4}$, 7

6. Use the Zero Location Theorem to determine which interval of $P(x)=-x^4+2x^3+x^2-5x+7$ has a zero.

   a. (0, 1) b. (3, 4) c. (-3, -2) d. (-2, -1).

7. Find the zeros of the function given by the equation $P(x)=(x^2-16)^2(3x-4)(x+5)^4$.

   a. -5, -4, $\frac{4}{3}$, 4 b. -5, $\frac{3}{4}$, 16 c. -5, -4, 4 d. -8, $\frac{4}{3}$, 5, 8

8. Use the Rational Zero Theorem to list the possible rational zeros for the function given by the equation $P(x) = 2x^4 - 7x^3 + 2x^2 - 5x + 12$.

a. $\pm1, \pm2, \pm3, \pm4, \pm6, \pm12$

b. $\pm1, \pm2, \pm3, \pm4, \pm6, \pm12, \pm\frac{1}{2}, \pm\frac{3}{2}$

c. $\pm1, \pm2$

d. $\pm1, \pm2, \pm\frac{1}{2}, \pm\frac{1}{3}, \pm\frac{1}{4}, \pm\frac{1}{6}, \pm\frac{1}{12}, \frac{2}{3}$

9. Find the smallest positive integer and the largest negative integer that are upper and lower bounds for the polynomial given by the equation $P(x) = 3x^4 + 5x^3 + 2x - 1$.

a. upper bound: 2, lower bound: -1
b. upper bound: 2, lower bound: 1
c. upper bound: 1, lower bound: -2
d. upper bound: 1, lower bound: -1

10. Use Descartes' Rule of Signs to state the number of possible positive and negative real zeros of $P(x) = -2x^5 - x^3 + x^2 + 5x - 11$.

a. positive real zeros: 4 or 2 or 0, negative real zeros: 1
b. positive real zeros: 2 or 0, negative real zeros: 1
c. positive real zeros: 2 or 0, negative real zeros: 3 or 1
d. positive real zeros: 3 or 1, negative real zeros 2 or 0

11. Find the zeros of the polynomial given by the equation $P(x) = x^3 - 4x^2 + x + 6$.

a. 1, 2, 3 b. -3, -2, 1 c. -3, -2, -1 d. -1, 2, 3

12. Given that 1 + 3i is a zero of $P(x) = x^4 - 2x^3 + 15x^2 - 10x + 50$, find the remaining zeros.

a. -5i, 5i, 1 - 3i,

b. -5, 5, 1-3i

c. $-\sqrt{5}, \sqrt{5}$, i - 3i

d. -2i, 2i, 1 - 3i

13. Find all the zeros of the polynomial $P(x) = x^4 + 2x^3 - 3x^2 - 4x - 12$.

a. -2, 1, 2, 3

b. -2, 3, $\frac{1}{2} \pm \frac{\sqrt{7}}{2}i$

c. -3, 2, $-\frac{1}{2} \pm \frac{\sqrt{7}}{2}i$

d. -3, 2, $\frac{-1 \pm \sqrt{7}}{2}$

14. Find a polynomial of lowest degree that has real coefficients and zeros 1, -2, and 0.

a. $x^3 + 2x^2 - x$ b. $x^3 - x^2 + 2x$

c. $x^4 + 2x^2 - 2x$ d. $x^3 + x^2 - 2x$

15. Find all vertical asymptotes of the rational function given by the equation $f(x) = \dfrac{3x^2 + 4x + 1}{2x^2 + 9x + 9}$.

a. No vertical asymptotes   b. $x = -\frac{3}{2}, x = -3$

c. $x = -\frac{1}{3}, x = -1$ d. $x = \frac{3}{2}, x = 3$

16. Find all horizontal asymptotes of the rational function given by the equation. $f(x) = \dfrac{6x^2 + 2x - 9}{3x^2 + 15x}$.

a. $y = 2$ b. $y = \frac{3}{2}, y = -\frac{1}{3}$

c. $y = 0, y = -5$ d. no horizontal asymptotes

17. Sketch the graph of $f(x) = \dfrac{x^2 - 2}{x^2 + x - 6}$.

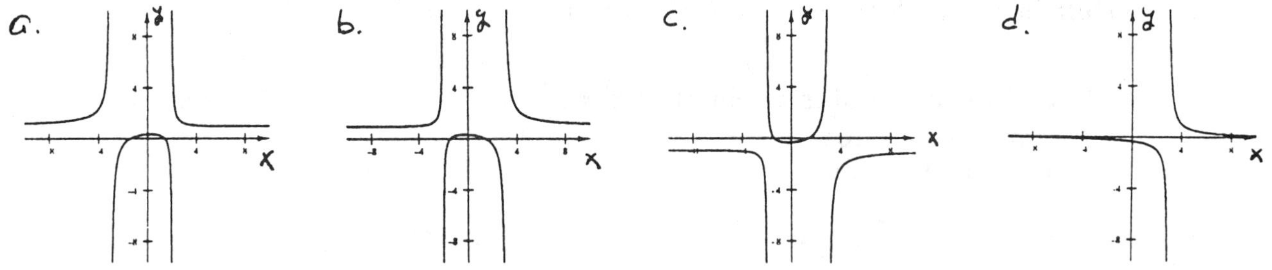

18. Sketch the graph of $f(x)=\frac{4x^2}{x^2+2}$.

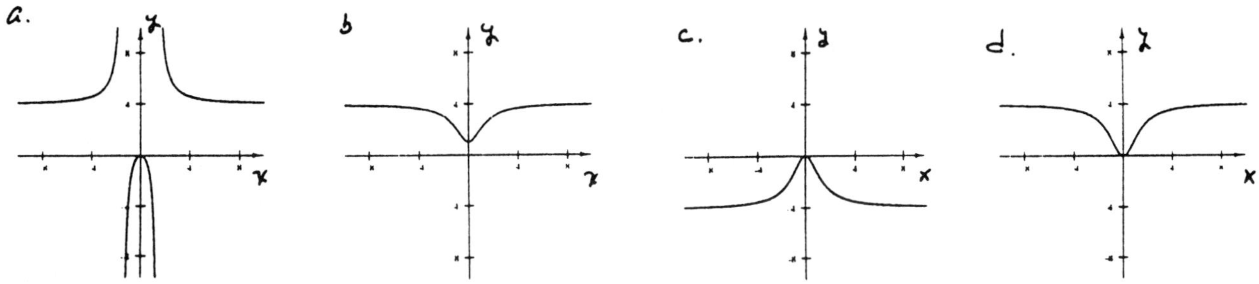

19. Sketch. the graph of $f(x)=\frac{3x^2-1}{x^2+1}$.

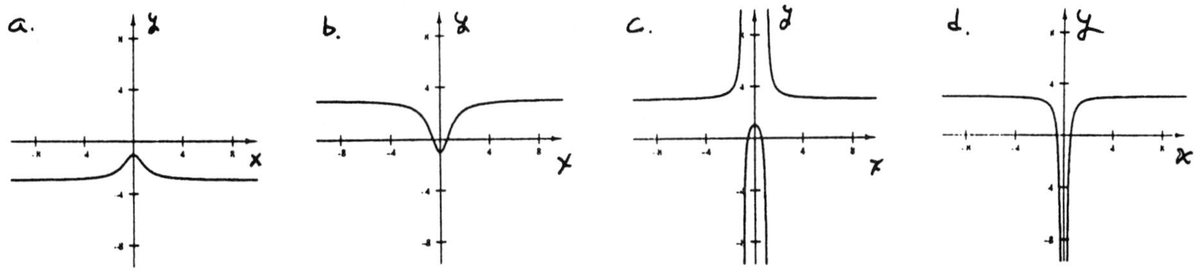

20. Approximate the zero in the interval -2 < x < -1 of the polynomial $P(x)=x^2-5x-8$. State your answer as the midpoint of the interval determined by three iterations of the bisection method.

a. -1.625   b. -1.375   c. -1.125   d. -1.875

1. Use synthetic division to divide $(2x^4 + 3x^3 - 17x^2 - 36) \div (x + 4)$.

a. $2x^3 + 5x^2 + 3x - 12 + \frac{12}{x+4}$

b. $2x^3 - 5x^2 + 3x - 12 + \frac{12}{x+4}$

c. $2x^3 - 11x^2 + 27x - 4 + \frac{4}{x+4}$

d. $2x^3 - 11x^2 + 27x - 4 + \frac{4}{x+4}$

2. Use the Remainder Theorem to find $P(-3)$ if $P(x) = -2x^4 - 6x^3 + 5x^2 + 8x - 11$.

a. 334 b. –314 c. 58 d. 10

3. Use synthetic division and the factor theorem to determine which of the following is a factor of $x^4 + 2x^3 - 3x^2 - 8x - 4$.

a. $x + 2$ b. $x - 2$ c. $x + 4$ d. $x - 1$

4. Examine the leading term of the function given by the equation $P(x) = -x^7 + 3x^6 - 5x + 1$ and determine the far-left and far-right behavior of the graph of the polynomial function.

a. Up to the far left and up to the far right
b. Up to the far left and down to the far right
c. Down to the far left and down to the far right
d. Down to the far left and up to the far right

5. Find the real zeros of the polynomial function given by the equation $12x^3 - 25x^2 + 12x = 0$.

a. $0, -\frac{4}{3}, -\frac{3}{4}$ b. $0, 3, 4$ c. $0, \frac{2}{3}, -\frac{3}{4}$ d. $0, \frac{4}{3}, \frac{3}{4}$

6. Use the Zero Location Theorem to determine which interval of $P(x) = -3x^5 + 4x^3 - 5x^2 + x - 9$ has a zero.

a. $(-2, -1)$ b. $(1, 2)$ c. $(-1, 0)$ d. $(3, 4)$

7. Find the zeros of the function given by the equation $P(x) = x^2(4x^2 - 9)^2(2x + 5)$.

a. $0, -\frac{5}{2}, \frac{2}{3}$ b. $0, -\frac{5}{2}, -\frac{3}{2}, \frac{3}{2}$ c. $0, -\frac{2}{3}, -\frac{5}{2}, \frac{2}{3}$ d. $0, -\frac{9}{4}, \frac{9}{4}, \frac{5}{2}$

8. Use the Rational Zero Theorem to list the possible rational zeros for the function given by the equation $P(x) = 4x^3 - 21x^2 + 7x - 10$.

a. $\pm 1, \pm 2, \pm 5, \pm 10$

b. $\pm 1, \pm 2, \pm 4$

c. $\pm 1, \pm 2, \pm 5, \pm 10, \pm\frac{1}{2}, \pm\frac{1}{4}, \pm\frac{5}{2}, \pm\frac{5}{4}$

d. $\pm 1, \pm 2, \pm 4, \pm\frac{1}{2}, \pm\frac{1}{5}, \pm\frac{1}{10}, \pm\frac{2}{5}, \pm\frac{4}{5}$

9. Find the smallest positive integer and the largest negative integer that are upper and lower bounds for the polynomial given by the equation $P(x) = 3x^4 + 4x^3 - 5x^2 + x - 9$.

a. upper bound: 3, lower bound: –2
b. upper bound: 2, lower bound: –3
c. upper bound: 1, lower bound: –3
d. upper bound: 2, lower bound: –4

10. Use Descartes' Rule of Signs to state the number of possible positive and negative real roots of $P(x) = 3x^5 + 4x^4 - 3x^3 - x^2 + 4$.

a. positive roots: 3, negative roots: 2
b. positive roots: 3 or 1, negative roots: 2 or 0
c. positive roots: 2 or 0, negative roots: 3 or 1
d. positive roots: 4 or 2 or 0: negative roots: 3 or 1

11. Find the zeros of the polynomial given by the equation $P(x) = 6x^3 - 23x^2 + 25x - 6$.

a. $-\frac{3}{2}, \frac{2}{3}, 1$

b. $-\frac{3}{2}, \frac{1}{3}, 1$

c. $-\frac{1}{3}, \frac{2}{3}, 2$

d. $\frac{1}{3}, \frac{3}{2}, 2$

12. Given that $-i$ is a zero of $P(x) = x^4 + x^3 + 2x^2 + x + 1$, find the remaining zeros.

a. $i, -\frac{1}{2} \pm \frac{\sqrt{3}}{2}i$

b. $i, -1, 1$

c. $i, \frac{-1 \pm \sqrt{3}}{2}$

d. $i, \frac{1}{2} \pm \frac{\sqrt{5}}{2}i$

13. Find all the zeros of the polynomial $P(x) = x^5 + 2x^3 - 7x^2 - 10x$.

a. $0, -2, -1, 1, 5$

b. $0, -1, 2, \frac{1 \pm \sqrt{21}}{2}$

c. $0, -1, 2, -\frac{1}{2} \pm \frac{\sqrt{19}}{2}i$

d. $-1, 2, \frac{-1 \pm \sqrt{19}}{2}$

14. Find a polynomial of lowest degree that has real coefficients and zeros $1 - \sqrt{3}$, -3 and 0.

a. $x^4 + 5x^3 + 4x^2 - 6x$

b. $x^4 + x^3 - 8x^2 - 6x$

c. $x^3 + x^2 - 8x - 6$

d. $x^4 - x^3 + 8x^2 + 6x$

15. Find all vertical asymptotes of the rational function given by the equation $f(x) = \frac{x^2 + 11}{x^2 - 2x - 15}$.

a. $x = \sqrt{11}$

b. $x = 1$

c. $x = -5, x = 3$

d. $x = -3, x = 5$

16 Find all horizonal asymptotes of the rational function given by the equation

$f(x) = \frac{3x^2 + 4x + 1}{5x^2 + 10x}$.

a. $y = \frac{3}{5}$

b. $y = -\frac{1}{3}, \ y = -1$

c. $y = 0, y = -2$

d. $y = \frac{5}{3}$

17. Sketch the graph of $f(x) = \frac{x^2 - 1}{x^2 + 3x - 4}$.

a.

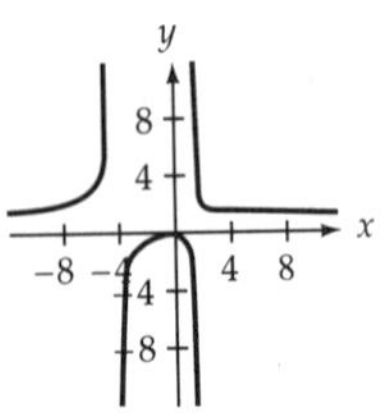

b.

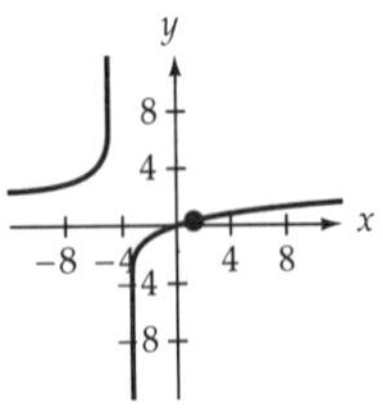

c.

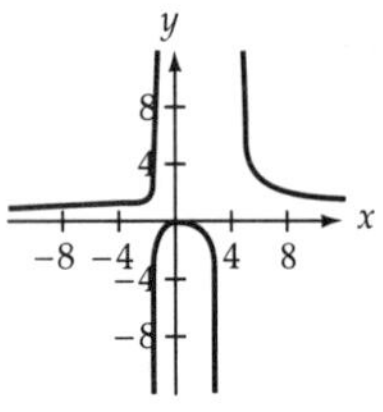

d.

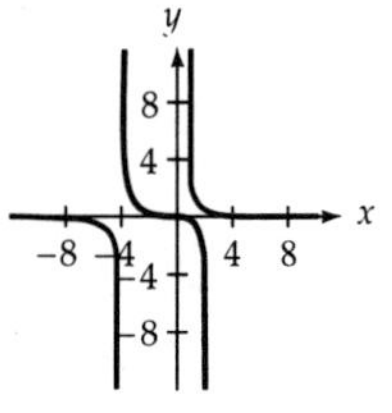

18. Sketch the graph of $f(x) = \dfrac{x^2 - 9}{x^2 + 2}$.

a.
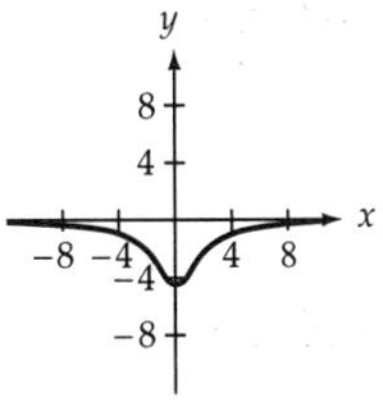

b.
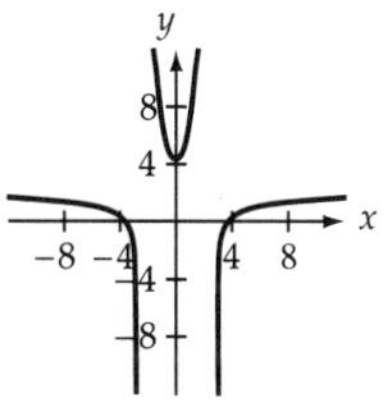

c.
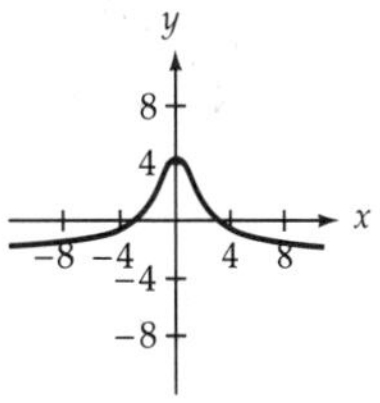

d.
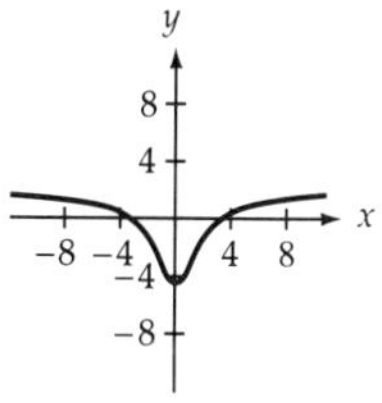

19. Sketch the graph of $f(x) = \dfrac{3x^2 + x - 2}{x^2 - 4x}$.

a.
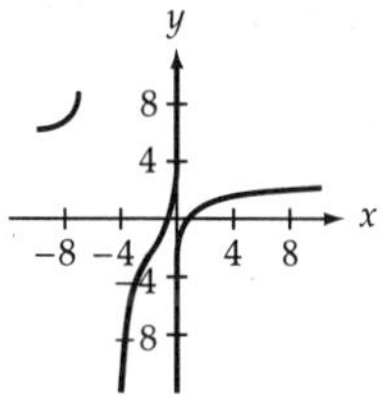

b.
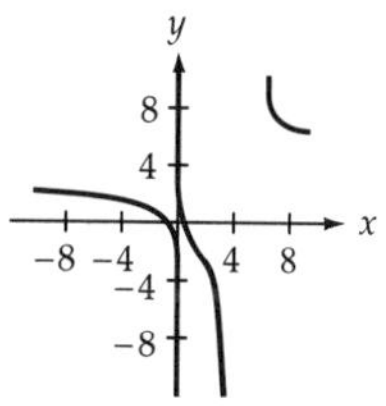

c.
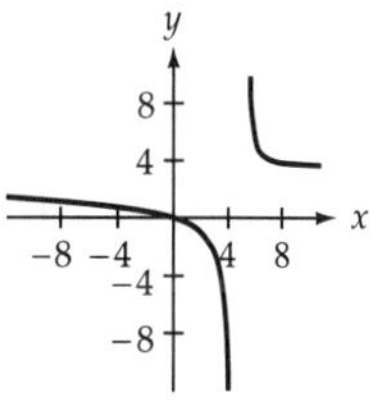

d.
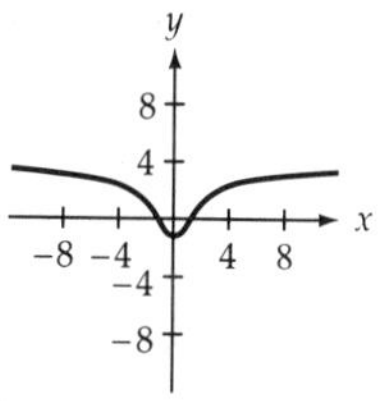

20. Approximate the zero in the interval $2 < x < 3$ of the polynomial $P(x) = x^2 - 5x - 8$ to within one-tenth of a unit.

a. 2.5 b. 2.8 c. 2.7 d. 2.9

1. Use synthetic division to divide $(3x^4 + 8x^3 + 16x + 9) \div (x + 3)$

a. $3x^3 + x^2 - x + 19 + \frac{-48}{x+3}$

b. $3x^2 + x - 19 + \frac{-32}{x+3}$

c. $3x^3 - x^2 + 3x + 7 + \frac{-12}{x+3}$

d. $3x^3 - 3x^2 + x + 19 + \frac{16}{x+3}$

2. Use the Remainder Theorem to find P(–3) if $P(x) = 2x^4 - 2x^2 + 7x - 8$.

a. –209 b. 130 c. 151 d. 115

3. Use synthetic division and the factor theorem to determine which of the following is a factor of $x^4 - 31x^2 - 180$.

a. $x - 6$ b. $x + 12$ c. $x - 18$ d. $x - 9$

4. Examine the leading term of the function given by the equation $P(x) = -3x^4 + 2x^2 + 16x - 9$ and determine the far-left and far-right behavior of the graph of the polynomial function.

a. up to the far left, up to the far right
b. down to the far left, down to the far right
c. up to the far left, down to the far right
d. down to the far left, up to the far right

5. Find the real zeros of the polynomial function given by the equation $x^4 - x^3 + x - 1 = 0$

a. $-1, 1, \frac{-1 \pm \sqrt{3}}{2}$

b. $-1, 1, \frac{-1 \pm \sqrt{5}}{2}$

c. $-1, 1, -\frac{1}{2} \pm \frac{\sqrt{3}}{2}i$

d. $-1, 1$

6. Use the Zero Location Theorem to determine which interval of $P(x) = 7x^3 + 4x^2 - x + 15$ has a zero.

a. (–1, 0) b. (1, 2) c. (–2, –1) d. (3, 4)

7. Find the zeros of the function given by the equation $P(x) = x^3(x^2 - 1)^2 (x - 2)^4$.

 a. 0, –1, 2 b. 0, 1, 2 c. 0, –2, –1, 1, 2 d. 0, –1, 1, 2

8. Use the rational Zero Theorem to list the possible rational zeros for the function given by the equation $P(x) = 3x^4 - 3x^3 + 4x^2 - 8$.

 a. $\pm 1, \pm 2, \pm 4, \pm 8$

 b. $\pm 1, \pm 2, \pm 4, \pm 8, \pm \frac{1}{3}, \pm \frac{2}{3}, \pm \frac{4}{3}, \pm \frac{8}{3}$

 c. $\pm 1, \pm 3$

 d. $\pm 1, \pm 3, \pm \frac{1}{2}, \pm \frac{1}{4}, \pm \frac{1}{8}, \pm \frac{3}{2}, \pm \frac{3}{4} \pm \frac{3}{8}$

9. Find the smallest positive integer and the largest negative integer that are upper and lower bounds for the polynomial given by the equation $P(x) = 6x^3 - 5x^2 - 13x + 12$.

 a. upper bound: 2, lower bound: –2
 b. upper bound: 3, lower bound: –1
 c. upper bound: 2, lower bound: –3
 d. upper bound: 4, lower bound: –2

10. Use Descartes' Rule of Signs to state the number of possible positive and negative real zeros of $P(x) = -3x^5 + 3x^4 - 2x^2 - 8x + 9$.

 a. positive roots: 3, negative roots: 2
 b. positive roots: 3 or 1, negative roots: 2 or 0
 c. positive roots: 4 or 2, negative roots: 2 or 0
 d. positive roots: 3 or 1, negative roots: 0

11. Find the zeros of the polynomial given by the equation $P(x) = 2x^3 + x^2 + 8x + 4$.

 a. $-\frac{1}{2}, -2, 2$ b. $-\frac{1}{2}, \frac{1}{2}, 4$ c. $-\frac{1}{2}, -2i, 2i$ d. $\frac{1}{2}, \frac{1}{2} \pm \frac{1}{2}i$

12. Given that 2i is a zero of $P(x) = 3x^4 - 10x^3 + 20x^2 - 40x + 32$, find the remaining zeros.

 a. $\frac{2}{3}, 4, -2i$

 b. $\frac{4}{3}, 2, -2i$

 c. $\frac{4}{3}, 2$

 d. $-2, \frac{3}{4}, -2i$

13. Find all the zeros of the polynomial $P(x) = x^4 + 6x^3 + 14x^2 + 15x - 36$.

a. $-1, 4, \frac{-3 \pm 3\sqrt{3}}{2}$

b. $1, -4, -\frac{3}{2} \pm \frac{3\sqrt{3}}{2} i$

c. $-2, 2, -\frac{3}{2} \pm \frac{3\sqrt{3}}{2} i$

d. $1, 4, \frac{-3 \pm 3\sqrt{3}}{2}$

14. Find a polynomial of lowest degree that has real coefficients and zeros $-5i$, 8, -3.

a. $x^4 - 5x^3 + 125x - 600$

b. $x^4 - 5x^3 + x^2 - 125x - 600$

c. $x^4 + 5x^3 - 49x^2 - 25x + 120$

d. $x^4 - 25x^2 - 125x + 120$

15. Find all the vertical asymptotes of the rational function given by the equation

$$f(x) = \frac{x^2 + 5}{x^2 + x - 20}.$$

a. $x = \sqrt{5}, x = -\sqrt{5}$

b. $x = 1$

c. $x = -4, x = 5$

d. $x = -5, x = 4$

16. Find all horizontal asymptotes of the rational function given by the equation

$$f(x) = \frac{12x^2 - 7x + 1}{9x^2 + 3x}.$$

a. $y = \frac{4}{3}$

b. $y = \frac{1}{3}, y = \frac{1}{4}$

c. $y = 0, y - \frac{1}{3}$

d. $y = \frac{3}{4}$

17. Sketch the graph of $f(x) = \frac{x^2 + 1}{x^2 + x - 6}$.

a.

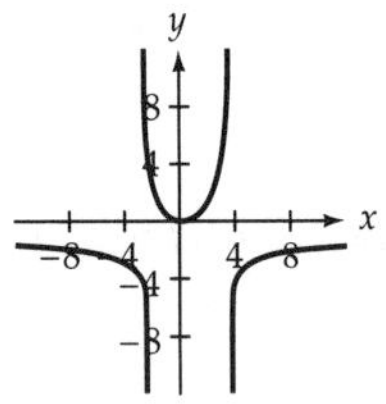

b.

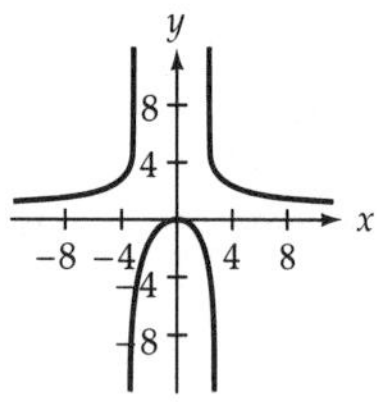

c.

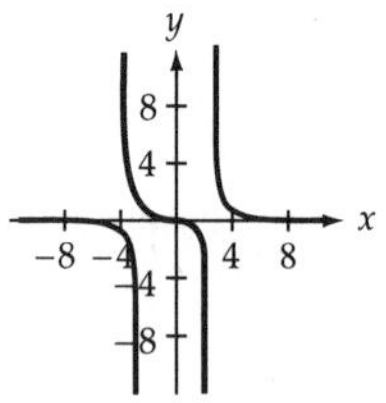

d.

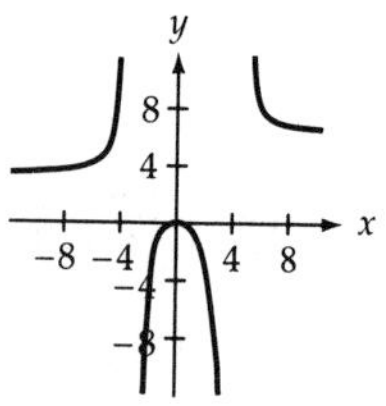

18. Sketch the graph of $f(x) = \dfrac{2x^2}{(x+1)^2}$.

a.

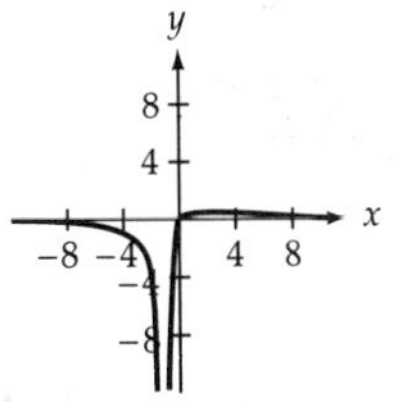

b.

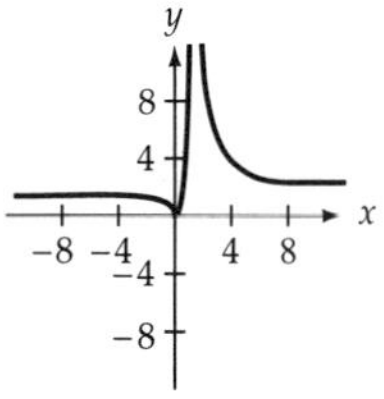

c.

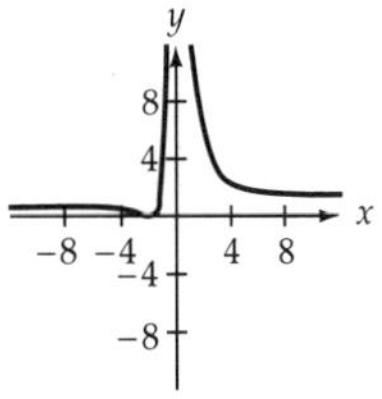

d.

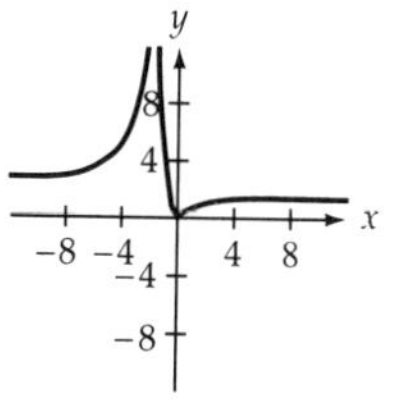

19. Sketch the graph of $f(x) = \dfrac{x^2 + 2x + 1}{x - 2}$.

a.

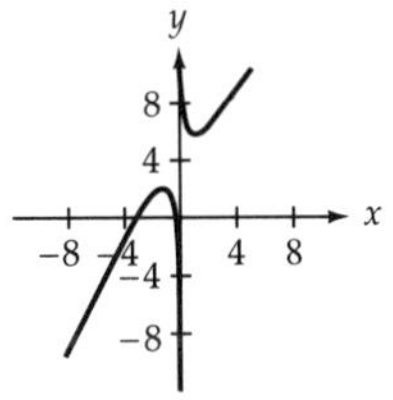

b.

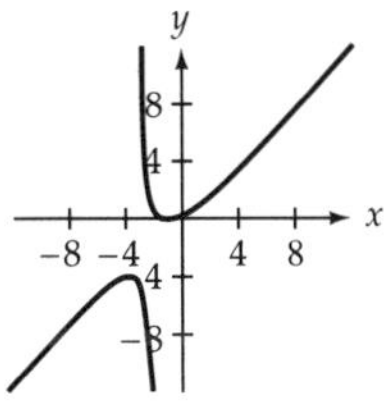

c.

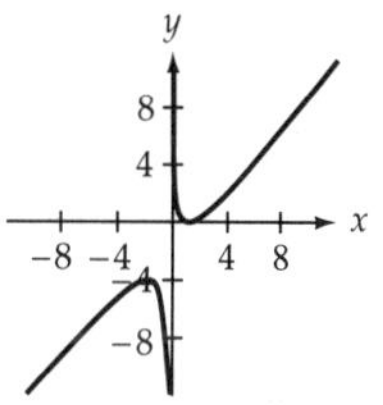

d.

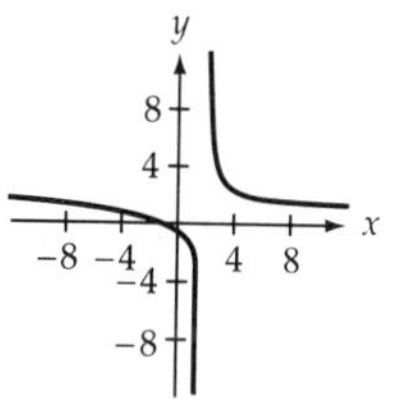

20. Approximate the zero in the interval $0 < x < 1$ of the polynomial $P(x) = -2x^2 + 7x - 3$. State your answer as the midpoint of the interval determined by two iterations of the bisecting method.

a. 0.25 b. 0.625 c. 0.125 d. 0.375

1. Given that $f(x)=5^x$, use a calculator to evaluate $f(\sqrt{7})$ accurate to six significant digits.

2. Given that $f(x)=e^{x/3}$, use a calculator to evaluate f(4.3) to six significant figures.

3. Graph $f(x)=2^{-3x/2}$.

4. Graph $f(x)=3^x$.

5. Graph $f(x)=e^{-x}$.

6. Write $\log_b(2x-5)=c$ in exponential notation.

7. Write $5^a=b$ in logarithmic form.

8. Write $\log_b \dfrac{x^5y^2}{z^2}$ in terms of the logarithms of a, b, and c.

9. Write $\log_b \dfrac{yz^4}{\sqrt[3]{x^2}}$ in terms of the logarithms of x, y, and z.

10. Write $\log_b(3x-4)-\log_b\sqrt{2x+5}$ as a single logarithm.

11. Given $\log_b 2=0{,}2891$ and $\log_b 5=0.6712$, evaluate $\log_b 10$.

12. Use the change of base formula and a calculator to approximate $\log_5 15$ to five significant digits.

13. Graph $f(x) = \log(x-2)$.

14. Graph $f(x) = \log_2 x - 1$.

15. Graph $f(x) = -\ln(x-2)$.

16. Solve $3^{4x-3} = \frac{1}{81}$.

17. Solve $3^x = 7.2$.

18. Solve $\log(11x) - \log(x+7) = 1$.

19. Find the balance if $25,000 is invested at an annual rate of 6.5% for 4 years, compounded continuously.

20. The number of people $P(t)$ exposed to a rumor in hours is given by the function $P(t) = 120{,}000\left(1 - e^{-0.0004t}\right)$. Find the number of hours until 40 percent of the population of 120,000 people have heard the rumor.

1. Given that $f(x)=2.3^x$, use a calculator to evaluate $f(\sqrt{5})$ accurate to six significant digits.

2. Given that $f(x)=e^{\sqrt{x}}$, use a calculator to evaluate $f(1.9)$ to six significant figures.

3. Graph $f(x)=4^x$.

4. Graph $f(x)=\left(\frac{1}{3}\right)^{2x}$.

5. Graph $f(x)=e^{x+1}$.

6. Write $\log_b(2x-7)=s$ in exponential form.

7. Write $4^{\sqrt{3x}}=y$ in logarithmic form.

8. Write $\log_b \frac{x^5z^2}{y}$ in terms of logarithms of x, y, and z.

9. Write $\log_b \frac{\sqrt[3]{x^2}y}{\sqrt{z}}$ in terms of x, y, and z.

10. Write $\log_{10}(3x+1)-9\log_{10}x$ as a single logarithm.

11. Given $\log_b 2 \approx 0.2038$ and $\log_b 5 \approx 0.4732$, evaluate $\log_b 10$.

12. Use the change-of-base formula and a calculator to approximate $\log_6 30$ to five significant digits.

13. Graph $f(x) = \log_3(x-1)$.

14. Graph $f(x) = -2 + \log_4 x$.

15. Graph $f(x) = -\ln x + 1$.

16. Solve $2^{3x-7} = \frac{1}{32}$.

17. Solve $3^x = 42$.

18. Solve $\log(3-303) - \log(2x-7) = 3$.

19. If $23,000 is invested at an annual interest rate of 7.6 percent compounded annually, find the balance after 8 years.

20. Use $N(t) = N_o(0.5)^{t/1660}$ to estimate the percentage of radium $\left({}^{226}Ra\right)$ that remains in a sample after 5000 years.

1. Given that $f(x) = 3.2^x$, use a calculator to evaluate $f(\pi)$ accurate to six significant digits.

2. Given that $f(x) = e^{x/5}$, use a calculator to evaluate f(4.6) to six significant digits.

3. Graph $f(x) = 4^{-x/2}$.

4. Graph $f(x) = \left(\frac{2}{3}\right)^x$.

5. Graph $f(x) = e^{-x/3}$.

6. Write $\log_b(4 - 3x) = r$ in exponential notation.

7. Write $7^{(x-1)} = y$ in logarithmic form.

8. Write $\log_b \frac{x^2y}{z^7}$ in terms of logarithms of a, b, and c.

9. Write $\log_b \frac{\sqrt[4]{x^3z}}{y^2}$ in terms of logarithms of x, y, and z.

10. Write $\log_4 x - \log_4(2x - 3) + \log_4(x + 2)$ as a single logarithm.

11. Given $\text{lob}_b 2 \approx 0.2314$ and $\log_b 5 \approx 0.5372$, evaluate $\log_b 10$.

12. Use the change of base formula and a calculator to approximate $\log_5 20$ to five significant digits.

13. Graph $f(x) = \log_4(x-2)$.

14. Graph $f(x) = 2 + \log_3 x$.

15. Graph $f(x) = -\ln(x+1)$.

16. Solve $\left(\frac{2}{3}\right)^x = \frac{16}{81}$.

17. Solve $e^{x-1} = 9$.

18. Solve $\text{Log}(\log x) = \frac{1}{2}$.

19. Find the balance is $8,500 is invested at an annual rate of 6.7 percent for 20 years compounded continuously.

20. After a race, a runner's pulse rate R in beats per minute decreases according to the function $R(t) = 145e^{-0.092t}$ $(0 \le t \le 15)$ where t is measured in minutes. How long, to the nearest minute, after the end of the race will the runner's pulse rate be 90 beats per minute?

1. Given that $f(x)=8^x$, use a calculator to evaluate $f(\pi)$ accurate to six significant digits.

   a. 9488.53 b. 687.291 c. 25.1327 d. 867.921

2. Given that $f(x)=e^{-x}$, use a calculator to evaluate $f(0.8)$ to six significant digits.

   a. 2.22554 b. 2980.96 c. o.496076 d. 0.449329

3. Graph $f(x)=\left(\frac{1}{3}\right)^x$.

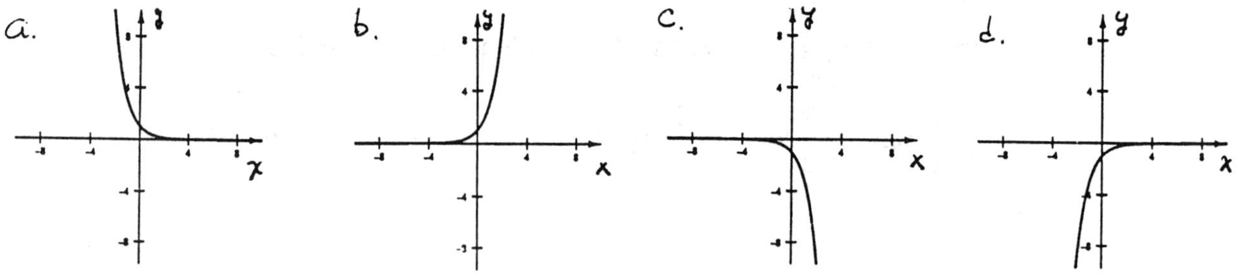

4. Graph $f(x)=e^x-1$.

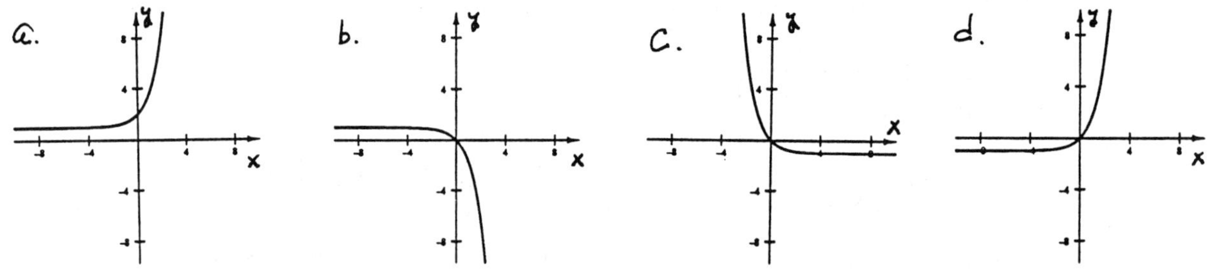

5. Graph $f(x)=2^x+1$.

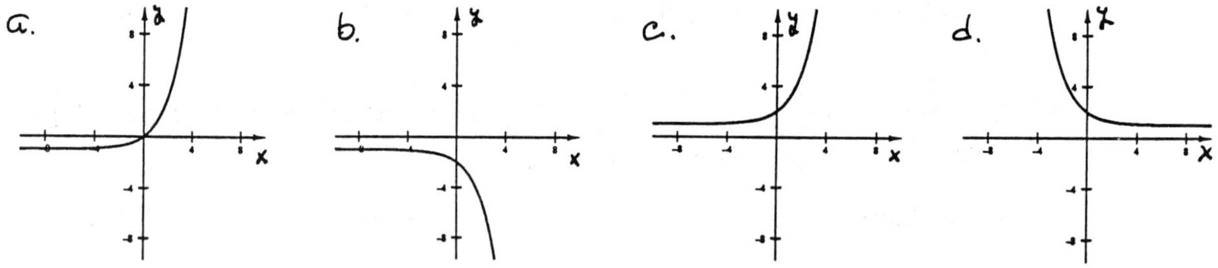

6. Write $\log_5(3x)=y-2$ in exponential form.

   a. $5^{y-2}=3x$ b. $(3x)^{y-2}=5$

   c. $(3x)^5=y-2$ d. $(y-2)^5=3x$

7. Write $4^{x-2} = y$ in logarithmic form.

a. $\log_y 4 = x - 2$

b. $\log_4(x-2) = y$

c. $\log_4 y = x - 2$

d. $\log_y(x-2) = 4$

8. Write $\log_4 \frac{xy^3}{z^4}$ in terms of logarithms of x, y, and z.

a. $3(\log_4 x + \log_4 y) - \log_4 z$

b. $\log_4 x + 3\log_4 y - 4\log_4 z$

c. $\log_4 x - 3\log_4 y - 4\log_4 z$

d. $\log_4 x + \log_4 3y - \log_4 4z$

9. Write $\log_5 \frac{x^{2/3}}{\sqrt[3]{y^2 z}}$ in terms of logarithms of z, y, and z.

a. $\frac{2}{3}\log_5 x - \frac{2}{3}\log_5 y + \frac{1}{3}\log_5 z$

b. $\frac{2}{3}\log_5 x - 3\log_5 y^2 z$

c. $\frac{3}{2}\log_5 x - (3\log_5 y - \log_5 z)$

d. $\frac{2}{3}\log_5 x - \frac{2}{3}\log_5 y - \frac{1}{3}\log_5 z$

10. Write $\log(2x-3) - \log x - \log(x-4)$ as a single logarithm.

a. $\log\frac{(2x-3)(x-4)}{x}$

b. $\log\frac{x(2x-3)}{x-4}$

c. $\log\frac{2x-3}{x(x-4)}$

d. $\log x(2x-3)(x-4)$

11. Given $\log_b 2 \approx 0.0270$ and $\log_b 5 \approx 0.6275$, evaluate $\log_b 10$.

a. 0.0162 b. 0.5645 c. 0.6005 d. 0.6545

12. Use the change-of-base formula and a calculator to approximate $\log_3 15$ to five significant digits.

a. 0.40568 b. 2.4650 c. 1.1761 d. 0.47712

13. Graph $f(x) = \log_5(x+2)$.

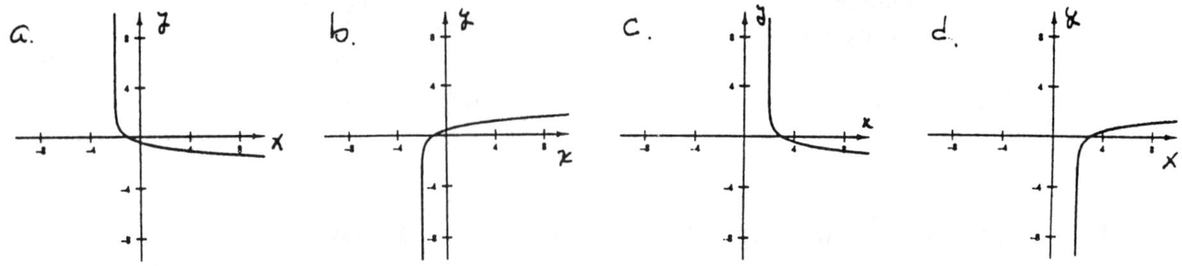

14. Graph $f(x) = 1 + \log_3 x$.

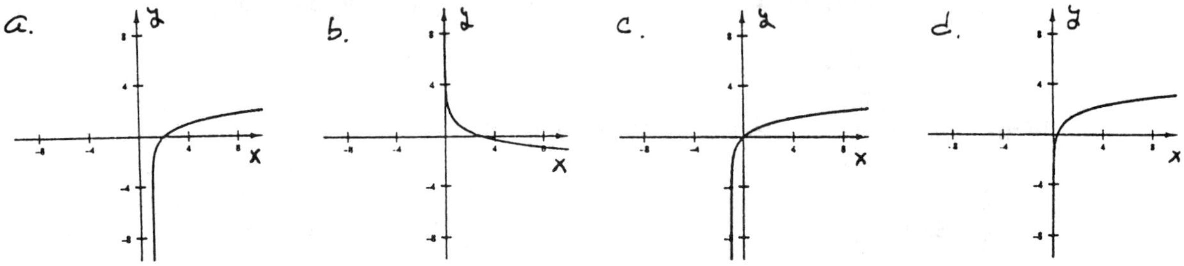

15. Graph $f(x) = \ln(x-5)$.

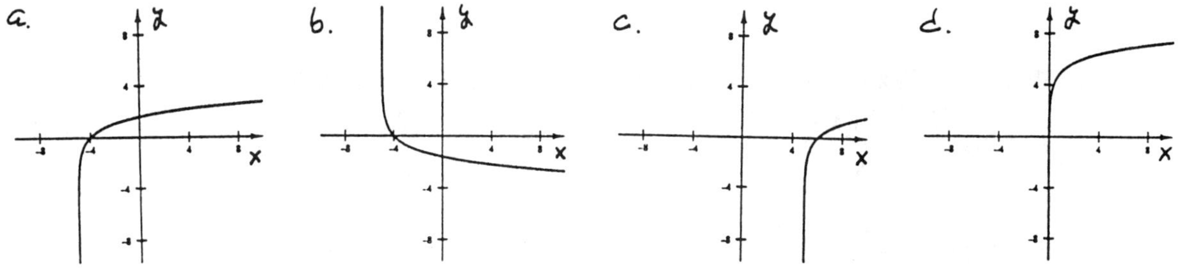

16. Solve $2^{3x-4} = \frac{1}{32}$.

a. $-\frac{1}{3}$ b. 3 c. $\frac{1}{3}$ d. $\frac{8}{3}$

17. Solve $7^x = 120$.

a. 2.8571 b. 2.4603 c. 2.2710 d. 2.5044

18. Solve $\ln 5 + \ln(x-6) = \ln(x+2)$.

a. $\frac{3}{2}$ b. 1 c. 7 d. 8

19. Find the balance if $22,000 is invested at an annual rate of 6.8 percent for 12 years compounded continuously.

a. $39,157.42
b. $49,751.59
c. $62,352.16
d. $29,462.11

20. The scrap value S of a product with an expected life of n years is given by $S(n) = P(1 - r)^n$, Where P is the original purchase price of the product and r is the annual rate of depreciation. A company car is purchased for $22,000 and is expected to last 5 years. What is its scrap value if the car depreciates at a rate of 18 percent per year?

a. $8651,28
b. $8156.28
c. $50,330.67
d. $7516.28

1. Given that $f(x) = 0.8^x$, use a calculator to evaluate $f(3.5)$ accurate to six significant digits.

   a. 0.457947 b. 0.572433 c. 2.72430 d. 0.366356

2. Given that $f(x) = e^{x/2}$, use a calculator to evaluate $f(\pi)$ to six significant digits.

   a. 11.5704 b. 4.73911 c. 4.81048 d. 23.1407

3. Graph $f(x) = 3^{2-x}$.

a. 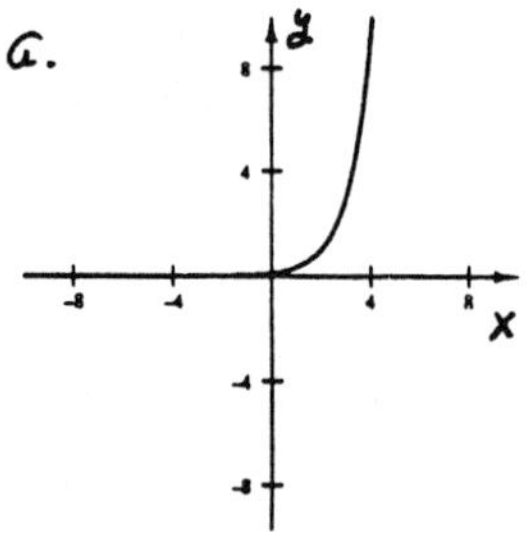

b. 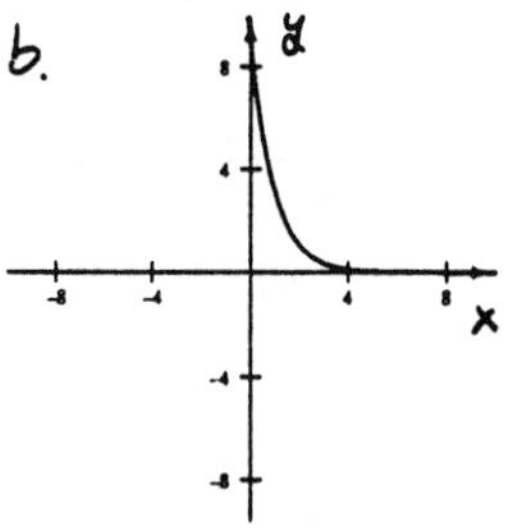

c. 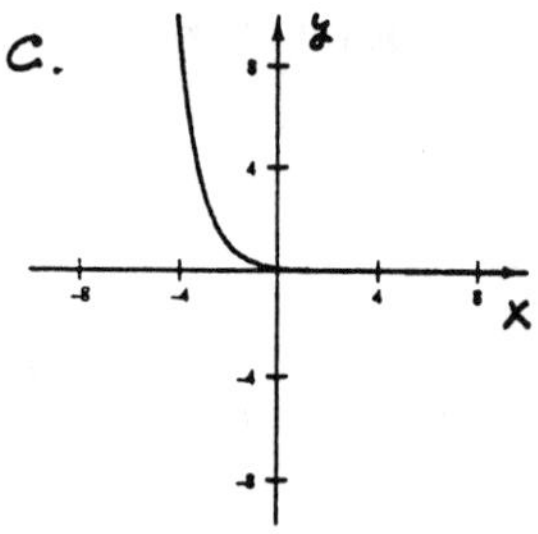

d. 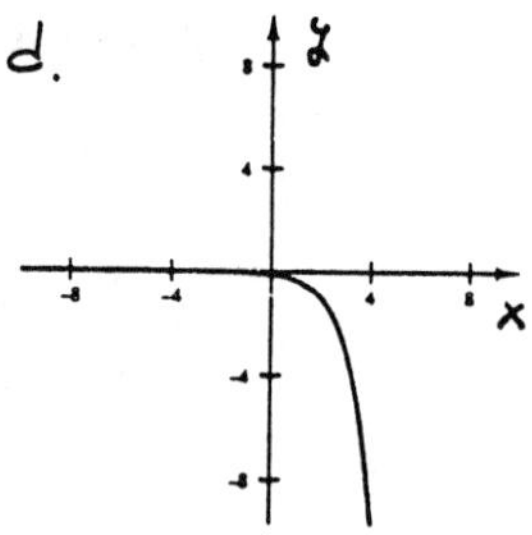

4. Graph $f(x) = \left(\frac{3}{2}\right)^{-x}$.

a. 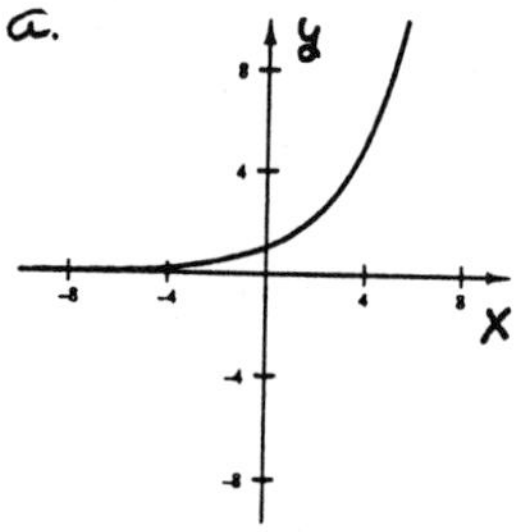

b. 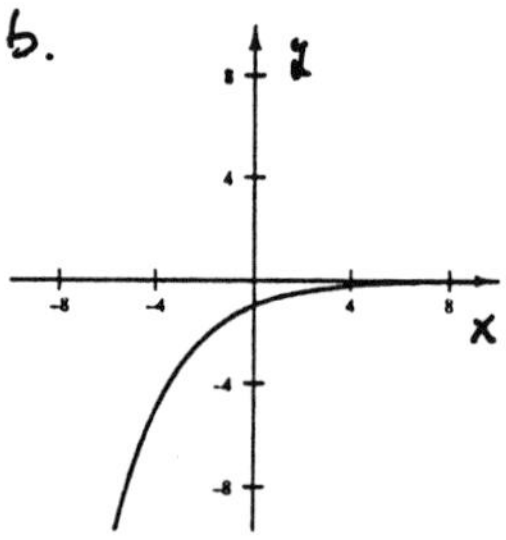

c. 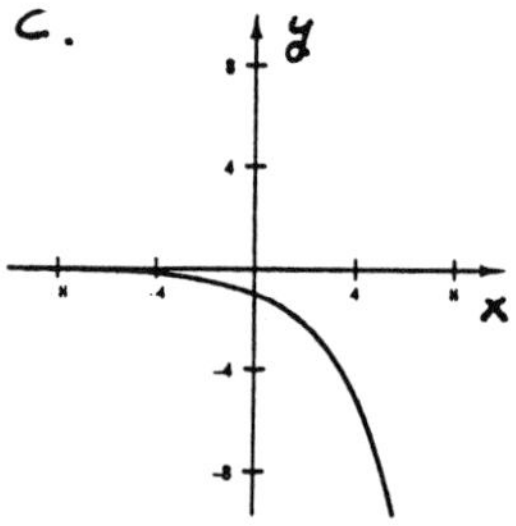

d. 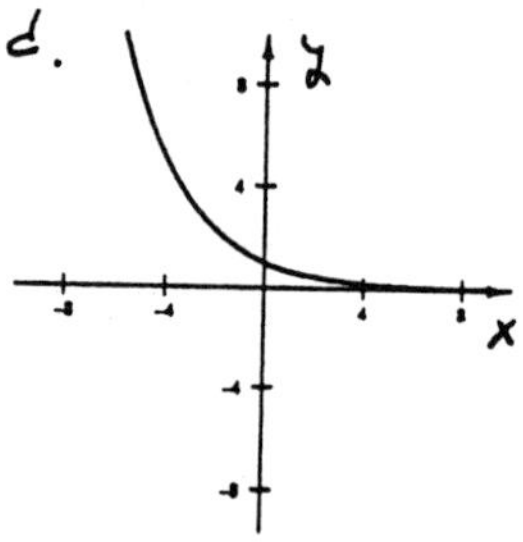

5. Graph $f(x) = -\frac{1}{3}e^x$.

a. 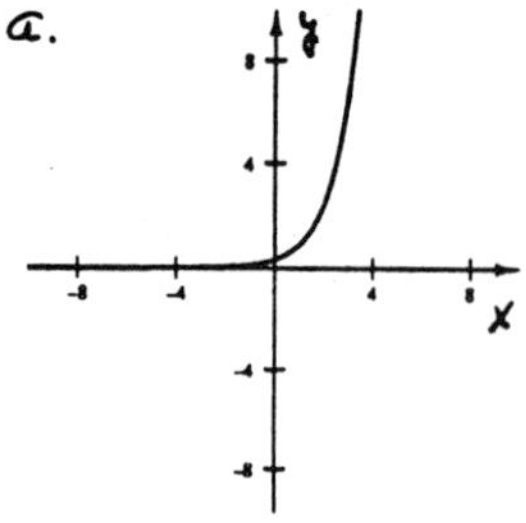

b. 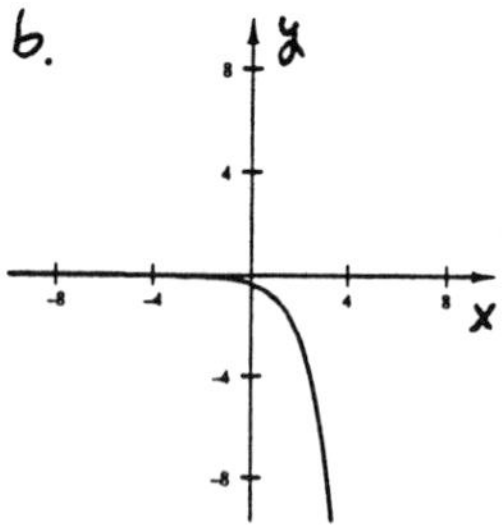

c. 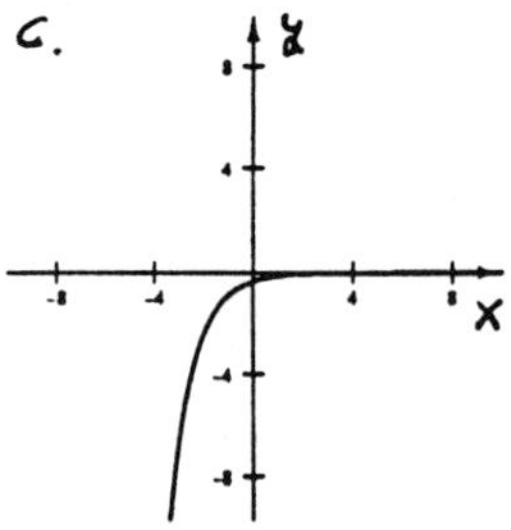

d. 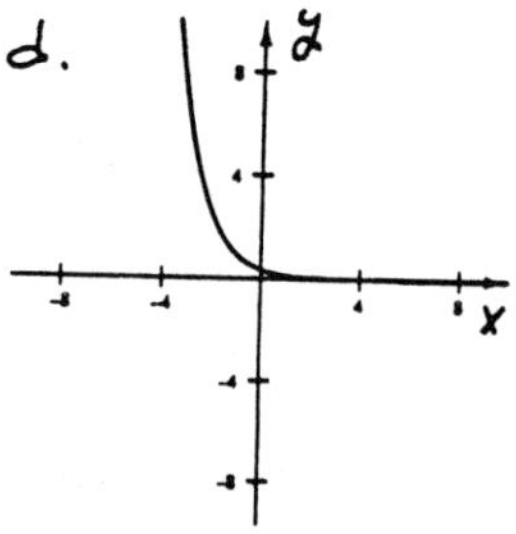

6. Write $\log_6(3x - 4) = y$ in exponential form.

   a. $y^6 = 3x - 4$ b. $6^y = 3x - 4$ c. $6^{3x-4} = y$ d. $y^{3x-4} = 6$

7. Write $5^{x+1} = y$ in logarithmic form.

a. $\log_5(x+1) = y$ b. $\log_y(x+1) = 5$

c. $\log_y 5 = x+1$ d. $\log_5 y = x+1$

8. Write $\log_b \frac{x^5 z}{y^3}$ in terms of the logarithms of x, y, and z.

a. $5\log_5 x + \log_b z - 3\log_b y$ b. $5\log_b x - \log_b z - 3;og_b y$

c. $5(\log_b x + \log_b z) - 3\log_b y$ d. $\log_b 5x - (\log_b z + \log_b 3y)$

9. Write $\log_b \frac{\sqrt[5]{x^3 z^7}}{y^{2/3}}$ in terms of logarithms of x, y, and z.

a. $15\log_b x + 35\log_b z - 5\log_b y$ b. $5(\log_b x^3 + \log_b z^7) - \frac{2}{3}\log_b y$

c. $\frac{3}{5}\log_b x + \frac{7}{5}\log_b z - \frac{2}{3}\log_b y$ d. $\frac{5}{3}\log_b x + \frac{5}{7}\log_b z - \frac{2}{3}\log_b y$

10. Write $\log_4(x-2) + \log_4 2x - \log_4(3x-7)$ as a single logarithm.

a. $\log_4 5$ b. $\log_4 \frac{2x(x-2)}{3x-7}$

c. $\log_4 \frac{x-2}{2x(3x-7)}$ d. $\log_4 2x(x-2)(3x-7)$

11. Given $\log_b 3 \approx 0.5647$ and $\log_b 5 \approx 0.8271$ find $\log_b \frac{5}{3}$.

a. -0.2624 b. 0.2624 c. 0.4671 d. -0.4671

12. Use the change-of-base formula and a calculator to approximate $\log_9 144$ to five significant digits.

a. 2.2619 b. 0.44214 c. 1.2041 d. 2.6219

13. Graph $f(x) = \log_7(x+3)$.

a. 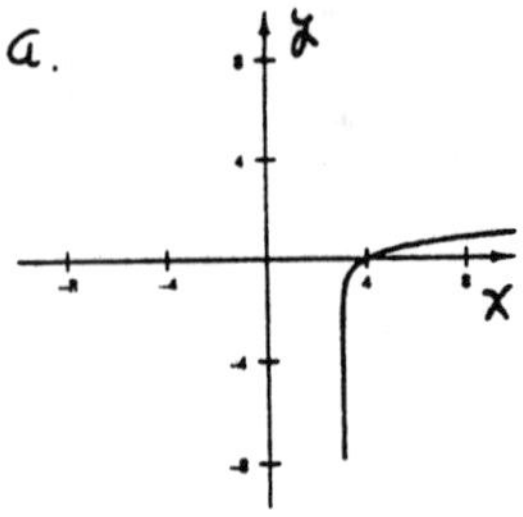

b. 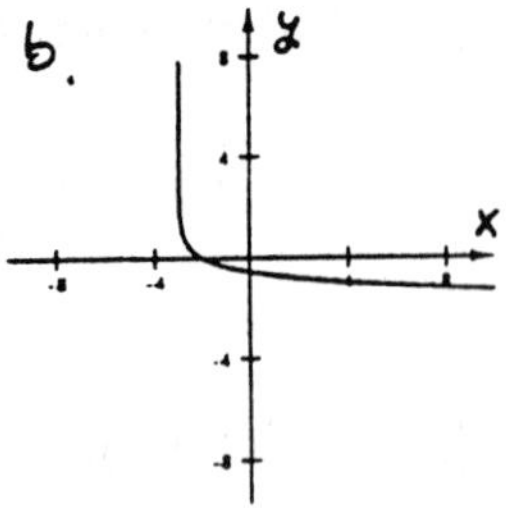

c. 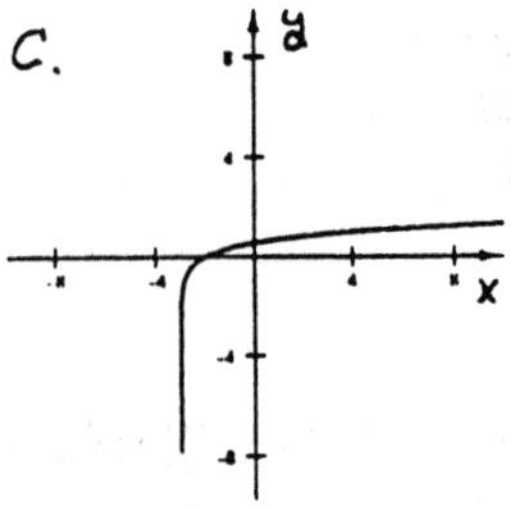

d. 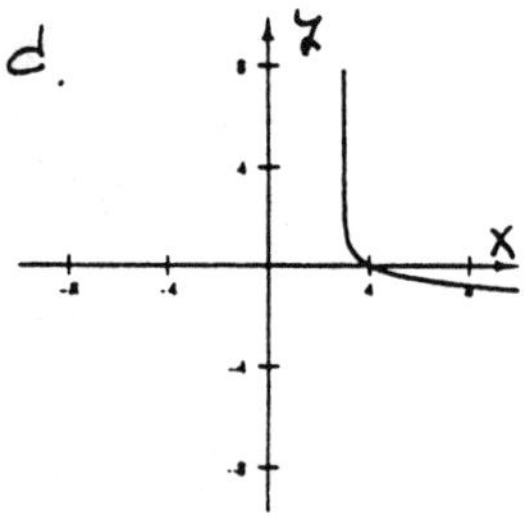

14. Graph $f(x) = 2 + \log_4 x$.

a. 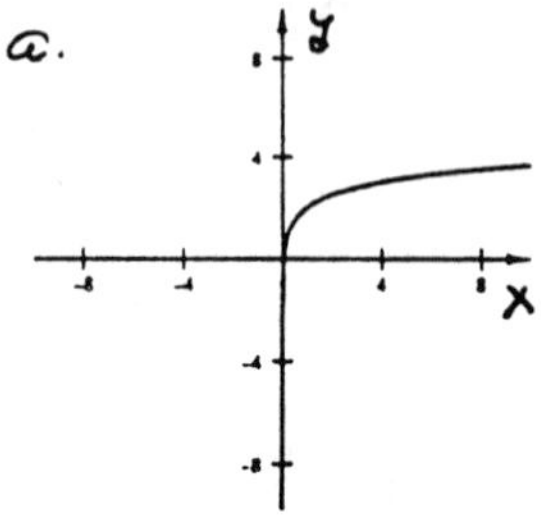

b. 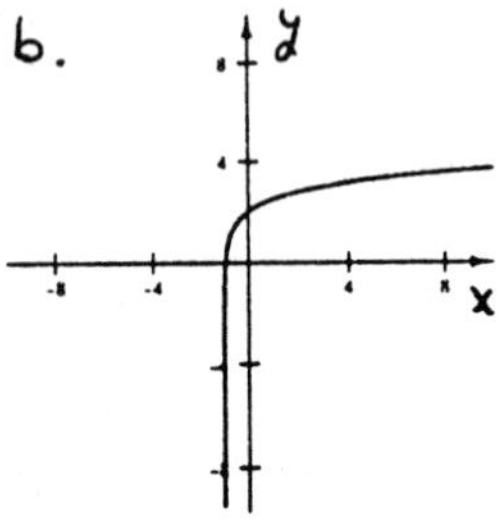

c. 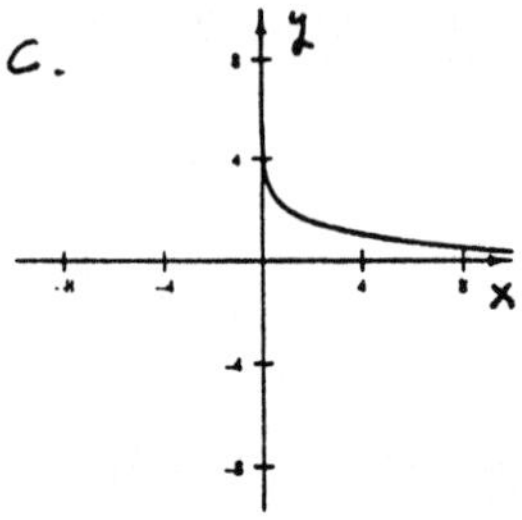

d. 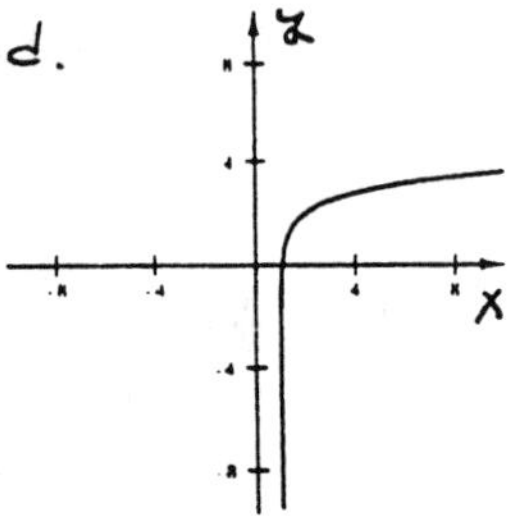

15. Graph $f(x) = -\ln(x-2)$.

a. 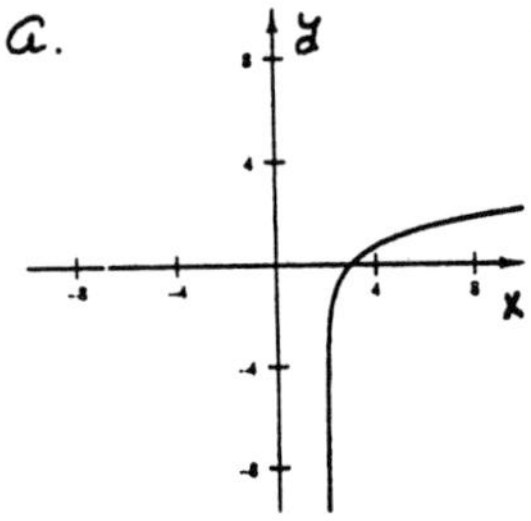

b. 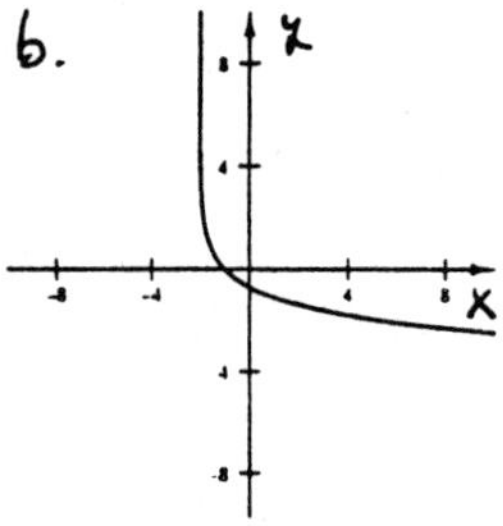

c. 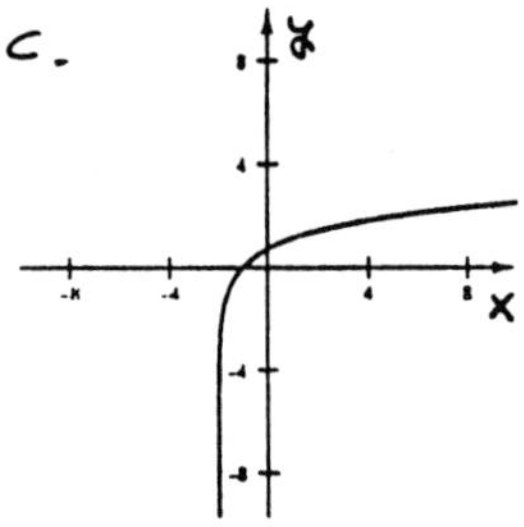

d. 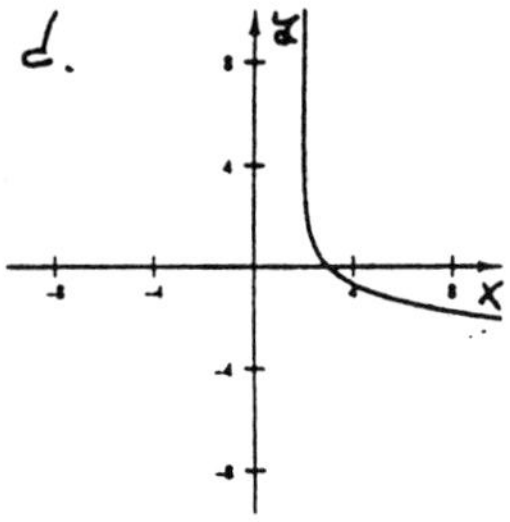

16. Solve $4^{2-3x} = \frac{1}{32}$.

a. -1   b. $\frac{7}{3}$ c. $\frac{3}{2}$ d. $-\frac{1}{6}$

17. Solve $6^x = 32$.

a. 0.72700   b. 1.9343   c. 5.3333   d. 1.3925

18. Solve $\ln(2x-3) - \ln x = \ln(1-4x)$.

a. $-\frac{3}{4}, 1$ b. $-1, \frac{1}{2}$ c. no solution   d. 1

19. Find the balance if \$12,000 is invested an annual rate of 5.5 percent for 9 years, compounded continuously.

a. \$19,429.13
b. \$19685.98
c. \$17,944.13
d. \$21,685.28

20. Find the exponential growth function for a city whose population was 75,000 in 1980 and 125,000 in 1990.

a. $N(t) = 75{,}000e^{-0.5108t}$
b. $N(t) = 75{,}000e^{0.05108t}$
c. $N(t) = 75{,}000e^{10t}$
d. $N(t) = 75{,}000e^{-10t}$

1. Given that $(x)=9^{x}$, use a calculator to evaluate $f\left(\sqrt{0.8}\right)$, accurate to six significant digits.

a. 5.79955 b. 7.13675 c. 8.80498 d. 25.4558

2. Given that $f(x)=e^{\sqrt{x}}$, use a calculator to evaluate $f(5)$, accurate to six significant digits.

a. 148.413 b. 8.91248 c. 9.35647 d. 6.05974

3. Graph $f(x)=\left(\frac{2}{3}\right)^{-x}$.

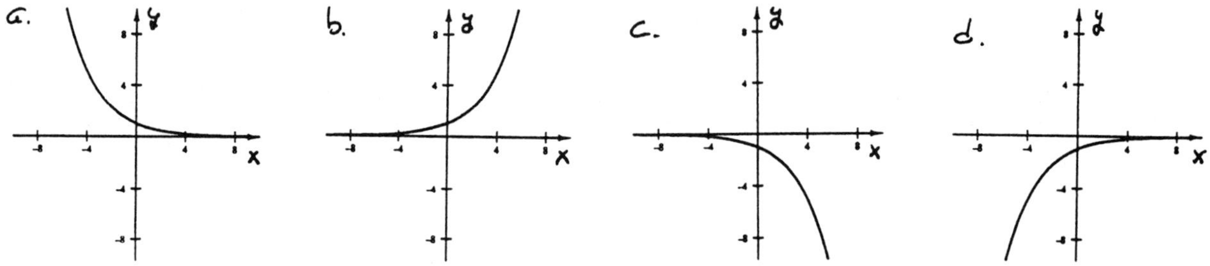

4. Graph $f(x)=3^{-x}$.

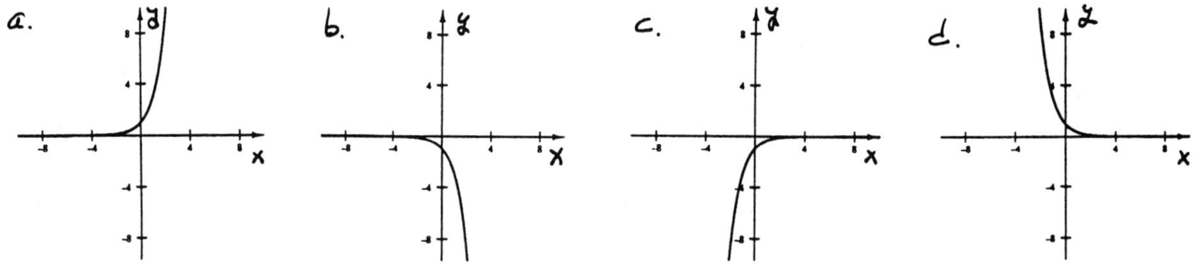

5. Graph $f(x)=\left(\frac{1}{e}\right)^{-2x}$.

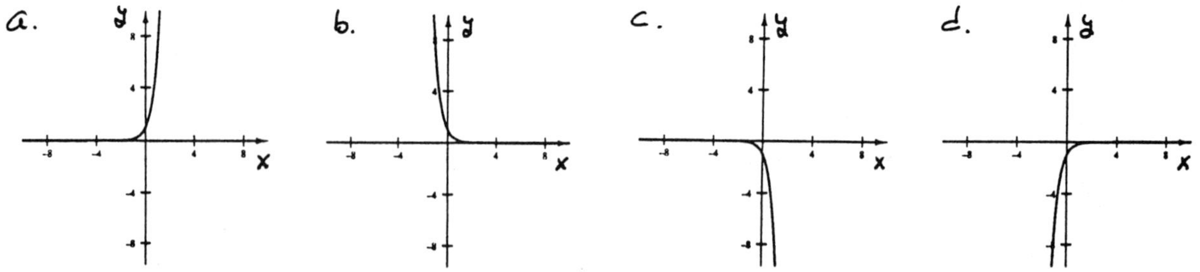

6. Write $\log_7(2x-3)=2y$ in exponential form.

a. $7^{2y}=2x-3$ b. $7^{2x-3}=2y$ c. $(2y)^7=2x-3$ d. $(2x-3)^7=2y$

7. Write $7^{2x-1} = 3y$ in logarithmic form.

a. $\log_7(2x-1) = 3y$ b. $\log_7 3y = 2x-1$

c. $\log_{3y}(2x-1) = 7$ d. $\log_{3y} 7 = 2x-1$

8. Write $\log_3 \frac{x^2y^5}{z^3}$ in terms of logarithms of x, y, and z.

a. $2\log_3 x + 5\log_3 y - 3\log_3 z$ b. $\log_3(2x+5y) - \log_3 3z$

c. $\frac{1}{2}\log_3 z + \frac{1}{5}\log_3 y - \frac{1}{3}\log_3 z$ c. $2\log_3 z - 5\log_3 y + 3\log_3 z$

9. Write $\log_8 \frac{\sqrt[3]{x^6y}}{z^{4/3}}$ in terms of logarithms of x, y, and z.

a. $6\log_8 x + \frac{1}{3}\log_8 y - \frac{3}{4}\log_8 z$ b. $2\log_8 x + \frac{1}{3}\log_8 y - \frac{4}{3}\log_8 z$

c. $3\log_8 6x + \frac{1}{3}\log_9 y - \frac{1}{3}\log 4z$ d. $\frac{1}{3}\log_8 x - 3\log_8 y - \frac{4}{3}\log_8 z$

10. Write $\log_5(2x-3) - 4\log_5 x + \log_5(x+4)$ in terms of logarithms of x, y, and z.

a. $\log_5 \frac{2x-3}{x^4(x+4)}$ b. $\log_5 \frac{(2x-3)(x+4)}{-4x}$

c. $\log_5 \frac{2x-3}{4x(x+4)}$ d. $\log_5 \frac{(2x-3)(x+4)}{x^4}$

11. Given $\log_7 2 \approx 0.3562$, $\log_7 3 \approx 0.5646, \approx$ and $\log_7 5 \approx 0.8271$, evaluate $\log_7 30$.

a. 1.7479 b. 1.4797 c. 0.16634 d. 1.9497

12. Use the change of base formula and a calculator to approximate $\log_9 50$ to five significant digits.

a. 0.56166 b. 1.7804 c. 1.6212 d. 1.8740

13. Graph $f(x) = \log(x-3)$.

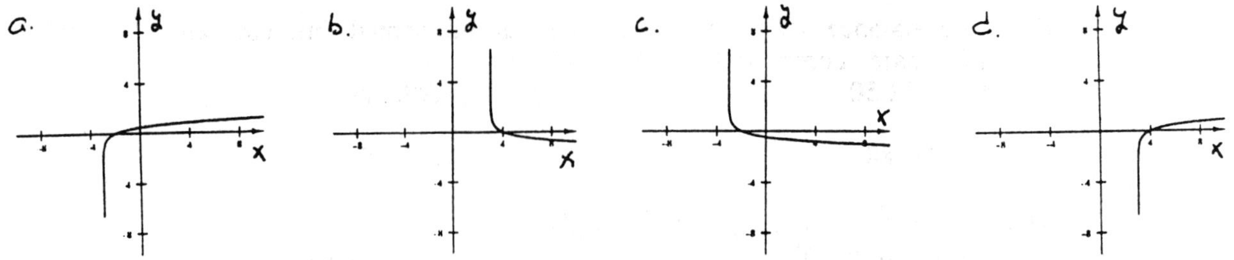

14. Graph $f(x) = 1 - \log_2 x$.

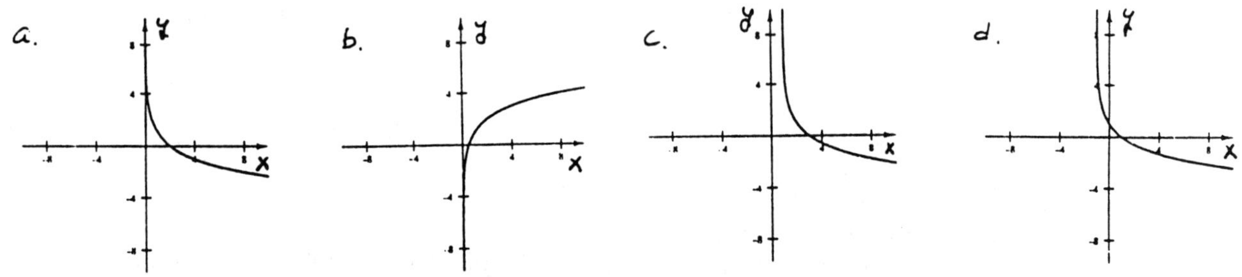

15. Graph $f(x) = \frac{1}{2}\ln(x-1)$.

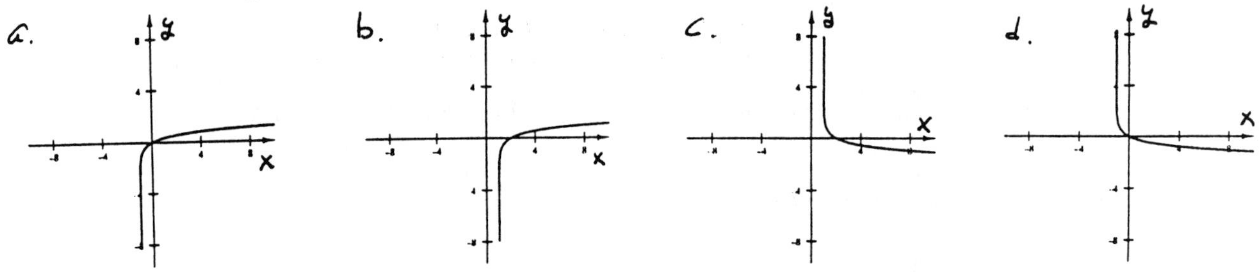

16. Solve $3^{2-5x} = \frac{1}{27}$.

a. $-\frac{1}{5}$ b. -5 c. 1 d. -1

17. Solve $12^x = 187$.

a. 1.02462 b. 2.10515 c. 2.27039 d. 2.4622

18. Solve $\log(8x) - \log(x-5) = 1$.

a. $\frac{10 \pm \sqrt{102}}{4}$ b. $\frac{25}{9}$ c. $-\frac{25}{9}$ d. 25

19. Find the balance if \$7500 is invested at an annual interest rate of 8 percent for 15 years compounded continuously.
    a. \$14,062.50
    b. \$24,900.88
    c. 29,400.88
    d. \$21062.50

20. Use $N(t) = N_0(0.5)^{t/5730}$ to estimate the amount of carbon in a bone that is 10,000 years old.
    a. 3%
    b. 67%
    c. 25%
    d. 30%

1. Convert 210° to exact radian measure.

2. Find the supplement of the angle whose radian measure is $\frac{7}{12}\pi$. Express your answer in terms of π.

3. Find the length of an arc that subtends a central angle of 105° in a circle with radius of 12 inches.

4. A wheel is rotating at 8 revolutions per second. Find the angular speed in radians per second.

5. A wheel with a diameter of 26 inches is rotating at 15 radians per second. Find the linear speed of a point on the edge of the wheel.

6. If θ is an acute angle of a right triangle and $\tan\theta = \frac{3}{7}$, find $\sec\theta$,

7. Find the value of $\sec 36^\circ$ to the nearest ten thousandths.

8. Find the exact value of $\sin\frac{11\pi}{6}\cos\frac{\pi}{3} - \tan\frac{5\pi}{4}$.

9. Find $W\left(\frac{5\pi}{4}\right)$.

10. Express $\frac{\csc^2\theta - 1}{\csc^2\theta}$ in terms of a single trigonometric function.

11. State the period of $y = -3\sin\frac{2}{3}x$.

12. State the amplitude, period, and phase shift for the equation $y = -2\sin\left(3x - \frac{\pi}{6}\right)$.

13. Graph $y = 2\cos\frac{3}{2}x$.

14. Graph $y = 2\csc 2x$.

15. Graph $y = 2\tan\left(\frac{1}{2}x - \frac{\pi}{2}\right)$.

16. Graph $y = -2\sin\left(2x - \frac{\pi}{4}\right) + 1$.

17. Graph $1 - \cos 2x$.

18. Graph $y = \sin x - \cos\frac{x}{2}$.

19. Find the height of a building if the angle of elevation to the top of the building changes from $20°$ to $35°$ as the observer moves a distance of 60 feet toward the building.

20. Write the equation for simple harmonic motion, given that the amplitude is 4 inches and the period is 2 seconds. Assume maximum displacement when $t = 0$.

1. Convert 240° to exact radian measure.

2. Find the supplement of the angle whose measure is $82^\circ 16'$.

3. Find the length of an arc that subtends a central angle of 225° in a circle with radius of 15 mentimeters.

4. A hard disk lin a computer rotates at 2400 rpm. Find the angular speed in radians per second.

5. The 14 inch wheel of a moving automobile is rotating at 14 revolutions per second. Assuming that the wheel is not skidding, find the speed of the automobile.

6. If θ is an acute angle of a right triangle and $\cot\theta = \frac{5}{3}$, find $\csc\theta$.

7. Find the value of cot 37° to the nearest ten thousandths.

8. Find the exact value of $\cos\frac{2\pi}{3}\cot\frac{\pi}{3} - \sin\frac{7\pi}{4}$.

9. Find $W\left(\frac{\pi}{6}\right)$.

10. Express $\frac{1}{1-\sec t} + \frac{1}{1+\sec t}$ in terms of a single trigonometric function.

11. State the period of $y = -2 \cot\frac{\pi}{4}x$.

12. State the amplitude, period, and phase shift for the equation $y = -\sin(3x + \pi)$.

13. Graph $y = -2\sin 2x$.

14. Graph $y = -3\csc \frac{x}{2}$.

15. Graph $y = \tan\left(x - \frac{\pi}{4}\right)$.

16. Graph $y = 3\cos\left(3x - \frac{\pi}{3}\right) + 2$.

17. Graph $y = 1 + \sin 2x$.

18. Graph $y = 2\sin x + \cos x$.

19. If a car climbs a hill that has a constant angle of 4° for a distance of 0.3 miles, how many vertical feet has it climbed?

20. Write the equation for simple harmonic motion, given that the amplitude is 12 centimeters and the period is 1.5 seconds. Assume maximum displacement when $t = 0$.

1. Convert 315° to exact radian measure.

2. Find the supplement of the angle whose radian measure is $\frac{8}{15}\pi$. Express your answer in terms of π.

3. Find the measure in radians of the central angle of a cirle with radius 27 centimeters an d arc length 12 centimeters.

4. A wheel is rotating at 40 revolutions per minute. Find the angular speed in radians per second.

5. A wheel with a diameter of 5 feet is rotating at 25 radians per second. Find the linear speed of a point on the edge of the wheel.

6. If θ is an acute angle of a right triangle and $\tan\theta = 3$, find $\sec\theta$.

7. Find the value of $\sec 123^\circ$ to the nearest ten thousandths.

8. Find the exact value of $\sin 315^\circ \cot 135^\circ + \cos 120^\circ$.

9. Find $W\left(\frac{13\pi}{6}\right)$.

10. Express $\frac{\cos^2 t}{\tan^2 t}$ in terms of the sine function.

11. State the period of $y = -3\cot\frac{\pi}{3}x$.

12. State the amplitude, period, and phase shift for the equation $y = \frac{2}{3}\sin\left(5x + \frac{\pi}{2}\right)$.

13. Graph $y = \frac{1}{2}\sin 4x$.

14. Graph $y = 2\sec 4x$.

15. Graph $y = \sin\left(\frac{1}{2}x - \pi\right)$.

16. Graph $y = -3\cos\left(x - \frac{\pi}{2}\right) - 5$.

17. Graph $y = -1 + \sin 2x$.

18. Graph $y = \sin x + \cos\frac{x}{2}$.

19. The angle of elevation from point A to the top of a tree is 30 degrees. At point B, 6 meters from A and on a line through the base of the tree and A, the angle of elevation is 37 degrees. Find the height of the tree.

20. Write the equation for simple harmonic motion, given that the amplitude is 8 feet and the period is 2.5 seconds. Assume maximum displacement when t = 0.

1. Convert 48° to exact radian measure.

a. $\frac{3\pi}{8}$ b. $\frac{2\pi}{15}$ c. $\frac{4\pi}{15}$ d. $\frac{7\pi}{15}$

2. Convert 153.31° to the DMS system to the nearest minute.

a. 153°31' b. 153°51' c. 153°46' d. 153°19'

3. Find the supplement of the angle whose radian measure is $\frac{5}{9}\pi$. Express your answer in terms of π.

a. $\frac{5}{9}\pi$ b. $\frac{\pi}{3}$ c. $\frac{13}{9}\pi$ d. $\frac{4}{9}\pi$

4. Find the length of an arc that subtends a central angle of 105° in a circle of radius 8 inches.

a. 4π in. b. $\frac{14}{3}\pi$ in. c. $\frac{96}{7}\pi$ in. d. $\frac{7}{2}\pi$ in.

5. A wheel with a diameter of 40 inches is rotating at 22 radians per second. Find the linear speed of a point on the edge of the wheel. Express the answer in ft/s.

a. 880 ft/s b. 73.3 ft/s c. 36.7 ft/s d. 23.5 ft/s

6. If θ is an acute angle of a right triangle and $\tan\theta = 8$, find $\cos\theta$.

a. $\frac{1}{\sqrt{65}}$ b. $\frac{1}{3\sqrt{7}}$ c. $\frac{8}{\sqrt{65}}$ d. $\frac{1}{65}$

7. Find the value of csc 141° to the nearest ten thousandths.

a. 0.6293 b. 1.5890 c. -1.2868 d. 1.3570

8. Find the exact value of $\tan\frac{5\pi}{3}\cos\frac{7\pi}{6}+\sin\frac{2\pi}{3}$.

a. $-\frac{3+2\sqrt{3}}{4}$ b. $\frac{3+2\sqrt{3}}{4}$ c. $\frac{3+\sqrt{3}}{2}$ d. 1

9. Find $W\left(\frac{2\pi}{3}\right)$.

a. $\left(\frac{\sqrt{3}}{2}, -\frac{1}{2}\right)$ b. $\left(-\frac{\sqrt{2}}{2}, \frac{\sqrt{2}}{2}\right)$ c. $\left(-\frac{\sqrt{2}}{2}, \frac{\sqrt{3}}{2}\right)$ d. $\left(-\frac{1}{2}, \frac{\sqrt{3}}{2}\right)$

10. Express $\frac{\sec^2 t}{\cot^2 t}$ in terms of the sine function.

a. $\frac{\sin^2 t}{1 - 2\sin^2 t + \sin^4 t}$ b. $\frac{\sin^2 t}{1 - \sin^4 t}$

c. $\frac{1 - \sin^4 t}{\sin^2 t}$ d. $\sin^4 t$

11. State the period of $y = -5\tan\frac{\pi}{2}x$.

a. 2 b. π c. $\frac{2}{\pi}$ d. $\frac{1}{2}$

12. State the amplitude, period, and phase shift for the equation $y = \frac{1}{2}\sin(4x - \pi)$.

a. $\frac{1}{2}, \frac{\pi}{4}, \frac{\pi}{4}$ b. $-\frac{1}{2}, \frac{\pi}{2}, \frac{\pi}{2}$ c. $\frac{1}{2}, \frac{\pi}{2}, \frac{\pi}{4}$ d. $\frac{1}{2}, 4, \pi$

13. Graph $y = \frac{2}{3}\sin 6x$.

a.
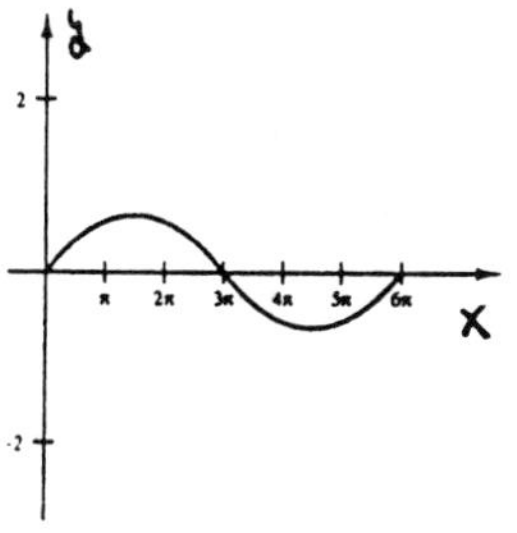

b.
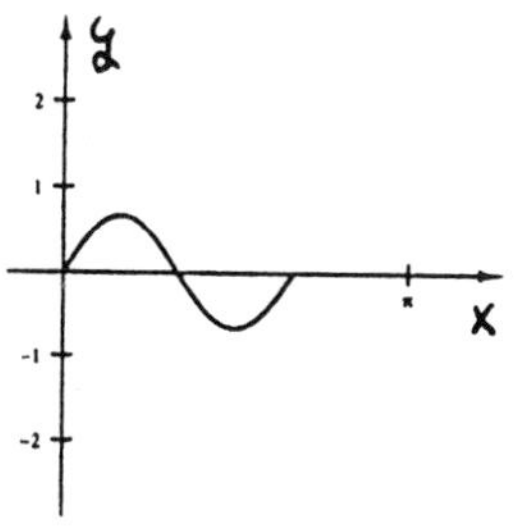

c.
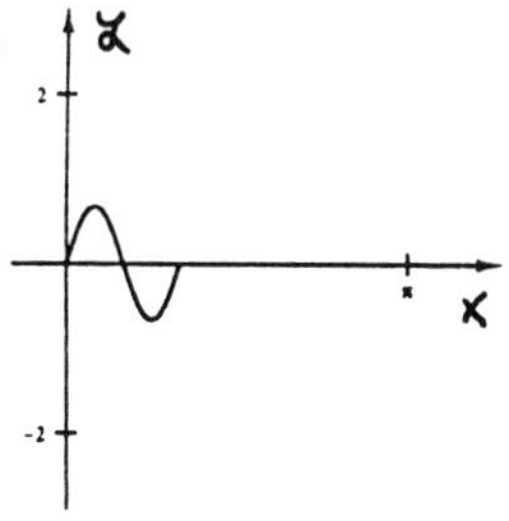

d.
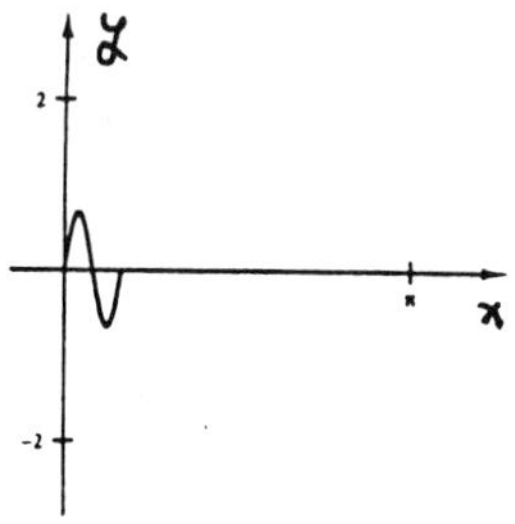

14. Graph $y = \sec 2x$.

a. 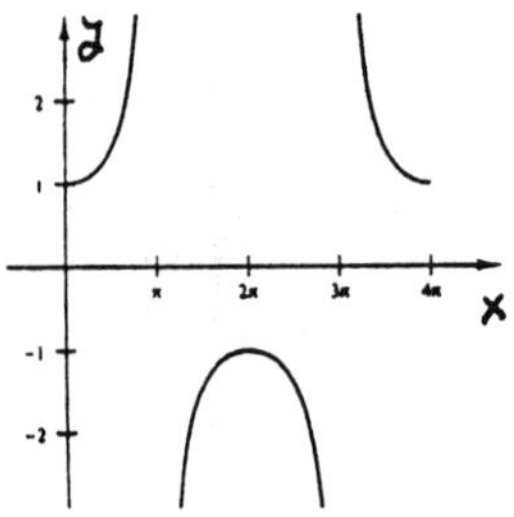

b. 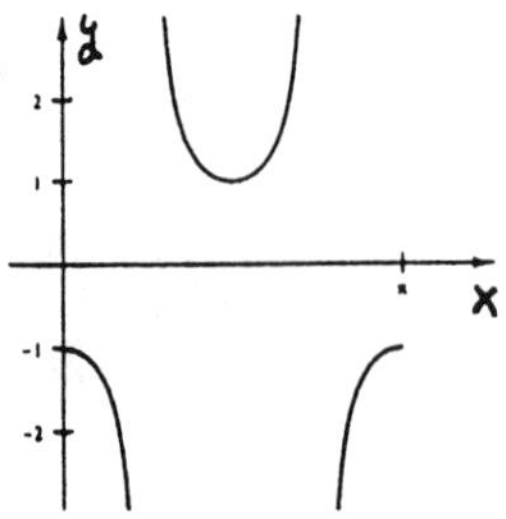

c. 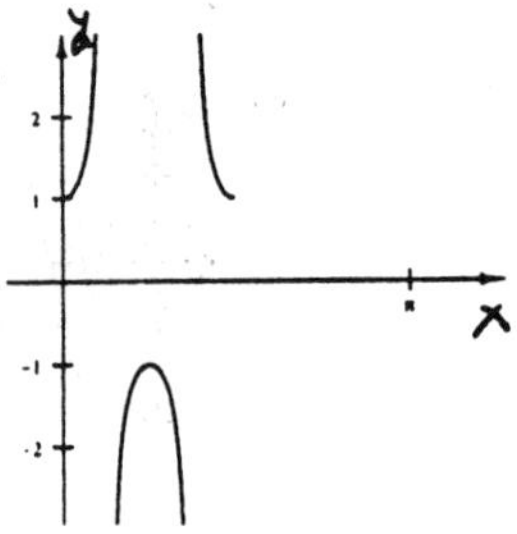

d. 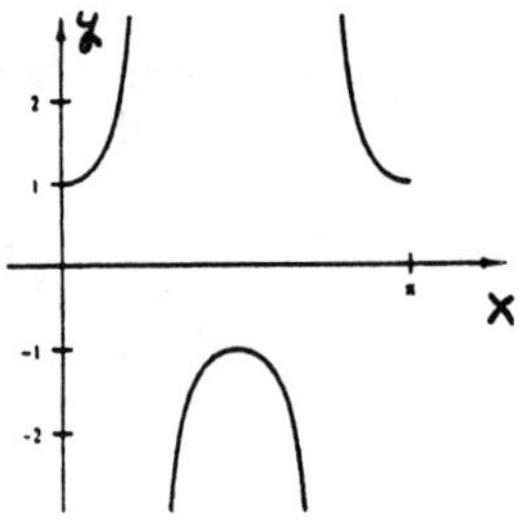

15. Graph $y = -2\tan\frac{x}{2}$.

a. 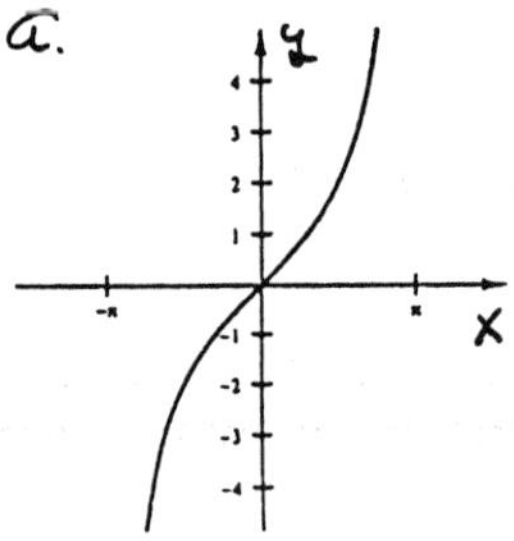

b. 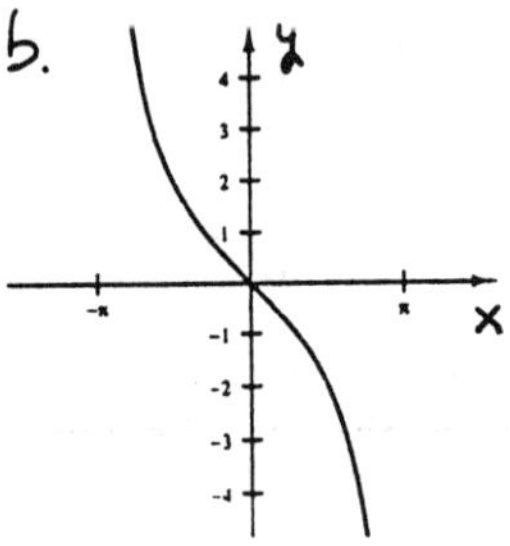

c. 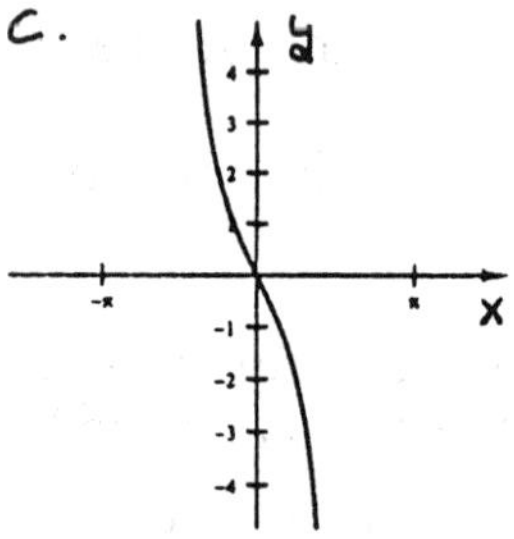

d. 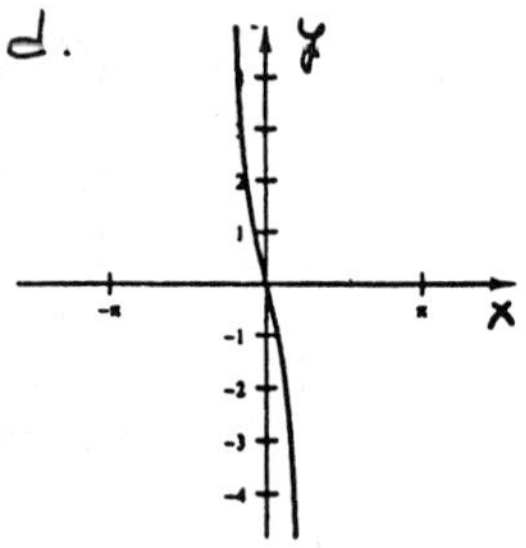

16. Graph $y = \sin\left(\frac{1}{4}x + \pi\right)$.

(a) 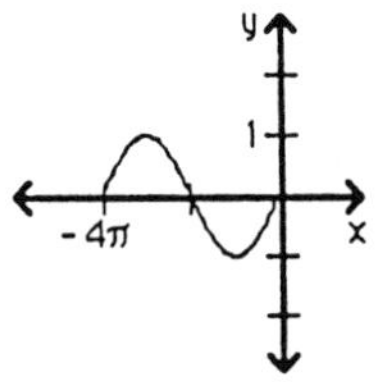

(b) 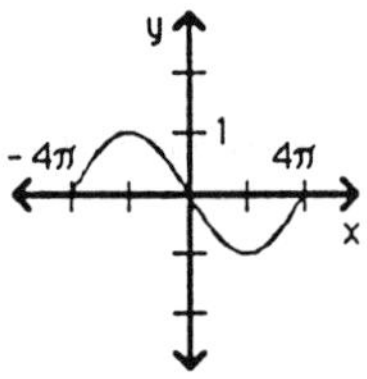

(c) 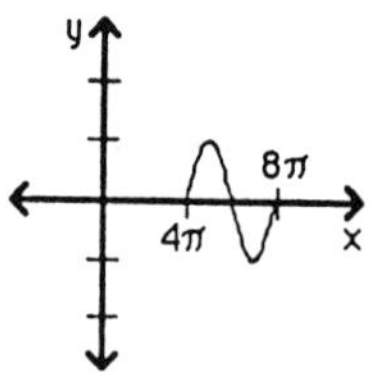

(d) 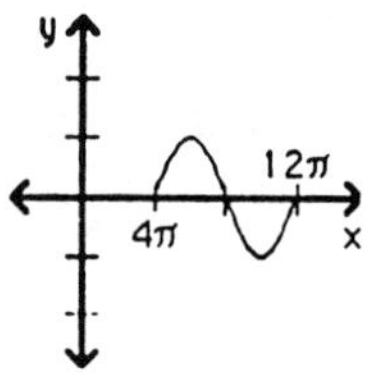

17. Graph $y = 1 + \text{cox}\frac{x}{2}$.

a. 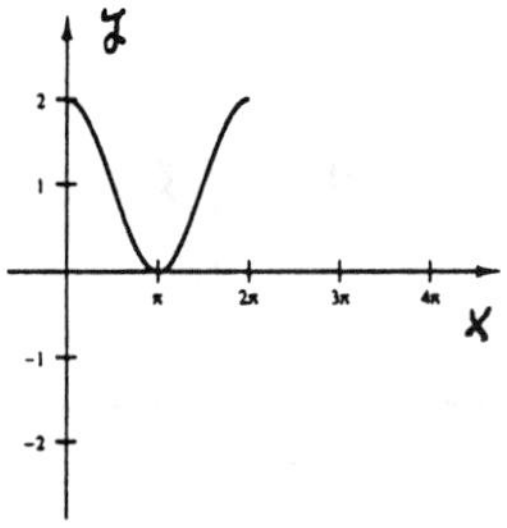

b. 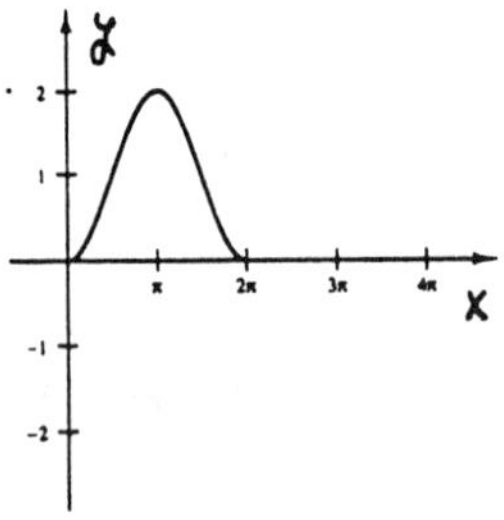

c. 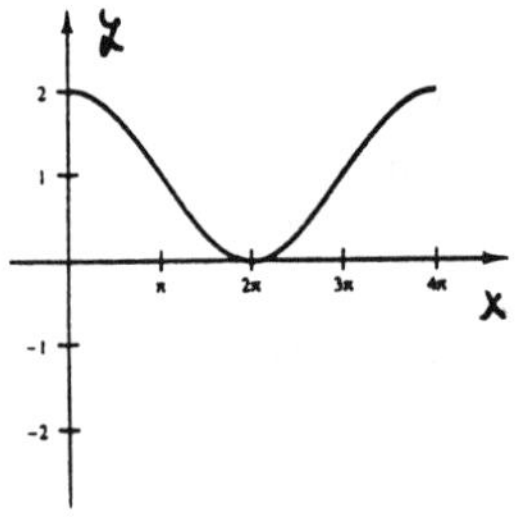

d. 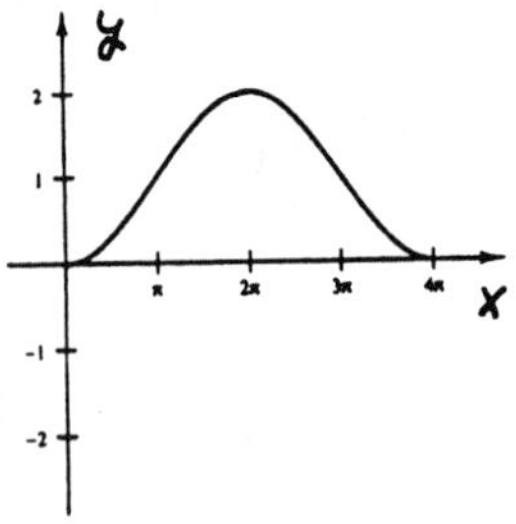

18. Graph $y = 2\sin x + \cos\frac{x}{2}$.

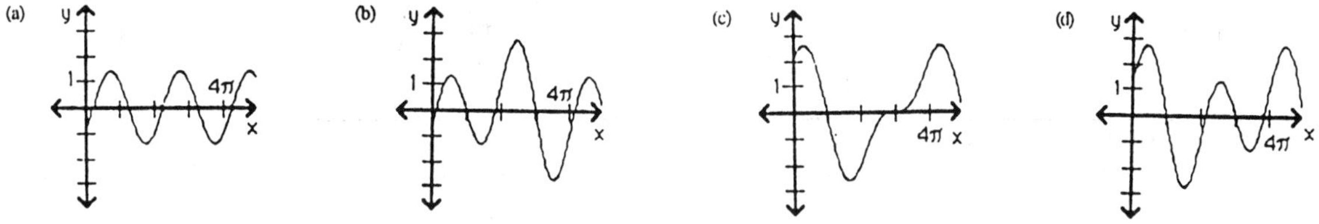

19. A tree casts a shadow of 7.5 feet when the angle of elevation of the sun is 38°. Find the height of the tree.

a. 14 ft b. 4.0 ft c. 5.9 ft d. 12 ft

20. Write an equation for simple harmonic motion given that the amplitude is 5 centimeters and the period is 4 seconds. Assume maximum displacement at time = 0.

a. $y = 4\cos\frac{2\pi t}{5}$ b. $y = 5\cos\frac{\pi t}{2}$ c. $y = 5\cos\frac{\pi t}{4}$ d. $y = 5\sin\frac{\pi t}{4}$

1. Convert 42°52' to decimal degrees.

a. 42.54° b. 42.9° c. 42.09° d. 42.06°

2. Convert $\frac{11\pi}{15}$ to exact degree measure.

a. 121° b. 132° c. 144° d. 96°

3. Find the measure in radians of the central angle of a circle with radius 15 centimeters and arc length 24 centimeters.

a. $\frac{5}{8}$ b. $\frac{8\pi}{5}$ c. $\frac{4}{5}$ d. $\frac{8}{5}$

4. A wheel is rotating at 80 rpm. Find the angular speed in radians per second.

a. $\frac{3\pi}{8}$ radians/s b. $\frac{8\pi}{3}$ radians/s

c. $\frac{4\pi}{3}$ radians/s d. 80π radians/s

5. If θ is an acute angle of a right triangle and $\sin\theta = \frac{2}{3}$, find $\tan\theta$.

a. $\frac{2\sqrt{5}}{5}$ b. $\frac{\sqrt{5}}{2}$ c. $\frac{2\sqrt{11}}{11}$ d. $\frac{\sqrt{5}}{2}$

6. If $\sin\theta = -\frac{\sqrt{3}}{2}$, $\pi \le \theta \le \frac{3\pi}{2}$, find $\sec\theta$.

a. $-\frac{2\sqrt{3}}{3}$ b. $\frac{1}{2}$ c. 2 d. –2

7. Find the value of sec 13° to the nearest ten thousandths.

a. 0.9744 b. 1.0263 c. 1.2063 d. 1.0042

8. Find the exact value of $\sin 270° \cos 135° - \tan 300°$.

   a. $-\dfrac{\sqrt{2} - 2\sqrt{3}}{4}$ b. $\dfrac{\sqrt{2} - 2\sqrt{3}}{2}$ c. $\dfrac{\sqrt{2} + 2\sqrt{3}}{2}$ d. $\dfrac{-\sqrt{2} - \sqrt{3}}{2}$

9. Find $W\left(\dfrac{7\pi}{6}\right)$

   a. $\left(-\dfrac{1}{2}, -\dfrac{\sqrt{3}}{2}\right)$ b. $(-\sqrt{3}, -1)$ c. $\left(-\dfrac{\sqrt{3}}{2}, -\dfrac{1}{2}\right)$ d. $(-1, -\sqrt{3})$

10. Express $\dfrac{1}{1 - \sec t} + \dfrac{1}{1 + \sec t}$ in terms of the cotangent function.

   a. $-2\cot^2 t$ b. $2\cot^2 t$ c. $\dfrac{1}{2\cot^2 t}$ d. $1 - \cot^2 t$

11. State the period of $y = \dfrac{3}{2} \cot \pi x$

   a. 1 b. π c. $\dfrac{3}{2}$ d. $\dfrac{\pi}{2}$

12. State the amplitude, period, and phase shift for the equation $y = -\dfrac{5}{2}\sin(2\pi x - 2) + 1$.

   a. $\dfrac{5}{2}, 2, \pi$ b. $\dfrac{5}{2}, 1, \pi$ c. $\dfrac{5}{2}, 1, \dfrac{1}{\pi}$ d. $-\dfrac{5}{2}, 1, \pi$

13. Graph $y = -2\sin\dfrac{x}{4}$.

   a.

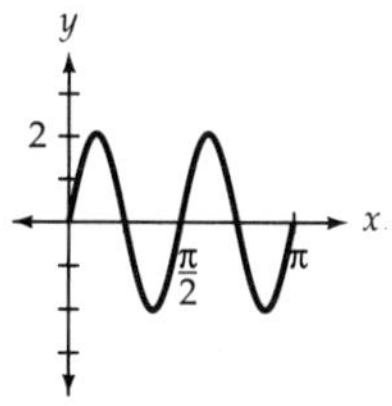

   b.

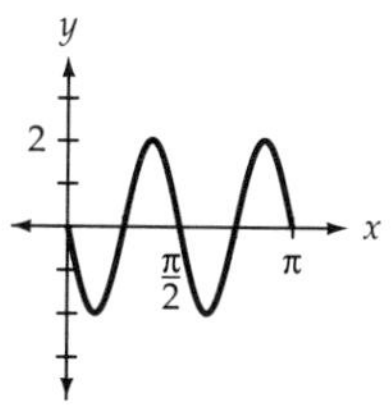

   c.

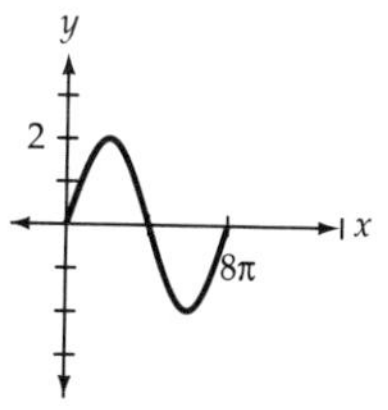

   d.

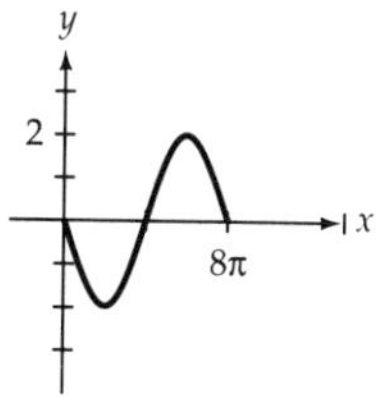

14. Graph $y = -5\csc \frac{x}{3}$.

a.
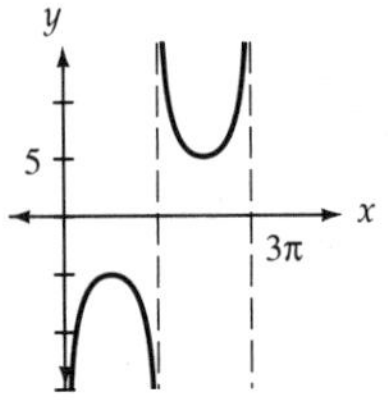

b.
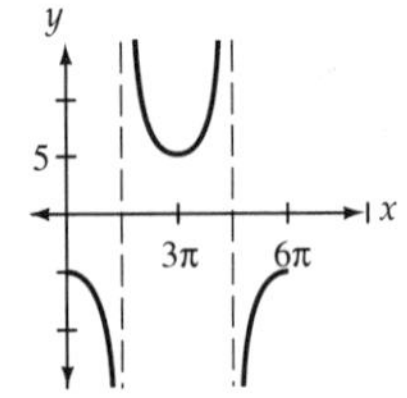

c.
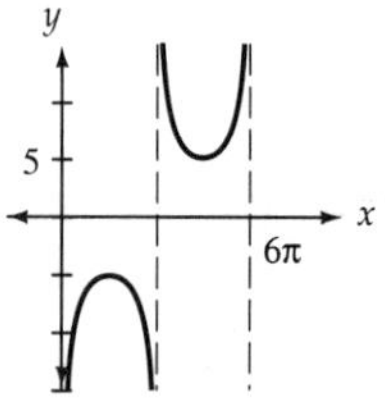

d.
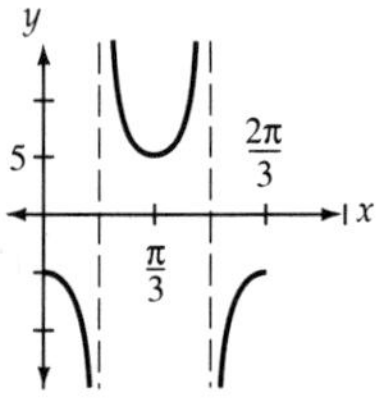

15. Graph $y = -2\tan \frac{x}{3}$.

a.
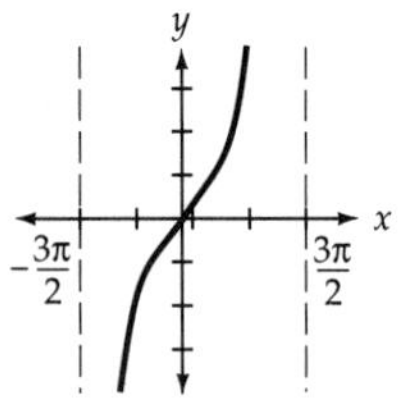

b.
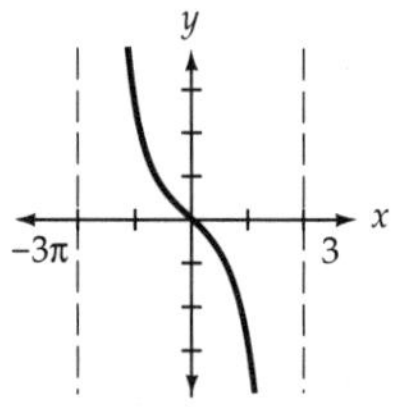

c.
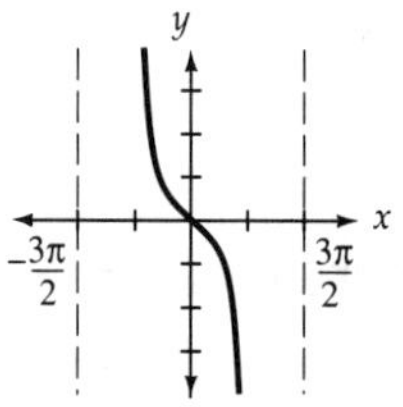

d.
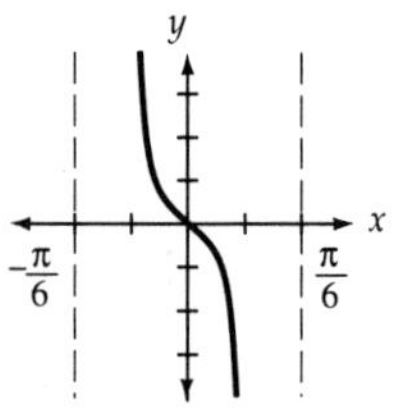

16. Graph $y = -2\cos(x - \pi) - 3$.

a.
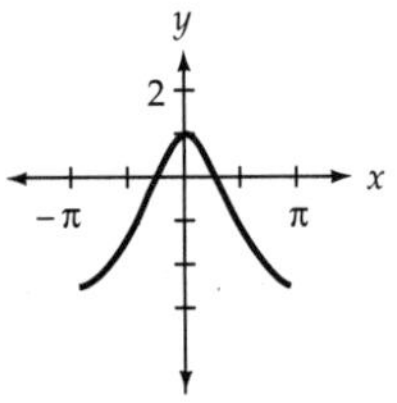

b.
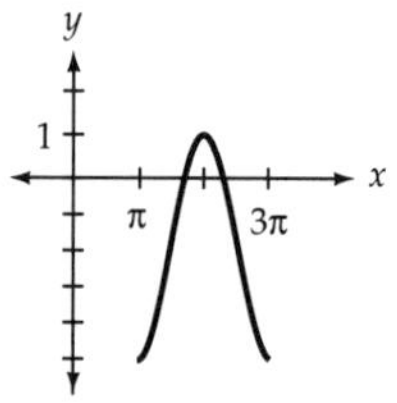

c.
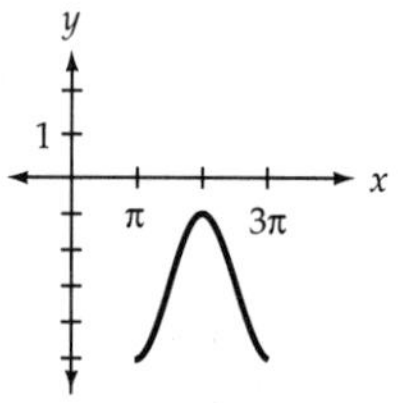

d.
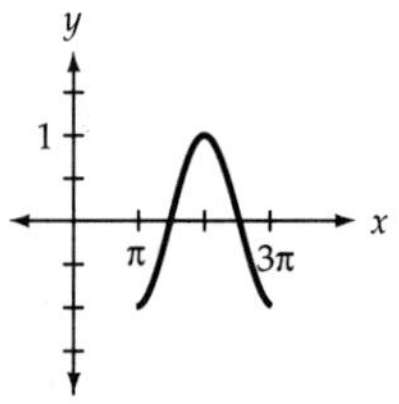

17. Graph $y = 1 + 2\sin \frac{x}{2}$.

a.
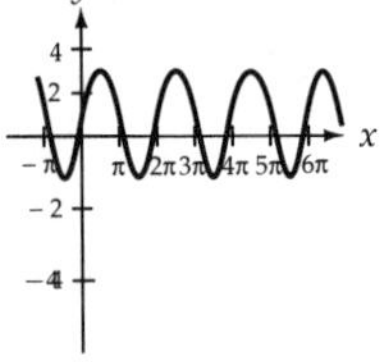

b.
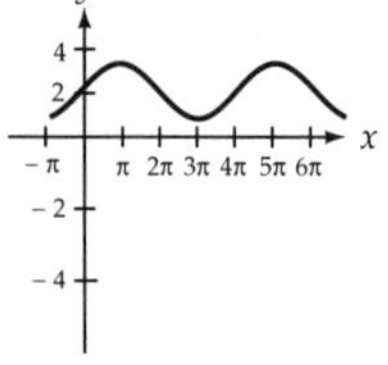

c.
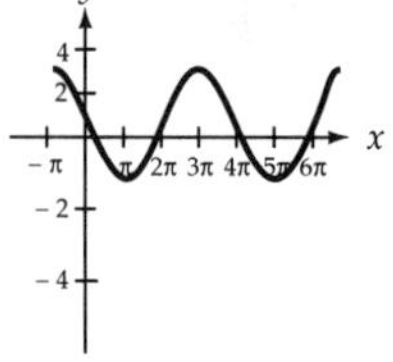

d.
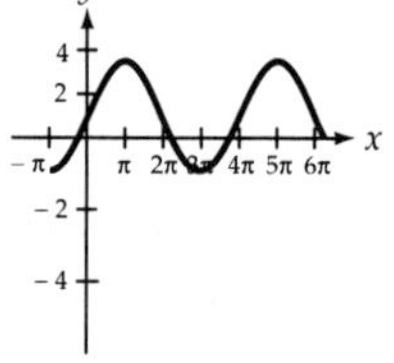

18. Graph $y = \frac{1}{2}x + \sin x$.

a.
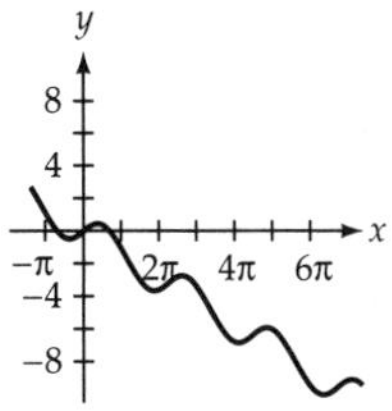

b.
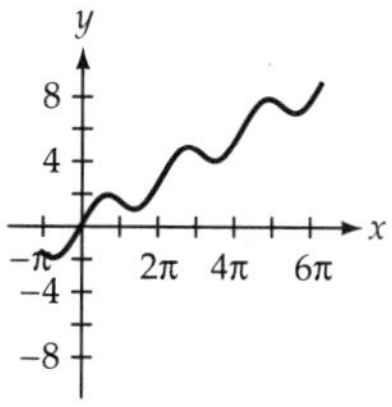

c.
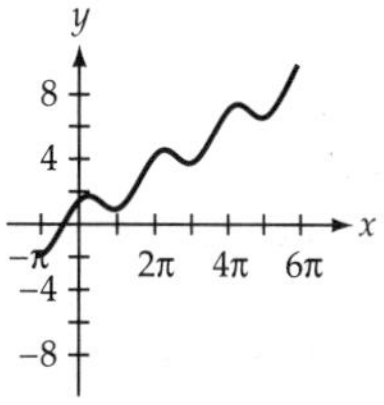

d.
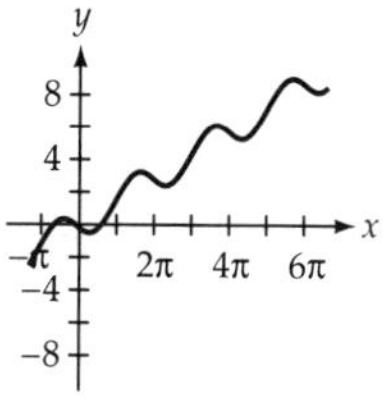

19. A pulley with a radius of 30 inches uses a belt to drive a pulley with a radius of 18 inches. Find the angle through which the smaller pulley turns as the 30-inch pulley makes one revolution.

a. $\frac{3\pi}{10}$ in. b. $\frac{5\pi}{3}$ in. c. $\frac{10\pi}{3}$ in. d. $\frac{3\pi}{5}$ in.

20. Write the equation for simple harmonic motion given that the amplitude is 11 centimeters and the period is 2 seconds. Assume zero displacement at $t = 0$.

a. $y = 11\sin\pi t$ b. $y = 11\sin\frac{\pi}{2}t$ c. $y = 11\cos 2\pi t$ d. $y = 11\sin\frac{\pi}{4}t$

1. Convert the decimal degree measure 112.24° to the DMS system to the nearest minute.

a. 112°14' b. 112°24' c. 112°34' d. 112°18'

2. Find the supplement of the angle whose radian measure is $\frac{5\pi}{12}$. Express your answer in terms of π.

a. $\frac{5\pi}{12}$ b. $\frac{\pi}{12}$ c. $\frac{7\pi}{12}$ d. $\frac{5\pi}{9}$

3. Find the equivalent number of radians in $3\frac{3}{4}$ revolutions.

a. 15π b. $\frac{15\pi}{2}$ c. $\frac{15\pi}{4}$ d. 30π

4. Find the length of an arc that subtends a central angle of 75° in a circle of radius 20 inches.

a. $\frac{25}{3}$ in. b. $\frac{\pi}{24}$ in. c. 48π in. d. $\frac{25\pi}{3}$ in.

5. If θ is an acute angle of a right triangle and $\tan\theta = 5$, find $\sin\theta$.

a. $\frac{\sqrt{26}}{5}$ b. $\frac{5\sqrt{26}}{26}$ c. $\frac{5\sqrt{6}}{12}$ d. $\frac{\sqrt{26}}{26}$

6. If $\cos\theta = \frac{1}{2}$, $270° \leq \theta \leq 360°$, find $\cot\theta$.

a. $\frac{1}{2}$ b. -2 c. $-\frac{\sqrt{3}}{3}$ d. $-\frac{2\sqrt{3}}{3}$

7. Find the value of csc 72° to the nearest ten thousandths.

a. 0.9511 b. -1.5015 c. 1.0515 d. 1.5015

8. Find the exact value of $\tan\frac{\pi}{3}\cos\frac{3\pi}{4} - \sin\frac{7\pi}{6}$

a. $\frac{-\sqrt{6}+1}{2}$ b. $\frac{\sqrt{6}+\sqrt{3}}{2}$ c. $\frac{\sqrt{6}-\sqrt{4}}{4}$ d. $\frac{\sqrt{3}+1}{2}$

9. Find $W\left(\frac{5\pi}{6}\right)$.

a. $\left(\frac{1}{2},-\frac{\sqrt{3}}{2}\right)$ b. $\left(-\frac{1}{2},-\frac{\sqrt{3}}{2}\right)$ c. $\left(-\frac{\sqrt{3}}{2},\frac{1}{2}\right)$ d. $\left(\frac{\sqrt{2}}{2},\frac{1}{2}\right)$

10. Express $\frac{1}{1-\cos t}+\frac{1}{1+\cos t}$ in terms of the cosecant function.

a. $2\csc^2 t$ b. $\frac{2}{\csc^2 t}$ c. $1-\csc^2 t$ d. $1+\csc^2 t$

11. State the period of $y=\frac{1}{2}\tan 8x$.

a. $\frac{\pi}{8}$ b. 8π c. $\frac{\pi}{4}$ d. 4π

12. State the amplitude, period, and phase shift for the equation $y=-\frac{3}{4}\sin\left(3x-\frac{\pi}{2}\right)$.

a. $\frac{3}{4},\frac{2\pi}{3},\frac{\pi}{6}$ b. $\frac{3}{4},\frac{3\pi}{2},\frac{\pi}{3}$ c. $\frac{4}{3},\frac{2\pi}{3},-\frac{\pi}{6}$ d. $\frac{3}{4},6\pi,-\frac{\pi}{2}$

13. Graph $y=-5\cos\frac{x}{4}$.

(a)
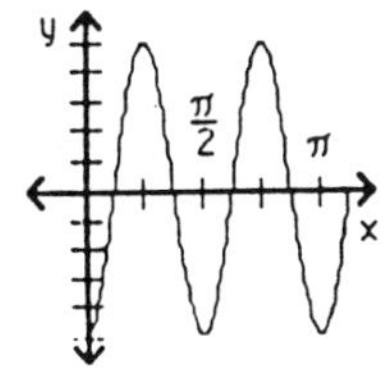

(b)
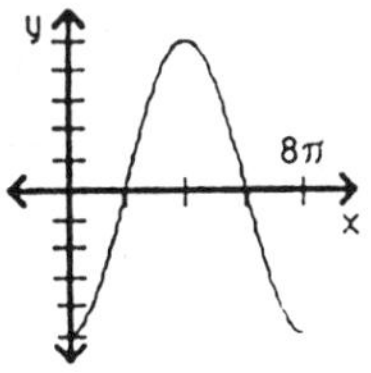

(c)
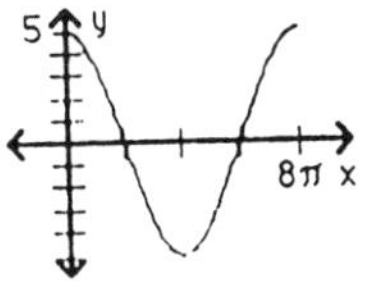

(d)
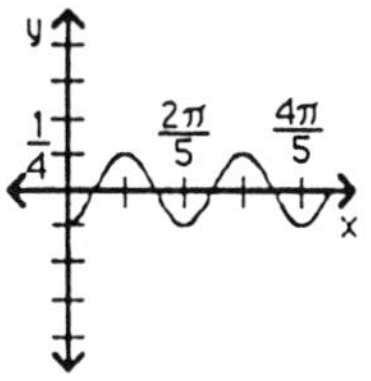

14. Graph $y = -\csc\frac{x}{4}$.

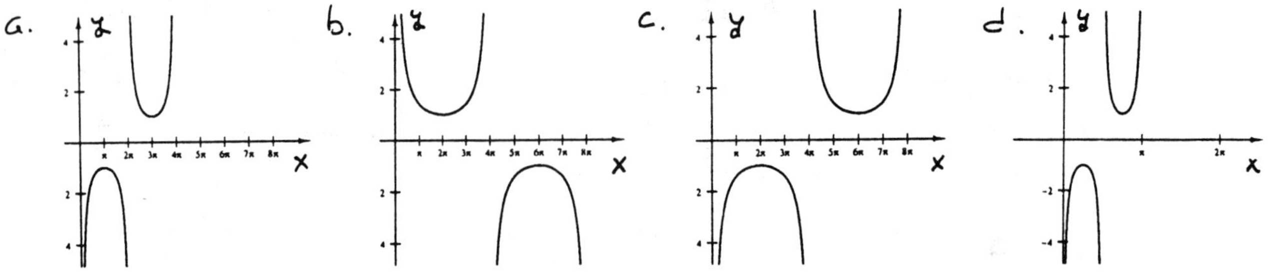

15. Graph $y = 2\tan 4x$.

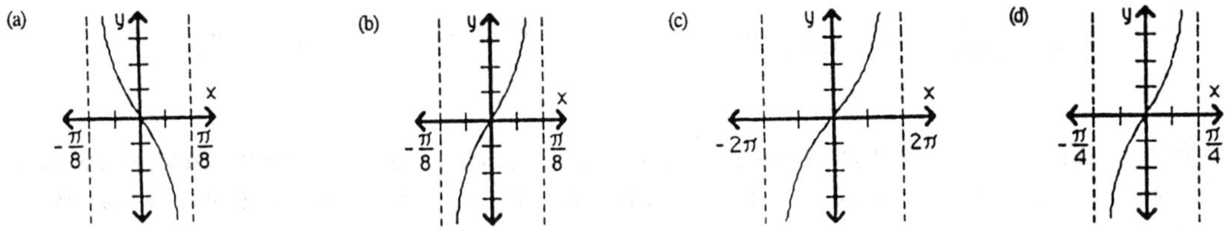

16. Graph $y = \frac{3}{5}\sin\left(x + \frac{\pi}{2}\right) + 4$.

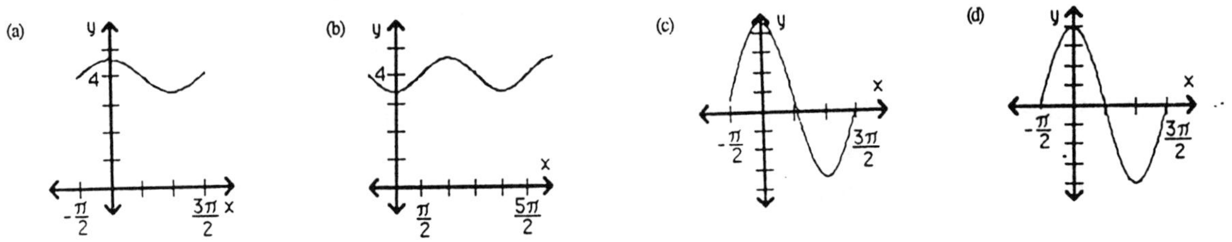

17. Graph $y = 2 - 3\cos\frac{x}{2}$.

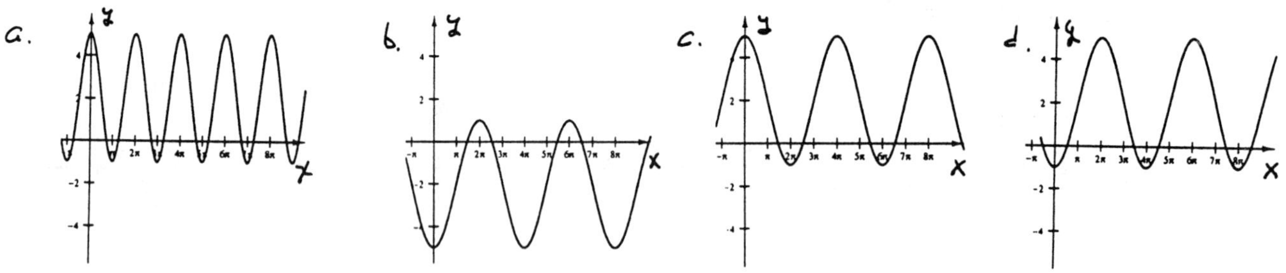

18. Graph $y = x - \sin x$.

(a)

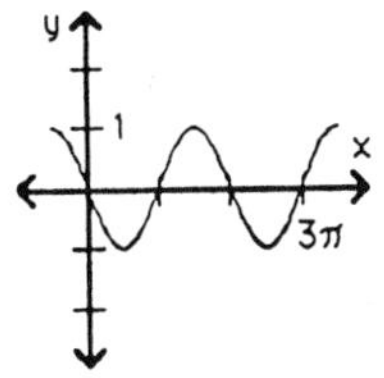

(b)

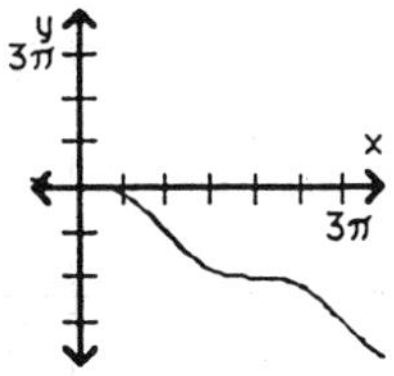

(c)

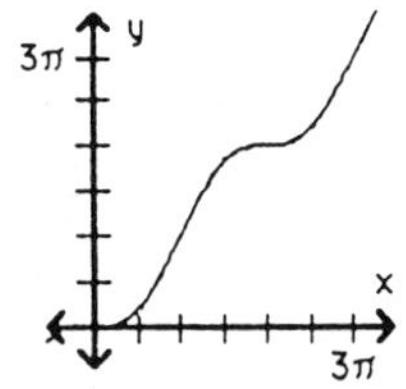

(d)

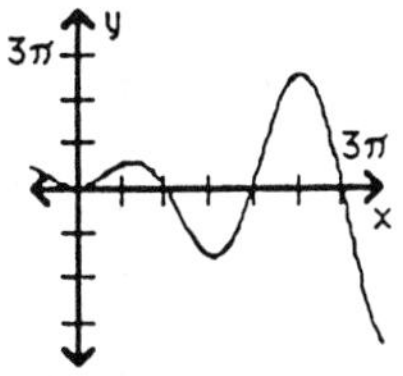

19. The angle of elevation from point A to the top of a building is 27°. At point B, 50 feet from A and on a line through the base of the building and A, the angle of elevation is 40°. Find the height of the building.

a. 25.5 ft b. 64.9 ft c. 55.6 ft d. 73.2 ft

20. Write the equation for simple harmonic motion, given that the amplitude is 4 feet and the period is 3 seconds. Assume maximum displacement at time $t = 0$.

a. $y = 4\cos 4\pi t$ b. $y = 4\cos\frac{3}{2}\pi t$

c. $y = 4\cos\frac{2}{3}\pi t$ d. $y = 4\sin 4\pi t$

1. Verify the identity $\cos x \cot x + \sin x = \csc x$.

2. Verify the identity $\dfrac{\cos x \tan x + \sin x}{\cos x} = 2 \tan x$.

3. Verify the identity $\dfrac{\cos^2 x}{1 - \sin x} = 1 + \sin x$.

4. Verify the identity $\dfrac{\csc x \tan x - \sec x}{\sec x} = 0$.

5. Find the exact value of $\cos 75^\circ$.

6. Given $\sin \alpha = \dfrac{5}{13}$ in quadrant II and $\cos \beta = -\dfrac{7}{25}$ in quadrant III, find $\sin(\alpha - \beta)$.

7. Verify the identity $\tan(2\pi - \theta) = -\tan \theta$.

8. Write $2 \sin \dfrac{x}{2} \cos \dfrac{x}{2}$ in terms of a single trigonometric function.

9. Find the exact value of $\cos 2\theta$ given that $\sin \theta = \dfrac{5}{13}$ in quadrant II.

10. Verify the identity $\cot \dfrac{x}{2} = \csc x + \cot x$.

11. Verify the identity $\dfrac{2\tan\frac{1}{2}x}{1+\tan^2\frac{1}{2}x} = \sin x$.

12. Find the exact value of $\cos 15^\circ \cos 75^\circ$.

13. Write $y = \frac{1}{2}\sin x - \frac{\sqrt{3}}{2}\cos x$ in the form $y = k\sin(x+\alpha)$, where α is measured in radians.

14. Find the exact degree value for $\sin^{-1}\left(-\frac{1}{2}\right)$.

15. Find the approximate radian measure for $\cos^{-1}(-0.2347)$.

16. Evaluate $\sin\left(\tan^{-1}\frac{15}{8}\right)$.

17. Graph $y = \cos^{-1}(x+1)$.

18. Solve the equation $6 + 6\sin x = 9$, where $0^\circ \le x < 360^\circ$.

19. Solve the equation $2\sin^2 x - 5\cos x + 5 = 0$, where $0^\circ \le x < 360^\circ$.

20. Solve the equation $4\sin x\cos x - 2\sqrt{3}\sin x - 2\cos x + \sqrt{3} = 0$, where $0 \le x < 2\pi$.

1. Verify the identity $\frac{\sin x}{1+\sin x} - \frac{\sin x}{1-\sin x} = -2\tan^2 x$.

2. Verify the identity $\tan x + \cot x = \sec x \csc x$.

3. Verify the identity $\sec^2 x \sin^2 x - \sec^2 x = -1$.

4. Verify the identity $\csc x(\cos x + \sin x) = \cot x + 1$.

5. Find the exact value of $\sin 255^\circ$.

6. Given $\sin\alpha = -\frac{3}{5}$ in quadrant III and $\cos\beta = \frac{8}{17}$ in quadrant I, find $\cos(\alpha+\beta)$.

7. Verify the identity $\sin\left(\frac{3\pi}{2}+\theta\right) = -\cos\theta$.

8. Write $\cos 5x \cos 3x - \sin 5x \sin 3x$ in terms of a single trigonometric function.

9. Find the exact value of $\sin 2\theta$ given that $\cos\theta = -\frac{5}{13}$ in quadrant II.

10. Verify the identity $\frac{1-\tan^2 \frac{x}{2}}{1+\tan^2 \frac{x}{2}} = \cos x$.

11. Verify the identity $\sin 2x = \dfrac{2\tan x}{1+\tan^2 x}$.

12. Find the exact value of $\cos 15^\circ \sin 105^\circ$.

13. Write $y = \sqrt{3}\sin x + \cos x$ in the form $y = k\sin(x+\alpha)$, where α is measured in radians.

14. Find the exact degree value for $\sin^{-1}\left(-\dfrac{\sqrt{2}}{2}\right)$.

15. Find the approximate radian measure for $\tan^{-1} 1.4222$.

16. Evaluate $\cos\left(\sin^{-1}\dfrac{7}{25}\right)$.

17. Graph $y = \sin^{-1} x + 2$.

18. Solve the equation $2 + \sqrt{3}\tan x - 1 = 0$, where $0^\circ \le x < 360^\circ$.

19. Solve the equation $2\sin^2 x - 11\cos x - 11 = 0$, where $0 \le x < 2\pi$.

20. Solve the equation $4\sin x\cos x - 2\sqrt{2}\sin x + 2\cos x - \sqrt{2} = 0$, where $0^\circ \le x < 360^\circ$.

1. Verify the identity $(1+\tan x)^2 = \sec^2 x + 2\tan x$.

2. Verify the identity $\tan^2 x \sin^2 x - \tan^2 x = -\sin^2 x$.

3. Verify the identity $\dfrac{1+\csc x}{\sec x} = \cos x + \cot x$.

4. Verify the identity $\tan^2 x - \sin^2 x = \tan^2 x \sin^2 x$.

5. Find the exact value of $\cos\left(\dfrac{5\pi}{4} + \dfrac{\pi}{3}\right)$.

6. Given $\sin\alpha = -\dfrac{3}{5}$ in quadrant III and $\cos\beta = -\dfrac{12}{13}$ in quadrant II. find $\cos(\alpha - \beta)$.

7. Verify the identity $\sin\left(\dfrac{\pi}{3} + x\right) = \dfrac{\sqrt{3}}{2}\cos x + \dfrac{1}{2}\sin x$.

8. Write $\sin 2\theta \cos\theta + \cos 2\theta \sin\theta$ in terms of a single trigonometric function.

9. Find the exact value of $\sin 2\theta$ given that $\cos\theta = \dfrac{3}{5}$ and θ is in quadrant IV.

10. Verify the identity $2\sec x \sin^2 \dfrac{x}{2} = \sec x - 1$.

11. Verify the identity $\cos 2x + 2\sin^2 x = 1$.

12. Find the exact value of $\cos 255^\circ \cos 75^\circ$.

13. Write $y = -\sqrt{3}\sin x + \cos x$ in the form $y = k\sin(x + \alpha)$, where α is measured in radians.

14. Find the exact degree value for $\cos^{-1}\left(-\frac{\sqrt{2}}{2}\right)$.

15. Find the approximate radian measure for $\sin^{-1} 0.4772$.

16. Evaluate $\cos^{-1}\left(\tan\frac{7\pi}{6}\right)$.

17. Graph $y = \cos^{-1}(x+2)$.

18. Solve the equation $4\sqrt{3} + 2\cos x = 3\sqrt{3}$, where $0^\circ \le x < 360^\circ$.

19. Solve the equation $2\sin^2 x - 3\cos x + 3 = 0$, where $0^\circ \le x < 360^\circ$.

20. Solve the equation $\sec x - 2\cos x - \tan x = 0$, where $0 \le x < 2\pi$.

1. Simplify the expression $\sec x - \sin x \tan x$.

a. $\sec^2 x$ b. cos x c. 1 - sin x d. $\cos^2 x$

2. Simplify the expression $\dfrac{\cos x \tan x - \sin x}{\sin x}$.

a. sin x -1 b. sin x c. 0 d. $\csc^2 x$

3. Simplify the expression $\dfrac{\sin x}{1+\sin x} - \dfrac{\sin x}{1-\sin x}$.

a. $-2\sin^2 x$ b. -2 c. $\sec^2 x$ d. $-2\tan^2 x$

4. Simplify the expression $\sin x + \csc x \cos^2 x$.

a. csc x b. sec x c. sin x + cos x d. 2 sin x

5. Find the exact value of $\sin\left(\frac{7\pi}{4} - \frac{\pi}{3}\right)$.

a. $\dfrac{-\sqrt{2}-\sqrt{6}}{4}$ b. $\dfrac{-\sqrt{2}+\sqrt{6}}{4}$ c. $\dfrac{\sqrt{2}+\sqrt{6}}{4}$ d. $\dfrac{\sqrt{2}-\sqrt{6}}{4}$

6. Given $\cos\alpha = -\frac{7}{25}$ in quadrant III and $\cos\beta = \frac{8}{17}$ in quadrant I, find $\cos(\alpha - \beta)$.

a. $-\dfrac{304}{425}$ b. $-\dfrac{416}{425}$ c. $\dfrac{304}{425}$ d. $-\dfrac{56}{425}$

7. Simplify the expression $\sin(30^\circ + \theta) - \cos(60^\circ + \theta)$.

a. $\frac{1}{2}\cos\theta$ b. $\frac{\sqrt{3}}{2}\sin\theta$ c. $\sqrt{3}\sin\theta$ d. $-\frac{1}{2}\cos\theta$

8. Write $\cos^2 4\theta - \sin^2 4\theta$ as a single trigonometric function.

a. $\cos 4\theta$ b. 1 c. $\sin 8\theta$ d. $\cos 8\theta$

9. Find the exact value of $\sin\frac{\alpha}{2}$ given $\cos\alpha = -\frac{7}{24}$ in quadrant III.

a. $\frac{\sqrt{93}}{12}$ b. $\frac{\sqrt{51}}{12}$ c. $-\frac{\sqrt{93}}{12}$ d. $-\frac{\sqrt{51}}{12}$

10. Simplify the expression $\cos 2\theta + 2\sin^2\theta$.

a. $\cos\theta - \sin\theta$ b. 1 c. 0 d. $\cos\theta + \sin\theta$

11. Write $2\cos^2 4\theta - 1$ as a single trigonometric expression.

a. $\cos 2\theta$ b. $\cos 8\theta$ c. 1 d. $\sin 8\theta$

12. Find the exact value of $\cos 195^\circ \sin 75^\circ$.

a. $\frac{2+\sqrt{3}}{4}$ b. $\frac{-2-\sqrt{3}}{4}$ c. $\frac{-2+\sqrt{3}}{2}$ d. $\frac{-2-\sqrt{3}}{2}$

13. Write $y = -\frac{1}{4}\sin x - \frac{\sqrt{3}}{4}\cos x$ in the form $y = k\sin(x+\alpha)$, where the measure of α is in degrees.

a. $y = \frac{1}{2}\sin(x + 240^\circ)$ b. $y = \frac{1}{2}\sin(x + 210^\circ)$

c. $y = 2\sin(x + 240^\circ)$ d. $y = \frac{1}{4}\sin(x + 120^\circ)$

14. Find the exact degree value for $\sin^{-1}\frac{\sqrt{3}}{2}$.

a. 30° b. 60° c. 120° d. -60°

15. Find the approximate radian measure for $\tan^{-1} 2.4107$.

a. 1.178 b. 2.4107 c. -0.897 d. 0.415

16. Evaluate $\sin\left(\cos^{-1}\frac{7}{25}\right)$.

a. $\frac{25}{7}$ b. $\frac{25}{24}$ c. $\frac{7}{24}$ d. $\frac{24}{25}$

17. Graph $y = \sin^{-1}(x+1)$.

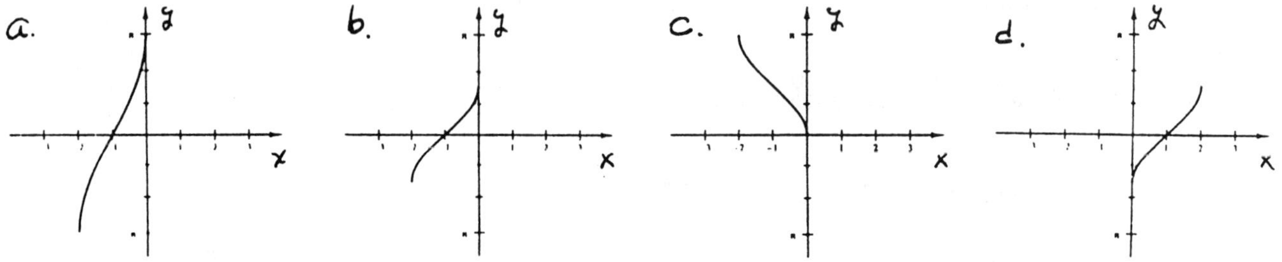

18. Solve the equation $4 + 6\tan x = -2$, where $0^\circ \le x < 360^\circ$.

a. 45°, 225° b. 45°, 315° c. 135°, 315° d. 150°, 330°

19. Solve the equation $2\sin^2 x + 9\cos x + 9 = 0$, where $0 \le x < 2\pi$./

a. $\frac{3\pi}{2}$ b. π c. $0, \frac{\pi}{2}$ d. 0

20. Solve the equation $4\cos^2\frac{x}{2} = 1$, where $0 \le x < 2\pi$.

a. $\frac{2\pi}{3}, \frac{4\pi}{3}$ b. $\frac{\pi}{3}, \frac{2\pi}{3}$ c. $\frac{\pi}{3}, \frac{2\pi}{3}, \frac{\pi 4}{3}$ d. $\frac{\pi}{3}$

1. Simplify the expression $\frac{1 + \sin x}{\cos x}$

a. $\cos x$ b. $1 + \tan x$ c. $\sec x + \sin x$ d. $\sec x + \tan x$

2. Simplify the expression $(\cot x + 1)(\cot x - 1) + 2$.

a. $\cot^2 x + 3$ b. $\csc^2 x + 2$ c. $\csc^2 x$ d. $2 \csc^2 x$

3. Simplify the expression $\tan^2 x + \cos^2 x + \frac{1}{\csc^2 x}$.

a. $\cos^2 x$ b. $\sec^2 x$ c. $1 - \sec^2 x$ d. $-\sec^2 x$

4. Simplify the expression $\sec x + \tan x$.

a. $\frac{1 + \sin x}{\cos x}$ b. $\frac{1}{\sin^2 x}$ c. $\frac{1 - \sin x}{\sin x}$ d. $\frac{1 + \tan x}{\sec x}$

5. Find the exact value of $\cos\left(\frac{7\pi}{4} - \frac{2\pi}{3}\right)$.

a. $\frac{-\sqrt{2} + \sqrt{6}}{4}$ b. $\frac{\sqrt{2} + \sqrt{6}}{4}$ c. $\frac{-\sqrt{2} - \sqrt{6}}{4}$ d. $\frac{\sqrt{6} - \sqrt{2}}{4}$

6. Given $\sin \alpha = \frac{15}{17}$ in quadrant I and $\cos \beta = \frac{4}{5}$ in quadrant IV, find $\cos(\alpha + \beta)$.

a. $\frac{77}{85}$ b. $-\frac{77}{85}$ c. $-\frac{13}{85}$ d. $\frac{13}{85}$

7. Simplify the expression $\cos\left(\frac{3\pi}{2} + \theta\right)$.

a, $\sin \theta$ b. $\cos \theta$ c. $-\sin \theta$ d. $-\cos \theta$

8. Write $\frac{2 \tan 5\theta}{1 - \tan^2 5\theta}$ as a single trigonometric function.

a. $\tan 10\theta$ b. $\tan 5\theta$ c. $\tan \frac{5\theta}{2}$ d. $-\tan 5\theta$

9. Write $1 - 2\sin^2\frac{\alpha}{2}$ as a single trigonometric function.

a. $\sin 2\alpha$ b. $\cos\alpha$ c. $\cos 2\alpha$ d. $\cos\frac{\alpha}{2}$

10. Find the value of $\sin\frac{\alpha}{2}$ given $\cos\alpha = \frac{2}{3}$ in quadrant IV.

a. $-\frac{\sqrt{5}}{6}$ b. $\frac{\sqrt{6}}{6}$ c. $-\frac{\sqrt{6}}{3}$ d. $-\frac{5\sqrt{6}}{6}$

11. Simplify the expression $\frac{\tan x + \sin x}{2\tan x}$.

a. $\cos^2\frac{x}{2}$ b. $\frac{1 - \cos x}{2}$ c. $\sin^2\frac{x}{2}$ d. $\cos^2 x$

12. Find the exact value of $\cos 105^\circ \sin 15^\circ$.

a. $\frac{2 - \sqrt{3}}{4}$ b. $\frac{\sqrt{3} - 1}{4}$ c. $\frac{\sqrt{3} - 2}{4}$ d. $\frac{-\sqrt{3} - 1}{4}$

13. Write $y = 5\sqrt{2}\sin x - 5\sqrt{2}\cos x$ in the form $y = k\sin(x + \alpha)$, where α is measured in radians.

a. $y = 10\sin(x + 135^\circ)$ b. $y = 100\sin(x + 135^\circ)$

c. $y = 5\sqrt{2}\sin(x + 315^\circ)$ d. $y = 10\sin(x + 315^\circ)$

14. Find the exact degree value for $\sin^{-1}\left(\frac{\sqrt{3}}{2}\right)$.

a. -60° b. 60° c. 120° d. 240°

15. Find the approximate radian measure for $\sin^{-1}(-0.3349)$.

a. -0.3349 b. -0.3415 c. -19.566 d. 0.4315

16. Evaluate $\cos\left(\sin^{-1}\frac{\sqrt{2}}{2}\right)$.

a. $\frac{\sqrt{3}}{2}$ b. $\frac{\sqrt{2}}{2}$ c. $\frac{1}{2}$ d. $-\frac{\sqrt{2}}{2}$

17. Graph $y = \cos^{-1}(x - 2)$.

a. 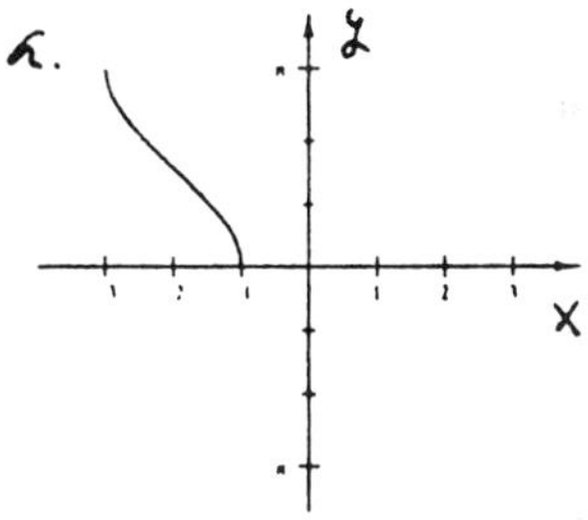

b. 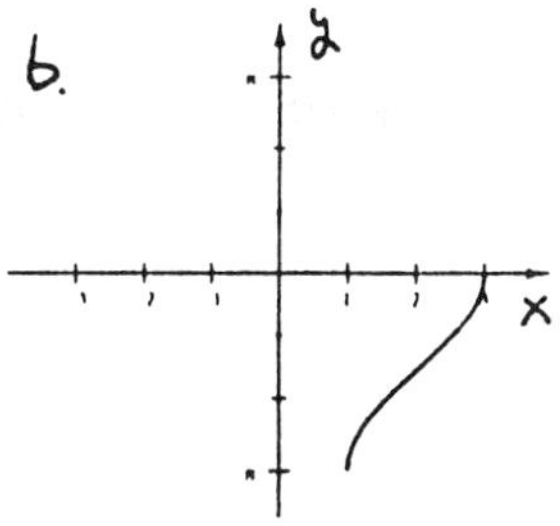

c. 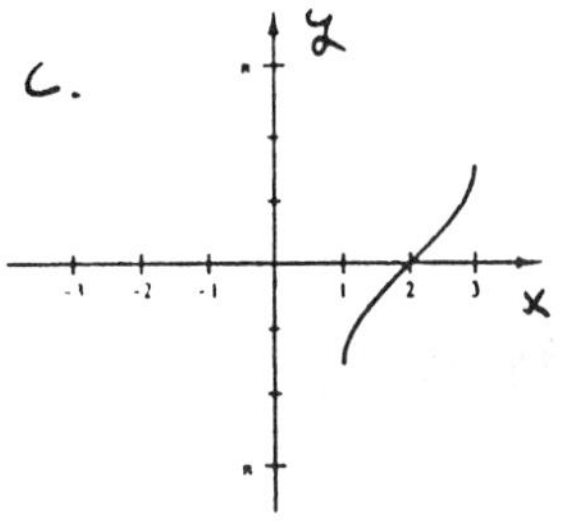

d. 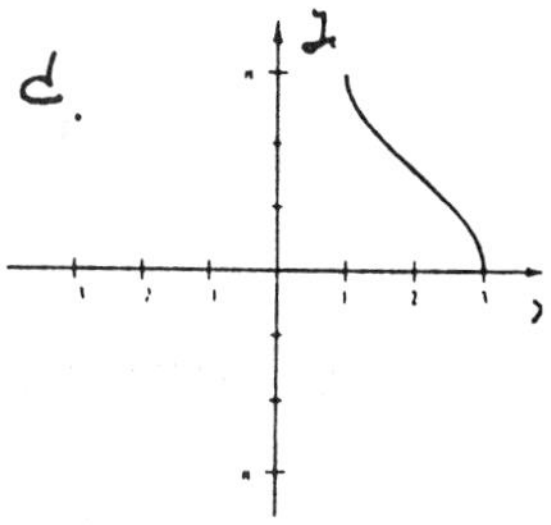

18. Solve the equation $4 \sin^2 x = 3$, where $0^\circ \le x < 360^\circ$.

a. 60°, 300°

b. 120°, 240°

c. 30°, 150°, 210°, 330°

d. 60°, 120°, 240°, 300°

19. Solve the equation $\sin 2x - \cos x = 0$, where $0^\circ \le x < 360^\circ$.

a. 30°, 90°, 180°, 330°

b. 30°, 90°, 150°, 270°

c. 60°, 90°, 180°, 300°

d. 60°, 120°, 150°, 180°

20. Solve the equation $2 \sin^3 x = \sin x$ where $0 \le x < 2\pi$.

a. $0, \frac{\pi}{4}, \frac{3\pi}{4}, \pi, \frac{5\pi}{4}, \frac{7\pi}{4}$

b. $0, \frac{\pi}{4}, \frac{3\pi}{4}, \pi$

c. $\frac{\pi}{4}, \frac{3\pi}{4}, \frac{5\pi}{4}, \frac{7\pi}{4}$

d. $0, \pi$

Chapter 7 Chapter Test Form F

1. Simplify the expression $\left(1 + \sqrt{2}\sin x\right)\left(1 - \sqrt{2}\sin x\right)$.

a. $2\sin^2 x$ b. $\cos 2x$ c. $1 + 2\sin^2 x$ d. $1 + \cos^2 x$

2. Simplify the expression $\frac{\cos x}{1 + \cos x} - \frac{\cos x}{1 - \cos x}$.

a. 0 b. $2\cot x \sec x$ c. $-2\cot^2 x$ d. $2\cot x$

3. Simplify the expression $\frac{1 - \cos x}{\sin x}$.

a. $\sin x$ b. $-\sin x$ c. $\csc x - \cos x$ d. $\csc x - \cot x$

4. Simplify the expression $\tan\left(x + \frac{\pi}{4}\right)$.

a. $\frac{\tan x}{1 - \tan x}$ b. $-\frac{\tan x + 1}{\tan x}$ c. $\frac{\tan x + 1}{1 - \tan x}$ d. 1

5. Find the exact value of $\cos 255^\circ$.

a. $\frac{-\sqrt{6} + \sqrt{2}}{4}$ b. $\frac{-\sqrt{6} - \sqrt{2}}{4}$ c. $\frac{-\sqrt{6} + 2}{4}$ d. $\frac{\sqrt{6} + \sqrt{2}}{2}$

6. Given $\cos\alpha = -\frac{7}{25}$ in quadrant II and $\sin\beta = -\frac{12}{13}$ in quadrant IV, find $\sin(\alpha + \beta)$.

a. $-\frac{2}{325}$ b. $\frac{323}{325}$ c. $\frac{36}{325}$ d. $-\frac{253}{325}$

7. Simplify the expression $\cos\left(\frac{3\pi}{2} - \theta\right)$.

a. $\sin\theta$ b. $-\sin\theta$ c. $\cos\theta$ d. $-\cos\theta$

8. Write $2\sin 5\theta \cos 5\theta$ in terms of a single trigonometric function.

a. $\sin 5\theta$ b. $\sin 10\theta$ c. $\sin\frac{5}{2}\theta$ d. $\cos 10\theta$

9. Find the exact value of $\sin \frac{\alpha}{2}$ given that $\cos \alpha = -\frac{3}{5}$ and α is in quadrant II.

a. $\frac{24}{25}$ b. $\frac{2\sqrt{5}}{5}$ c. $\frac{\sqrt{5}}{10}$ d. $\frac{\sqrt{5}}{5}$

10. Simplify the expression $\sin^2 \frac{\theta}{2} \cos^2 \frac{\theta}{2}$.

a. $\frac{\cos^2 \theta}{4}$ b. $4 \sin^2 \theta$ c. $\frac{\sin^2 \theta}{4}$ d. $4 \cos^2 \theta$

11. Simplify the expression $\sec^2 x \cos 2x$.

a. $1 - \tan^2 x$ b. $\cos^2 x - 1$ c. $1 + \sin^2 x$ d. $\frac{\tan x + 1}{\tan x - 1}$

12. Find the exact value of $\sin 225^\circ \sin 15^\circ$.

a. $\frac{-\sqrt{3} - 1}{4}$ b. $\frac{-\sqrt{3} + 1}{4}$ c. $\frac{-\sqrt{3} + 1}{2}$ d. $\frac{\sqrt{3} - 1}{2}$

13. Write $y = -4 \sin x + 4\sqrt{3} \cos x$ in the form $y = k \sin(x + \alpha)$, where the measure of α is in degrees.

a. $y = 8 \sin(x + 30^\circ)$ b. $y = 8 \sin(x + 240^\circ)$

c. $y = 8 \sin(x + 150^\circ)$ d. $y = 8 \sin(x + 120^\circ)$

14. Find the exact degree value for $\sin^{-1}\left(-\frac{\sqrt{2}}{2}\right)$.

a. 120° b. 135° c. -45° d. 45°

15. Find the approximate degree measure for $\tan^{-1}(-3.4221)$.

a. -73.7° b. -1.28 c. 73.7° d. 37.7°

16. Evaluate $\tan\left(\tan^{-1} \frac{2}{3}\right)$.

a. $\frac{3}{2}$ b. 33.69 c. $\frac{2}{3}$ d. $\frac{4\sqrt{13}}{13}$

17. Graph $y = \sin^{-1}\left(\frac{x}{2} - 1\right)$.

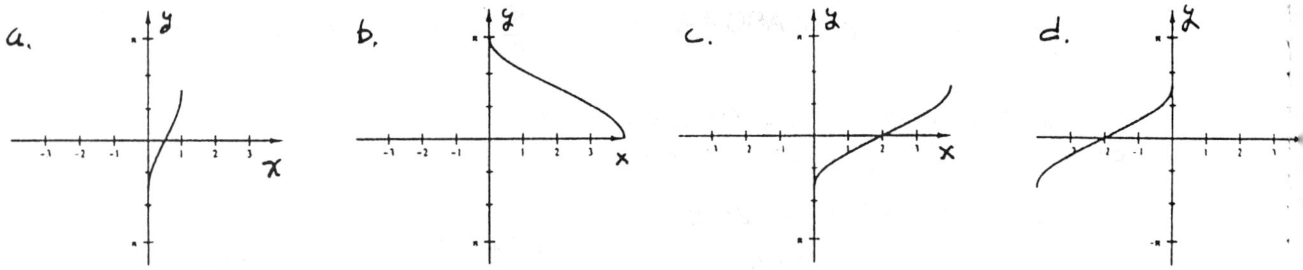

18. Solve the equation $2\sin^2 x - 3\sin x + 1 = 0,\ 0 \le x < 2\pi$.

a. $\frac{\pi}{6}, \frac{\pi}{2}, \frac{5\pi}{6}$ b. $\frac{\pi}{3}, \frac{\pi}{2}, \frac{3\pi}{2}$ c. $\frac{\pi}{6}, \frac{\pi}{3}, \frac{5\pi}{3}$ d. $\frac{\pi}{6}, \frac{\pi}{2}, \frac{5\pi}{6}, \frac{3\pi}{2}$

19. Solve the equation $\sin^2 x - \cos^2 x + 1 = 0,\ 0 \le x < 2\pi$.

a. $0, \pi$ b. $\frac{\pi}{2}, \frac{3\pi}{2}$ c. $0, 2\pi$ d. $\frac{\pi}{2}$

20. Solve the equation $2\sin^2 x - \sqrt{3}\sin x - 2\sin x + \sqrt{3} = 0,\ 0^\circ \le x < 360^\circ$.

a. $60^\circ, 90^\circ$ b. 90°

c. $60^\circ, 90^\circ\ 120^\circ$ d. $60^\circ, 90^\circ, 120^\circ, 240^\circ, 300^\circ$

1. Solve triangle ABC if A = 62°, B = 51°, and c = 19.

2. Find B in triangle ABC if A = 34°, b = 7, and a = 11.

3. In triangle ABC, B = 71°, a = 4, and c = 7. Find side b.

4. In triangle ABC, a = 26, b = 15, and c =17. Find Angle A.

5. Given angle C = 61°, side a = 55, and side b = 29. Find the area of triangle ABC.

6. Given angle A = 35°·, angle C = 80°, and side a = 15, find the area of triangle ABC.

7. Given side a = 12, side b = 22, and side c = 18, find the area of triangle ABC.

8. A vector has a magnitude of 16 and direction of 240°. Write an equivalent vector in the form $v = a_1 i + a_2 j$.

9. Find 2u – 9v given the vectors u = 6i – 7j and v = i + 11j.

10. Find the dot product of u = –4i + j and v = 3i + 7j.

11. Find the angle to the nearest degree between the vectors u = ⟨2, 3⟩ and v = ⟨6, 4⟩.

12. Write z = –6 + 6i in trigonometric form.

13. Write 3 cis 225° in standard form.

14. Multiple: 5 cis 155° • 8 cis 55°. Write the answer in standard form.

15. Divide: $\frac{45 \text{ cis } 75°}{9 \text{ cis } 145°}$. Write the answer in trigonometric form.

16. Find $(1 - i)^8$. Write the answer in standard form.

17. Find the three cube roots of 8i. Write your answer in standard form.

18. A motorist drove 86 miles at a heading of 146°. The motorist then drove 74 miles at a heading of 236°. Find the distance of the motorist from the starting point.

19. One ship leaves a port at 10:00 A.M. traveling at 14 mph at a heading of 48°. At 12:00 noon another ship leaves the port traveling at 20 mph at a heading of 160°. Find the distance between the ships at 2:00 P.M.

20. A triangular commercial piece of real estate is priced at $12.50 per square foot. Find the cost of the lot measuring 90 ft by 140 ft by 120 ft.

1. Solve triangle ABC if B = 65°, C = 38°, and b = 24.

2. Find C in triangle ABC if A = 56°, c = 14 and a = 8.0.

3. In triangle ABC, B = 43°, a = 8, and c = 14. Find side b.

4. In triangle ABC, a = 22, b = 18, and c = 14. Find angle B.

5. Given angle B = 22°, side a = 34, and side b = 45, find the area of triangle ABC.

6. Given A = 110°, angle C = 42°, and side a = 25, find the area of triangle ABC.

7. Given side a = 60, side b = 72, and side c = 54. find the area of triangle ABC.

8. A vector has magnitude 8.0 and direction 70°. Write an equivalent vector in the form $\mathbf{v} = a_1\mathbf{i} + a_2\mathbf{j}$.

9. Find $4\mathbf{u} - 6\mathbf{v}$ given the vectors $\mathbf{u} = 3\mathbf{i} - 7\mathbf{j}$ and $\mathbf{v} = 5\mathbf{i} + 2\mathbf{j}$.

10. Find the dot product of $\mathbf{u} = -3\mathbf{i} + 2\mathbf{j}$ and $\mathbf{v} = 5\mathbf{i} - \mathbf{j}$.

11. Find the angle to the nearest degree between the vectors $\mathbf{u} = \langle -2, 5 \rangle$ and $\mathbf{v} = \langle -1, 4 \rangle$

12. Write $z = -5\sqrt{3} + 5i$ in trigonometric form.

13. Write $z = 4(\cos 210° + i \sin 210°)$ in standard form.

14. Multiply: 6 cis 85° · 9 cis 312°. Write the answer in standard form.

15. Divide: $\frac{12 \text{ cis } 335°}{24 \text{ cis } 122°}$. Write the answer in trigonometric form.

16. Find $(1 - \sqrt{3}\,i)^8$. Write the answer in standard form.

17. Find the square roots of -64i. Write your answer in standard form.

18. A pilot is flying at a heading of 120° at 250 mph. A 55 mph wind is blowing from the southwest at a h of 36°. Find the ground speed and course of the plane.

19. A 45-pound force keeps a 550-pound object from sliding down an inclined ramp. Find the angle of th

20. Two ships left a port at the same time. One ship traveled at a speed of 15 mph at a heading of 55°. The ship traveled at a speed of 22 mph at a heading of 115°. Find the distance between the two ships after hours of travel.

Chapter 8 Chapter Test Form C

1. Solve triangle ABC if $A = 105°$, $B = 22°$, and $c = 18$.

2. Find A in triangle ABC if $B = 66°$, $a = 56$, and $b = 108$.

3. In triangle ABC, $B = 39°$, $a = 56$, and $c = 42$. Find side b.

4. In triangle ABC, $a = 42$, $b = 32$, and $c = 55$. Find angle C.

5. Given angle $B = 112°$, side $a = 12$, and side $c = 15$, find the area of triangle ABC.

6. Given angle $A = 44°$, angle $B = 82°$, and side $c = 25$, find the area of triangle ABC.

7. Given side $a = 9$, side $b = 12$, and side $c = 6$, find the area of triangle ABC.

8. A vector has a magnitude of 75 and direction of 320°. Write an equivalent vector in the form $\mathbf{v} = a_1\mathbf{i} + a_2\mathbf{j}$.

9. Find $-5\mathbf{u} + 8\mathbf{v}$ given the vectors $\mathbf{u} = \langle -4, 5\rangle$ and $\mathbf{v} = \langle 6, -3\rangle$.

10. Find the dot product of $\mathbf{u} = -3\mathbf{i} - 8\mathbf{j}$ and $\mathbf{v} = -4\mathbf{i} + 6\mathbf{j}$.

11. Find the angle to the nearest degree between the vectors $\mathbf{u} = \langle 3, 5\rangle$ and $\mathbf{v} = \langle -6, 2\rangle$

12. Write $z = 2\sqrt{2} - 2i$ in trigonometric form.

13. Write 10 cis 120° in standard form.

14. Multiply: 5 cis 330° · 3 cis 225°. Write the answer in standard form.

15. Divide: $\frac{15 \text{ cis } 25°}{25 \text{ cis } 140°}$. Write the answer in standard form.

16. Find $(2\sqrt{3} \ -2i)^6$. Write the answer in standard form.

17. Find the square roots of -36i. Write the answer in standard form.

18. A motorist drove 55 miles at a heading of 112°. The motorist then drove 35 miles at a heading of 202°. Find the distance of the motorist from the starting point.

19. Find the number of acres in a pasture whose shape is a triangle measuring 560 yards by 800 yards by 640 yards. (An acre is 4840 square yards.)

20. Two fire lookouts are located 15 miles apart. Lookout A is at a bearing of N35°W of lookout B. A fire was sighted at a bearing of N55°E from A and N30°E from B. Find the distance of the fire from lookout A.

1. Solve triangle ABC if $A = 53°$, $B = 79°$, and $c = 15$.

a. $C = 48°$, $a = 13$, $b = 22$

b. $C = 48°$, $a = 16$, $b = 20$

c. $C = 48°$, $a = 20$ $b = 16$

d. $C = 48°$, $a = 13$, $b = 20$

2. Find B in triangle ABC if $A = 34°$, $b = 16$, and $a = 19$.

a. 40°, 140°,

b. 28°, 152°

c. 28°

d. No solution

3. In triangle ABC, $B = 44°$, $a = 29$, and $c = 17$. Find side b.

a. 39

b. 420

c. 150

d. 21

4. In triangle ABC, $b = 30$, $a = 23$, and $c = 18$. Find angle A.

a. 129°

b. 29°

c. 52°

d. 151°

5. Given angle $A = 52°$, side $c = 96$, and side $b = 72$, find the area of triangle ABC.

a. 3560

b. 2700

c. 3600

d. 2840

6. Given angle $A = 64°$, angle $C = 23°$, and side $b = 15$, find the area of triangle ABC.

a. 35

b. 52

c. 40

d. 46

7. Given side $a = 23$, side $b = 29$, and side $c = 12$, find the area of triangle ABC.

a. 260

b. 130

c. 78

d. 147

8. A vector has a magnitude of 15 and direction 230°. Write an equivalent vector in the form $\mathbf{v} = a_1\mathbf{i} + a_2\mathbf{j}$.

a. $\mathbf{v} = 9.6\mathbf{i} - 11\mathbf{j}$

b. $\mathbf{v} = -9.6\mathbf{i} + 11\mathbf{j}$

c. $\mathbf{v} = -9.6\mathbf{i} - 11\mathbf{j}$

c. $\mathbf{v} = 9.6\mathbf{i} + 11\mathbf{j}$

9. Find $3\mathbf{u} - 7\mathbf{v}$ given the vectors $\mathbf{u} = 5\mathbf{i} + 11\mathbf{j}$ and $\mathbf{v} = 2\mathbf{i} - 3\mathbf{j}$.

a. $\mathbf{i} + 12\mathbf{j}$

b. $\mathbf{i} + 30\mathbf{j}$

c. $\mathbf{i} + 54\mathbf{j}$

d. $13\mathbf{i} + 30\mathbf{j}$

10. Find the dot product of **u** = 2i -7j and **v** = i + 8j.

a. -54 b. -13 c. 58 d. -12

11. Find the angle between the vectors **u** = ⟨3, -1⟩ and **v** = ⟨-4, 11⟩.

a. 38° b. 128° c. 142° d. 52°

12. Write $z = 6 - 6i\sqrt{3}$ in trigonometric form.

a. -12 cis 300° b. 144 cis 240° c. 12 cis 240° d. 12 cis 300°

13. Write 3 cis 180° in standard form.

a. -3i b. 3i c. 3 d. -3

14. Multiply: 10(cos 120° + i sin 120°) · 5(cos 90° + i sin 90°). Write the answer in standard form.

a. $25\sqrt{3} - i$ b. 50 - 50i c. $-25\sqrt{3} - 25i$ d. $5\sqrt{3} - 5i$

15. Divide: $\frac{60 \text{ cis } 240°}{15 \text{ cis } 170°}$. Write the answer in trigonometric form.

a. 4 cis 70° b. 4 cis 2° c. 45 cis 70° d. 45 cis 2°

16. Find $(\sqrt{3} - i)^4$. Write the answer in standard form.

a. $8 + 8i\sqrt{3}$ b. $-8 - 8i\sqrt{3}$ c. $-16 - 16i\sqrt{3}$ d. $-8 + 8i\sqrt{3}$

17. Find the square root of -25i.

a. $-5\sqrt{2} + 5i\sqrt{2}$

b. $-\frac{5\sqrt{2}}{2} + \frac{5\sqrt{2}}{2}i, \frac{5\sqrt{2}}{2} - \frac{5\sqrt{2}}{2}i$

c. $\frac{5\sqrt{2} - 5\sqrt{2}}{2}, \frac{5\sqrt{2} + 5\sqrt{2}}{2}$

d. $5\sqrt{2} - 5\sqrt{2}$

18. A 180-pound box is dragged 12 feet along a level floor. Find the work done if a force of 80-pounds at an angle If 32° is used.

a. 12,211 ft-lb b. 960 ft-lb c. 509 ft-lb d. 814 ft-lb

19. Two ships left a port at the same time. One ship traveled at a speed of 16 mph at a heading of 55°. The other ship traveled at a speed of 20 mph at a heading of 220°. Find the distance between the two ships after 4 hours of travel.

a. 124 mi b. 143 mi c. 155 mi d. 139 mi

20. Find the area of a parallelogram with sides of 12 ft and 16 ft ans with one angle of 66°.

a. 170 ft^2 b. 120 ft^2 c. 88 ft^2 d. 60 ft^2

Chapter 8 Chapter Test Form E

1. Solve triangle ABC if A = 82°, B = 70°, and c = 23.

   a. C = 28°, a = 49, b = 46
   b. C = 28°, a = 49, b = 22
   c. C = 28°, a = 32, b = 46
   d. C = 28°, a = 32, b = 46

2. Find B in triangle ABC if A = 23°, b = 19, and a = 11.

   a. no solution
   b. 42°
   c. 42°, 138°
   d. 35°

3. In triangle ABC, B = 86°, a = 18, and c = 12. Find side b.

   a. 22
   b. 21
   c. 438
   d. 19

4. In triangle ABC, a = 52, b = 17, and c = 40. Find angle A.

   a. 90°
   b. 127°
   c. 53°
   d. 37°

5. Given triangle C = 78°, side a = 49, and side b = 56, find the area of triangle ABC.

   a. 1340
   b. 571
   c. 285
   d. 2680

6. Given angle B = 35°, angle C = 72°, and side b = 18, find the area of triangle ABC.

   a. 214
   b. 257
   c. 305
   d. 269

7. Given side a = 22, side b = 30, and side c = 36, find the area of triangle ABC.

   a. 420
   b. 329
   c. 382
   d. 297

8. A vector has a magnitude of 26 and direction 310°. Write an equivalent vector in the form $\mathbf{v} = a_1\mathbf{i} + a_2\mathbf{j}$.

   a. $\mathbf{v} = 279\mathbf{i} + 136\mathbf{j}$
   b. $\mathbf{v} = 7\mathbf{i} - 20\mathbf{j}$
   c. $\mathbf{v} = 17\mathbf{i} - 20\mathbf{j}$
   d. $\mathbf{v} = 20\mathbf{i} - 17\mathbf{j}$

9. Find 3**u** - 7**v** given the vectors **u** = 5**i** + 11**j** and **v** = 2**i** - 3**j**.

a. **i** + 12**j** b. **i** + 30**j** c. **i** + 54**j** d. 13**i** + 30**j**

10. Find the dot product of **u** = ⟨4, 1⟩ and **v** = ⟨-2, 5⟩.

a. 13 b. 3 c. - 3 d. - 13

11. Find the angle to the nearest degree between the vectors **u** = 5**i** + 8**j** and **v** = -2**i** + 9**j**.

a. 45° b. 135° c. 71° d. 19°

12. Write z = 5i in trigonometric form.

a. 5 cos 270° b. 5 cis 90° c. -5 cis 90° d. 5 cis 180°

13. Write z = 3(cos 240° + i sin 240°) in standard form.

a. $-\frac{3}{2} + \frac{3\sqrt{3}}{2}i$ b. $-\frac{3\sqrt{3}}{2} - \frac{3}{2}i$ c. $-\frac{3}{2} - \frac{3\sqrt{3}}{2}i$ d. $-\frac{\sqrt{3}}{2} - \frac{1}{2}i$

14. Multiply: 16 cis 70° · 3 cis 120°. Write the answer in trigonometric form.

a. 48 cis 310° b. 48 cis 190° c. 13 cis 310° d. 13 cis 190°

15. Divide: $\frac{85 \text{ cis } 220°}{17 \text{ cis } 80°}$. Write the answer in standard form.

a. -3.8302 - 3.2139i b. 34 - 58.8897i

c. -3.8302 + 3.2139i d. -3.2139 + 3.8302i

16. Find $(-1 + i)^8$. Write the answer in standard form.

a. 16 b. 16i c. - 16 d. - 16i

17. Find the square roots of -144i. Write the answer in standard form.

a. $-6\sqrt{2} + 6i\sqrt{2}$, $-6\sqrt{2} - 6i\sqrt{2}$ b. $-6\sqrt{2} - 6i\sqrt{2}$, $6\sqrt{2} + 6i\sqrt{2}$

c. $6\sqrt{2} - 6i\sqrt{2}$, $-6\sqrt{2} - 6i\sqrt{2}$ d. $-6\sqrt{2} + 6i\sqrt{2}$, $6\sqrt{2} - 6i\sqrt{2}$

18. A motorist drove 52 miles at a heading of 152°. The motorist then drove 90 miles at a heading of 242°. Find the distance of the motorist from the starting point.

a. 142 mi b. 104 mi c. 73 mi d. 68 mi

19. A triangular industrial piece of real estate is priced at $7.40 per square foot. Find the cost of the lot measuring 420 feet by 510 feet by 360 feet.

a. $666,298 b. $552,958 c. $77,724 d. $432,560

20. On ship leaves a port at 12:00 noon traveling at 15 mph at a heading of 310°. At 1:00 P.M. another ship leaves the port traveling at 24 mph at a heading of 70°. Find the distance between the ships at 4:00 P.M..

a. 94 mi b. 114 mi c. 124 mi d. 76 mi

1. Solve triangle ABC if A = 35°, B = 78°, and c = 83.

a. C = 67°, a = 88, b = 52
b. C = 67°, a = 52, b = 88
c. C = 67°, a = 64, b = 77
d. C = 67°, a = 77, b = 64

2. Find A in triangle ABC if C = 125°, a = 18, and c = 42.

a. 21°
b. 15°
c. 28°
d. 42°

3. In triangle ABC A = 32°, b = 66, and c = 42. Find side a.

a. 26
b. 32
c. 38
d. 44

4. In tgriangle ABC, a = 37, b = 18, and c = 23. Find angle A.

a. 129°
b. 29°
c. 52°
d. 151°

5. Given angle C = 85°, side a = 27, and side b = 19, find the area of triangle ABC.

a. 22
b. 45
c. 256
d. 511

6. Given angle C = 105°, angle B = 19°, and side c = 54, find the area of triangle ABC.

a. 315
b. 382
c. 512
d. 407

7. Given side a = 10, side b = 14, and side c = 18, find the area of triangle ABC.

a. 35
b. 70
c. 46
d. 82

8. A vector has a magnitude of 31 and a direction 130°. Write an equivalent vector in the form $\mathbf{v} = a_1\mathbf{i} + a_2\mathbf{j}$.

a. -20i - 24j
b. 24i - 20j
c. -20i + 24j
d. 67i - 122j

9. Find 6**u** - 15**v** given the vectors **u** = 8i - 2j and **v** = i - 9j.

a. 33i - 57j
b. 53i - 21j
c. 53i + 33j
d. 43i + 33j

10. Find the dot product of **u** = **i** -5**j** and **v** =6**i** + 4**j**.

a. -14 b. 14 c. -1 d. 26

11. Find the angle to the nearest degree between the vectors **u** = ⟨-1, 4⟩ and **v** = ⟨6, 2⟩.

a. 4° b. 42° c. 68° d. 86°

12. Write $z = 4\sqrt{3} - 4i$ in trigonometric form.

a. 8 cos 330° b. 8 cis 180° c. 8 cis 30° d. cis 300°

13. Write $z = 9(\cos \pi + i \sin \pi)$ in standard form.

a. -9 b. 9i c. -9i d. 9

14. Multiply: 2(cos 80° + i sin 80°) · 5(cos 120° + i sin 120°). Write the answer in trigonometric form.

a. 10 cis 200° b. 7 cis 40° c. 10 cis 40° d. 7 cis 200°

15. Divide: $\frac{15 \text{ cis } 170°}{3 \text{ cis } 120°}$. Write the answer in standard form.

a. 3.8302 + 3.2139i b. -3.2139 - 3.8302i

c. 3.2139 + 3.8302i d. -3.8302 - 3.2139i

16. Find $(\sqrt{3} + 1)^6$. Write the answer in standard form.

a. 64 b. -64 c. 32 d. -32

17. Find the square roots of -100i. Write your answer in standard form.

a. $-5\sqrt{2} - 5i\sqrt{2}$, $-5\sqrt{2} - 5i\sqrt{2}$ b. $-5\sqrt{2} + 5i\sqrt{2}$, $5\sqrt{2} + 5i\sqrt{2}$
c. $-5\sqrt{2} + 5i\sqrt{2}$, $5\sqrt{2} - 5i\sqrt{2}$ c. $-5\sqrt{2} - 5i\sqrt{2}$, $5\sqrt{2} + 5i\sqrt{2}$

18. A motorist drove 95 miles at a heading of 139°. The motorist then drove 82 miles at a heading of 229°. Find the distance of the motorist from the starting point.

a. 48 mi b. 177 mi c. 64 mi d. 125 mi

19. A plane is flying at an airspeed of 300 mph at a heading of 140°. A wind of 50 mph is blowing from the west. Find the ground speed of the plane.

a. 312 mph b. 318 mph c. 328 mph d. 334 mph

20. A triangular commercial piece of real estate is priced at $8.60 per square foot. Find the cost of the lot measuring 230 feet by 190 feet by 300 feet.

a. $207,678 b. $167,988 c. $187,898 d. $257,788

Chapter 9 Chapter Test Form A

1. Find the vertex, focus, and directrix of the parabola given by the equation $x^2 = -\frac{1}{8}y$.

2. Find the vertex, focus, and directrix of the parabola given by the equation $x^2 - 10x - 8y + 33 = 0$.

3. Find the equation in standard form of the parbola with directrix $x = 5$ and focus $(11, -7)$.

4. Sketch the graph of the parabola with focus $(0, 3)$ and directrix $y = 5$.

5. Find the vertices and foci of the ellipse given by the equation $\frac{x^2}{25} + \frac{y^2}{49} = 1$.

6. Sketch the graph of $\frac{x^2}{100} + \frac{y^2}{16} = 1$.

7. Find the vertices and foci of the ellipse given by the equation $x^2 + 9y^2 - 24x + 18y + 9 = 0$.

8. Find the equation in standard form of the ellipse with center at $(3, 8)$, foci $\left(3 + \sqrt{11}, 8\right)$, $\left(3 - \sqrt{11}, 8\right)$, and minor axis of length 10.

9. Find the eccentricity of the ellipse given by the equation $16x^2 + 25y^2 = 100$.

10. Graph $\frac{x^2}{81} - \frac{y^2}{4} = 1$.

11. Find the vertices, foci, and asymptotes of the hyperbola given by the equation $\frac{x^2}{25} - \frac{y^2}{100} = 1$.

12. Graph $4x^2 - y^2 + 24x - 2y + 31 = 0$.

13. Find the vertices and foci of the hyperbola given by the equation $\frac{(y + 5)^2}{25} - \frac{(x - 3)^2}{4} = 1$.

14. Find the equation in standard form of the hyperbola with vertices at (0, 3) and (-4, 3) and foci $\left(-2 + \sqrt{15}, 3\right), \left(-2 - \sqrt{15}, 3\right)$.

15. Graph $r = 6$.

16. Graph $r = 3 + 3 \cos \theta$.

17. Graph $r = 6 \sin 3\theta$.

18. Find the rectangular coordinates of the point whose polar coordinates are $\left(3, \frac{5\pi}{4}\right)$.

19. Find the rectangular form of $r - r \sin \theta = 9$.

20. Find the polar form of $3x^2 + 3y^2 = 25$.

Chapter 9 Chapter Test Form B

1. Find the vertex, focus, and directrix of the parabola given by the equation $x + \frac{1}{2}y = 0$.

2. Find the vertex, focus, and directrix of the parabola given by the equation $4x - y^2 + 14y - 37 = 0$.

3. Find the equation in standard form of the parabola with directrix $y = -7$ and focus $(-3, -3)$.

4. Graph the parabola with focus $(-6, 0)$ and directrix $x = 0$.

5. Find the vertices and foci of the ellipse given by the equation $\frac{x^2}{81} + \frac{y^2}{16} = 1$.

6. Graph $\frac{x^2}{25} + \frac{y^2}{64} = 1$.

7. Find the vertices and foci of the ellipse given by the equation $2x^2 + y^2 + 16x - 14y + 77 = 0$.

8. Graph $x^2 + 9y^2 - 6x + 18y + 9 = 0$.

9. Find the eccentricity of the ellipse given by the equation $49x^2 + 45y^2 = 245$.

10. Graph $\frac{y^2}{9} - \frac{x^2}{49} = 1$.

11. Find the vertices, foci, and asymptotes of the hyperbola given by the equation $\frac{x^2}{36} - \frac{y^2}{81} = 1$.

12. Graph $-x^2 + 4y^2 - 6x - 16y + 3 = 0$.

13. Find the vertices and foci of the hyperbola given by the equation $\frac{(x+9)^2}{225} - \frac{(y-3)^2}{100} = 1$.

14. Find the equation in standard form of the hyperbola with vertices at $(7, 0)$ and $(7, -8)$ and foci $\left(7, -4 + \sqrt{31}\right)$, $\left(7, -4 - \sqrt{31}\right)$.

15. Graph $r = 3$.

16. Graph $r = 5 - 5 \sin \theta$.

17. Graph $r = 4 \cos 3\theta$.

18. Find the rectangular coordinates of the point whose polar coordinates are $\left(14, \frac{5\pi}{3}\right)$.

19. Find the polar form of $y = 8$.

20. Find the rectangular form of $r + r \cos \theta = 2$.

1. Find the vertex, focus, and directrix of the parabola given by the equation $3y^2 = 2x$.

2. Find the vertex, focus, and directrix of the parabola given by the equation $x^2 - 8x + 2y + 16 = 0$.

3. Find the equation in standard form of the parabola with directrix $y = 2$ and focus $(2, 0)$.

4. Graph the parabola with focus $(2, 0)$ and directrix $x = 5$.

5. Find the vertices and foci of the ellipse given by the equation $\frac{x^2}{12} + \frac{y^2}{36} = 1$.

6. Graph $\frac{x^2}{9} + \frac{y^2}{4} = 1$.

7. Find the vertices and foci of the ellipse given by the equation $4x^2 + 9y^2 - 16x + 36y + 16 = 0$.

8. Find the equation in standard form of the ellipse with center $(2, 0)$, foci $(6, 0)$ and $(-2, 0)$, and minor axis of length 4.

9. Find the eccentricity of the ellipse given by the equation $16x^2 + 4y^2 = 49$.

10. Graph $\frac{y^2}{4} - \frac{x^2}{49} = 1$.

11. Find the vertices, foci, and asymptotes given by the equation $\frac{x^2}{36} - \frac{y^2}{25} = 1$.

12. Graph $x^2 + 2x - 4y^2 + 16y = 31$.

13. Find the vertices, foci, and asymptotes of the hyperbola given by the equation $\frac{(y-3)^2}{25} - \frac{(x+2)^2}{4} = 1$.

14. Find the equation in standard form of the hyperbola with vertices at $\left(-5, 4 + 2\sqrt{5}, -5, 4 - 2\sqrt{5}\right)$ and foci $(-5, 10)$ and $(-5, -2)$.

15. Graph $r = 5$.

16. Graph $r = 2 - 2\cos\theta$.

17. Graph $r = 2\sin 2\theta$.

18. Find the polar coordinates of the point whose rectangular coordinates are $(-2, 3)$.

19. Find the rectangular form of $r = 2\cos\theta$.

20. Find the polar form of $3x + 2y = 4$.

1. Find the vertex, focus, and directrix of the parabola given by the equation $2y^2 = -3x$.

 a. vertex: (0, 0), focus: $\left(\frac{3}{8}, 0\right)$, directrix: $x = -\frac{3}{8}$
 b. vertex: (0, 0), focus: $\left(-\frac{3}{8}, 0\right)$, directrix: $x = \frac{3}{8}$
 c. vertex: (0, 0), focus: (3, 0), directrix: $x = -3$
 d. vertex: (0, 0), focus: (−3, 0), directrix: $x = 3$

2. Find the vertex, focus, and directrix of the parabola given by the equation $x^2 + 8x - 16y + 80 = 0$.

 a. vertex: (4, −4), focus: (4, 0), directrix: $y = -8$
 b. vertex: (−4, 4), focus: (−4, 20), directrix: $y = -12$
 c. vertex: 4, −4), focus: (8, −4), directrix: $x = 0$
 d. vertex: (−4, 4), focus: (−4, 8), directrix: $y = 0$

3. Find the equation in standard form of the parabola with directrix $x = 7$ and focus (−1, 3).

 a. $(y - 3)^2 = 12(x - 3)$
 b. $(y - 3)^2 = -16(x - 3)$
 c. $(x - 3)^2 = 3(y - 3)$
 d. $(y - 3)^2 = 3(x - 3)$

4. Graph the parabola with focus (0, 1) and directrix $y = 5$.

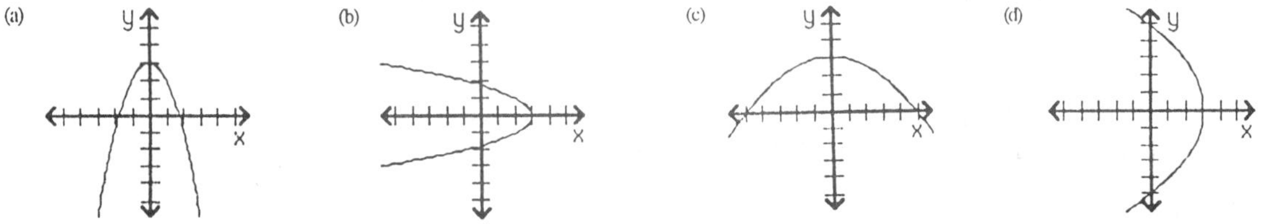

5. Graph $\frac{x^2}{16} + \frac{y^2}{36} = 1$,

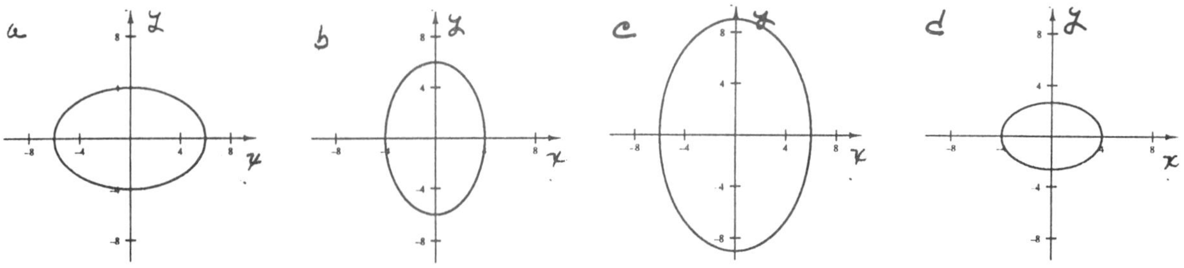

6. Find the vertices and foci of the ellipse given by the equation $5x^2 + y^2 + 60x + 4y + 159 = 0$.

a. vertices: $(-11, -2)$, $(-1, -2)$, foci: $\left(-6 + 2\sqrt{5}, -2\right)$, $\left(-6 - 2\sqrt{5}, -2\right)$

b. vertices: $(6, 7)$, $(6, -3)$, foci: $\left(6, 2 + 2\sqrt{5}\right)$, $\left(6, 2 - 2\sqrt{5}\right)$

c. vertices: $(6, 7)$, $(6, -3)$, foci: $\left(6 + 2\sqrt{5}, 2\right)$, $\left(6 - 2\sqrt{5}, 2\right)$

d. vertices: $(-6, 3)$, $(-6, -7)$, foci: $\left(-6, -2 + 2\sqrt{5}\right)$, $\left(-6, -2 - 2\sqrt{5}\right)$

7. Find the equation in standard form of the ellipse with center at $(-3, 6)$, foci $\left(-3 + \sqrt{15}, 6\right)$, $\left(-3 - \sqrt{15}, 6\right)$, and minor axis of length 12.

a. $\dfrac{(x + 3)^2}{159} + \dfrac{(y - 6)^2}{144} = 1$

b. $\dfrac{(x - 3)^2}{39} + \dfrac{(y + 6)^2}{24} = 1$

c. $\dfrac{(x + 3)^2}{51} + \dfrac{(y - 6)^2}{36} = 1$

d. $\dfrac{(x - 3)^2}{39} + \dfrac{(y - 6)^2}{24} = 1$

8. Graph $3x^2 + y^2 - 42x + 4y + 142 = 0$.

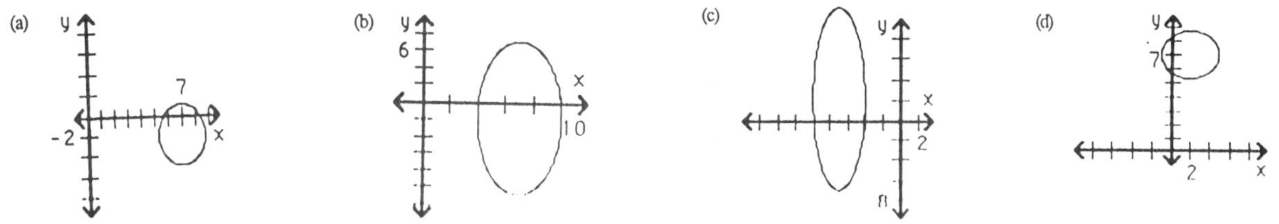

9. Find the eccentricity of the ellipse $25x^2 + 9y^2 = 400$

a. $\frac{4}{5}$

b. $\frac{5}{4}$

c. $\frac{3}{5}$

d. $\frac{25}{16}$

10. Graph $\dfrac{y^2}{36} - \dfrac{x^2}{49} = 1$

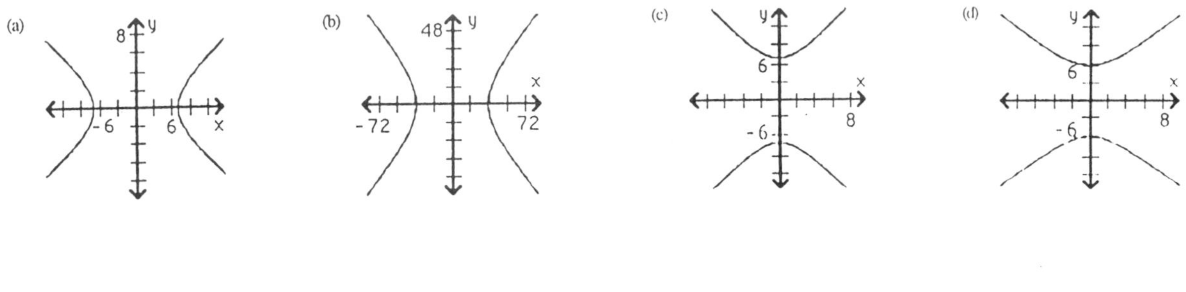

11. Find the vertices, foci, and asymptotes of the hyperbola given by the equation $\frac{y^2}{100} - \frac{x^2}{81} = 1$.

a. vertices: (0, 10), (0, −10), foci: (0, 9), (0, −9). Asymptotes: $y = \pm\frac{9}{10}x$

b. vertices: (9, 0), (−9, 0), foci: $\left(0, \sqrt{181}\right), \left(0, -\sqrt{181}\right)$, asymptotes: $y = \pm\frac{9}{10}x$

c. vertices: (9, 0), (−9, 0), foci: $\left(0, \sqrt{181}\right), \left(0, -\sqrt{181}\right)$, asymptotes: $y = \pm\frac{10}{9}x$

d. vertices: (0, 10), (0, −10), foci: $\left(0, \sqrt{181}\right), \left(0, -\sqrt{181}\right)$, asymptotes: $y = \pm\frac{10}{9}x$

12. Graph $x^2 - 16y^2 + 2x + 128y - 271 = 0$

(a)
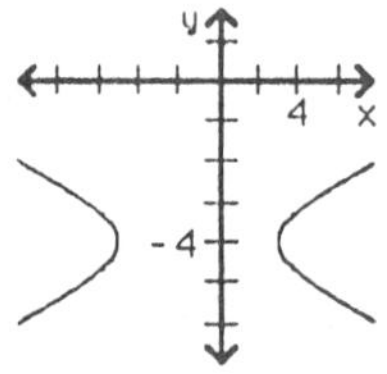

(b)
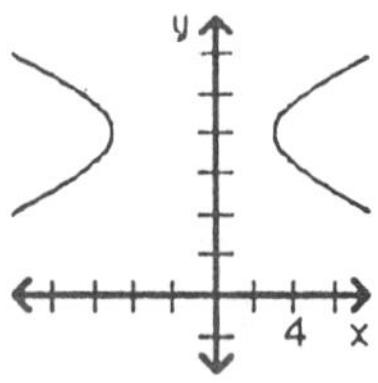

(c)
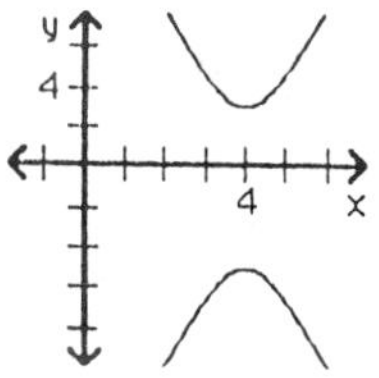

(d)
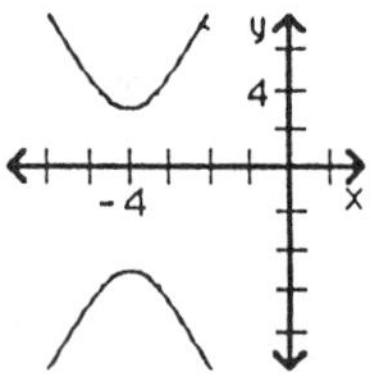

13. Find the vertices and foci of the hyperbola given by the equation $\frac{(x + 7)^2}{9} - \frac{(y + 1)^2}{1} = 1$.

a. vertices: (−4, −1), (−10, −1), foci: $\left(-7 + \sqrt{10}, -1\right), \left(-7 - \sqrt{10}, -1\right)$

b. vertices: (−7, 2), (−7, −4), foci: $\left(-7, -1 + \sqrt{10}\right), \left(-7, -1 - \sqrt{10}\right)$

c. vertices: (10, 1), (4, 1), foci: $\left(7 + \sqrt{10}, 1\right), \left(7 - \sqrt{10}, 1\right)$

d. vertices: (10, 1), (4, 1), foci: $\left(7 + 2\sqrt{2}, 1\right), \left(7 - 2\sqrt{2}, 1\right)$

14. Find the equation in standard form of the hyperbola with vertices at (9, 2). (11, 2) and foci $\left(10 + \sqrt{6}, 2\right), \left(10 - \sqrt{6}, 2\right)$.

a. $\frac{(y - 2)^2}{1} - \frac{(x - 2)^2}{5} = 1$

b. $\frac{(+ 10)^2}{1} - \frac{(y - 2)^2}{5} = 1$

c. $\frac{(x - 10)^2}{1} - \frac{(y - 2)^2}{5} = 1$

d. $\frac{(x + 10)^2}{1} - \frac{(y + 1)^2}{6} = 1$

15. Graph $r = 10$.

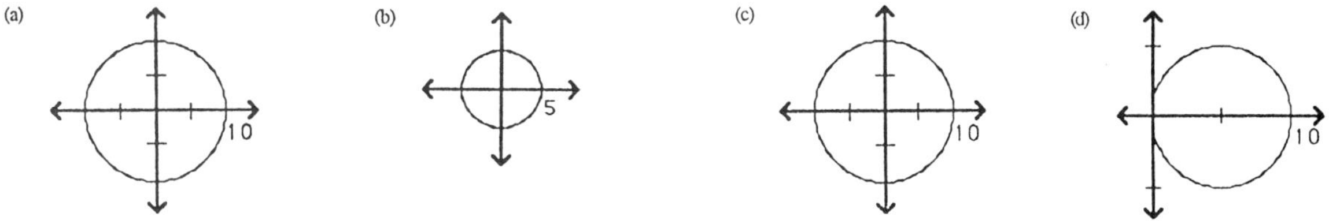

16. Graph $r = 4 + 4 \cos \theta$.

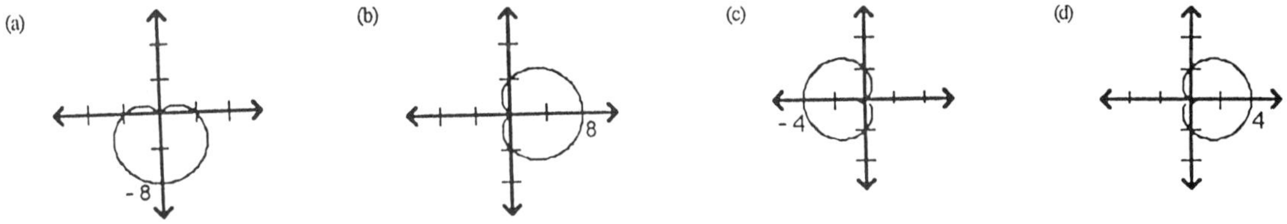

17. Graph $r = -8 \sin 3\theta$.

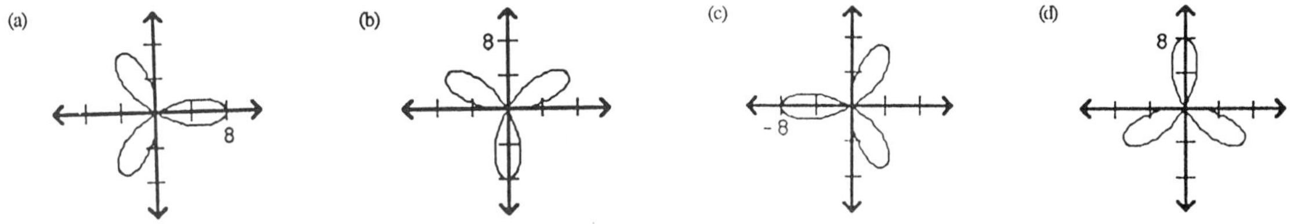

18. Find the polar coordinates of the point whose rectangular coordinates are $\left(-4\sqrt{3}, 4\right)$.

a. $\left(8, \frac{11\pi}{6}\right)$ b. $\left(4, \frac{7\pi}{6}\right)$ c. $\left(8, \frac{5\pi}{6}\right)$ d. $\left(8, \frac{7\pi}{6}\right)$

19. Find the rectangular form of the equation $r = 11 \sin \theta$.

a. $x^2 + y^2 = 11$ b. $x^2 = 11y$ c. $y^2 = 11x$ d. $x^2 + y^2 = 11y$

20. Find the polar form of $y^2 = 4x + 4$.

a. $r^2 \sin^2\theta = 4$

b. $r^2 \sin^2\theta - 4r \sin \theta = 4$

c. $r^2 \sin^2\theta - 4r \cos \theta = 4$

d. $r \sin \theta - 4r \cos \theta = 4$

1. Find the vertex, focus, and the directrix of the parabola given by the equation $x^2 = -6y$.

a. vertex: (0, 0), focus: $(0, \frac{3}{2})$, directrix: $y = -\frac{3}{2}$

b. vertex: (0, 0), focus: $(0, -\frac{3}{2})$, directrix: $y = \frac{3}{2}$

c. vertex: (0, 0), focus: $(0, \frac{2}{3})$, directrix: $y = -\frac{2}{3}$

d. vertex: (0, 0), focus: $(0, -\frac{2}{3})$, directrix: $y = \frac{2}{3}$

2. Find the vertex, focus, and the directrix given by the equation $24x - y^2 - 2y - 73 = 0$.

a. vertex: (3, -1), focus: (9, -1), directrix: x = -3
b. vertex: (-1, 3), focus: (5, 3), directrix: x = -7
c. vertex: (1, 3), focus: (1, 9), directrix: y = -3
d. vertex: (3, -1), focus: (c, 5), directrix: y = -7

3. Find the equation in standard form of the parabola with directrix y = 4 and focus (3, -6).

a. $(x - 3)^2 = 20(y + 1)$ b. $(y - 3)^2 = -5(x + 1)$

c. $(x - 3)^2 = -20(y + 1)$ d. $(x + 3)^2 = 20(y - 1)$

4. Graph the parabola with focus (-2, 0) and directrix x = 4.

(a)
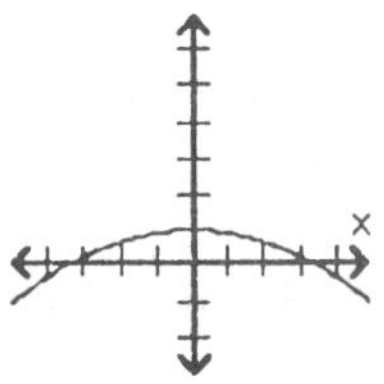

(b)
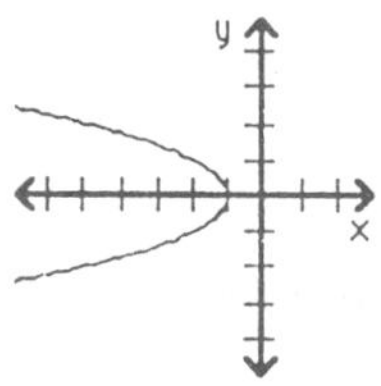

(c)
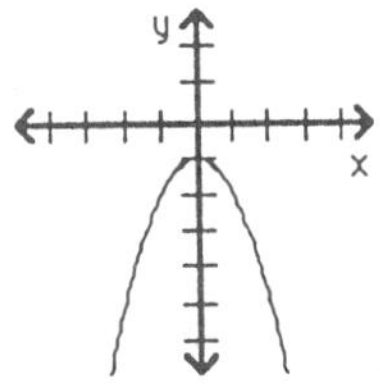

(d)
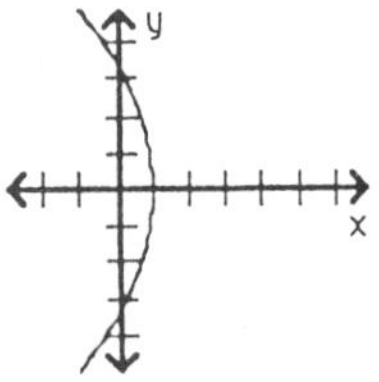

5. Graph $\frac{x^2}{121} + \frac{y^2}{9} = 1$.

(a)
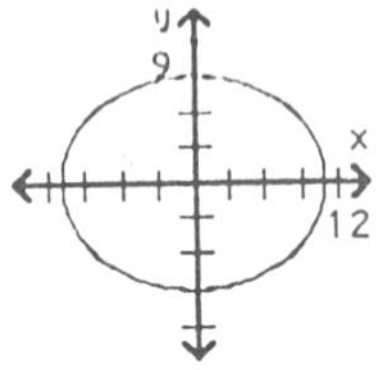

(b)
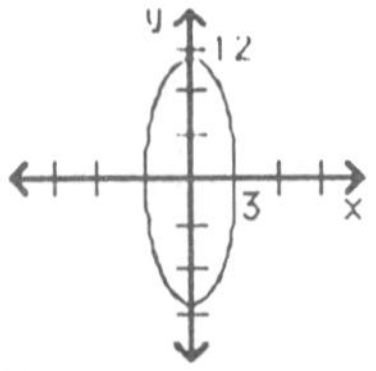

(c)
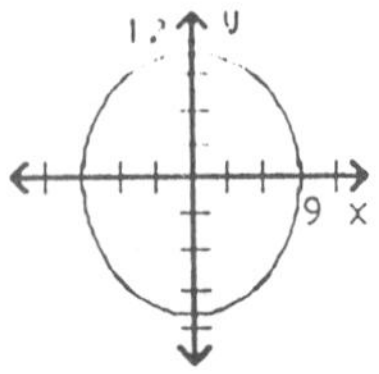

(d)
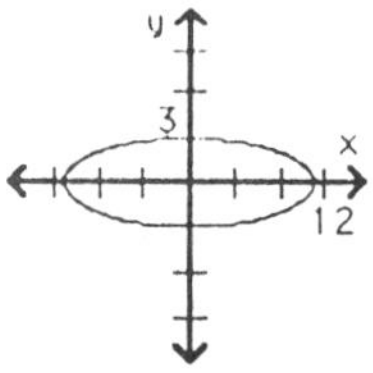

6. Find the vertices and foci of the ellipse given by the equation $x^2 + 2y^2 - 8x - 36y + 162 = 0$.

a. vertices: (0, 9), (8, 9), foci: $(4 + 2\sqrt{2}, 9)$, $(4 - 2\sqrt{2}, 9)$
b. vertices: (20, 9), (-12, 9), foci: (12, 9), (-4, 9)
c. vertices: (-8, -9), (0, -9), foci: $(-4 + 2\sqrt{2}, 9)$, $(-4 - 2\sqrt{2}, 9)$
d. vertices: (4, 13), (4, 5), foci: $(4, 9 + 2\sqrt{2})$, $(4, 9 - 2\sqrt{2})$

7. Find the equation in standard form of the ellipse with center (-2, 6), foci $(-2, 6 + \sqrt{14})$, $(-2, 6 - \sqrt{14})$, and minor axis of length 16.

a. $\frac{(x-2)^2}{64} + \frac{(y+6)^2}{78} = 1$ b. $\frac{(x+2)^2}{64} + \frac{(y-6)^2}{78} = 1$

c. $\frac{(x+2)^2}{32} + \frac{(y-6)^2}{46} = 1$ d. $\frac{(x+2)^2}{78} + \frac{(y-6)^2}{64} = 1$

8. Graph $x^2 + 4y^2 + 20x - 40y + 199 = 0$.

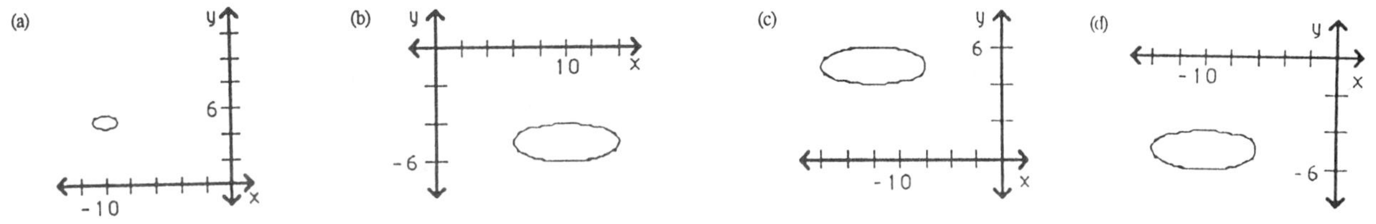

9. Find the eccentricity of the ellipse $81x^2 + 32y^2 = 162$.

a. $\frac{9}{7}$ b. $\frac{7}{9}$ c. $\frac{7}{4}$ d. $\frac{5}{9}$

10. Graph $\frac{y^2}{64} - \frac{x^2}{4} = 1$.

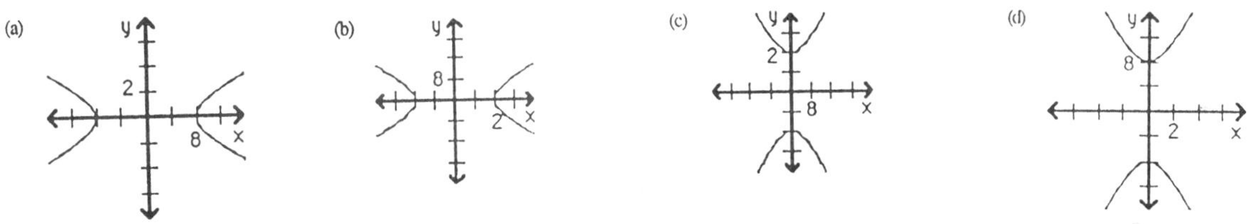

11. Find the equation in standard form of the hyperbola with vertices (3, 0), (-3, 0) and foci $(\sqrt{17}, 0)$. $(-\sqrt{17}, 0)$

a. $\frac{x^2}{3} - \frac{y^2}{14} = 1$ b. $\frac{x^2}{9} - \frac{y^2}{17} = 1$ c. $\frac{y^2}{8} - \frac{x^2}{9} = 1$ d. $\frac{x^2}{9} - \frac{x^2}{8} = 1$

12. Graph $-4x^2 + y^2 - 8x + 14y + 41 = 0$.

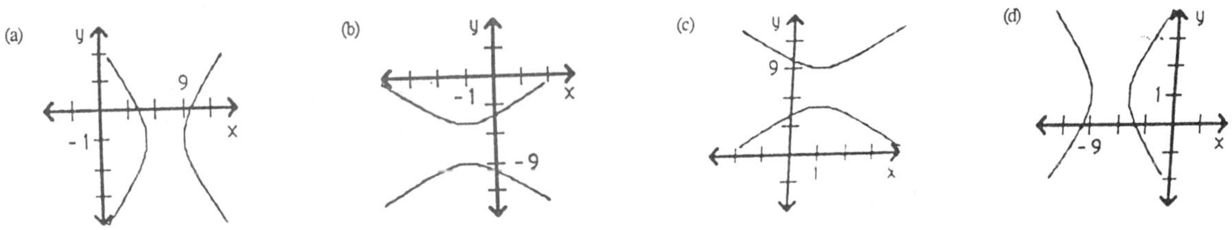

13. Find the vertices and foci of the hyperbola given by the equation $\frac{(y-8)^2}{16} - \frac{(x+3)^2}{1} = 1$.

a. vertices: (8, 1), (8, -7), foci: $(8, -3 + \sqrt{17})$, $(8, -3 - \sqrt{17})$
b. verices: (8, 1), (8, -7), foci: $(8 + \sqrt{17}, -3)$, $(8 - \sqrt{17}, -3)$
c. vertices: (-3, 12), (-3, 4), foci: $(-3, 8 + \sqrt{17})$, $(-3, 8 - \sqrt{17})$
d. vertices: (-3, 12), (-3, 4), foci: $(-3, 8 + \sqrt{15})$, $(-3, 8 - \sqrt{15})$

14. Find the equation in standard form of the hyperbola with vertices at (3, -5), (9, -5) and foci $(3 + \sqrt{21}, -5)$, $(3 - \sqrt{21}, -5)$.

a. $\frac{(x-6)^2}{9} - \frac{(y+5)^2}{15} = 1$ b. $\frac{(x+6)^2}{9} - \frac{(y-5)^2}{15} = 1$

c. $\frac{(x-6)^2}{3} - \frac{(y+5)^2}{18} = 1$ d. $\frac{(x+5)^2}{15} - \frac{(y-3)^2}{9} = 1$

15. Graph $r = 12$.

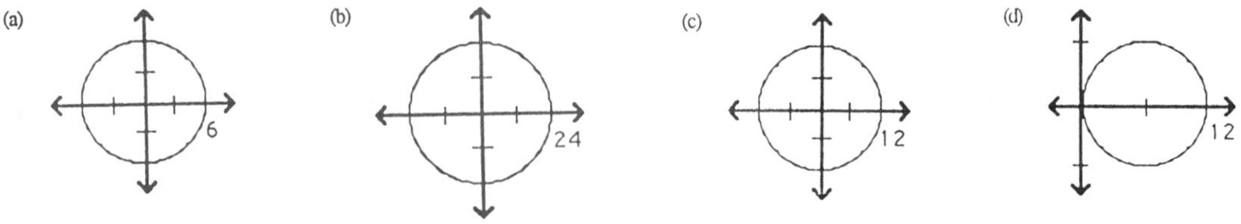

16. Graph $r = 7 + 7 \sin \theta$.

(a)

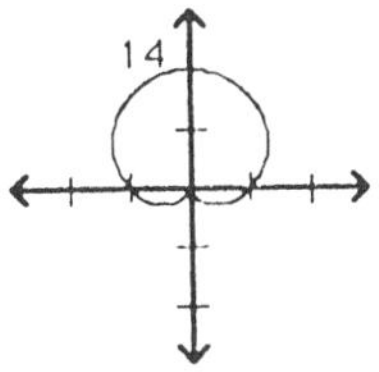

(b)

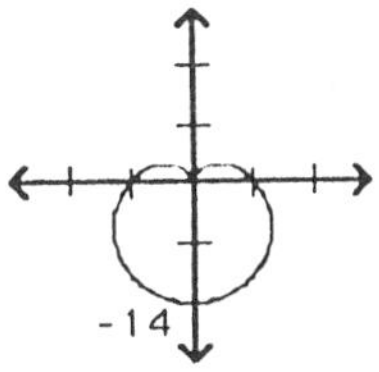

(c)

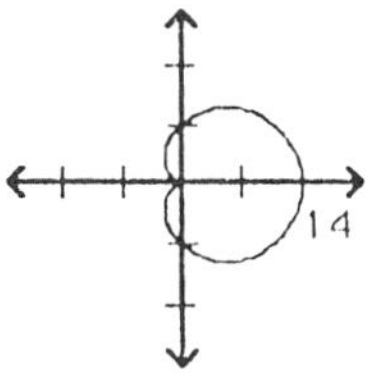

(d)

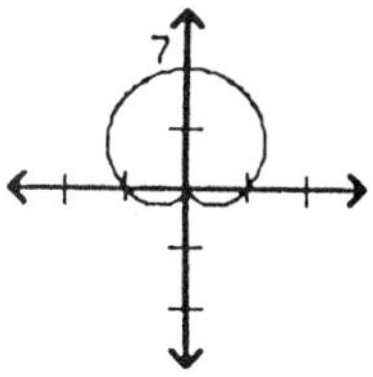

17. Graph $r = -2 \cos 3\theta$.

(a)

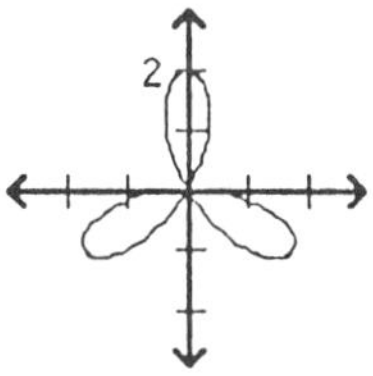

(b)

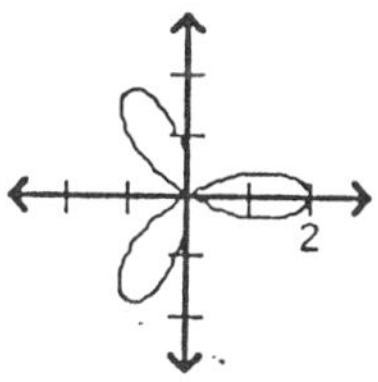

(c)

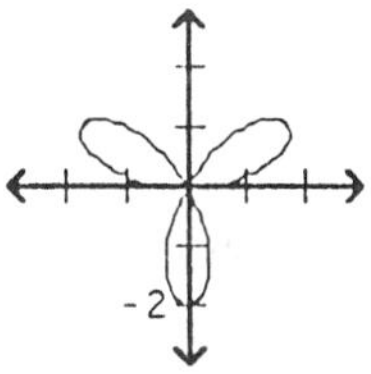

(d)

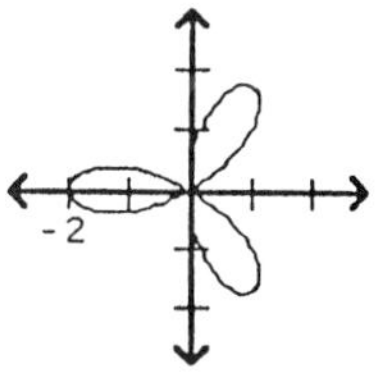

18. Find the rectangular coordinates of the point whose polar coordinates are $(-6, \frac{7\pi}{6})$.

a. $(-3, -3\sqrt{3})$ b. $(-3\sqrt{3}, -3)$ c. $(3, 3\sqrt{3})$ d. $(3\sqrt{3}, 3)$

19. Find the polar form of $x = 5$.

a. $r \cos \theta = 5$ b. $r \sin \theta = 5$ c. $r \cos \theta = 25$ d. $r \sin \theta = 25$

20. Find the rectangular form of $r = 2 \cos \theta$.

a. $x + y = 2$

b. $\sqrt{x^2 + y^2} = 2$

c. $x^2 - 2x + y^2 = 0$

d $x = 2y$

1. Find the vertex, focus, and directrix of the parabola given by the equation $6x = 4y^2$.

a. vertex: (0, 0), focus: $\left(-\frac{3}{8}, 0\right)$, directrix: $y = \frac{3}{8}$

b. vertex: (0, 0), focus: $\left(\frac{3}{8}, 0\right)$, directrix: $x = -\frac{3}{8}$

c. vertex: (0, 0), focus: $\left(-\frac{3}{8}, 0\right)$, directrix: $x = \frac{3}{8}$

d. vertex: (0, 0), focus: $\left(0, \frac{3}{8}\right)$, directrix: $y = -\frac{3}{8}$

2. Find the vertex, focus, and directrix of the parabola given by the equation $x^2 - 4x - 8y - 20 = 0$.

a. vertex: (–3, 2), focus: (–6, 2), directrix: $x = -5$
b. vertex: (4, –3), focus: (4, –6), directrix: $y = 1$
c. vertex: (–2, 3), focus: (–2, 1), directrix: $y = 5$
d. vertex: (2, –3), focus: (2, –1), directrix: $y = -5$

3. Find the equation in standard form of the parabola with directrix $y = -5$ and focus (4, 3).

a. $(x - 4)^2 = 16(y + 1)$

b. $(y - 3^2 = 4(x - 4)$

c. $(x + 4)^2 = 4(y + 1)$

d. $(x - 4)^2 = 4(y - 1)$

4. Graph the parabola with focus (–2, 1) and directrix $x = -4$.

a.
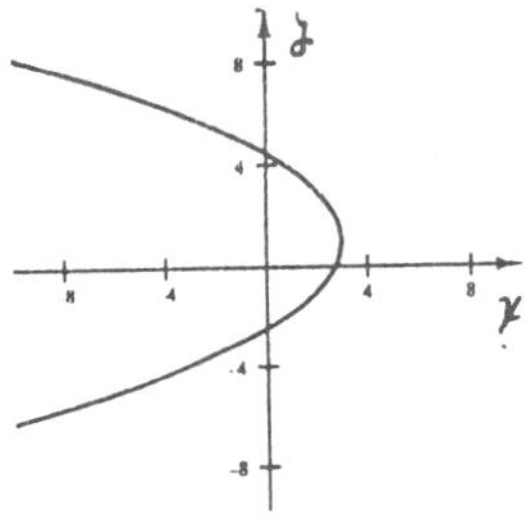

b.
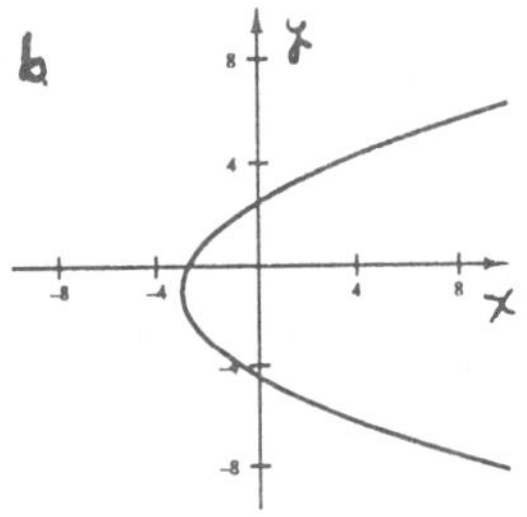

c.
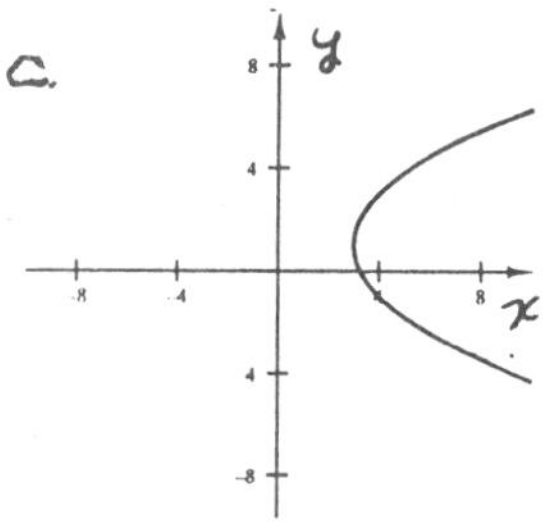

d.
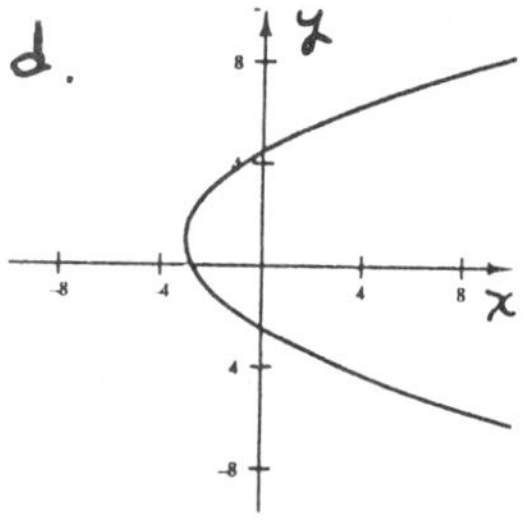

5. Find the vertices and foci of the ellipse given by the equation $\frac{x^2}{20} + \frac{y^2}{36} = 1$.

a. vertices: (6, 0), (–6, 0), foci: $\left(2\sqrt{5}, 0\right)$, $\left(-2\sqrt{5}, 0\right)$.

b. vertices: (0, 6), (–6, 0), foci: $\left(0, 2\sqrt{5}\right)$, $\left(0, -2\sqrt{5}\right)$.

c. vertices: (6, 0), (–6, 0), foci: (4, 0), (–4, 0).

d. vertices: (0, 6), (0, –6), foci: (0, 4), (0, –4).

6. Graph $\frac{x^2}{4} + \frac{y^2}{25} = 1$.

a. 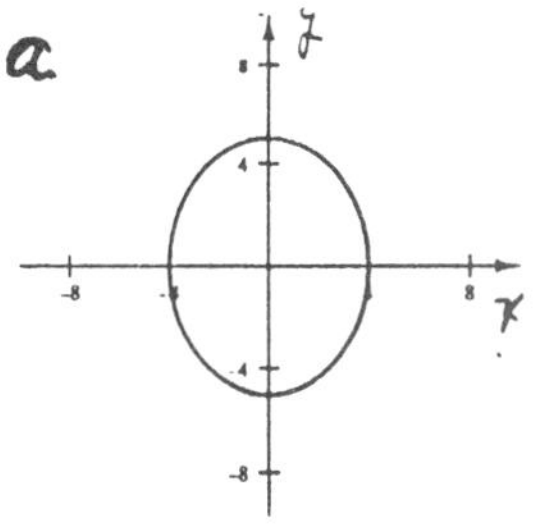

b. 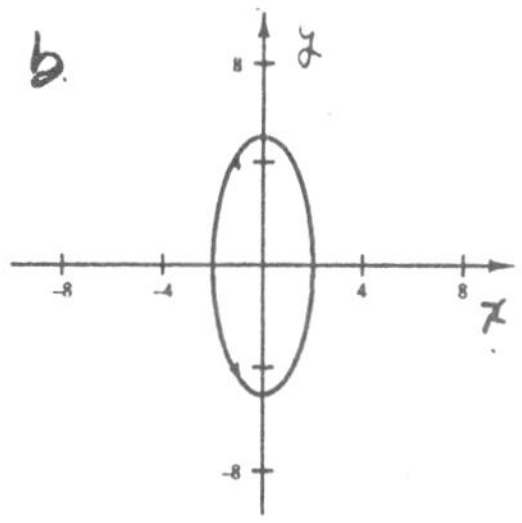

c. 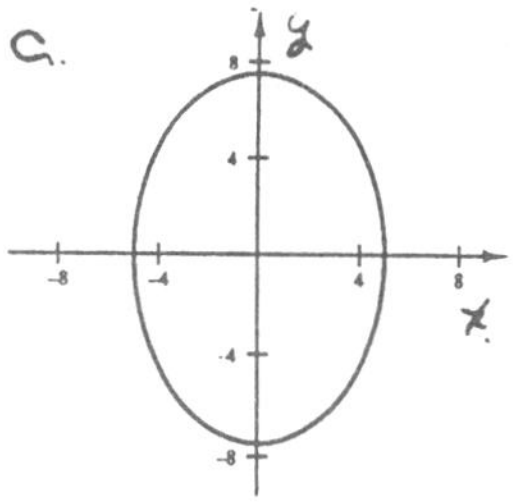

d. 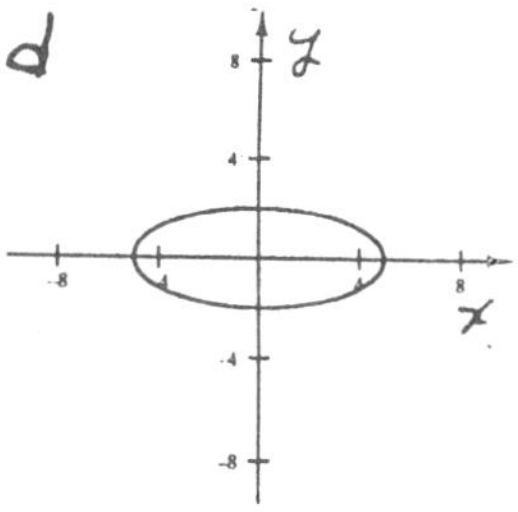

7. Find the vertices and foci of the ellipse given by the equation $x^2 + 16y^2 - 6x + 64y + 57 = 0$.

a. vertices: $(-1, -2)$, $(7, -2)$, foci: $\left(3 + \sqrt{15}, -2\right)$, $\left(3 - \sqrt{15}, -2\right)$

b. vertices: $(-2, -1)$, $(-2, 7)$, foci: $\left(-2, 3 + \sqrt{15}\right)$, $\left(-2, 3 - \sqrt{15}\right)$

c. vertices: $(1, -2)$, $(7, -2)$, foci: $\left(3 + \sqrt{15}, 2\right)$, $\left(3 - \sqrt{15}, 2\right)$

d. vertices: $(-1, -2)$, $(7, -2)$, foci: $\left(3 + \sqrt{17}, -2\right)$, $\left(3 - \sqrt{17}, -2\right)$

8. Find the equation in standard form of the ellipse with center $(-2, 3)$, foci $(-2, 7)$ and $(-2, -1)$, and minor axis of length 4.

a. $\frac{(x-2)^2}{20} + \frac{(y+3)^2}{4} = 1$

b. $\frac{(x+2)^2}{20} + \frac{(y-3)^2}{4} = 1$

c. $\frac{(x+2)^2}{4} + \frac{(y-3)^2}{20} = 1$

d. $\frac{(x-2)^2}{4} + \frac{(y+3)^2}{20} = 1$

9. Find the eccentricity of the ellipse given by the equation $25x^2 + 9y^2 = 225$.

a. $\frac{4}{5}$ b. $\frac{5}{4}$ c. $\frac{9}{16}$ d. $\frac{16}{25}$

10. Graph $\frac{y^2}{4} - \frac{x^2}{25} = 1$.

a. 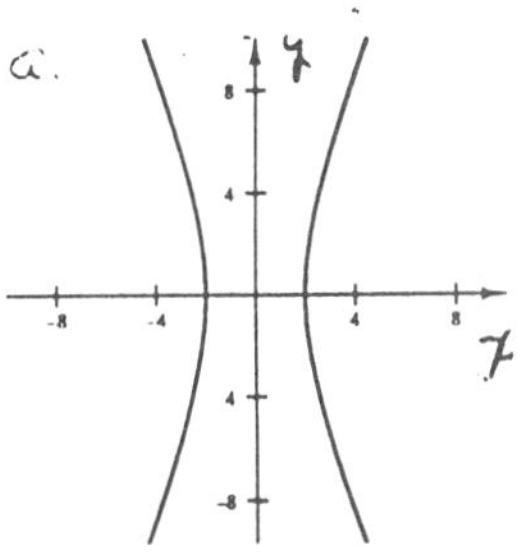

b. 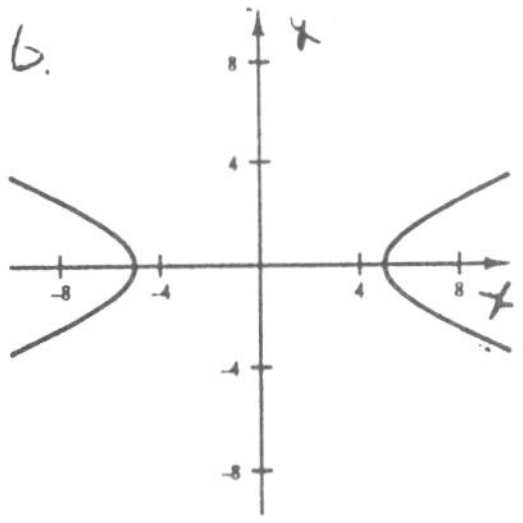

c. 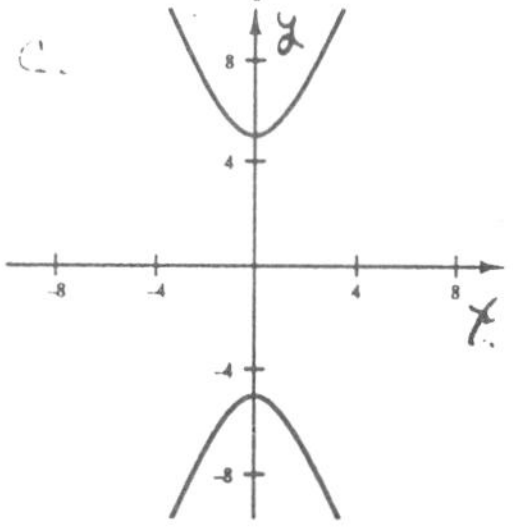

d. 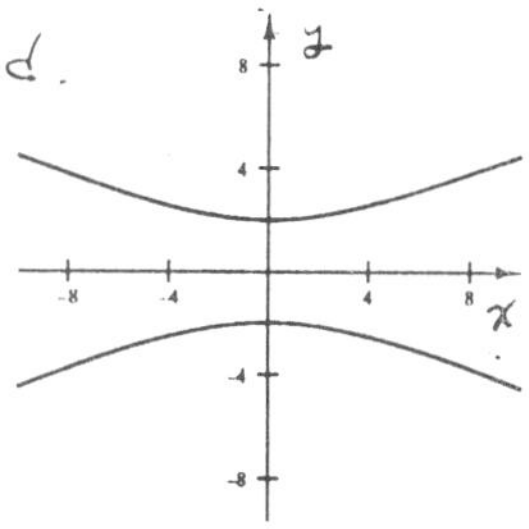

11. Find the vertices, foci, and asymptotes of the hyperbola given by the equation $\frac{x^2}{25} - \frac{y^2}{100} = 1$.

a. vertices: (0, 5), (0, −5), foci: $\left(0, 5\sqrt{3}\right)$, $\left(0, -5\sqrt{3}\right)$, $y = \pm 2x$

b. vertices: (5, 0), (−5, 0), foci: $\left(5\sqrt{5}, 0\right)$, $\left(-5\sqrt{5}, 0\right)$, $y = \pm\frac{1}{2}x$

c. vertices: (5, 0), (−5, 0), foci: $\left(5\sqrt{3}, 0\right)$, $\left(-5\sqrt{3}, 0\right)$, $y = \pm 2x$

d. vertices: (5, 0), (−5, 0), foci: $\left(5\sqrt{5}, 0\right)$, $\left(-5\sqrt{5}, 0\right)$, $y = \pm 2x$

12. Graph $4y^2 - 8y - x^2 - 4x = 16$.

a.
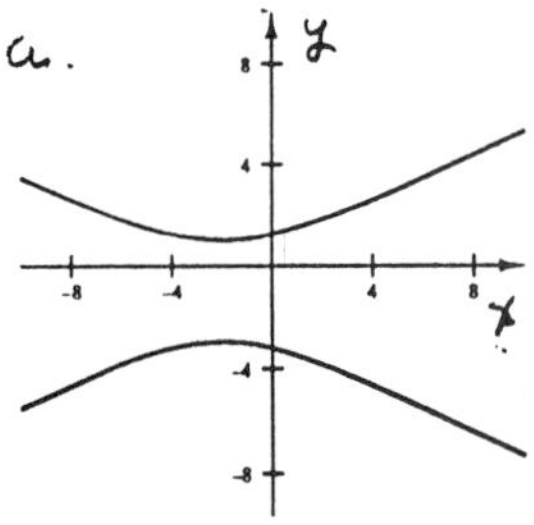

b.
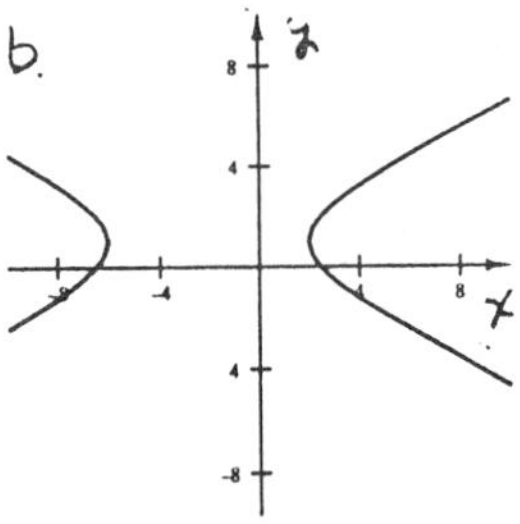

c.
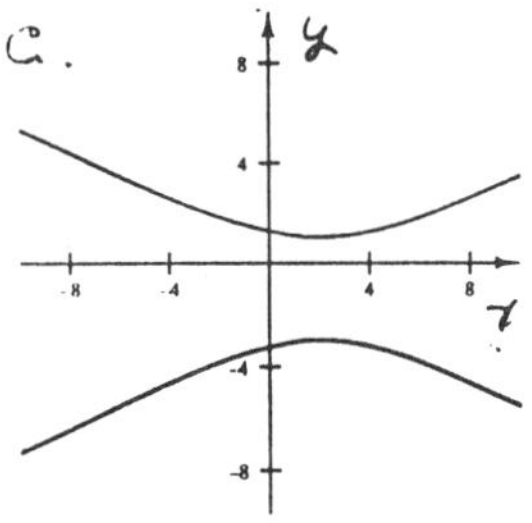

d.
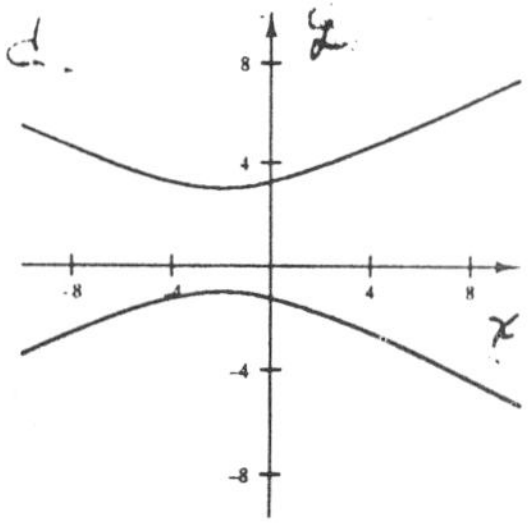

13. Find the vertices and foci of the hyperbola given by the equation $\frac{(x-3)^2}{16} - \frac{(y+4)^2}{36} = 1$.

a. vertices: (−4, 7), (−4, −1), foci: $\left(-2, 4 + 2\sqrt{13}\right)$, $\left(-2, 4 - 2\sqrt{13}\right)$

b. vertices: (7, −4), (−4, −1), foci: $\left(4 + 2\sqrt{5}, -4\right)$, $\left(4 - 2\sqrt{5}, -4\right)$

c. vertices: (10, −6), (−4, −6), foci: $\left(3 + 2\sqrt{5}, -6\right)$, $\left(3 - 2\sqrt{5}, -6\right)$

d. vertices: (7, −4), (−1, −4), foci: $\left(3 + 2\sqrt{13}, -4\right)$, $\left(3 - 2\sqrt{13}, -4\right)$

14. Find the equation in standard form of the hyperbola with vertices at (1, 4) and (−5, 4) and foci (3, 4) and (−7, 4).

a. $\frac{(x+2)^2}{9} - \frac{(y-4)^2}{16} = 1$

b. $\frac{(x-2)^2}{9} - \frac{(y+4)^2}{16} = 1$

c. $\frac{(x-2)^2}{16} - \frac{(y-4)^2}{9} = 1$

d. $\frac{(y-4)^2}{16} - \frac{(x+2)^2}{9} = 1$

15. Graph $r = 5$.

a. 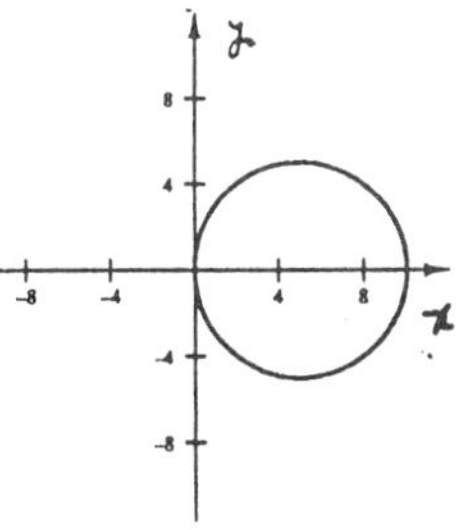

b. 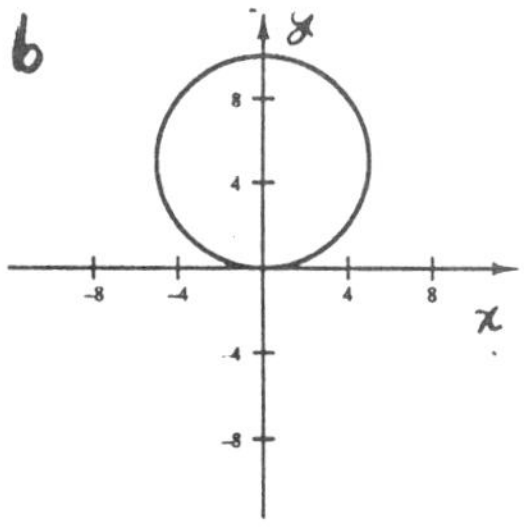

c. 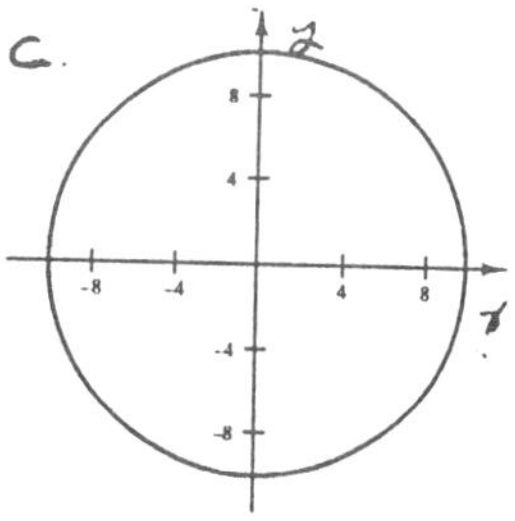

d. 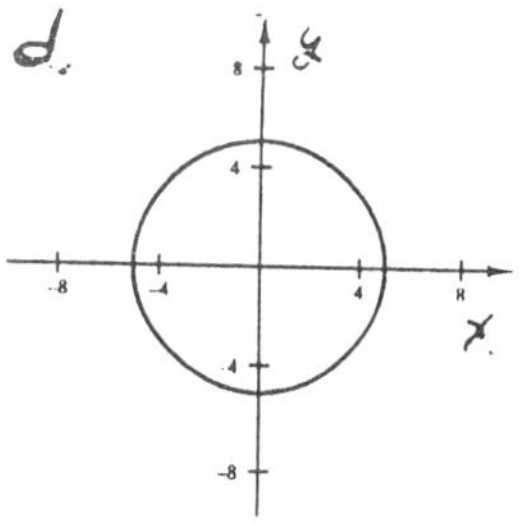

16. Graph $r = -2 \sin \theta$.

a. 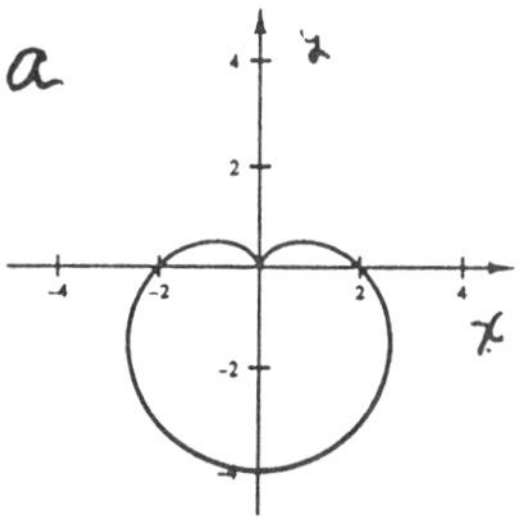

b. 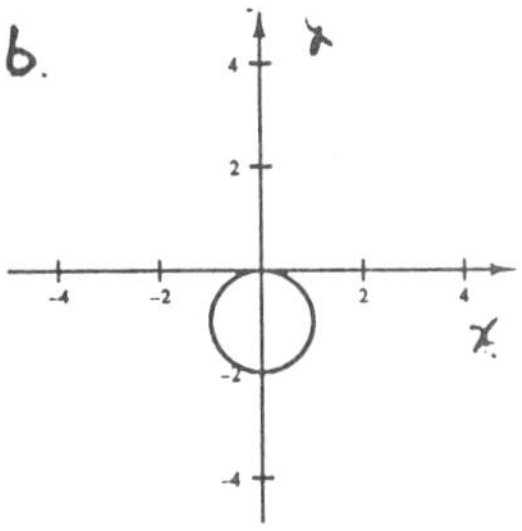

c. 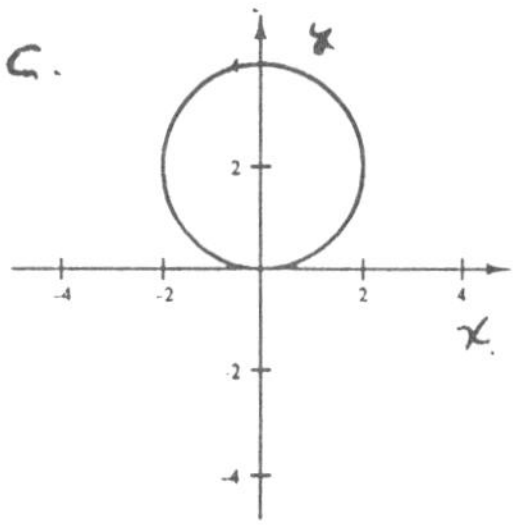

d. 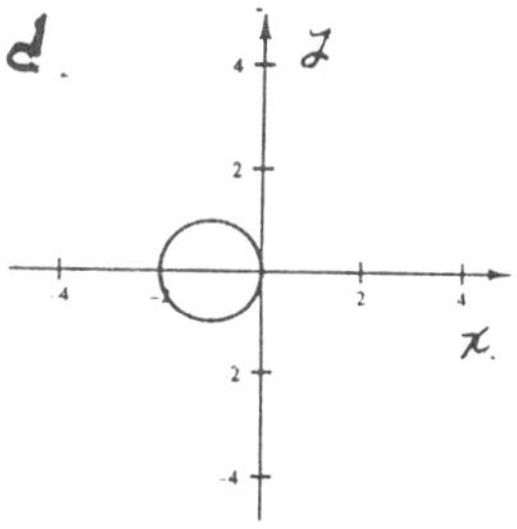

17. Graph $r = 3 - \sin \theta$.

a. 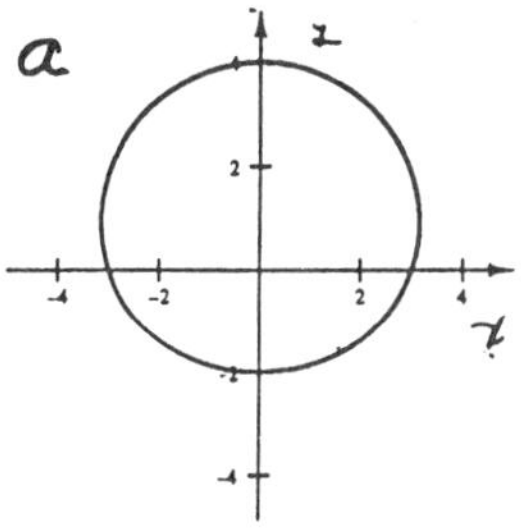

b. 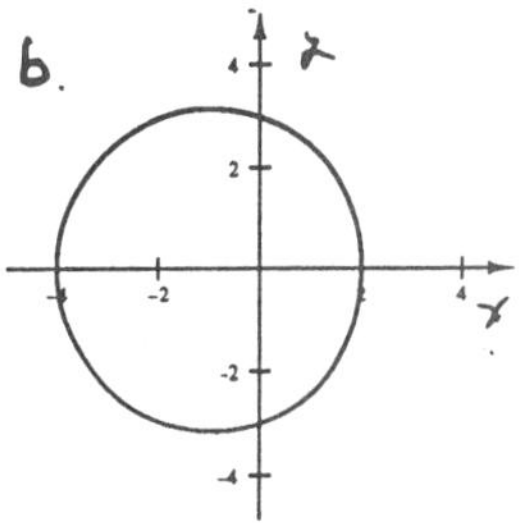

c. 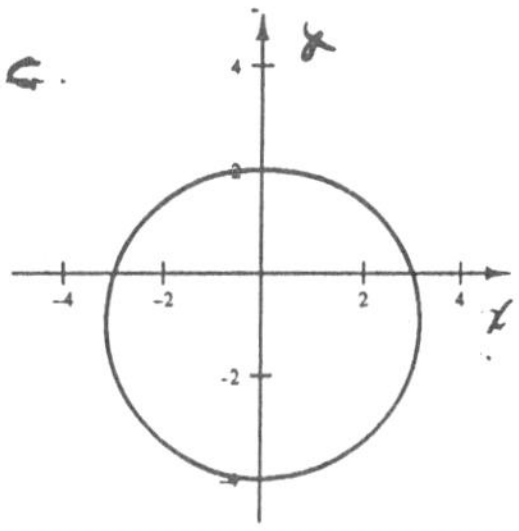

d. 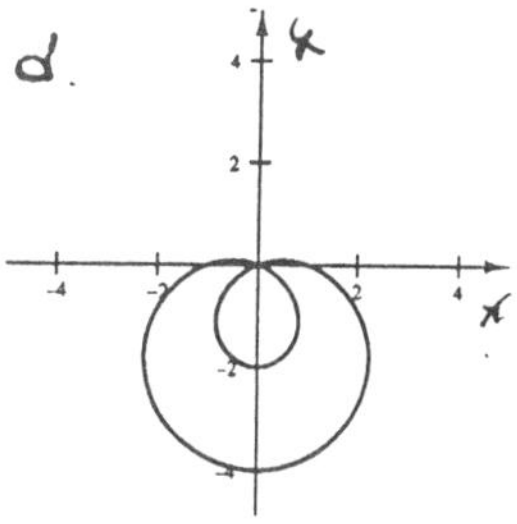

18. Find the polar coordinates of the point whose rectangular coordinates are $(-5, 5)$.

a. $\left(5, \frac{3\pi}{4}\right)$ b. $\left(5\sqrt{2}, \frac{3\pi}{4}\right)$ c. $\left(-5, \frac{3\pi}{4}\right)$ d. $\left(5\sqrt{2}, \frac{5\pi}{4}\right)$

19. Find the polar form of $x^2 - y^2 = 1$.

a. $r \cos \theta = 1$ b. $r^2 \cos 2\theta = 1$ c. $r^2 \sin 2\theta = 1$ d. $r^2 \sin^2 \theta = 1$

20. Find the rectangular form of $r + r \sin \theta = 4$.

a. $x^2 = 16 - 8y$ b. $y^2 = x^2 - 16$ c. $x^2 = y - 4$ d. $\sqrt{x^2 + y^2} = 4$

1. Solve the system of equations: $\begin{cases} 4x - 3y = 20 \\ y = 5x - 3. \end{cases}$

2. Solve the system of equations: $\begin{cases} 2x + y = 5 \\ 6x + 3y = 7. \end{cases}$

3. Solve the system of equations: $\begin{cases} x + 2y - 3z = 0 \\ 5x - y + 9z = 14 \\ 2x + 4y - 6z = 0. \end{cases}$

4. Solve the system of equations: $\begin{cases} 4x + 3y - z = 7 \\ 5x - 2y - 2z = 13 \\ 2z - 5y - 3z = 5. \end{cases}$

5. Solve the homogeneous system of equations: $\begin{cases} x + 3y + 2z = 0 \\ 5x + 7y + 4z = 0 \\ 10x + 2y - z = 0. \end{cases}$

6. Solve the system of equations: $\begin{cases} y = x^2 + 4x - 9 \\ y = -x - 3. \end{cases}$

7. Solve the system of equations: $\begin{cases} x^2 - 25y^2 = 25 \\ x^2 + y^2 = 25. \end{cases}$

8. Solve the system of equation: $\begin{cases} 8x + y - 3z = 24 \\ 2x + y + 3z = 6. \end{cases}$

9. Graph $y < 4x - 3$.

10. Graph $y \geq x^2 - 7x - 18$.

11. Graph the solution set of the system of inequalities: $\begin{cases} 5x - 3y \leq 15 \\ x + 2y > 8. \end{cases}$

12. Graph the solution set of the system of inequalities: $\begin{cases} y < 6x + 2 \\ y \geq x^2 + 4x - 5. \end{cases}$

13. Graph the solution set of the system of inequalities: $\begin{cases} x + y \leq 8 \\ 2x - y \leq 12 \\ x \geq 0,\ y \geq 0. \end{cases}$

14. Find the partial fraction decomposition of $\frac{6x - 2}{(x - 3)(x + 5)}$.

15. Find the partial fraction decomposition of $\frac{7x^2 + 16x + 2}{x(x + 1)^2}$.

16. Find the partial fraction decomposition of $\frac{2x^2 + 5x - 35}{(3x + 1)(x^2 + 9)}$.

17. Find an equation of the circle that passes through the points (–1, 6), (1, 4), and (2, 9).

18. A motorboat traveled a distance of 48 miles in 2 hours while traveling with the current. Against the current, the same trip took 3 hours. Find the rate of the motorboat in calm water and the rate of the current.

19. A chemist has two salt solutions. One solution is 20 percent salt and the other solution is 45 percent salt. How many liters of each must be used to produce 100 liters of a solution that is 37.5 percent salt?

20. Minimize the objective function $C = 2x + 9y$ with constraints $2x + 5y \le 35$, $x + y \ge 2$, $y \ge 5x - 20$, $x \ge 0$, $y \ge 0$.

1. Solve the system of equations: $\begin{cases} 4x - 5y = 17 \\ y = 2x - 7. \end{cases}$

2. Solve the system of equations: $\begin{cases} 4x - 7y = -13 \\ 5x - 3y = 1. \end{cases}$

3. Solve the system of equations: $\begin{cases} x + y + z = 0 \\ 2x - 3y + z = -11 \\ 3x + 5y + 2z = 17. \end{cases}$

4. Solve the system of equations: $\begin{cases} x - 2y + z = 5 \\ 2x + 3y - 2z = -8 \\ x + 5y - 3z = 7. \end{cases}$

5. Solve the system of equations: $\begin{cases} 2x + y - 3z = 6 \\ 5x - 3y - 2z = 37. \end{cases}$

6. Solve the homogeneous system of equations:

$$\begin{cases} 4x - 2y - 3z = 0 \\ 7x - y + z = 0 \\ x + y + 3z = 0. \end{cases}$$

7. Solve the system of equations: $\begin{cases} x^2 + y^2 = 6 \\ y^2 = 5x. \end{cases}$

8. Solve the system of equations: $\begin{cases} x^2 - 49y^2 = 49 \\ x^2 + y^2 = 49. \end{cases}$

9. Graph $5x - 2y \geq 10$.

10. Graph $(x + 7)^2 + (y - 3)^2 \leq 4$.

11. Graph the solution set of the system of inequalities: $\begin{cases} 2x + 7y \geq 14 \\ x - 4y < 8. \end{cases}$

12. Graph the solution set of the system of inequalities: $\begin{cases} y < 8x + 4 \\ y \geq x^2 + 6x - 7. \end{cases}$

13. Graph the solution set of the system of inequalities: $\begin{cases} 2x + 3y \geq 12 \\ x + 2y \leq 12 \\ x \leq 8,\ y \leq 4. \end{cases}$

14. Find the partial fraction decomposition of: $\dfrac{7x + 15}{x(2x + 5)}$.

15. Find the partial fraction decomposition of $\dfrac{7x^2 - 29x + 27}{3(x - 3)^2}$.

16. Find the partial fraction decomposition of $\dfrac{18x^2 + 23x + 4}{(5x + 2)(x^2 + 1)}$.

17. Find the equation of the circle that passes through the points (6, 3), (13, -4), and (-4, 3).

18. A canoeist can row 10 miles with the current in 2 hours. Rowing against the current, it takes the canoeist 5 hours to travel the same distance. Find the rate of the canoeist in calm water and the rate of the current.

19. A silversmith has two alloys. The first alloy is 35 percent silver, the second is 70 percent silver. How many grams of each should be mixed to product 100 grams of an alloy that is 56 percent silver?

20. Maximize the objective function $C = 5x + y$ with constraints $x + y \geq 3$, $5x + 6y \leq 66$, $y \geq 6x - 30$. $x \geq 0$, $y \geq 0$.

1. Solve the system of equations: $\begin{cases} 3x + 2y = -8 \\ y = -3x + 2. \end{cases}$

2. Solve the system of euations: $\begin{cases} 3x - 7y = 11 \\ 2x + 5y = -12. \end{cases}$

3. Solve the system of equations: $\begin{cases} 2x + 3y - z = -6 \\ -2x + y - 5z = -12 \\ 3x + y - 7z = -4. \end{cases}$

4. Solve the system of equations: $\begin{cases} -x + 5y + z = -27 \\ 2x - y + z = -2 \\ x + y + z = 17. \end{cases}$

5. Solve the system of equations: $\begin{cases} 2x - 3y + z = 5 \\ 5x - 7y - 3z = -12. \end{cases}$

6. Solve the system of equations: $\begin{cases} 3x + 2y - 5z = 0 \\ x - 4y + 2z = 0 \\ -x + 18y - 13z = 0. \end{cases}$

7. Solve the system of equations: $\begin{cases} x^2 + y^2 = 4 \\ y^2 = 3x. \end{cases}$

8. Solve the system of equaitons: $\begin{cases} y = x - 2 \\ y = x^2 - 2x - 6. \end{cases}$

9. Graph $3x - 5y \geq 4$.

10. Graph $x - y^2 < 0$.

11. Graph the solution set: $\begin{cases} 2x + 5y \leq 10 \\ x - 3y > 6. \end{cases}$

12. Graph the solution set: $\begin{cases} 2x + 3y < 4 \\ y \leq x^2 + 3x - 4. \end{cases}$

13. Graph the solution set: $\begin{cases} x - 3y \leq 8 \\ 3x - y \leq 12 \\ x - y \geq -4. \end{cases}$

14. Find the partial fraction decomposition of $\dfrac{2x - 3}{(x - 4)(x + 2)}$.

15. Find the partial fraction decomposition of $\dfrac{2x^2 + 3x - 2}{x^2(x - 2)}$.

16. Find the partial fraction decomposition of $\dfrac{8x^2 + 5x - 5}{(3x + 1)(x^2 - 3)}$.

17. Find the equation of the circle that passes through the points (6, 3), (13, -4), and (-4, 3).

18. Flying with the wind, a plane traveled 1500 miles in 5 hours. Flying against the wind, the plane traveled the same distance in 6 hours. Find the rate of the plane in calm air and the rate of the wind.

19. A coin bank contains only dimes and quarters. The value of the coins is \$5.45. If the dimes were quarters and the quarters were dimes, the value of the coins would be \$4.70. Find the original number of dimes and quarters in the coin bank.

20. Maximize the objective function $C = 4x + 2y$ with constraints $2x + 3y \leq 18$, $4x + y \leq 16$, $x \geq 0$, and $y \geq 0$.

1. Solve the system of equations: $\begin{cases} 2x - 3y = 15 \\ y = \frac{2}{3}x - 10. \end{cases}$

a. Inconsistent b. $-\frac{15}{4}, -\frac{25}{2}$ c. $\frac{15}{4}, -\frac{15}{2}$ d. $\frac{45}{4}, -\frac{5}{2}$

2. Solve the system of equations: $\begin{cases} x - 3y = 12 \\ 2x - 6y = 24. \end{cases}$

a. (6, -2) b. (-3, -5) c. Dependent d. Inconsistent

3. Solve the system of equations: $\begin{cases} x + y - 4z = 6 \\ 2x - 5y + 3z = 2 \\ x - 6y + 7z = -4. \end{cases}$

a. (8, 2, 1) b. (7, 3, 1) c. Dependent d. Inconsistent

4. Solve the system of equations: $\begin{cases} x + 5y - z = 4 \\ 3x + 4y + z = 4 \\ x - y + z = 0. \end{cases}$

a. (6, −2, −8) b. (2, 0, −2) c. Dependent d. Inconsistent

5. Solve the system of equations: $\begin{cases} x - y + z = -6 \\ 2x - 3y + 4z = -12. \end{cases}$

a. (2c, c, c − 6) b. (c, 2c, c − 6) c. (c − 6, 2c, c) d. (c − 6, c, 2c)

6. Solve the homogeneous system of equations: $\begin{cases} 3x - 2y + z = 0 \\ x - 3y + 2z = 0 \\ 2x + y - z = 0. \end{cases}$

a. $\left(7c, \frac{7}{5}c, c\right)$ b. $\left(-\frac{1}{7}c, \frac{7}{5}c, c\right)$ c. $\left(\frac{29}{7}c, \frac{5}{7}c, c\right)$ d. $\left(\frac{1}{7}c, \frac{5}{7}c, c\right)$

7. Solve the system of equations: $\begin{cases} y = 3x + 9 \\ y = x^2 - 2x - 5. \end{cases}$

a. (7, 12)
b. (7, 12), (−2, −15)
c. (7, 30), (−2, 3)
d. (7, −12), (2, −3)

8. Solve the system of equations: $\begin{cases} x^2 - 100y^2 = 100 \\ x^2 + y^2 = 100. \end{cases}$

a. (0, 10), (0, −10)
b. (10, 0)
c. No real solution
d. (10, 0), (−10, 0)

9. Graph $3x - 4y \geq 12$.

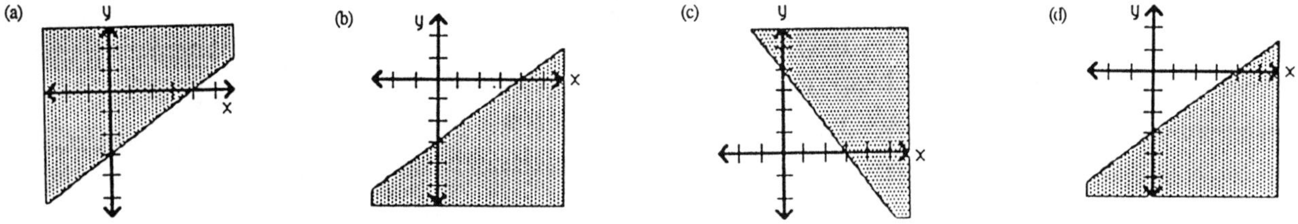

10. Graph $\frac{x^2}{4} + \frac{y2}{16} < 1$.

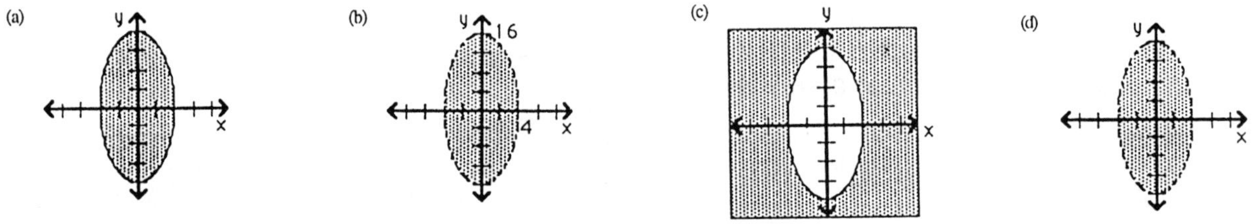

11. Graph the solution set: $\begin{cases} 3x - 8y \leq 24 \\ x + 5y > 10. \end{cases}$

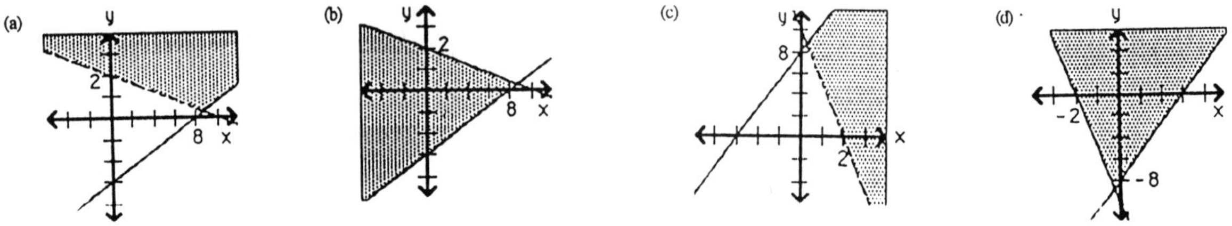

12. Graph the solution set: $\begin{cases} 2x + 3y \leq 18 \\ 2x - 3y < 6 \\ x \geq 0,\ y \geq 0. \end{cases}$

a. 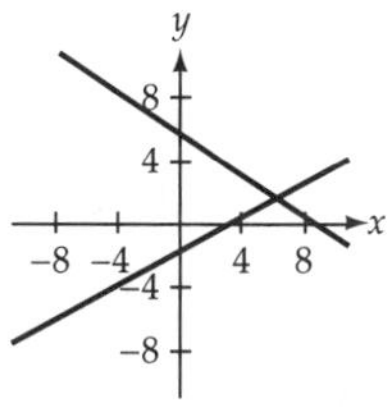

b. 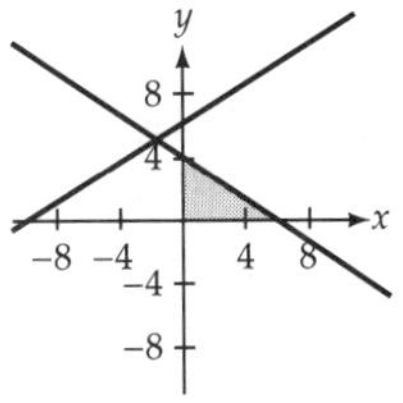

c. 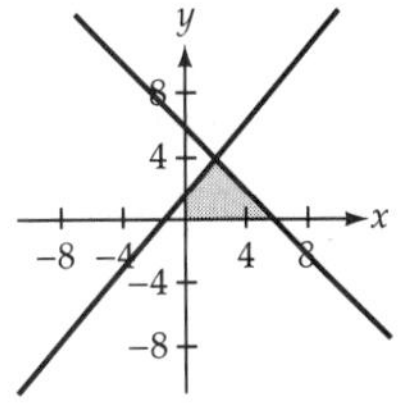

d. 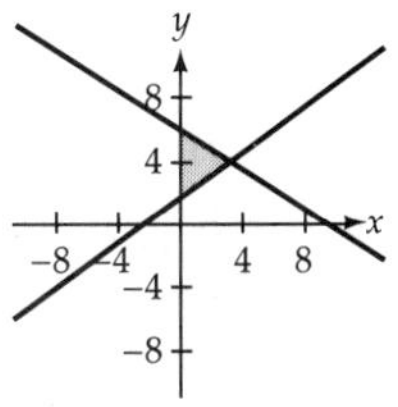

13. Graph the solution set: $\begin{cases} 3x - 4y \leq -8 \\ y \geq x^2 + 3x - 1. \end{cases}$

a. 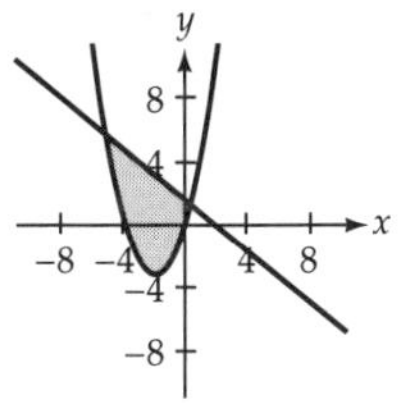

b. 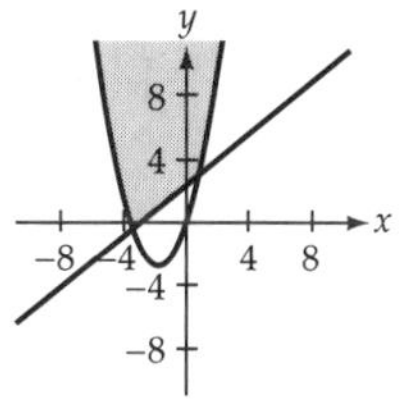

c. 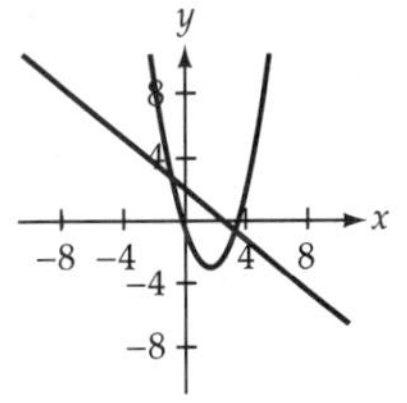

d. 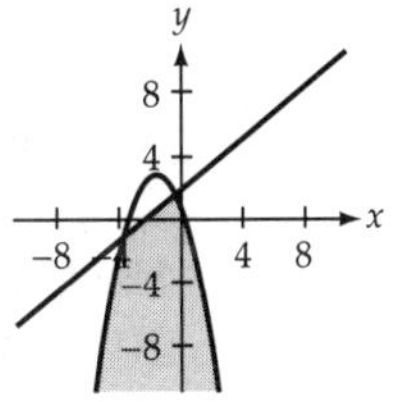

14. Find the partial fraction decomposition of $\dfrac{12x + 2}{(x - 4)(x + 6)}$.

a. $\dfrac{5}{x - 4} + \dfrac{-7}{x + 6}$ b. $\dfrac{7}{x - 4} + \dfrac{5}{x + 6}$ c. $\dfrac{5}{x - 4} + \dfrac{7}{x + 6}$ d. $\dfrac{7}{x - 4} + \dfrac{-5}{x + 6}$

15. Find the partial fraction decomposition of $\dfrac{3x^2 + 19x + 16}{x(x + 4)^2}$.

a. $\dfrac{1}{x} + \dfrac{3}{x} - \dfrac{5}{(x + 4)^2}$

b. $\dfrac{1}{x} + \dfrac{2}{x + 4} + \dfrac{3}{(x + 4)^2}$

c. $\dfrac{1}{x} + \dfrac{2}{x + 4} - \dfrac{5}{(x + 4)^2}$

d. $\dfrac{3}{x} + \dfrac{-5}{(x + 4)^2}$

16. Find the partial fraction decomposition of $\dfrac{3x^2 + 5x - 4}{(x - 1)(x^2 + 2)}$.

a. $\dfrac{6}{x - 1} + \dfrac{x + 6}{x^2 + 1}$

b. $\dfrac{6}{x - 1} - \dfrac{x + 6}{x^2 + 1}$

c. $\dfrac{2}{x - 1} - \dfrac{x - 6}{x^2 + 1}$

d. $\dfrac{2}{x - 1} + \dfrac{x + 6}{x^2 + 1}$

17. Find the equation of the circle that passes through the points (4, 10), (1, 1) and (8, 2).

a. $x^2 + 4x + y^2 - 5y + 25 = 0$ b. $x^2 - 8x + y^2 - 10y + 16 = 0$

c. $x^2 - 8x + y^2 - 5y + 25 = 0$ d. $x^2 + 4x + y^2 - 10y + 16 = 0$

18. A motorboat traveled a distance of 72 miles in 3 hours while traveling with the current. Against the current the same trip took 4.5 hours. Find the rate of the motorboat in calm water and the rate of the current.

a. motorboat: 20 mph, current: 4 mph b. motorboat: 16 mph, current: 2 mph

c. motorboat: 18 mph, current: 4 mph d. motorboat: 20 mph, current: 2 mph

19. A silversmith has two alloys. The first alloy is 28 percent silver, the second is 60 percent silver. How many grams of each should be mixed to produce 80 grams of an alloy that is 52 percent silver?

a. 15 g of 28% silver, 65 g of 60% silver
b. 20 g of 28% silver, 60 g of 60% silver
c. 25 g of 28% silver, 55 g of 60% silver
d. 30 g of 28% silver, 50 g of 60% silver

20. Maximize the objective function $C = 3x + 2y$ with constraints $x + 5y \le 40$, $7x - 2y \le 21$, $3x + y \ge 3$, $x \ge 0$, $y \ge 0$.

a. 24 b. 16 c. 29 d. 31

Chapter 10 Chapter Test Form E

1. Solve the system of equations: $\begin{cases} 7x - 3y = 1 \\ y = 8x + 11. \end{cases}$

   a. (-2, -5) b. (2, 27) c. Dependent d. Inconsistent

2. Solve the system of equations: $\begin{cases} 4x + 5y = 11 \\ 5x + 3y = -9. \end{cases}$

   a. (-6, 7) b. (9, -5) c. Dependent d. Inconsistent

3. Solve the system of equations: $\begin{cases} 3x - y + 2z = 15 \\ x + 2y + 3z = 5 \\ -2x + y - 3z = -16. \end{cases}$

   a. (2, 3, -3) b. (3, -2, 3) c. (2, -3, 3) d. (1, -3, 3)

4. Solve the system of equations: $\begin{cases} -x + 2y + 3z = 1 \\ x - y + z = -4 \\ -3x + 2y - 7z = 15. \end{cases}$

   a. (0, 2, -1) b. (2, 2, -4) c. Dependent d. Inconsistent

5. Solve the system of equations: $\begin{cases} 4x + 3y - 11z = 8 \\ x + y - 3z = 3. \end{cases}$

   a. (2c - 1, c, c + 4) b. (c - 1, 2c + 4, c)

   c. (2c - 1, c + 4, c) d. (2c + 1, c + 4, c)

6. Solve the homogeneous system of equations: $\begin{cases} 2x + y + z = 0 \\ 4x - 5y + z = 0 \\ 3x - 2y + z = 0. \end{cases}$

   a. $\left(-\frac{3c}{7}, -\frac{c}{7}, c\right)$ b. $\left(\frac{c}{7}, \frac{3c}{7}, c\right)$ c. (-c, 3c, c) d. (3c, -7c, c)

7. Solve the system of equations: $\begin{cases} y \quad x^2 + 5x - 3 \\ y = 8x - 5. \end{cases}$

a. (–1, –13), (2, –21)

b. (1, 3), (2, 11)

c. No real solution

d. (1, 3)

8. Solve the system of equations: $\begin{cases} x^2 - 144y^2 = 144 \\ x^2 + y^2 = 144. \end{cases}$

a. No real solution

b. (0, 12)

c. (12, 0), (–12, 0)

d. (0, 12), (0, –12)

9. Graph $2x - 3y \leq 6$.

a.

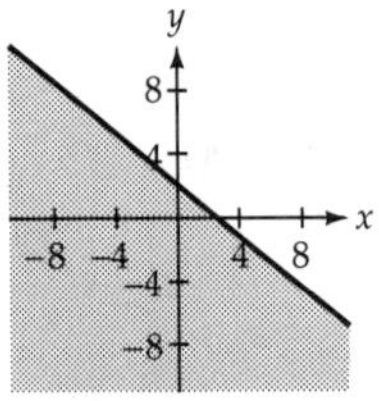

b.

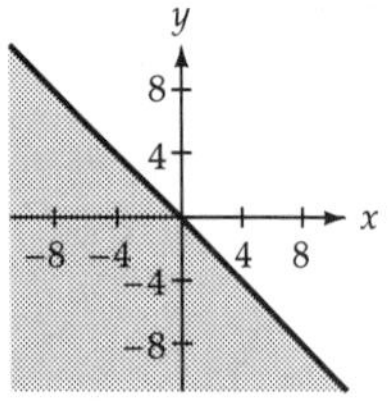

c.

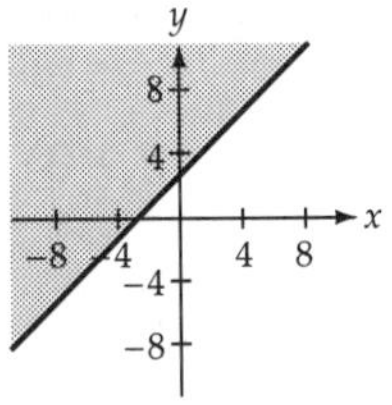

d.

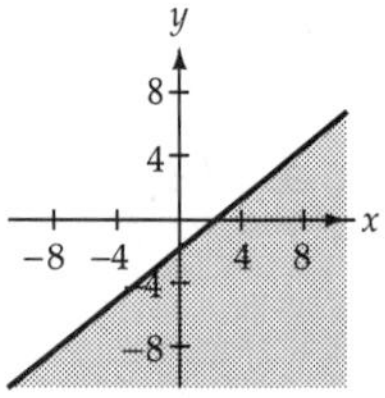

10. Graph $y \geq x^2 - 3x - 4$.

a.

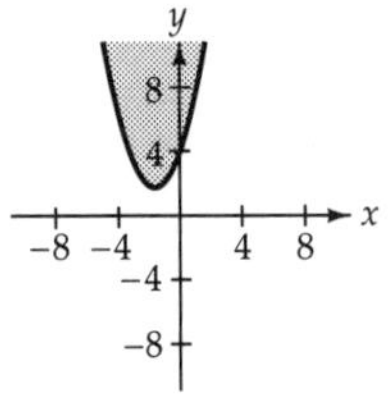

b.

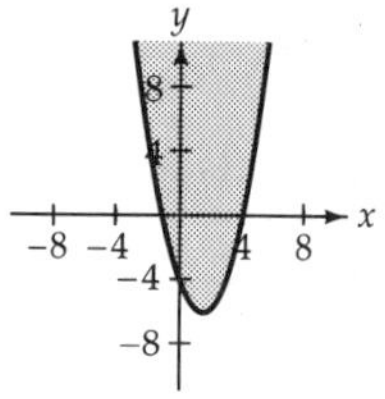

c.

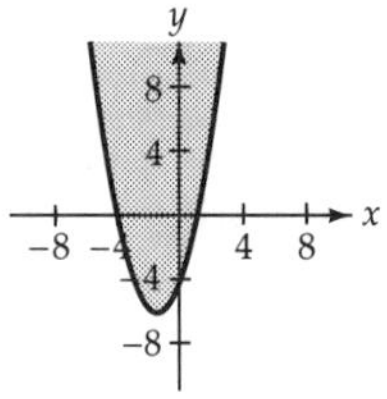

d.

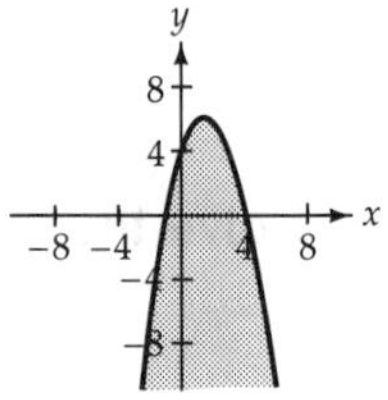

11. Graph the solution set: $\begin{cases} 4x + 5y \leq 20 \\ x - 3y < 9. \end{cases}$

a.

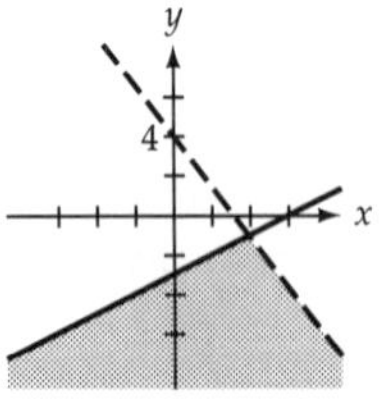

b.

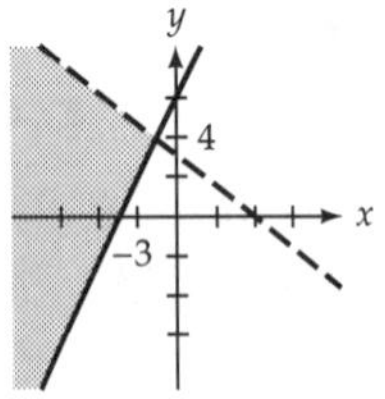

c.

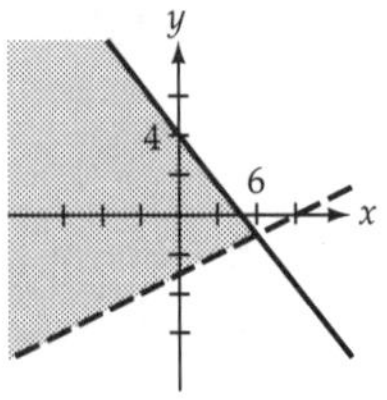

d.

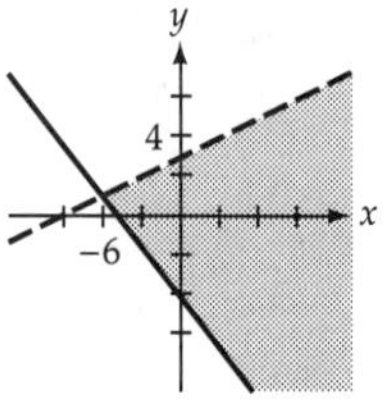

12. Graph the solution set: $\begin{cases} y > 2x + 6 \\ y \geq x^2 + 8x - 9. \end{cases}$

a. 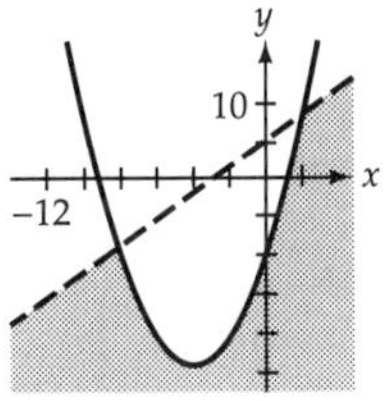

b. 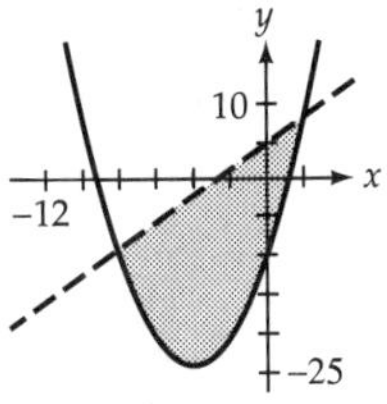

c. 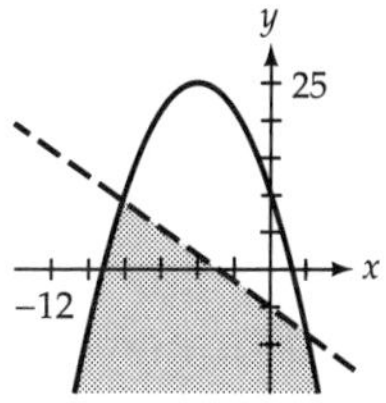

d. 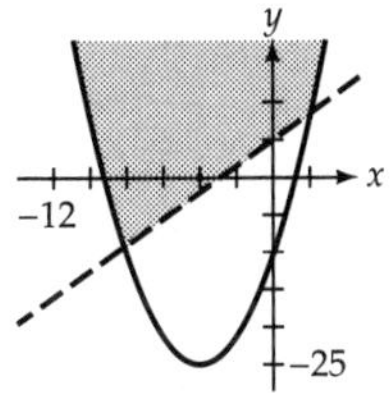

13. Graph the solution set: $\begin{cases} -3x + y \geq -6 \\ -x + 2y \geq 8 \\ x \geq 0,\ y \geq 0. \end{cases}$

a. 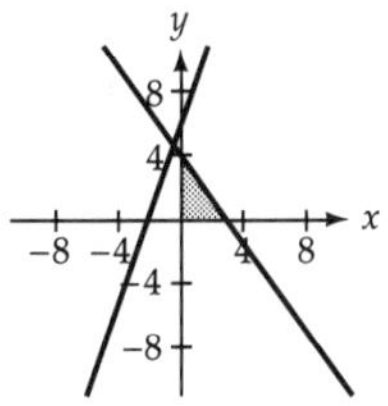

b. 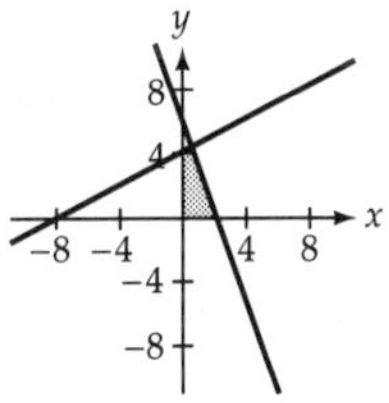

c. 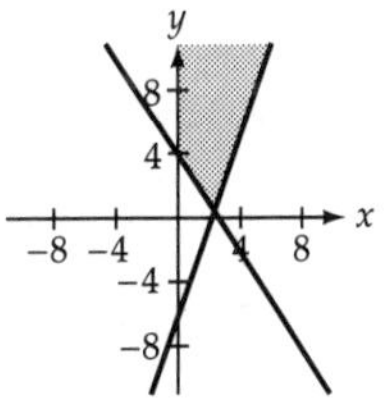

d. 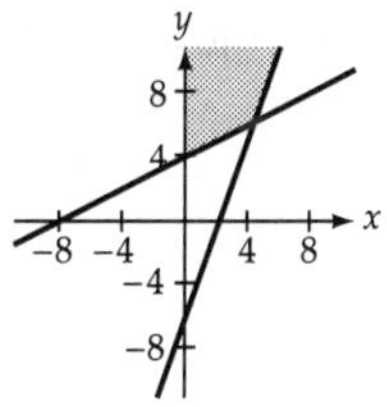

14. Find the partial fraction decomposition of $\dfrac{14x + 2}{(x - 5)(x + 7)}$.

a. $\dfrac{6}{x - 5} + \dfrac{8}{x + 7}$ b. $\dfrac{8}{x - 5} + \dfrac{6}{x + 7}$ c. $\dfrac{6}{x - 5} + \dfrac{-8}{x + 7}$ d. $\dfrac{-6}{x - 5} + \dfrac{8}{x + 7}$

15. Find the partial fraction decomposition of $\dfrac{3x^2 - 2x + 1}{x(x + 1)^2}$.

a. $\dfrac{2}{x} + \dfrac{1}{x + 1} + \dfrac{6}{(x + 1)^2}$

b. $\dfrac{6}{x} - \dfrac{2}{x + 1} - \dfrac{1}{(x + 1)^2}$

c. $\dfrac{1}{x} + \dfrac{2}{x + 1} - \dfrac{6}{(x + 1)^2}$

d. $\dfrac{2}{x} - \dfrac{6}{x + 1} + \dfrac{1}{(x + 1)^2}$

16. Find the partial fraction decomposition of $\dfrac{12x^2 + 19x - 53}{(7x - 1)(x^2 + 25)}$.

a. $\dfrac{-2}{7x - 1} + \dfrac{2x + 3}{x^2 + 25}$ b. $\dfrac{2}{7x - 1} + \dfrac{2x - 3}{x^2 + 25}$ c. $\dfrac{2}{7x - 1} + \dfrac{2x + 3}{x^2 + 25}$ d. $\dfrac{-2}{7x - 1} + \dfrac{2x - 3}{x^2 + 25}$

17. Find the equation of the form $y = ax^2 + bx + c$ whose graph passes through the points (-1, 7), (2, 10), and (1, 5).

a. $y = 2x^2 - x + 4$ b. $y = 2x^2 + x + 4$

c. $y = x^2 + 2x - 4$ c. $y = 2x^2 - x - 4$

18. During a race, a high-performance sailboat traveled 54 miles with the wind in 3 hours. The return trip, against the wind, took 6 hours. Find the rate of the boat in calm water and the rate of the wind.

a. sailboat: 12 mph, wind: 6 mph b. sailboat: 14 mph, wind: 4 mph

c. sailboat: 15 mph, wind: 3 mph d. sailboat: 13.5 mph, wind: 4.5 mph

19. A silversmith has two alloys. The first alloy is 18 percent silver, the second is 54 percent silver. How many grams of each should be mixed to produce 120 grams of an alloy that is 39 percent silver?

a. 30 g 18% silver, 90 g 54% silver b. 40 g 18% silver, 80 g 54% silver

c. 50 g 18% silver, 70 g 54% silver d. 60 g 18% silver, 60 g 54% silver

20. Minimize the objective function $C = 2x + 5y$ with constraints $3x + 4y \geq 12$, $2x + 7y \leq 49$, $5x - y \leq 30$, $x \geq 0$, $y \geq 0$.

a. 15 b. 20 c. 19 d. 8

1. Solve the system of equations: $\begin{cases} 4x - 3y = -3 \\ \quad\quad y = 6x - 13 \end{cases}$

a. (3, –31) b. (3, 5) c. Dependent d. Inconsistent

2. Solve the system of equations: $\begin{cases} 5x - 2y = 6 \\ 3x + 5y = 16. \end{cases}$

a. (3, 2) b. (–2, 3) c. (2, 2) d. (–2, –3)

3. Solve the system of equations: $\begin{cases} 4x - 5y + \quad = 23 \\ x + y + \quad = 2 \\ x + 5y + 2z = -5. \end{cases}$

a. (5, –1, –2) b. (3, –2, 1) c. Dependent d. Inconsistent

4. Solve the system of equations: $\begin{cases} x + y + \quad = -3 \\ 2x + 7y - 9z = -1 \\ x + 6y - 10z = -2. \end{cases}$

a. (5, –3, –5) b. (–4, 1, 0) c. Dependent d. Inconsistent

5. Solve the system of equations: $\begin{cases} 2x + y - 3z = 2 \\ 4x - 5y + z = 5. \end{cases}$

a. $\left(c + \frac{15}{4}, c - \frac{1}{7}, c\right)$ b. $\left(2c + 7, c - \frac{1}{7}, c\right)$

c. $\left(c - \frac{7}{2}, c + \frac{15}{4}, c\right)$ d. (c + 7, c – 7, c)

6. Solve the homogeneous system of equations: $\begin{cases} x - y + 2z = 0 \\ 2x - 6y + 3z = 0 \\ x + 3y + 3z = 0. \end{cases}$

a. (4c, c, 9c) b. (9c, c, 4c) c. (c, 9c, –4c) d. (9c, c, –4c)

7. Solve the system of equations $\begin{cases} x^2 + x^2 = 10 \\ y^2 = 9x. \end{cases}$

a. No real solution

b. $(10, 3\sqrt{10}), (10, -3\sqrt{10})$

c. (1, 3), (1, –3)

d. $(1, \sqrt{3}), (1, -\sqrt{3})$

8. Solve the system of equations: $\begin{cases} x^2 - y^2 = 9 \\ x^2 + y^2 = 9. \end{cases}$

a. (0, 0)

b. (3, 0), (–3,0)

c. (0, 3), (0, –3)

d. (3, 0)

9. Graph $y > 2x - 6$.

a.

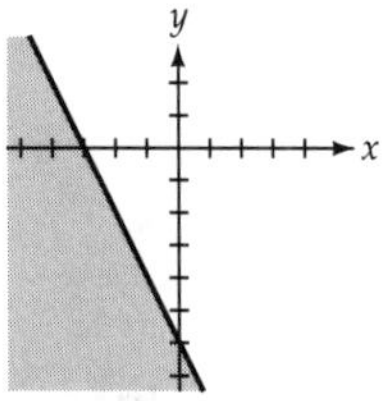

b.

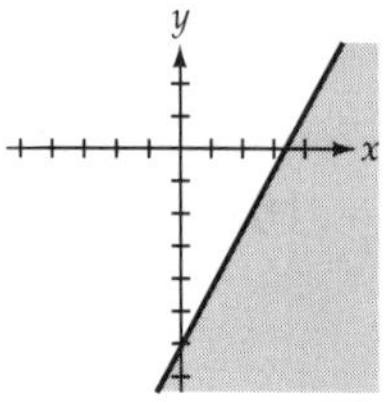

c.

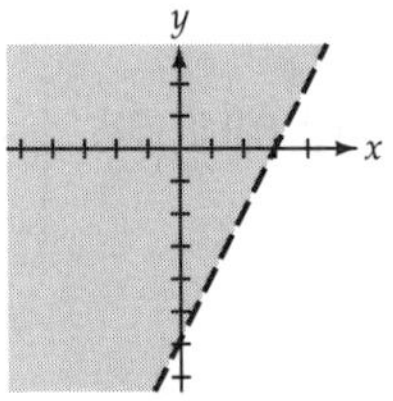

d.

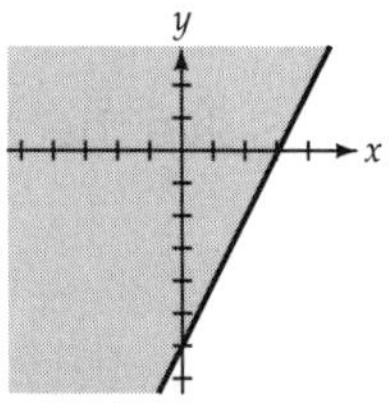

10. Graph $y \le x^2 - 2x + 1$.

a.

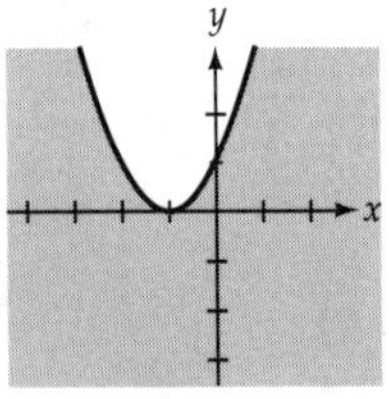

b.

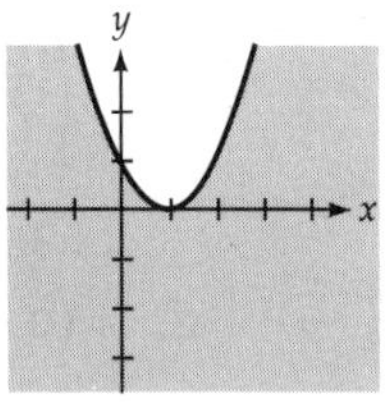

c.

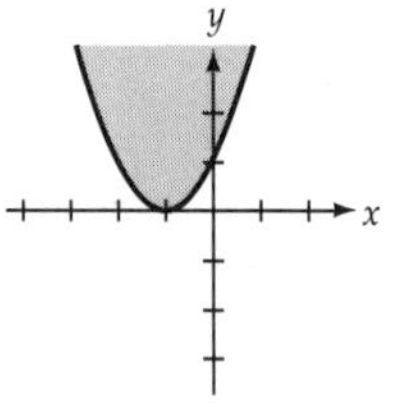

d.

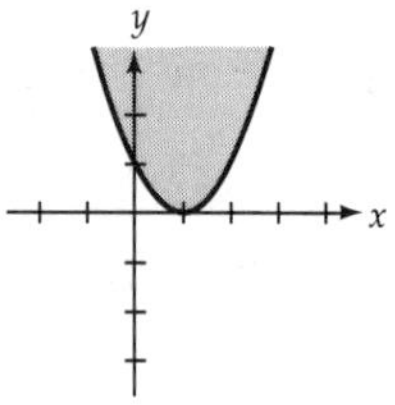

11. Graph the solution set: $\begin{cases} x^2 + y^2 \le 9 \\ x^2 - y^2 \ge 3. \end{cases}$

a.
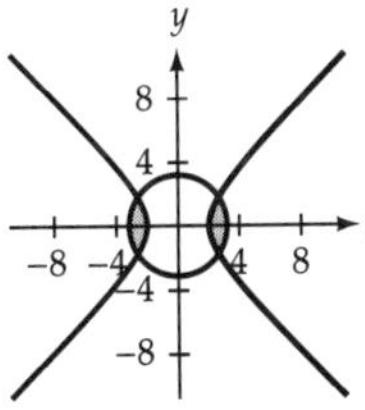

b.
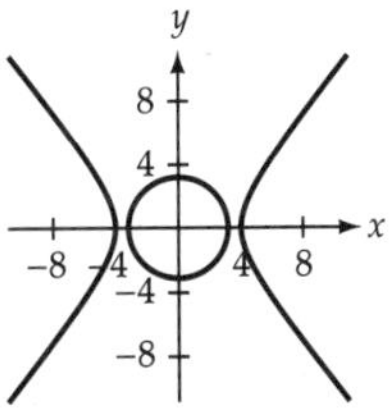

c.
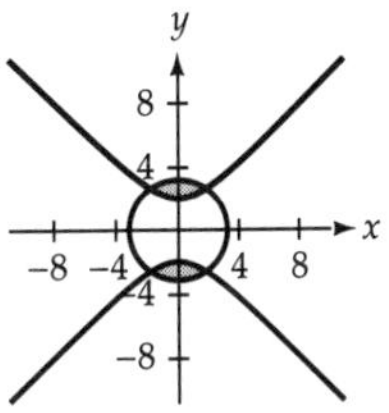

d.
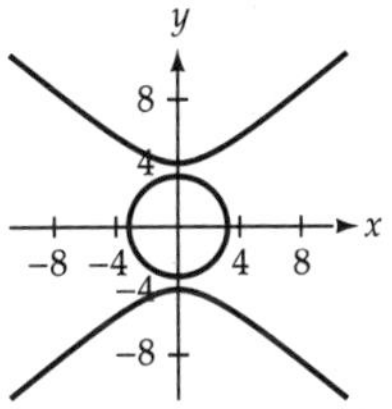

12. Graph the solution set: $\begin{cases} y < 4x - 8 \\ y \ge x^2 - 10x - 11. \end{cases}$

a.
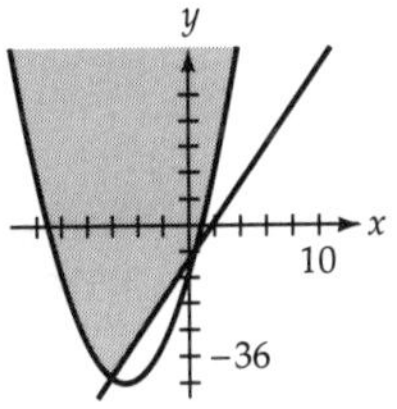

b.
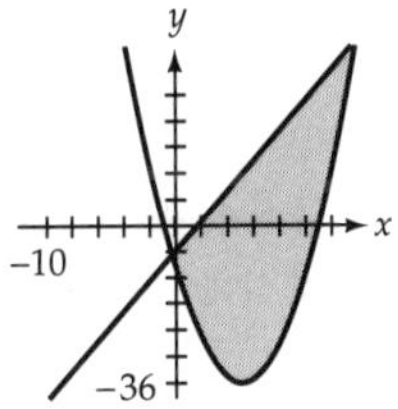

c.
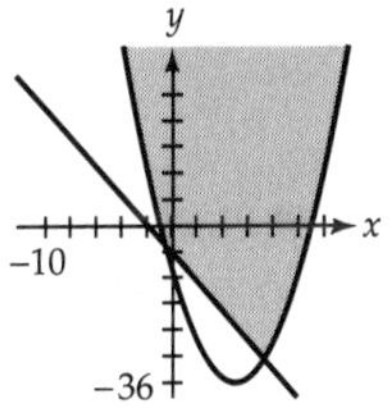

d.
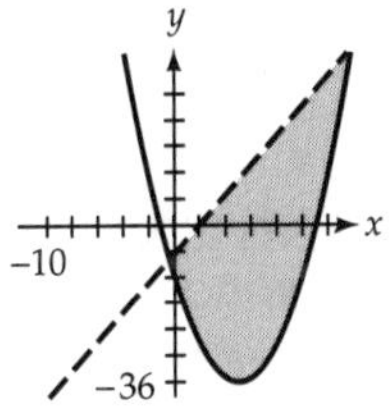

13. Graph the solution set: $\begin{cases} 4x + 5y \le 20 \\ 3x - 4y \le 12 \\ x \ge 0,\ y \ge 0. \end{cases}$

a.
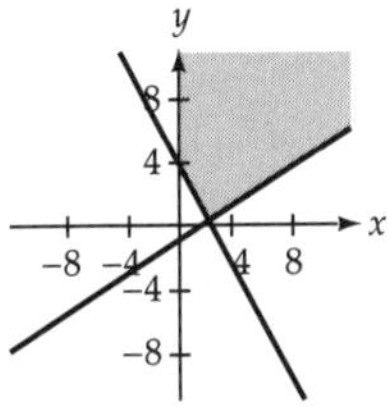

b.
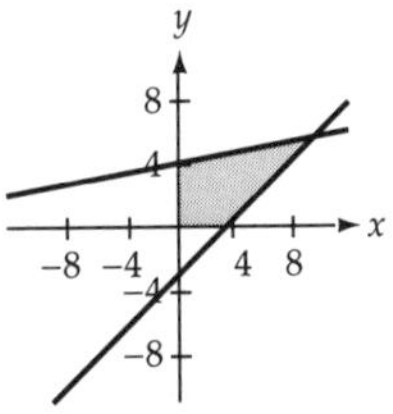

c.
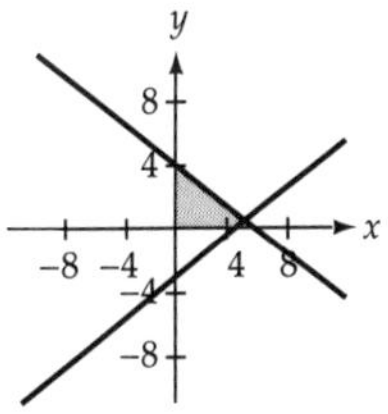

d.
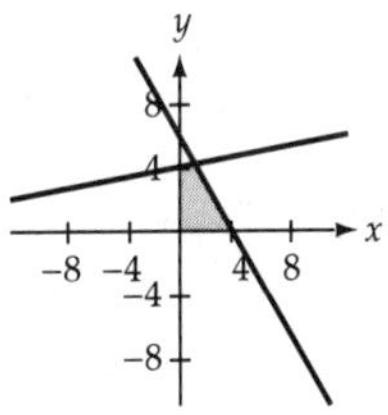

14. Find the partial fraction decomposition of $\dfrac{x - 12}{x(3x - 2)}$.

a. $\dfrac{3}{x} - \dfrac{13}{3x - 2}$ b. $\dfrac{6}{x} - \dfrac{17}{3x - 2}$ c. $\dfrac{17}{x} - \dfrac{6}{3x - 2}$ d. $\dfrac{3}{x} + \dfrac{13}{3x - 2}$

15. Find the partial fraction decomposition of $\dfrac{14x^2 - 29x + 24}{x(x - 2)^2}$.

a. $\dfrac{6}{x} + \dfrac{-8}{x - 2} + \dfrac{11}{(x - 2)^2}$ b. $\dfrac{6}{x} + \dfrac{8}{x - 2} + \dfrac{11}{(x - 2)^2}$

16. Find the partial fraction decomposition of $\frac{16x^2 + 21x + 3}{(6x + 5)(x^2 + 1)}$.

a. $\frac{-1}{6x + 5} + \frac{3x + 1}{x^2 + 1}$

b. $\frac{2}{6x + 5} + \frac{3x + 1}{x^2 + 1}$

c. $\frac{-2}{6x + 5} + \frac{3x - 1}{x^2 + 1}$

d. $\frac{2}{6x + 5} + \frac{3x + 1}{x^2 + 1}$

17. Find an equation of the form $y = ax^2 + bx + c$ whose graph passes through the points (1, 4), (–1, 0), and (–2, 7).

a. $y = 3x^2 - 2y - 1$

b. $y = 3x^2 + 2x + 1$

c. $y = 2x^2 - 3x + 1$

d. $y = 3x^2 + 2x - 1$

18. During a storm a small corporate jet traveled 750 miles in 3 hours with the wind. Against the wind, the same trip took 5 hours. Find the rate of the plane in calm air and the rate of the wind.

a. plane: 175 mph, wind: 75 mph

b. plane: 200 mph, wind: 50 mph

c. plane: 150 mph, wind: 100 mph

d. plane: 225 mph, wind: 25 mph

19. A goldsmith has two gold alloys. The first alloy is 25 percent gold and the second alloy is 55 percent gold. How many grams of each should be mixed to produce 40 grams of an alloy that is 43 percent gold?

a. 24 g of 55% gold, 16 g of 25% gold

b. 16 g of 55% gold, 24 g of 25% gold

c. 18 g of 55% gold, 22 g of 25% gold

d. 25 g of 55% gold, 15 g of 25% gold

20. Maximize the objective function $C = 3x + 5y$ with constraints $2x + 3y \le 90$, $4x + y \le 60$, $x \ge 0$, and $y \ge 0$.

a. 135

b. 147

c. 150

d. 45

1. Write the augmented matrix, coefficient matrix, and constant matrix for the system of equations $\begin{cases} 3x - 4y + z = -1 \\ 2x - 3y - 4z = 17. \\ x - 5z = 4 \end{cases}$

2. Write a system of equations that is equivalent to the augmented matrix

$$\begin{bmatrix} -2 & 5 & -2 & 3 & 1 \\ 4 & -2 & 3 & 1 & -4 \\ -1 & -4 & 2 & 3 & -1 \end{bmatrix}.$$

In Exercises 3 to 5, solve the system of equations by using the Gauss–Jordan method.

3. $\begin{cases} 2x + y + z = 3 \\ x + 2y - z = 12 \\ x + y - 2z = 11 \end{cases}$

4. $\begin{cases} 2x + 5y - 3z = 3 \\ 3x + 2y + z = 10 \\ x - y + 2z = 5 \end{cases}$

5. $\begin{cases} x + 2y - z + w = 6 \\ -2x + 2y + 2z + 2w = 14 \\ 2x - y + z + w = -2 \end{cases}$

In Exercises 6 to 14, let $A = \begin{bmatrix} -1 & 2 & 4 \\ 4 & -2 & 2 \end{bmatrix}$, $B = \begin{bmatrix} 1 & 0 & 1 \\ -3 & 2 & 2 \\ 3 & 1 & 4 \end{bmatrix}$, and $C = \begin{bmatrix} 5 & 0 & 2 \\ 4 & 3 & -1 \\ 1 & -1 & 0 \end{bmatrix}$.

Perform each possible operation. If an operation is not possible, so state.

6. $-5A$

7. $A + C$

8. $2B - 4C$

9. AB

10. CB

11. $BC - CB$

12. C^2

13. A^2

14. B^{-1}

15. Find the minor and cofactor of B_{32} for matrix B.

16. Find the determinant of B by expanding by cofactors of row 2.

17. Find the determinant of C by using elementary row operations.

18. Solve the system of equations by using inverse matrix methods: $\begin{cases} 2x + 7y = 4 \\ 3x + 8y = 1. \end{cases}$

19. Find the value of y for the following system of equations by using Cramer's Rule.

$$\begin{cases} x + y + z = 1 \\ 2x - y + 3z = -2 \\ -x + y - z = 1 \end{cases}$$

20. In a laboratory, scientists are feeding rats a special diet that is to contain 34 grams of protein, 82 grams of carbohydrates, and 57 grams of fat. The following table shows the protein, carbohydrate, and fat amounts in one serving of foods A, B, and C.

	Protein	Carbohydrate	Fat
Food A	12	1	19
Food B	6	3	10
Food C	1	18	2

How many servings of each food are necessary to meet the goals of the specific diet?

Chapter 11 Chapter Test Form B

1. Write the augmented matrix, coefficient matrix, and constant matrix for the system of equation $\begin{cases} 2x - 3y + z = 3 \\ -x + 4y = -2 \\ 3x - 5y + 2z = 17 \end{cases}$.

2. Write a system of equations that is equivalent to the augmented matrix $\begin{bmatrix} 1 & 2 & 4 & 5 & 1 \\ -2 & 3 & -1 & 4 & -7 \\ 2 & -3 & 4 & -1 & 3 \end{bmatrix}$.

In Exercises 3 to 5, solve the system of equations by using the Gauss-Jordan method.

3. $\begin{cases} 2x + y + z = 1 \\ x + y + z = 0 \\ x + y - 2z = -6 \end{cases}$

4. $\begin{cases} 2x + 3y - z = 5 \\ 3x + 4y - z = 9 \\ x - y + 2z = 10 \end{cases}$

5. $\begin{cases} -x + y + 2z + w = 3 \\ 3x - y + z + 2w = 27 \\ -2x + 3y - 2z - w = -15 \end{cases}$

In Exercises 6 to 17, let $A = \begin{bmatrix} 6 & 0 & -3 \\ 1 & 2 & 9 \end{bmatrix}$, $B = \begin{bmatrix} 4 & 0 & 5 \\ -3 & 2 & -3 \\ 6 & 2 & 9 \end{bmatrix}$, and $C = \begin{bmatrix} 1 & 0 & 1 \\ 4 & 2 & 3 \\ -4 & 3 & -5 \end{bmatrix}$.

Perform each possible operation. If an operation is not possible, so state.

6. $-4B$
7. $A + B$
8. $3B - 2C$
9. AC
10. CB
11. $AC - CB$
12. A^2
13. B^2
14. B^{-1}

15. Find the minor and cofactor of B_{12} for matrix B.

16. Find the determinant of B by expanding by cofactors of column 3.

17. Find the determinant of C by using elementary row operations.

18. Solve the system by using inverse matrix methods: $\begin{cases} x + 2y + z = 4 \\ 2x - y + 3z = -1 \\ 3x + 2y - z = 0. \end{cases}$

19. Find the value of y for the following system of equations by using Cramer's Rule.

$$\begin{cases} x + y + z = 9 \\ 2x - y + 2z = 16 \\ -x + 2y - 3z = -12 \end{cases}$$

20. In a laboratory, scientists are feeding rats a special daily diet that is to contain 37 grams of protein, 45 grams of carbohydrates, and 55 grams of fat. The following table shows the protein, carbohydrates, and fat amounts in one serving of foods A, B, and C.

	Protein	Carbohydrate	Fat
Food A	4	6	9
Food B	3	7	2
Food C	11	1	6

How many servings of each food are necessary to meet the goals of the specified diet?

1. Write the augmented matrix, coefficient matrix, and constant matrix for the system of equations $\begin{cases} x - 3y + z = -4 \\ 2x - 3y - 5z = 2 \\ x + 7x - 8z = -1 \end{cases}$.

2. Write a system of equations that is equivalent to the augmented matrix $\begin{bmatrix} 2 & 1 & -2 & 3 & 4 \\ -1 & 2 & 3 & 1 & 5 \\ 5 & 4 & -1 & 2 & 3 \end{bmatrix}$.

In Exercises 3 to 5, solve the system of equations by using the Gauss-Jordan method.

3. $\begin{cases} x - 2y + z = 6 \\ 2x + y - z = -3 \\ -x + 4y + 2z = 15 \end{cases}$

4. $\begin{cases} 2x - y + z = -2 \\ x + 2y - z = 6 \\ 3x + y - 2z = 0 \end{cases}$

5. $\begin{cases} x + y - z + w = -1 \\ 2x - y + z - 2w = 6 \\ 3x - 2y + 2z - w = 13 \end{cases}$

In Exercises 6 to 17, let $A = \begin{bmatrix} 1 & -2 & 4 \\ -1 & 3 & 2 \end{bmatrix}$, $B = \begin{bmatrix} 1 & -2 & 3 \\ 2 & -1 & 0 \\ 0 & 2 & 1 \end{bmatrix}$, $C = \begin{bmatrix} 1 & 1 & -1 \\ 1 & 0 & -2 \\ 0 & 2 & 3 \end{bmatrix}$.

Perform each possible operation. If an operation is not possible, so state.

6. $-3C$

7. $A + C$

8. $2B - 3C$

9. AB

10. BC

11. $BC - CB$

12. A^2

13. C^2

14. B^{-1}

15. Find the minor and cofactor of B_{12} for matrix B.

16. Find the determinant of B by expanding by cofactors of column 2.

17. Find the determinant of C by using elementary row operations.

18, Solve the system of equations by using inverse matrix methods:

$$\begin{cases} x + y + z = -4 \\ x - 2z = 0 \\ 2y + 3z = -7. \end{cases}$$

19. Find the value of y for the following system of equations by using Cramer's Rule.

$$\begin{cases} -2x + z = 2 \\ 5y + 12z = 9 \\ y + z = 1. \end{cases}$$

20. A simplified economy has three major industries: manufacturing, agriculture, and service. The input-output matrix for this economy is:

$$\begin{bmatrix} .12 & .18 & .08 \\ .06 & .10 & .14 \\ .20 & .15 & .10 \end{bmatrix}$$

Set up, but do not solve, a matrix equation that when solved will determine the gross output needed to satisfy consumer demand of \$60 million of manufacture, \$80 million of agriculture, and \$45 million of service.

1. Use elementary row operations to write the matrix in echelon form: $\begin{bmatrix} 4 & 3 & 2 & 7 \\ -4 & 2 & 3 & 8 \\ 5 & -1 & -1 & 2 \end{bmatrix}$

a. $\begin{bmatrix} 1 & 0 & 2 & 10 \\ 0 & 1 & 3 & 3 \\ 0 & 0 & 1 & -6 \end{bmatrix}$ b. $\begin{bmatrix} 1 & 1 & 4 & 10 \\ 0 & 1 & 6 & 3 \\ 0 & 0 & 1 & 6 \end{bmatrix}$ c. $\begin{bmatrix} 1 & 0 & -2 & -5 \\ 0 & 1 & 6 & 3 \\ 0 & 0 & 1 & 6 \end{bmatrix}$ d. $\begin{bmatrix} 1 & -4 & -3 & -5 \\ 0 & 1 & 1 & 3 \\ 0 & 0 & 1 & 6 \end{bmatrix}$

2. Solve the system of equations by the Gauss-Jordan method: $\begin{cases} x - 3y + 2z = 13 \\ 2x + y - 3z = -2 \\ 4x - 4y + 9z = -7. \end{cases}$

a. (–2, –1, 3) b. (4, 1, –3) c. (–2, –7, –3) d. No solution

3. Solve the system of equations by the Gauss-Jordan method: $\begin{cases} 3x + 2y + z = 20 \\ 2x + y + z = 13 \\ x + 2y - z = 8 \end{cases}$

a. (c – 6, c + 1, c) b. (6 – c, c + 1, c) c. (6 – c, c – 1, c) d. No solution

4. Given A = $\begin{bmatrix} 7 & -2 \\ -3 & 5 \end{bmatrix}$ and B = $\begin{bmatrix} -1 & 2 \\ -8 & 6 \end{bmatrix}$, find 2A + 7B.

a. $\begin{bmatrix} 7 & 10 \\ -62 & 52 \end{bmatrix}$ b. $\begin{bmatrix} 0 & 10 \\ -62 & 52 \end{bmatrix}$ c. $\begin{bmatrix} 7 & 10 \\ 50 & 47 \end{bmatrix}$ d. $\begin{bmatrix} 63 & -18 \\ -27 & 45 \end{bmatrix}$

5. Find the product $\begin{bmatrix} 5 & -1 \\ 2 & 3 \\ 0 & 7 \end{bmatrix}\begin{bmatrix} 8 & 4 & -1 \\ 0 & 2 & 6 \end{bmatrix}$.

a. $\begin{bmatrix} 40 & 18 & -11 \\ 16 & 14 & 16 \\ 0 & 14 & 42 \end{bmatrix}$ b. $\begin{bmatrix} 40 & 16 & 0 \\ 18 & 14 & 14 \\ -11 & 16 & 42 \end{bmatrix}$ c. $\begin{bmatrix} 36 & 21 \\ 28 & 5 \\ 28 & -7 \end{bmatrix}$ d. Undefined

6. Find the product $\begin{bmatrix} 8 & -1 & 9 \\ 4 & 2 & 5 \end{bmatrix} \begin{bmatrix} 6 & 2 & 0 \\ 1 & -3 & 4 \end{bmatrix}$.

a. $\begin{bmatrix} 52 & -8 & 20 \\ 24 & -6 & 0 \end{bmatrix}$ b. $\begin{bmatrix} 46 & 45 \\ 28 & 18 \end{bmatrix}$ c. $\begin{bmatrix} 48 & 4 \\ -2 & -6 \\ 0 & 20 \end{bmatrix}$ d. Undefined

7. Given A = $\begin{bmatrix} 4 & 0 & 5 \\ -3 & 2 & -3 \\ 6 & 1 & 0 \end{bmatrix}$, find A^2.

a. $\begin{bmatrix} 16 & 0 & 25 \\ 9 & 4 & 9 \\ 36 & 1 & 0 \end{bmatrix}$ b. $\begin{bmatrix} 46 & 5 & 20 \\ -36 & 1 & -21 \\ 21 & 2 & 27 \end{bmatrix}$ c. $\begin{bmatrix} 46 & 5 & 20 \\ -36 & 2 & 27 \\ 1 & 21 & -21 \end{bmatrix}$ d. $\begin{bmatrix} 16 & 0 & 5 \\ 9 & 1 & -21 \\ 21 & 2 & 27 \end{bmatrix}$

8. Find the system of equations that is equivalent to the matrix equation

$$\begin{bmatrix} 9 & 4 & 1 \\ -5 & 0 & 11 \\ 2 & 6 & 3 \end{bmatrix} \begin{bmatrix} x \\ y \\ z \end{bmatrix} = \begin{bmatrix} 1 \\ 9 \\ -8 \end{bmatrix}.$$

a. $\begin{cases} 9x - 5y + 2z = 1 \\ 4x + 6z = 9 \\ x + 11y + 3z = -8 \end{cases}$

b. $\begin{cases} 9x - 5y + 2z = 1 \\ 4x + 6y = 9 \\ x + 11y + 3z = -8 \end{cases}$

c. $\begin{cases} 9x + 4y + z = 1 \\ -5x + 11z = 9 \\ 2x + 6y + 3z = -8 \end{cases}$

d. $\begin{bmatrix} 9x + 4y + z = 1 \\ -5x + 11y = 9 \\ 2x + 6y + 3z = -8 \end{bmatrix}$

9. Find the inverse of the matrix $\begin{bmatrix} 4 & 9 \\ 3 & 7 \end{bmatrix}$.

a. $\begin{bmatrix} -7 & 9 \\ 3 & -4 \end{bmatrix}$ b. $\begin{bmatrix} 7 & 9 \\ 3 & 4 \end{bmatrix}$ c. $\begin{bmatrix} 7 & -9 \\ -3 & 4 \end{bmatrix}$ d. No inverse

10. Find the inverse of the matrix $\begin{bmatrix} 1 & 4 & 6 \\ 2 & 8 & 9 \\ 3 & 12 & 11 \end{bmatrix}$.

a. $\begin{bmatrix} -1 & -4 & -6 \\ -2 & -8 & -9 \\ -3 & -12 & -11 \end{bmatrix}$ b. $\begin{bmatrix} 11 & 2 & 3 \\ 4 & 8 & 12 \\ 6 & 9 & 1 \end{bmatrix}$ c. $\begin{bmatrix} 6 & \frac{1}{3} & \frac{2}{3} \\ 0 & 1 & 3 \\ 7 & 6 & -4 \end{bmatrix}$ d. No inverse

11. Find the inverse of the matrix $\begin{bmatrix} 1 & 0 & 1 \\ 4 & 2 & 3 \\ -4 & 3 & -5 \end{bmatrix}$.

a. $\begin{bmatrix} -1 & 0 & -1 \\ -4 & -2 & -3 \\ 4 & -3 & 5 \end{bmatrix}$ b. $\begin{bmatrix} -19 & 3 & -2 \\ 8 & -1 & 1 \\ 20 & -3 & 2 \end{bmatrix}$ c. $\begin{bmatrix} 1 & 4 & -4 \\ 0 & 2 & 3 \\ 1 & 3 & -5 \end{bmatrix}$ d. No inverse

12. Solve the system of equations by using inverse matrix methods: $\begin{cases} 8x + 3y = 5 \\ 7x + 5y = 21. \end{cases}$

a. (2, 7) b. (7, –2) c. (–2, 7) d. Inconsistent

13. Solve the system of equations by using inverse matrix methods: $\begin{cases} 6x + 3y + 2z = 1 \\ 2x + y + z = 1 \\ 5x + 2y + 2z = 3. \end{cases}$

a. (1, –3, –2) b. (3, 1, –2) c. (1, –3, 2) d. Inconsistent

14. Given $A = \begin{bmatrix} 7 & 0 & 1 \\ 3 & -5 & 4 \\ 6 & 2 & -4 \end{bmatrix}$, find M_{32} and C_{12}.

a. $M_{32} = 25$, $C_{12} = 36$
b. $M_{32} = -31$, $C_{12} = 0$
c. $M_{32} = -34$, $C_{12} = 0$
d. $M_{32} = 25$, $C_{12} = -36$

15. Find the determinant of matrix $A = \begin{bmatrix} 8 & -4 \\ 2 & 11 \end{bmatrix}$.

a. 96
b. 82
c. –34
d. 94

16. Find the determinant of matrix $A = \begin{bmatrix} -7 & 7 & 1 \\ 4 & 0 & 6 \\ -1 & 3 & 2 \end{bmatrix}$.

a. 40
b. –40
c. 62
d. –62

17. Find the determinant of matrix $A = \begin{bmatrix} 5 & 4 & -6 & 8 \\ 0 & 0 & 3 & 0 \\ -1 & 0 & 4 & -5 \\ 1 & 1 & -5 & 1 \end{bmatrix}$

a. –1
b. –3
c. –5
d. –7

18. Solve the system of equations using Cramer's Rule: $\begin{cases} 5x - 6y = 3 \\ 2x + 7y = -4. \end{cases}$

a. $\left(-\frac{3}{47}, -\frac{26}{47}\right)$
b. $\left(\frac{3}{47}, -\frac{26}{47}\right)$
c. $\left(-\frac{3}{25}, -\frac{26}{25}\right)$
d. $\left(\frac{3}{25}, \frac{26}{25}\right)$

19. Solve the system of equations for y: $\begin{cases} x + 8z = 1 \\ x + 3z = 0 \\ 7y - 3z = 5. \end{cases}$

a. $\frac{4}{5}$
b. $\frac{22}{35}$
c. $-\frac{4}{5}$
d. $\frac{22}{7}$

20. A simplified economy has three major industries: mining, manufacturing, and transportation

The input-output matrix for this economy is $\begin{bmatrix} .13 & .19 & .9 \\ .07 & .11 & .05 \\ .18 & .12 & .10 \end{bmatrix}$. Set up, but do not solve, a matrix equation that when solved will determine the gross output needed to satisfy consumer demand of 80 million for mining, 55 million for manufacturing, and 12 million for transportation.

a. $\left(\begin{bmatrix} 1 & 0 & 0 \\ 0 & 1 & 0 \\ 0 & 0 & 1 \end{bmatrix} - \begin{bmatrix} .13 & .19 & .09 \\ .07 & .11 & .05 \\ .18 & .12 & .10 \end{bmatrix}\right)^{-1} \cdot \begin{bmatrix} 80 \\ 55 \\ 12 \end{bmatrix}$

b. $\begin{bmatrix} 1 & 0 & 0 \\ 0 & 1 & 0 \\ 0 & 0 & 1 \end{bmatrix} - \begin{bmatrix} .13 & .19 & .09 \\ .07 & .11 & .05 \\ .18 & .12 & .10 \end{bmatrix} \cdot \begin{bmatrix} 80 \\ 55 \\ 12 \end{bmatrix}$

c. $\begin{bmatrix} 1 & 0 & 0 \\ 0 & 1 & 0 \\ 0 & 0 & 1 \end{bmatrix} - \left(\begin{bmatrix} .13 & .19 & .09 \\ .07 & .11 & .05 \\ .18 & .12 & .10 \end{bmatrix}\right)^{-1} \cdot \begin{bmatrix} 80 \\ 55 \\ 12 \end{bmatrix}$

d. $\left(\begin{bmatrix} .13 & .19 & 0 \\ .07 & .11 & 0 \\ .18 & .12 & 1 \end{bmatrix} - \begin{bmatrix} 1 & 0 & 0 \\ 0 & 1 & 0 \\ 0 & 0 & 1 \end{bmatrix}\right)^{-1} \cdot \begin{bmatrix} 80 \\ 55 \\ 12 \end{bmatrix}$

1. Use elementary row operations to write the matrix in echelon form: $\begin{bmatrix} 5 & 2 & -3 & -4 \\ -5 & -1 & 2 & 7 \\ -4 & -2 & 1 & 4 \end{bmatrix}$.

a. $\begin{bmatrix} 1 & 0 & -2 & 0 \\ 0 & 1 & -1 & 3 \\ 0 & 0 & 5 & -10 \end{bmatrix}$

b. $\begin{bmatrix} 1 & 0 & -2 & 0 \\ 0 & 0 & 1 & 3 \\ 0 & 0 & 1 & 2 \end{bmatrix}$

c. $\begin{bmatrix} 2 & 0 & -4 & 0 \\ 0 & 1 & -1 & 3 \\ 0 & 0 & 1 & 2 \end{bmatrix}$

d. $\begin{bmatrix} 2 & 0 & -4 & 0 \\ 0 & 1 & -1 & 3 \\ 0 & 0 & 1 & 2 \end{bmatrix}$

2. Solve the system of equations by the Gauss-Jordan method: $\begin{cases} 2x + y + z = 0 \\ x + 2y - z = 9 \\ x - y + 2z = -9. \end{cases}$

a. (–1, 2, 4) b. (1, 2, 4) c. (2, 2, –4) d. dependent

3. Solve the system of equations by the Gauss-Jordan method: $\begin{cases} x + y + z = 1 \\ 2x - y + 3z = 3 \\ -2 + 2y - 2z = -2. \end{cases}$

a. $\left(\frac{1 - c}{3}, \frac{c - 1}{3}, c\right)$

b. $\left(\frac{4 - 4c}{3}, \frac{c - 1}{3}, c\right)$

c. $\left(\frac{c + 1}{3}, \frac{c - 1}{3}, c\right)$

d. $\left(\frac{c}{3}, -\frac{c}{3}, c\right)$

4. Given A = $\begin{bmatrix} 6 & -3 \\ 4 & 1 \end{bmatrix}$ and B = $\begin{bmatrix} -2 & -3 \\ 9 & -7 \end{bmatrix}$, find 2A + 3B.

a. $\begin{bmatrix} 18 & 3 \\ 35 & 23 \end{bmatrix}$ b. $\begin{bmatrix} 18 & 3 \\ 35 & -19 \end{bmatrix}$ c. $\begin{bmatrix} 6 & -3 \\ 19 & 35 \end{bmatrix}$ d. $\begin{bmatrix} 6 & -3 \\ 19 & 35 \end{bmatrix}$

5. Find the product $\begin{bmatrix} 8 & -1 \\ 7 & -4 \\ -3 & 0 \end{bmatrix}\begin{bmatrix} 11 & 2 \\ 0 & -1 \end{bmatrix}$.

a. $\begin{bmatrix} 88 & 35 & -6 \\ 0 & -12 & 0 \end{bmatrix}$ b. $\begin{bmatrix} 88 & 77 & -33 \\ 43 & 23 & -15 \\ 15 & 18 & -6 \end{bmatrix}$ c. $\begin{bmatrix} 88 & 43 & 15 \\ 77 & 23 & 18 \\ -33 & -15 & -6 \end{bmatrix}$ d. Undefined

6. Find the product $\begin{bmatrix} -2 & 4 & 6 \\ 3 & 1 & 5 \end{bmatrix}\begin{bmatrix} 0 & 3 & 5 \\ 1 & 0 & 6 \end{bmatrix}$.

a. $\begin{bmatrix} 0 & 12 & 30 \\ 3 & 0 & 30 \end{bmatrix}$ b. [3 12 60] c. $\begin{bmatrix} -2 & 12 & 11 \\ 3 & 1 & 11 \end{bmatrix}$ d. Undefined

7. Given A = $\begin{bmatrix} 4 & -3 & 0 \\ 2 & 0 & 1 \\ 8 & 5 & -6 \end{bmatrix}$, find A^2.

a. $\begin{bmatrix} 10 & 16 & -6 \\ -12 & -1 & -54 \\ -3 & -6 & 41 \end{bmatrix}$ b. $\begin{bmatrix} 10 & -12 & -3 \\ 16 & -1 & -6 \\ -6 & -54 & 41 \end{bmatrix}$

c. $\begin{bmatrix} 16 & 9 & 0 \\ 4 & 0 & 1 \\ 64 & 25 & 36 \end{bmatrix}$ d. Undefined

8. Find the system of equations that is equivalent to the matrix equation

$\begin{bmatrix} 8 & -5 & 2 \\ 3 & 0 & 1 \\ 4 & 9 & 6 \end{bmatrix}\begin{bmatrix} x \\ y \\ z \end{bmatrix} = \begin{bmatrix} 4 \\ 6 \\ -7 \end{bmatrix}$.

a. $\begin{cases} 8x + 3y + 4z = 4 \\ -5x + 9z = 6 \\ 2x + y + 6z = -7 \end{cases}$ b. $\begin{cases} 8x - 5y + 2z = 4 \\ 3x + \quad = 6 \\ 4x + 9y + 6z = -7 \end{cases}$

c. $\begin{cases} 8x + 3y + 4z = 4 \\ -5x + 9y = 6 \\ 2x + y + 6z = -7 \end{cases}$ d. $\begin{cases} 8x - 5y + 2z = 4 \\ 3x + z = 6 \\ 4x + 9y + 6z = -7 \end{cases}$

9. Find the inverse of the matrix $\begin{bmatrix} 2 & 5 \\ 5 & 12 \end{bmatrix}$.

a. $\begin{bmatrix} 12 & -5 \\ -5 & 2 \end{bmatrix}$ b. $\begin{bmatrix} -12 & 5 \\ 5 & -2 \end{bmatrix}$ c. $\begin{bmatrix} 5 & 12 \\ 2 & 5 \end{bmatrix}$ d. No inverse

10. Find the inverse of the matrix $\begin{bmatrix} 1 & 11 & 5 \\ 2 & 22 & 8 \\ 3 & 33 & -9 \end{bmatrix}$.

a. $\begin{bmatrix} 1 & 2 & 3 \\ 11 & 22 & 33 \\ 5 & 8 & -9 \end{bmatrix}$ b. $\begin{bmatrix} -1 & -11 & -5 \\ -2 & -22 & -8 \\ -3 & -33 & 9 \end{bmatrix}$ c. $\begin{bmatrix} 11 & 9 & 2 \\ -3 & -22 & 5 \\ 33 & 1 & -8 \end{bmatrix}$ d. No inverse

11. Find the inverse of the matrix $\begin{bmatrix} 1 & 0 & 2 \\ 5 & -3 & 6 \\ -5 & 2 & -8 \end{bmatrix}$.

a. $\begin{bmatrix} 6 & 2 & 3 \\ 5 & 1 & 2 \\ \frac{5}{2} & 1 & \frac{3}{2} \end{bmatrix}$ b. $\begin{bmatrix} 6 & 2 & 3 \\ 5 & 1 & 2 \\ -\frac{5}{2} & -1 & -\frac{3}{2} \end{bmatrix}$ c. $\begin{bmatrix} 6 & 2 & 3 \\ 5 & 1 & 2 \\ 5 & 2 & 3 \end{bmatrix}$ d. No inverse

12. Solve the system of equations by using inverse matrix methods: $\begin{cases} 7x + 3y = 5 \\ 6x + 5y = 14. \end{cases}$

a. (–1, 4) b. (–1, –4) c. (4, –1) d. Inconsistent

13. Solve the system of equations by using inverse matrix methods: $\begin{cases} 3x + y + 2z = 1 \\ 3x + 2y + 3z = -1 \\ 4x + y + 2z = 3. \end{cases}$

a. (2, –3, 1) b. (2, 1, –3) c. (2, 1, 3) d. Inconsistent

14. Given $A = \begin{bmatrix} 5 & 1 & 4 \\ 2 & 0 & 6 \\ -7 & 3 & -2 \end{bmatrix}$, find M_{32} and C_{12}.

a. $M_{32} = 22, C_{12} = -38$
b. $M_{32} = -38, C_{12} = 22$
c. $M_{32} = 22, C_{12} = 18$
d. $M_{32} = 22, C_{12} = -18$

15. Find the determinant of matrix $A = \begin{bmatrix} 6 & -3 \\ 4 & 5 \end{bmatrix}$.

a. 18
b. –18
c. 42
d. –42

16. Find the determinant of matrix $A = \begin{bmatrix} -5 & 5 & 2 \\ 0 & 3 & 1 \\ 7 & 2 & 8 \end{bmatrix}$.

a. –137
b. 103
c. –117
d. –103

17. Find the determinant of matrix $A = \begin{bmatrix} 2 & 4 & -3 & -9 \\ 0 & 0 & 5 & 0 \\ -1 & 0 & 2 & 6 \\ 1 & 1 & -3 & 1 \end{bmatrix}$

a. –3
b. –31
c. –125
d. 11

18. Solve the system of equations using Cramer's Rule: $\begin{cases} 8x + 5y = 2 \\ 3x - 7y = -4. \end{cases}$

a. $\left(-\frac{6}{41}, \frac{38}{41}\right)$
b. $\left(-\frac{6}{71}, \frac{38}{71}\right)$
c. $\left(\frac{6}{41}, \frac{38}{41}\right)$
d. $\left(-\frac{6}{71}, -\frac{38}{71}\right)$

19. Solve the system of equations for z: $\begin{cases} 5y + 3z = 1 \\ -5x + z = 2 \\ -7x + 4z = 3. \end{cases}$

a. $-\frac{1}{13}$
b. $\frac{1}{13}$
c. $\frac{2}{13}$
d. $-\frac{5}{13}$

20. In a laboratory, scientists are feeding rats a special daily diet that is to contain 35 grams of protein, 61 grams of carbohydrates, and 58 grams of fat. The following table shows the protein, carbohydrate, and fat amounts in one serving of foods A, B, and C.

	Protein	Carbohydrate	Fat
Food A	8	2	15
Food B	3	17	1
Food C	2	4	10

How many servings of each food are necessary to meet the goals of the specified diet?

a. 4 servings food A, 2 servings food B, 1 serving food C
b. 1 serving food A, 3 servings food B, 3 servings food C
c. 3 servings food A, 1 serving food B, 3 servings food C
d. 3 servings food A, 3 servings food B, 1 serving food C

1. Use elementary row operations to write the matrix in echelon form: $\begin{bmatrix} 6 & 2 & 3 & 10 \\ -6 & 1 & -2 & -9 \\ -5 & -1 & -2 & -8 \end{bmatrix}$.

a. $\begin{bmatrix} 1 & 1 & 1 & 2 \\ 0 & 1 & 1 & 1 \\ 0 & 0 & 1 & 2 \end{bmatrix}$ b. $\begin{bmatrix} 1 & 1 & 1 & 2 \\ 0 & 1 & -1 & 2 \\ 0 & 0 & 1 & 2 \end{bmatrix}$ c. $\begin{bmatrix} 1 & 1 & 1 & 2 \\ 0 & 1 & 1 & 1 \\ 0 & 0 & 1 & -2 \end{bmatrix}$ d. $\begin{bmatrix} 1 & 1 & 1 & 3 \\ 0 & 1 & 1 & 1 \\ 0 & 0 & 1 & 2 \end{bmatrix}$

2. Solve the system of equations by the Gauss-Jordan method: $\begin{cases} 2x - y + z = 4 \\ -x + 2y - z = 7 \\ x + 2y - 2z = 3. \end{cases}$

a. (–1, 2, 4) b. (–1, 4, 2) c. (2, –1, 4) d. (4, 2, –1)

3. Solve the system of equations by the Gauss-Jordan mehtod: $\begin{cases} 2x + y + z = 13 \\ x + 2y - z = 20 \\ x - y + 2z = -7. \end{cases}$

a. (2 + c, c + 9, c) b. (2 – c, c + 9, c) c. (c – 2, c + 9, –c) d. No solution

4. Given A = $\begin{bmatrix} 8 & 6 \\ -3 & -4 \end{bmatrix}$ and B = $\begin{bmatrix} -1 & 5 \\ -6 & 2 \end{bmatrix}$, find 6A + 3B.

a. $\begin{bmatrix} 47 & 41 \\ -24 & -22 \end{bmatrix}$ b. $\begin{bmatrix} 45 & 51 \\ -36 & -18 \end{bmatrix}$ c. $\begin{bmatrix} 45 & 51 \\ 0 & -18 \end{bmatrix}$ d. $\begin{bmatrix} 63 & 99 \\ -81 & -18 \end{bmatrix}$

5. Find the product $\begin{bmatrix} 4 & 2 \\ 8 & 0 \\ -1 & 3 \end{bmatrix}\begin{bmatrix} 6 & 0 & -3 \\ 1 & 2 & 9 \end{bmatrix}$.

a. $\begin{bmatrix} 26 & 48 & -3 \\ 4 & 0 & 6 \\ 6 & -24 & 30 \end{bmatrix}$ b. $\begin{bmatrix} 26 & 6 \\ -3 & 30 \end{bmatrix}$ c. $\begin{bmatrix} 26 & 4 & 6 \\ 48 & 0 & -24 \\ -3 & 6 & 30 \end{bmatrix}$ d. Undefined

6. Find the product $\begin{bmatrix} -8 & 1 & 6 \\ 2 & 3 & 5 \end{bmatrix}\begin{bmatrix} 4 & 2 & 0 \\ 0 & 7 & 1 \end{bmatrix}$.

a. $\begin{bmatrix} -32 & 2 & 0 \\ 0 & 21 & 5 \end{bmatrix}$ b. $[-32 \;\; 23 \;\; 5]$ c. $\begin{bmatrix} -32 & 23 & 5 \\ 8 & 13 & 1 \end{bmatrix}$ d. Undefined

7. Given A = $\begin{bmatrix} 8 & 5 & 0 \\ 2 & 0 & 1 \\ -4 & 3 & -9 \end{bmatrix}$, find A^2.

a. $\begin{bmatrix} 64 & 25 & 0 \\ 4 & 0 & 1 \\ 16 & 9 & 81 \end{bmatrix}$ b. $\begin{bmatrix} 74 & 12 & 5 \\ 40 & 13 & -9 \\ 5 & -9 & 84 \end{bmatrix}$ c. $\begin{bmatrix} 74 & 40 & 5 \\ 12 & 13 & -9 \\ 10 & -47 & 84 \end{bmatrix}$ d. Undefined

8. Find the system of equations that is equivalent to the matrix equation
$\begin{bmatrix} 7 & -1 & 3 \\ 5 & 0 & 2 \\ 3 & 1 & 6 \end{bmatrix}\begin{bmatrix} x \\ y \\ z \end{bmatrix} = \begin{bmatrix} -4 \\ -9 \\ 1 \end{bmatrix}$.

a. $\begin{cases} 7x - y + 3z = -4 \\ 5x + 2z = -9 \\ 3x + y + 6z = 1 \end{cases}$

b. $\begin{cases} 7x + 5y + 3z = -4 \\ -x + z = -9 \\ 3x + 2y + 6z = 1 \end{cases}$

c. $\begin{cases} 7x - y + 3z = -4 \\ 5x + 2y = -9 \\ 3x + y + 6z = 1 \end{cases}$

d. $\begin{cases} 7x + 5y + 3z = -4 \\ -x + y = -9 \\ 3x + 2y + 6z = 1 \end{cases}$

9. Find the inverse of the matrix $\begin{bmatrix} 7 & 16 \\ 4 & 9 \end{bmatrix}$.

a. $\begin{bmatrix} -9 & 4 \\ 16 & -7 \end{bmatrix}$ b. $\begin{bmatrix} -9 & 16 \\ 4 & -7 \end{bmatrix}$ c. $\begin{bmatrix} 9 & 16 \\ 4 & 7 \end{bmatrix}$ d. $\begin{bmatrix} -9 & 16 \\ 4 & 7 \end{bmatrix}$

10. Find the inverse of the matrix $\begin{bmatrix} 1 & 6 & 5 \\ 2 & 12 & -3 \\ 3 & 18 & 9 \end{bmatrix}$

a. $\begin{bmatrix} -1 & -6 & -5 \\ -2 & -12 & 3 \\ -3 & -18 & -9 \end{bmatrix}$ b. $\begin{bmatrix} -12 & 3 & 3 \\ 6 & -18 & 9 \\ 2 & 9 & -5 \end{bmatrix}$ c. $\begin{bmatrix} 1 & 2 & 3 \\ 6 & 12 & 18 \\ 5 & -3 & 9 \end{bmatrix}$ d. No inverse

11. Find the inverse of the matrix $\begin{bmatrix} 1 & 4 & 1 \\ 3 & 10 & 5 \\ -3 & -14 & -2 \end{bmatrix}$.

a. $\begin{bmatrix} 25 & -3 & 5 \\ -\frac{9}{2} & \frac{1}{2} & -1 \\ -6 & 1 & -1 \end{bmatrix}$ b. $\begin{bmatrix} 25 & -3 & 5 \\ -6 & 1 & -1 \\ -9 & 1 & -2 \end{bmatrix}$ c. $\begin{bmatrix} 25 & 5 & -3 \\ -6 & 1 & -1 \\ -\frac{9}{2} & \frac{1}{2} & -1 \end{bmatrix}$ d. No inverse

12. Solve the system of equations by using inverse matrix methods: $\begin{cases} 3x + 2y = 1 \\ 2x + y = 2. \end{cases}$

a. (–4, 3) b. (–3, 4) c. (1, –1) d. (3, –4)

13. Solve the system of equations by using inverse matrix methods: $\begin{cases} x + y - z = -4 \\ 2x - y + 3x = 0 \\ -x - 2y + 4z = 8. \end{cases}$

a. (1, 1, 6) b. (2, 1, –1) c. (–2, –1, 1) d. (2, 1, 3)

14. Given A = $\begin{bmatrix} 4 & 0 & 7 \\ 3 & 1 & -2 \\ -9 & 5 & -4 \end{bmatrix}$, find M_{32} and C_{12}.

a. $M_{32} = -29$, $C_{12} = 30$ b. $M_{32} = -29$, $C_{12} = -30$

c. $M_{32} = -13$, $C_{12} = -35$ d. $M_{32} = 13$, $C_{12} = 35$

15. Find the determinant of matrix A = $\begin{bmatrix} 7 & 5 \\ 4 & -9 \end{bmatrix}$

a. –7 b. –25 c. –83 d. –43

16. Find the determinant of matrix $A = \begin{bmatrix} 2 & 7 & -5 \\ 9 & 0 & 4 \\ -3 & 6 & 8 \end{bmatrix}$.

a. –906 b. –270 c. –605 d. 642

17. Find the determinant of matrix $A = \begin{bmatrix} 6 & 3 & -5 & 11 \\ 0 & 0 & 2 & 0 \\ -2 & 0 & 2 & -4 \\ 1 & 1 & -2 & 1 \end{bmatrix}$,

a. 0 b. 8 c. –2 d. –4

18. Solve the system of equations using Cramer's Rule: $\begin{cases} 5x - 6y = 3 \\ 2x + 7y = -4. \end{cases}$

a. $\left(-\frac{3}{47}, -\frac{26}{47}\right)$ b. $\left(-\frac{6}{7}, \frac{7}{14}\right)$ c. $\left(\frac{12}{47}, \frac{7}{47}\right)$ d. $\left(-\frac{12}{47}, -\frac{7}{47}\right)$

19. Solve the system of equations for z: $\begin{cases} -x + 3z = 1 \\ 3y - z = -1 \\ 2x + 3z = 2. \end{cases}$

a. $-\frac{4}{9}$ b. $-\frac{2}{9}$ c. $-\frac{5}{9}$ d. $\frac{5}{9}$

20. In a laboratory, scientists are feeding rats a special daily diet that is to contain 50 grams of protein, 62 grams of carbohydrates, and 49 grams of fat. The following table shows the protein, carbohydrate, and fat amounts in one serving of foods A, B, and C.

	Protein	Carbohydrate	Fat
Food A	23	2	6
Food B	1	18	5
Food C	12	3	14

How many servings of each food are necessary to meet the goals of the specified diet?

a. 1 serving food A, 3 servings food B, 2 serving food C
b. 1 serving food A, 2 servings food B, 3 servings food C
c. 1 serving food A, 3 servings food B, 3 servings food C
d. 2 servings food A, 3 servings food B, 1 serving food C

1. Find the third and seventh terms of the sequence $a_n = n(n - 7)$.

2. Find the sixth term of the sequence $a_n = \dfrac{(-1)^{n+1}}{2n+5}$.

3. Find the first three terms of the recursively defined sequence $a_1 = 1$, $a_n = 3n \bullet a_{n+1}$.

In Exercises 4 to 6, classify the sequence as an arithmetic sequence, a geometric sequence or neither.

4. $1, 7, 49, \ldots, 7^{n-1}, \ldots$

5. $-7, -2, 3, \ldots, -12 + 5n, \ldots$

6. $-\dfrac{1}{3}, -\dfrac{1}{9}, -\dfrac{1}{27}, \ldots, -\dfrac{1}{3^n}, \ldots$

In Exercises 7 to 9, find the indicated sum of the series.

7. $\displaystyle\sum_{i=1}^{4} 2i(i + 5)$

8. $\displaystyle\sum_{i=1}^{5} \frac{i}{2+i}$

9. $\displaystyle\sum_{n=1}^{5} 3^n$

10. Find the twenty-seventh term of the arithmetic sequence whose first three terms are $-4, -1, 2, \ldots$

11. Find the nth term of the geometric sequence $\dfrac{2}{3}, -\dfrac{4}{9}, \dfrac{8}{27}, \ldots$

12. Find the sum of the infinite geometric series given by $\displaystyle\sum_{n=1}^{\infty} \left(\frac{5}{6}\right)^n$

13. Write $0.\overline{17}$ as the quotient of integers in simplest form.

14. Use mathematical induction to prove $1 + 6 + 11 + 16 + 21 + \ldots + 5n - 4 = \frac{n(5n - 3)}{2}$.

15. Use mathematical induction to prove 3 is a factor of $4^n - 1$ for all positive integers.

16. Write the binomial expansion of $(x^3 - 4y)^5$.

17. Find the fifth term in the expansion $(4x - 3y^2)^{12}$.

18. Logs are stacked so that there are 40 logs on the bottom row, 39 logs on the second row, and so on, decreasing by one log each row. How many total logs are in a stack of eight rows?

19. A serial number consists of five characters. The first two characters are upper-case letters of the alphabet. The next two characters are selected from the digits 1 through 9. The last character is an upper-case letter of the alphabet. How many serial numbers are possible if no letter or number can be used twice in the same serial number?

20. One number is selected from the integers 25 through 35. Find the probability that the number is even or less than 29.

1. Find the third and seventh terms of the sequence $a_n = \frac{2n-3}{3n}$.

2. Find the seventh term of the sequence $a_n = (-1)^n + \frac{1}{n}$.

3. Find the first three terms of the recursively defined sequence $a_1 = -2$, $a_n = (-1)^n n^2 a_{n-1}$.

In Exercises 4 to 6, classify the sequence as an arithmetic sequence, a geometric sequence or neither.

4. $3, 6, 12, \ldots, 3(2^{n-1}), \ldots$

5. $\frac{1}{3}, \frac{1}{5}, \frac{1}{7}, \ldots, \frac{1}{2n+1}, \ldots$

6. $-2, 3, 8, \ldots, 5n-7, \ldots$

In Exercises 7 to 9, find the indicated sum of the series.

7. $\sum_{i=1}^{4} \left(\frac{1}{4}\right)^{i+1}$.

8. $\sum_{i=1}^{6} i^3$

9. $\sum_{i=1}^{5} \frac{(-1)^i}{2i}$

10. Find the nth term of the arithmetic sequence whose first three terms are $-3, 8, 19$.

11. Find the seventh term of the geometric sequence $-5, 20, -80, \ldots$

12. Find the sum of the infinite geometric series given by $\sum_{n=1}^{\infty} \left(\frac{2}{3}\right)^n$.

13. Write $0.\overline{35}$ as the quotient of integers in simplest form.

14. Use mathematical induction to prove $2 + 5 + 8 + 11 + \ldots + 3n - 1 = \frac{1}{2}n(3n + 1)$.

15. Use mathematical induction to prove $3^n < 3^{n+1}$ for all positive integers.

16. Write the binomial expansion of $(x^2 - 3y)^5$.

17. Find the seventh term in the expansion $(5x - 2y^3)^9$.

18. A contest offers twenty prizes. The first prize is $10,000, and each successive prize is $300 less than the preceding prize. What is the total amount of money distributed in prizes?

19. Nine people are to be seated in a row on a stage for an awards ceremony. Find the number of ways the group can be seated if the master of ceremonies must be seated in the middle?

20. Five cards are randomly selected from a deck of cards containing ten black cards and twelve red cards. What is the probability that two black cards and three red cards are selected?

1. Find the third and seventh terms of the sequence $a_n = \dfrac{n+5}{2n}$.

2. Find the eighth term of the sequence $a_n = \dfrac{(-1)^{n+1}3n}{2n-5}$.

3. Find the first three terms of the recursively defined sequence $a_1 = 2$, $a_n = n^2\,(a_{n+1})$.

In Exercises 4 to 6, classify the sequence as an arithmetic sequence, a geometric sequence or neither.

4. $1, 4, 7, \ldots, 3n-2, \ldots$

5. $1, \frac{1}{2}, \frac{1}{3}, \ldots, \frac{1}{n}, \ldots$

6. $\frac{4}{5}, -\frac{8}{15}, \frac{16}{45}, \ldots, \frac{4}{5}\left(-\frac{2}{3}\right)^{n-1}, \ldots$

In Exercises 7 to 9, find the indicated sum of the series.

7. $\displaystyle\sum_{i=1}^{4} (i^2 + i)$

8. $\displaystyle\sum_{n=1}^{4} \frac{n^2}{n+1}$

9. $\displaystyle\sum_{n=1}^{5} 5^n$

10. Find the sum of the first eighteen terms of the arithmetic sequence whose first three terms are $-4, -1, 2, \ldots$

11. Find the nth term of the geometric sequence $\frac{3}{5}, -\frac{9}{25}, \frac{27}{125}, \ldots$

12. Find the sum of the infinite geometric series given by $\displaystyle\sum_{n=1}^{\infty} \left(\frac{4}{7}\right)^n$. .

13. Write $0.\overline{47}$ as the quotient of integers in simplest form.

14. Use mathematical induction to prove $1 + 7 + 13 + 19 + 25 + \ldots + 6n - 5 = n(3n - 2)$.

15. Use mathematical induction to prove $1 + 4n \le 5^n - 1$ for all positive integers.

16. Write the binomial expansion of $(x + 2y^2)^5$.

17. Find the sixth term in the expansion $(3x + 2y^2)^9$.

19. A contest offers 15 prizes. The first prize is $20,000 and each successive prize is $400 less than the preceding prize. What is the total amount of money distributed in prizes?

19. Nine people attend a movie and all sit in the same row with no seats separating anyone in the group. Find the number of ways the group can sit together if two people in the group must sit side by side?

20. A ratings poll suggests that 35 percent of the viewing families watch the program ranked number 1 each week. If two families are chosen randomly, what is the probability that both families watched the program ranked number 1?

1. Find the third and seventh terms of the sequence $a_n = \frac{3n-2}{n}$.

a. $\frac{17}{3}, \frac{7}{19}$ b. $\frac{7}{3}, \frac{19}{7}$ c. $\frac{11}{3}, \frac{19}{7}$ d. $\frac{7}{3}, \frac{23}{7}$

2. Find the ninth term of the sequence $a_n = \frac{n^2}{n+3}$.

a. $\frac{9}{4}$ b. 3 c. $\frac{27}{4}$ d. $\frac{27}{2}$

3. Find the first three terms of the recursively defined sequence $a_1 = -2$, $a_n = (-1)^{n+1}(a_{n-1})$.

a. −2, 2, −2 b. −2, 2, 2 c. −2, −1, 0 d. −2, 2, −2

4. Which of the following are geometric sequences?
i. $3, 12, 27, \ldots, 3n^2, \ldots$ ii. $-\frac{1}{4}, -\frac{1}{16}, -\frac{1}{64}, \ldots, -\frac{1}{4^n}$ iii. $-3, -1, 1, \ldots, 2n-5, \ldots$

a. i b. i and ii c. iii d. ii

5. Which of the following are arithmetic sequences?
i. $3, 12, 27, \ldots, 3n^2, \ldots$ ii. $-\frac{1}{4}, -\frac{1}{16}, -\frac{1}{64}, \ldots, -\frac{1}{4^n}$ iii. $-3, -1, 1, \ldots, 2n-5, \ldots$

a. i b. ii and iii c. ii d. iii

6. Evaluate $\frac{12!}{6!}$.

a. 720 b. 6 c. 665,280 d. 55,440

7. Find the indicated sum of the series $\sum_{i=1}^{3} (-1)^i 5^i$.

a. 105 b. −105 c. −155 d. −125

8. Find the indicated sum of the series $\sum_{n=1}^{5} (-3)^n$

a. −726 b. 183 c. 61 d. −183

9. Find the nth term of the arithmetic sequence whose first three terms are 2, 9, 16.

a. $a_n = n + 7$ b. $a_n = 7n + 2$ c. $a_n = 7n - 5$ d. $a_n = 7n + 9$

10. Find the eighth term of the geometric sequence –6, 36, –216, ...

a. 1,679,616 b. –1,679,616 c. 279,936 d. –279,936

11. Find the nth term of the geometric sequence $\frac{3}{5}, -\frac{9}{25}, \frac{27}{125}$.

a. $\left(\frac{3}{5}\right)^{n-1}$ b. $\left(-\frac{3}{5}\right)^{n+1}$ c. $\left(-\frac{3}{5}\right)^{n-1}$ d. $-\frac{3}{5}\left(\frac{3}{5}\right)^{n-1}$

12. Find the sum of the infinite geometric series given by $\sum_{k=1}^{\infty}\left(\frac{3}{7}\right)^{k}$.

a. $\frac{3}{4}$ b. $\frac{3}{7}$ c. $\frac{4}{3}$ d. 2

13. Write $0.\overline{72}$ as the quotient of integers in simplest form.

a. $\frac{72}{99}$ b. $\frac{8}{11}$ c. $\frac{23}{33}$ d. $\frac{72}{999}$

14. State the induction hypothesis necessary to prove $\left(1 - \frac{1}{2}\right)\left(1 - \frac{1}{3}\right)\left(1 - \frac{1}{4}\right) + \ldots + \left(1 - \frac{1}{n+1}\right) = \frac{1}{n+1}$ using mathematical induction.

a. $\left(1 - \frac{1}{2}\right)\left(1 - \frac{1}{3}\right)\left(1 - \frac{1}{4}\right) + \ldots + \left(1 - \frac{1}{k+1+1}\right) = \frac{1}{k+1+1}$

b. $\left(1 - \frac{1}{2}\right)\left(1 - \frac{1}{3}\right)\left(1 - \frac{1}{4}\right) + \ldots + \left(1 - \frac{1}{k+1}\right) = \frac{1}{k+1}$

c. $S_1 = \frac{1}{2} = \frac{1}{1+1}$

d. $S_n = \left(1 - \frac{1}{2}\right)\left(1 - \frac{1}{3}\right)\left(1 - \frac{1}{4}\right) + \ldots + \left(1 - \frac{1}{k+1+1}\right)$

15. State the induction hypothesis necessry to prove $1 + 6n \le 7^n$ using mathematical induction.

a. $1 + 6n \le 7^n$ b. $1 + 6k \le 7^k$ c. $12 + 6(k + 1) \le 7^{k+1}$ d. $1 + 6(1) \le 7^1$

16. Write the binomial expansion of $(2x^2 - 3y)^5$.

a. $32x^{10} - 240x^8y + 720x^6y^2 - 1080x^4y^3 + 810x^2y^4 + 243y^5$
b. $32x^{10} - 120x^8y + 360x^6y^2 - 540x^4y^3 + 405x^2y^4 + 243y^5$
c. $32x^{10} - 240x^8y + 360x^6y^2 - 1080x^4y^3 + 405x^2y^4 + 243x^5$
d. $32x^{10} - 120x^8y + 720x^6y^2 - 540x^4y^3 + 405x^2y^4 + 243y^5$

17. Find the eighth term in the expansion $(5x + 2y^2)^{10}$.

a. $1{,}920{,}000x^3y^{14}$ b. $16{,}000x^3y^{14}$
c. $1{,}200x^3y^{14}$ d. $5{,}400x^3y^{14}$

18. The seating section in a theater has 35 seats in the first row, 39 seats in the second row, and so on, increasing by four seats each row for a total of twelve rows. How many seats are in the twelfth row, and how many seats are there in the section?

a. 79, 684 b. 79, 708 c. 83, 732 d. 83, 708

19. A committee of 5 is to be selected from a group of 6 men and 8 women. How many different committees are possible?

a. 240,240 b. 2,002 c. 726,485,760 d. 48,048

20. A state lottery game allows a person to choose six numbers from the integers 1 to 50. Repetitions of numbers are not allowed. If three or more numbers match the numbers chosen by the lottery, the player wins a prize. Find the probability that the player will match exactly three numbers.

a. 0.016669 b. 0.001667 c. 0.0001389 d. 0.00139

1. Find the third and sixth terms of the sequence $a_n = \left(-\frac{1}{5}\right)^n$.

a. $-\frac{1}{25}, -\frac{1}{625}$ b. $\frac{1}{125}, \frac{1}{625}$ c. $-\frac{1}{125}, \frac{1}{15{,}625}$ d. $-\frac{1}{125}, \frac{1}{3125}$

2. Find the sixth term of the sequence $a_n = (-1)^{n-1}\left(\frac{n^2}{5}\right)$.

a. $\frac{36}{5}$ b. -5 c. $-\frac{36}{5}$ d. 5

3. Find the first three terms of the recursively defined sequence $a_1 = 2$, $a_n = -2n(a_{n-1})$.

a. 2, –8, 48 b. 2, 8, 48 c. 2, –6, 36 d. 2, –4, 16

4. Which of the following are arithmetic sequences?

i. $-3, -12, -27, \ldots, -3n^2, \ldots$ ii. $1, -\frac{2}{3}, \frac{4}{9}, \ldots, \left(-\frac{2}{3}\right)^n, \ldots$ iii. $-1, 1, 3, \ldots, 2n-1 \ldots$

a. i b. ii c. iii d. i and iii

5. Which of the following are geometric sequences?

i. $-3, -12, -27, \ldots, -3n^2, \ldots$ ii. $1, -\frac{2}{3}, \frac{4}{9}, \ldots, \left(-\frac{2}{3}\right)^n, \ldots$ iii. $-1, 1, 3, \ldots, 2n-1, \ldots$

a. i b. ii c. iii d. i and ii

6. Evaluate $\frac{11!}{9!}$.

a. 2 b. 110 c. 3,628,800 d. 3

7. Find the indicated sum of the series $\sum_{i=1}^{5} (4i + 7)$.

a. 75 b. 102 c. 95 d. 27

8. Find the indicated sum of the series $\sum_{n=1}^{4} (-4)^n$.

a. 204 b. –204 c. –257 d. –255

9. Find the twenty-sixth term of the arithmetic sequence whose first three terms are –2, 3, 8.

a. 127 b. 128 c. 123 d. 132

10. Find the sum of the first twenty-one terms of the arithmetic sequence show first three terms are $4, \frac{9}{2}, 5$.

a. 185 b. 180 c. $194\frac{1}{4}$ d. 189

11. Find the ninth term of the geometric sequence –6, 36, –216, ...

a. –10,077,696 b. –1,679,616 c. 1,679,616 d. 10,077,696

12. Find the sum of the infinite geometric series given by $\sum_{n=1}^{\infty} \left(\frac{7}{9}\right)^n$.

a. $\frac{9}{2}$ b. $\frac{2}{9}$ c. $\frac{4}{7}$ d. $\frac{7}{2}$

13. Write $0.\overline{46}$ as the quotient of integers in simplest form.

a. $\frac{46}{9}$ b. $\frac{47}{99}$ c. $\frac{46}{99}$ d. $\frac{46}{999}$

14. State the induction hypothesis necessary to prove $1 + 12 + 23 + 34 + \ldots + 11n - 10 = \frac{n(11n-9)}{2}$ using mathematical induction.

a. $S_{k+1} = 1 + 12 + 23 + 34 + \ldots + 11(k+1) - 10 = \frac{(k+1)[11(k+1, -9)]}{2}$

b. $1 + 12 + 23 + 34 + \ldots + 11n - 10 = \frac{n(11n-9)}{2}$

c. $S_1 = 1 = \frac{1[11(1) - 9]}{2}$

d. $S_k = 1 + 12 + 23 + 34 + \ldots + 11k - 10 = \frac{k(11k-9)}{2}$

15. State the induction hypothesis necessary to prove that if $0 < a < 1$, then $a^{n+1} < a^n$ for all positive integers.

a. $a^{n+1} < a^n$ b. $a^{k+1} < a^k$ c. $a^{k+2} < a^{k+1}$ d. $a^{1+1} < a^1$

16. Write the binomial expansion of $(2x^4 - 3y)^5$.

a. $32x^{20} + 48x^{16}y + 24x^{12}y^2 + 108x^8y^3 + 160x^4y^4 + 243y^5$
b. $32x^{20} - 240x^{16}y + 720x^{12}y^2 - 1080x^8y^3 + 810x^4y^4 - 243y^5$
c. $32x^{20} + 240x^{16}y^2 + 720^{12}2y^2 - 1080x^8y^3 + 800x^4y^4 + 243y^5$
d. $32x^{20} - 48x^{16}y + 24x^{12}y^2 - 108x^8y^3 + 800x^4y^4 - 243y^3$

17. Find the fifth term in the expansion $(2x - 3y^2)^8$.

a. $145{,}152x^4y^8$ b. $90{,}720x^4y^8$ c. $145{,}152x^3y^5$ d. $136{,}080x^4y^8$

18. An exercise program calls for walking 15 minutes each day for a week. Each week thereafter, the amount of time spent walking increases by 10 minutes. In how many weeks, after the exercise program starts, will a person be walking 75 minutes per day?

a. 7 b. 6 c. 8 d. 9

19. In how many ways can a president, vice president, secretary, and treasurer be selected from a committee of twelve people?

a. 980 b. 11,880 c. 12,640 d. 14,560

20. From a deck of regular playing cards, a card is drawn at random. Find the probability that the card is red or a queen.

a. $\frac{15}{26}$ b. $\frac{7}{13}$ c. $\frac{1}{2}$ d. $\frac{19}{52}$

1. Find the third and seventh terms of the sequence $a_n = \frac{2n+3}{2n}$.

a. $\frac{11}{6}, \frac{17}{14}$ b. $\frac{11}{6}, \frac{19}{14}$ c. $\frac{3}{2}, \frac{17}{14}$ d. $\frac{3}{2}, \frac{19}{14}$

2. Find the eighth term of the sequence $a_n = \frac{2n+3}{2n+5}$.

a. $\frac{3}{2}$ b. $\frac{19}{21}$ c. $\frac{2}{3}$ d. $\frac{13}{15}$

3. Find the first three terms of the recursively defined sequence $a_1 = 3$, $a_n = -5n\,(a_{n-1})$.

a. 3, –30, 450 b. 3, 30, 450 c. 3, –15, 150 d. 3, –45, 450

4. Which of the following are arithmetic sequences?

i. $1, 18, 324, \ldots, 18^{n-1}, \ldots$ ii. $-5, 2, 9, \ldots, -12 + 7n, \ldots$ iii. $6, 12, 24, \ldots 6(2^{n-1}), \ldots$

a. i b. ii c. iii d. i and ii

5. Which of the following are geometric sequences?

i. $1, 18, 324, \ldots, 18^{n-1}, \ldots$ ii. $-5, 2, 9, \ldots, -12 + 7n, \ldots$ iii. $6, 12, 24, \ldots 6(2^{n-1}), \ldots$

a. i b. ii c. iii d. i and iii

6. Evaluate $\frac{9!}{5!}$.

a. 4 b. 15,120 c. 3024 d. $\frac{9}{5}$

7. Find the indicated sum of the series $\sum_{i=1}^{4} \left(\frac{1}{n}\right)^2$.

a. $\frac{217}{144}$ b. $\frac{61}{144}$ c. $\frac{65}{48}$ d. $\frac{205}{144}$

8. Find the indicated sum of the series $\sum_{i=1}^{5} (3 + 2^i)$.

a. 72 b. 77 c. 84 d. 90

9. Find the nth partial sum of the arithmetic sequence $a_n = 6n + 5$, $n = 8$.

a. 228 b. 140 c. 280 d. 256

10. Find the ninth term of the geometric sequence –4, 16, –64, …

a. –16,384 b. 66,560 c. –262,144 d. –1024

11. Find the nth term of the geometric sequence $-\frac{7}{3}, -\frac{49}{9}, \frac{343}{27}$.

a. $\left(\frac{7}{3}\right)^{n-1}$ b. $\left(-\frac{7}{3}\right)^{n}$ c. $\frac{7}{3}\left(\frac{7}{3}\right)^{n-1}$ d. $\frac{7}{3}\left(-\frac{7}{3}\right)^{n-1}$

12. Find the sum of the infinite geometric series given by $\sum_{n=1}^{\infty}\left(\frac{1}{11}\right)^{n}$.

a. $\frac{1}{10}$ b. $\frac{10}{11}$ c. $\frac{11}{10}$ d. 11

13. Write $0.\overline{56}$ as the quotient of integers in lowest form.

a. $\frac{56}{9}$ b. $\frac{56}{999}$ c. $\frac{5}{9}$ d. $\frac{56}{99}$

14. State the induction hypothesis necessary to prove $1 + 9 + 17 + \ldots + 8n - 7 = n(4n - 3)$ using mathematical induction.

a. $1 + 9 + 17 + \ldots + 8k - 7 = k(4k - 3)$
b. $1 + 9 + 17 + \ldots + 8(k + 1) - 7 = (k + 1)[4(k + 1) - 3]$
c. $S_1 = 1 = 1[4(1) - 3]$
d. $1 + 9 + 17 + \ldots + 8n - 7 = n(4n - 3)$

15. State the induction hypothesis necessary to prove $1 + 5n \le 6^n$ for all positive integers.

a. $1 + 5(1) \le 6^1$ b. $1 + 5(k + 1) \le 6^{k+1}$ c. $1 + 5k \le 6^k$ d. $1 + 5n \le 6^n$

16. Write the binomial expansion of $(4x^3 - y)^5$.

a. $1024x^{15} - 256x^{12}y + 64x^9y^2 - 16x^6y^3 + 4x^3y^4 + y^5$
b. $1024x^{15} + 256x^{12}y + 64x^9y^2 + 16x^6y^3 + 4x^3y^3 + y^5$
c. $1024x^{15} + 1280x^{12}y + 640x^9y^2 + 160x^3y^4 + 20x^3y^4 - y^5$
d. $1024x^{15} - 1280x^{12}y + 640x^9y^2 - 160x^6y^3 + 20x^3y^4 - y^5$

17. Find the fifth term in the expansion $(4x - 3y^3)^9$.

a. $-10{,}450{,}944x^5y^{12}$ b. $82{,}944x^5y^{12}$ c. $10{,}450{,}944x^5y^{12}$ d. $-126x^5y^{12}$

18. An automobile dealer offers three midsize cars. A customer selecting one of these cars must choose one of two different engines, one of four different colors, and one of three different interior packages. How many different selections can the customer make?

a. 36 b. 64 c. 72 d. 96

19. A committee of 7 is to be selected from a group of 3 men and 11 women. How many different committees are possible?

a. 3432 b. 17,297,280 c. 2,471,040 d. 2

20. Two numbers are drawn from the integers 1, 2, 3, 4, 5, 6 without replacement. Find the probability that the sum of the two numbers is greater than 9.

a. $\frac{1}{5}$ b. $\frac{1}{6}$ c. $\frac{1}{15}$ d. $\frac{2}{15}$

College Algebra and Trigonometry Test Answers

Chapter 1 Form A

1. Commutative Property of Addition

2. $\{2, 4, 9\}$

3. 4

4. 13

5. $-\frac{x^5}{108y^6}$

6. $\frac{x^{18}}{12y^{11}}$

7. 1.29×10^{-7}

8. $2x^4 - 5x^3 + 6x^2 - 23x + 20$

9. 185

10. $(7x - 2)(x - 3)$

11. $(3x - 2y)(a + b)$

12. $3x(x - y)(x^2 + xy + y^2)$

13. $(x + y)^2(x - y)(x^2 - xy + y^2)$

14. $\frac{(x-2)^2}{3x}$

15. $\frac{3x^2 + 4x - 8}{(x-2)(x+2)(x+4)}$

16. $\frac{x(x-2)}{(x-1)(x+2)}$

17. $\frac{x - y - 1}{x - y}$

18. $\frac{y^2}{3y - 2}$

19. $\frac{a}{b^{9/4}}$

20. $-2x\sqrt[3]{5}$

21. $\frac{2\sqrt[4]{a^3}}{a}$

22. $\frac{y\sqrt{y} + y\sqrt{2}}{y - 2}$

23. $-24 - 18i$

24. $-13 - 26i$

25. $-\frac{9}{13} + \frac{6}{13}i$

Chapter 1 Form B

1. $-8, 0.17, 0, 0.\overline{74}$

2. Commutative Property of Adition

3. -28

4. 18

5. $[-4, 1]$

6. $-\frac{24}{b^5}$

7. $-\frac{x^{16}}{486y^{11}}$

8. 2.34×10^9

9. $2x^2 - xy - y^2 + 6y - 8$

10. $-5a^2 - 16ab - 3b^2$

11. 7140 committees

12. $7(x - y)(x - 3y)$

13. $(2x + y)(a - 2b)$

14. $(2a - 3b)(x - y)(x + y)$

15. $\frac{4x^2 + 6xy + 9y^2}{2x - 3y}$

16. $\frac{(x - 4)(x + 1)}{(x + 2)(x + 3)(x - 1)}$

17. $\frac{(x - 1)(2x - 3)}{3x^2}$

18. $\frac{a}{a + 1}$

19. $a^{5/4}b^{7/6}$

20. $3ab^2\sqrt[3]{2a^2b}$

21. $a - 6\sqrt{a + 2} + 11$

22. $\frac{\sqrt{2x} + 2}{x - 2}$

23. $28 + 16i$

24. $-\frac{9}{10} + \frac{3}{10}i$

25. i

Chapter 1 Form C

1. $\sqrt{19}, \pi,$ 0.565565556...

2. Commutative Property of Multiplication

3. -10

4. 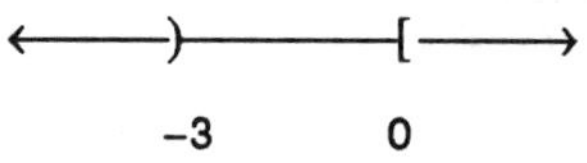

5. $-\frac{9a^2b}{2}$

6. $-\frac{9b^{11}}{32a^9}$

7. 0.00000061

8. $x^4 - 4x^2 + 12x - 9$

9. 110 ft

10. $(7x - 4)(x + 2)$

11. $x(x - 1)(2a - b)$

12. $xy(2x - 3y^2)(4x^2 + 6xy^2 + 9y^4)$

13. $(y + 1)(y^2 - y + 1)(x - 1)(x + 1)$

14. $\frac{x + 2}{(x - 4)(x + 4)(x - 3)}$

15. $\frac{x}{x - 3}$

16. $2xy^2z^3\sqrt[3]{12xy^2z^2}$

17. $\frac{x^2}{x^2 - 5x + 5}$

18. $\frac{b^{19/6}}{a^2}$

19. $9x + 12\sqrt{xy} + 4y$

20. $\frac{\sqrt[4]{3y^3}}{3}$

21. $-x^2y\sqrt{2y}$

22. $\frac{2x\left(\sqrt{2}+x\right)}{2-x^2}$

23. $9+3i$

24. $-\frac{12}{17}+\frac{3}{17}i$

25. -1

Chapter 1 Form D

1. c	2. b	3. d	4. a	5. b
6. d	7. d	8. a	9. b	10. c
11. a	12. b	13. c	14. d	15. a
16. d	17. c	18. a	19. b	20. d
21. a	22. b	23. c	24. a	25. a

Chapter 1 Form E

1. a	2. d	3. a	4. c	5. d
6. b	7. a	8. b	9. c	10. a
11. d	12. a	13. d	14. c	15. b
16. a	17. a	18. b	19. d	20. c
21. a	22. b	23. c	24. c	25. c

Chapter 1 Form F

1. d	2. b	3. c	4. a	5. d

6. c	7. a	8. c	9. a	10. b
11. a	12. c	13. d	14. a	15. b
16. c	17. b	18. c	19. d	20. a
21. c	22. b	23. c	24. b	25. c

Chapter 2 Form A

1. 6
2. 6
3. $\frac{ab}{b + a^2}$
4. $4 \pm \sqrt{42}$
5. $\frac{7}{8} + \frac{5\sqrt{7}}{8}i$ or $\frac{7}{8} - \frac{5\sqrt{7}}{8}i$
6. $1, -2, 1 + i\sqrt{3}, 1 - i\sqrt{3}$
7. no solution
8. $\frac{81}{16}$
9. 16
10. $x \leq 9$
11. $(-\infty, -4] \cup (-1, 1]$
12. $-\frac{7}{2} \leq x \leq 1$
13. $1, \frac{19}{3}$
14. $-4 < x < 1$
15. $x \leq -\frac{24}{5}$ or $x \geq 2$
16. 31 ft
17. \$5500 at 7.5%
\$2500 at 5.2%
18. 40 mph
19. 36 h
20. 240 ~~oz~~ g.

Chapter 2 Form B

1. $-\frac{24}{7}$
2. $\frac{9}{7}$
3. $\frac{12 + 6b}{2 + 3b}$

4. $\frac{3}{2} + \frac{\sqrt{47}}{2}i$ or $\frac{3}{2} - \frac{\sqrt{47}}{2}i$

5. $\frac{5 \pm \sqrt{13}}{6}$

6. $-2, 3, -1 + i\sqrt{3}, -1 - i\sqrt{3}$

7. 3

8. $\frac{3}{4}, \frac{7}{3}$

9. $\frac{35}{2}$

10. $x \geq \frac{7}{15}$

11. $(-6, -1] \cup [8, \infty)$

12. $1 < x \leq 5$

13. $-\frac{1}{5}, 1$

14. $-\frac{8}{5} \leq x \leq 2$

15. $x < -1$ or $x > \frac{11}{3}$

16. 15 mph

17. 900 ml

18. 60 minutes

19. $3800

20. 117 or more miles per day

Chapter 2 Form C

1. $\frac{9}{8}$

2. 3

3. $\frac{ab}{b^2 - a}$

4. $\frac{9 \pm \sqrt{65}}{2}$

5. $-2, 3$

6. $\frac{3}{2}, 2, -1 + i\sqrt{3}, -1 - i\sqrt{3}$

7. $6 + 2\sqrt{7}$

8. $\frac{1}{512}, 125$

9. $-2\sqrt{2}, 2\sqrt{2}, -2i, 2i$

10. $x \leq 1$

11. $[-4, 1)$

12. $-\frac{7}{5} < x \leq 2$

13. $-1, 9$

14. $\frac{3}{2} \leq x \leq \frac{7}{2}$

15. $x < -1$ or $x > \frac{13}{3}$

16. 30 mph

17. 4.2 h

18. 5 ft

19. $1\frac{2}{3}$ liter

20. 8200 \$9 tickets, 3800 \$15 tickets

Chapter 2 Form D

1. c	2. d	3. a	4. b	5. a
6. d	7. c	8. b	9. b	10. c
11. a	12. b	13. b	14. a	15. a
16. a	17. b	18. b	19. c	20. b

Chapter 2 Form E

1. a	2. d	3. d	4. b	5. d
6. c	7. a	8. d	9. b	10. a
11. d	12. b	13. c	14. b	15. a
16. d	17. a	18. b	19. a	20. d

Chapter 2 Form F

1. c	2. a	3. b	4. d	5. a
6. b	7. c	8. b	9. c	10. b
11. a	12. d	13. a	14. c	15. b
16. d	17. b	18. b	19. a	20. c

College Algebra and Trigonometry Test Answers

Chapter 3 Form A

1. $\left(-\frac{1}{2}, \frac{3}{2}\right)$

2. x–intercepts: $\left(-\sqrt{2}, 0\right), \left(\sqrt{2}, 0\right)$
   y–intercept: $(0, -4)$

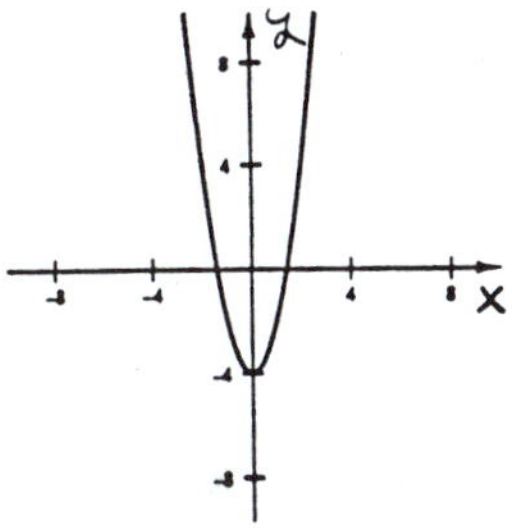

3.

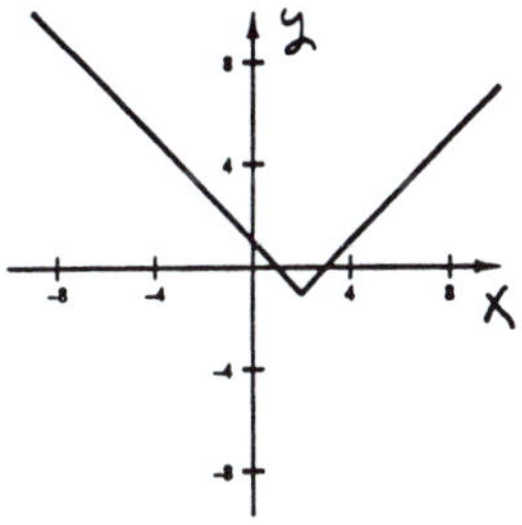

4. center: $(-5, 3)$, radius: 4

5. $-3 \le x \le 3$

6. $v = 16{,}000 - 3000t, \quad 0 \le t \le 4$

7. increasing: $(-\infty, -1]$
   constant: none
   decreasing: $[-1, \infty)$

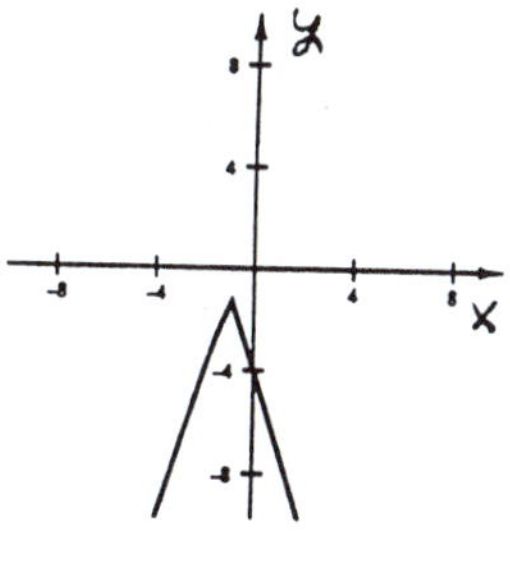

8. Domain: all real numbers
   Range: $\{y \mid y \le -2\}$

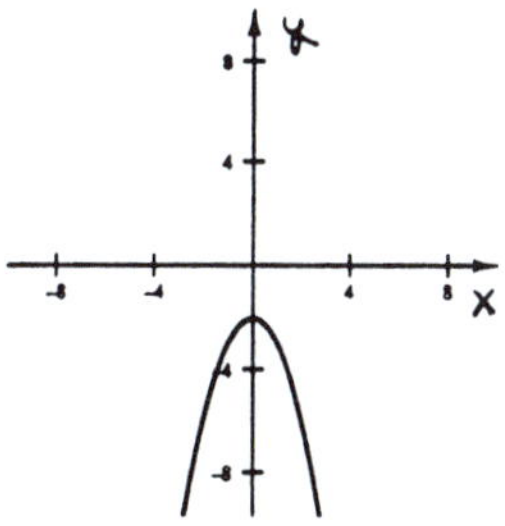

9.

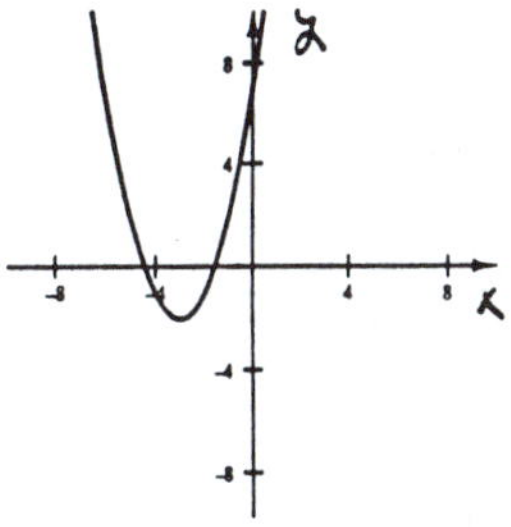

10.

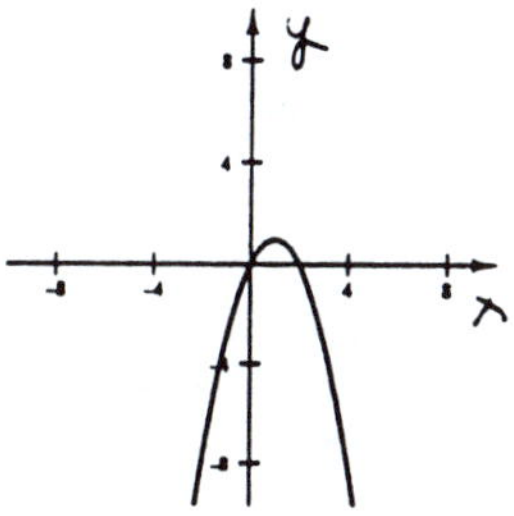

11. b

12. $y = -\frac{2}{5}x + \frac{21}{5}$

13.1minimum: $\frac{23}{8}$

14. a) $\frac{x^2 + 4x - 6}{x^2 + 3}$
b) $\frac{4x-9}{x^2+3}$

15. $2x + h + 8$

16. $4x^2 + 24x + 35$

17. $f(x) = x^2 + 4$, domain: $x \geq 0$, range: $y \geq 4$
$f^{-1}(x) = \sqrt{x - 4}$, domain $x \geq 4$, range: $y \geq 0$

18. $f^{-1}(x) = 2x - 4$

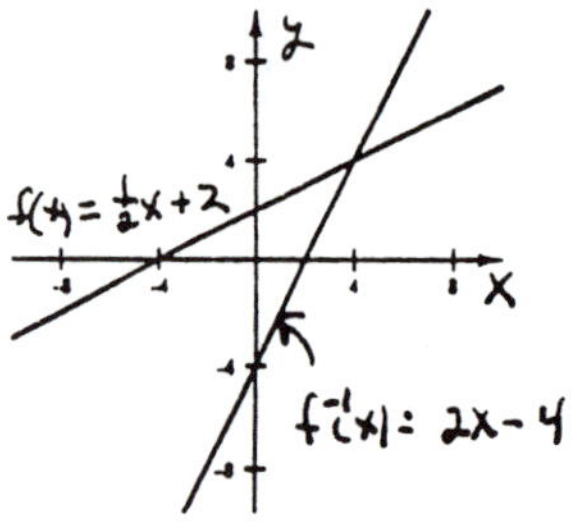

19. 3

20. 25.9 decibels

Chapter 3 Form B

1. $\left(-1, \frac{17}{2}\right)$

2. x–intercept: (1, 0)

y–interceopt: (0, 3)

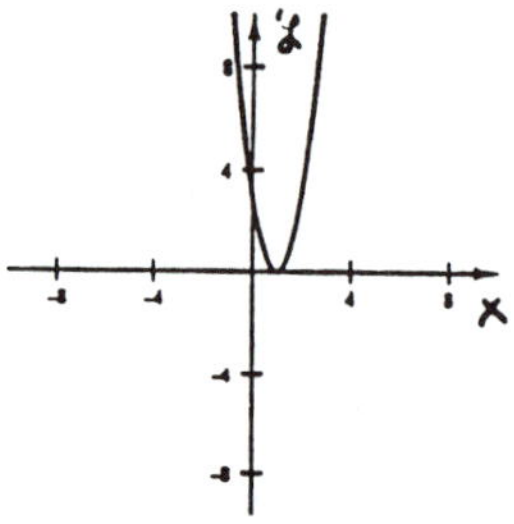

3.

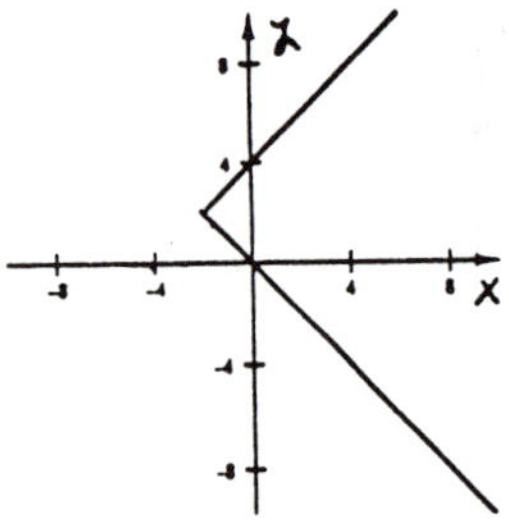

4. center: (1, −9)
radius: 7

5. $x \leq -5$ or $x \geq 5$

6. a. $C = 4000 + 50x$
b. $R = 87.50x$
c. $P = 37.50x - 4000$

7. a. increasing: none
b. constant: $(-\infty, 0] \cup [2, \infty)$
c. decreasing: $[0, 2]$

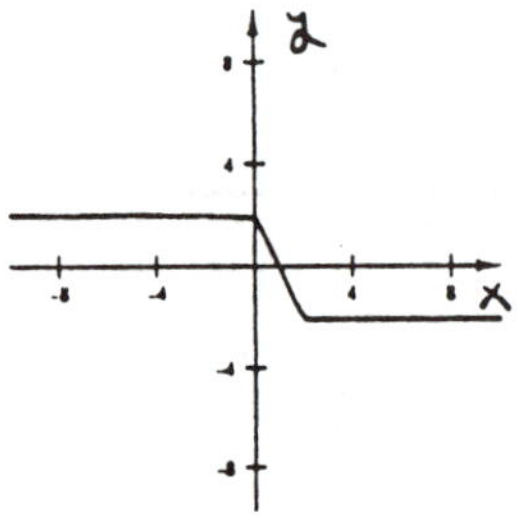

8. Domain: real numbers
Range: real numbers

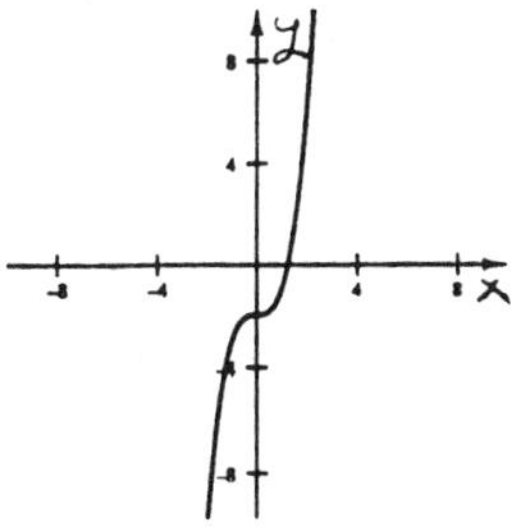

9. b, c

10.

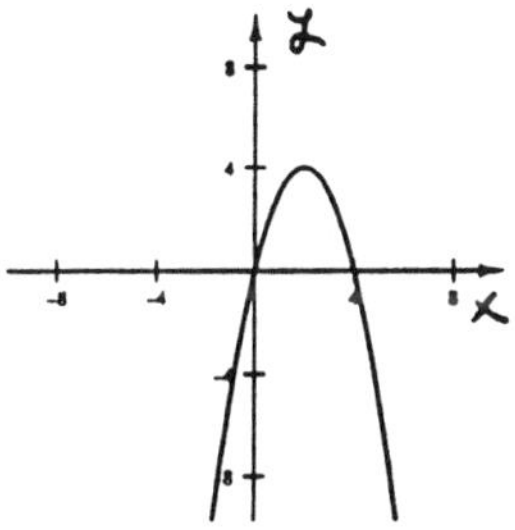

11. a

12. $y = -\frac{5}{2}x - \frac{7}{2}$

13. minimum: −30

14. $(g - f)(x) = -x^2 + 3x + 2$ $(fg)(x) = x^3 + x^2 - 5x + 3$

15. $2x + h + 15$

16. −59

17. $f(x) = x^2 - 4$, domain: $x \geq 0$, range: $y \geq -4$

$f^{-1}(x) = \sqrt{x + 4}$, domain: $x \geq -4$, range: $y \geq 0$

18. $f^{-1}(x) = \frac{1}{3}x - \frac{1}{3}$

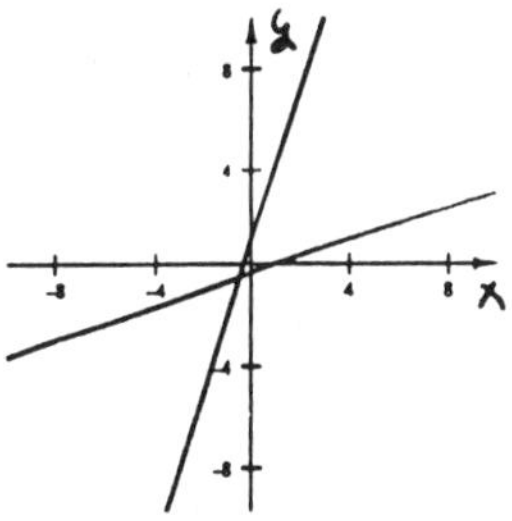

19. 3.75 in.

20. 100 m x 100 m

Chapter 3 Form C

1. 12.08

2. b, c

3.

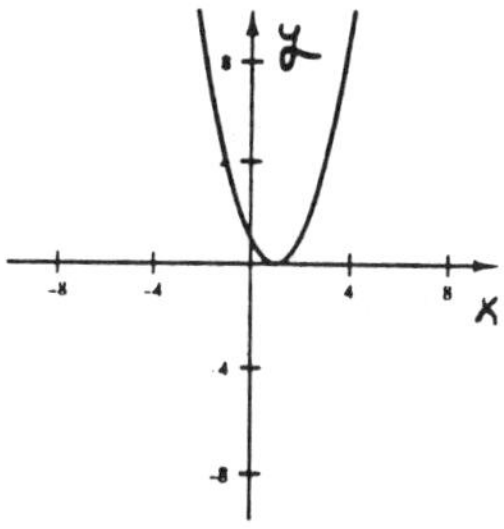

4. center: $(2, -1)$
radius: $2\sqrt{2}$

5. $x \geq -6$

6. $v = -15{,}000t + 450{,}000,$
$0 \leq t \leq 30$

7.

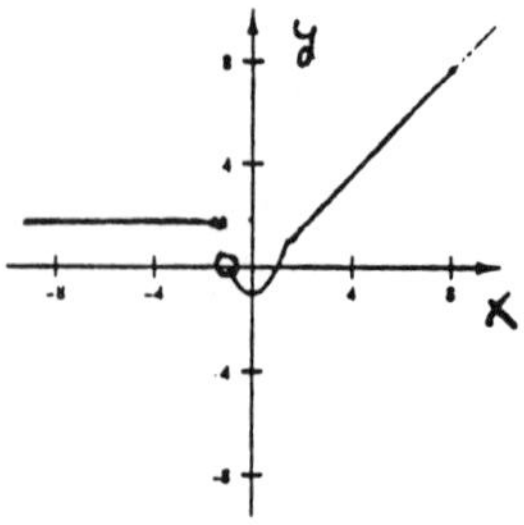

8. Domain: all reals
Range: $[2, \infty)$

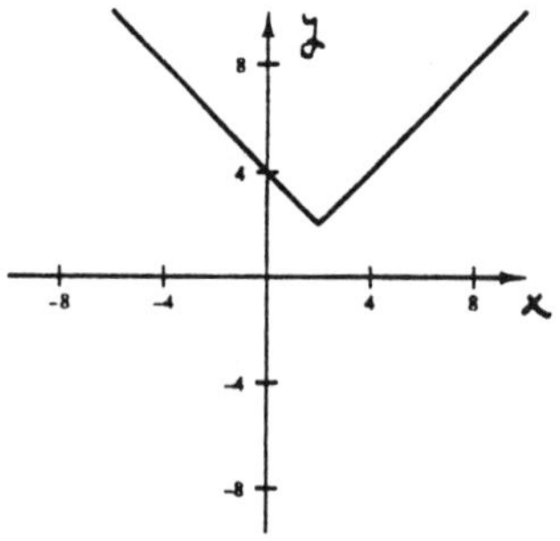

9.

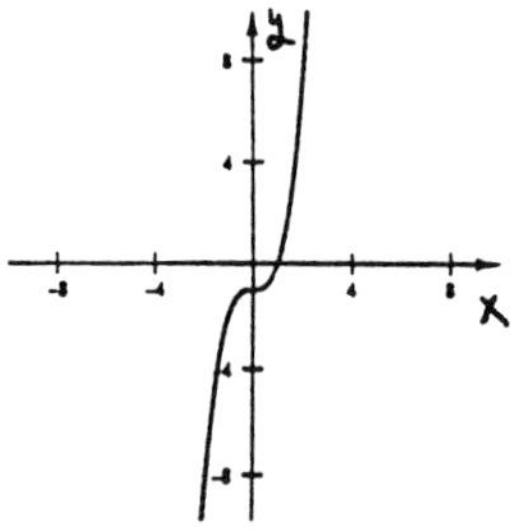

10.

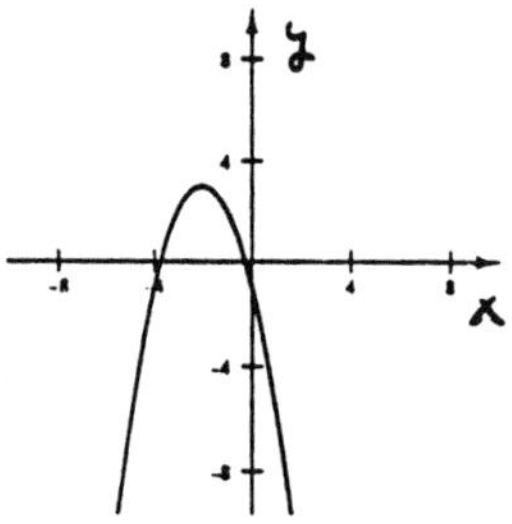

11. c

12. $y = \frac{2}{5}x - \frac{19}{5}$

13. minimum: $2\frac{7}{8}$

14. $(f - g)(x) = x^2 - x - 1$
$(fg)(x) = 2x^4 - x^3 + 14x^2 - 5x + 20$

15. -6

16. $-24x^2 + 11$

17. $f(x) = \frac{7x}{x + 5}$, Domain: all real except $x \neq -5$,
Range: all real except $y \neq 7$
$f^{-1}(x) = \frac{5x}{7 - x}$, Domain: all reals except $x \neq 7$
Range: all real except $y \neq -5$

18. $f^{-1}(x) = \frac{1}{3}x + \frac{4}{3}$

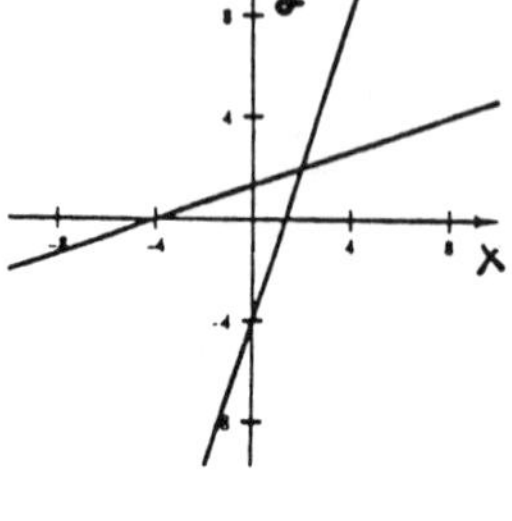

19. 5.5 amps

20. $d = \sqrt{81 + 144t^2}$

Chapter 3 Form D

1. b	2. a	3. c	4. c	5. c
6. b	7. b	8. d	9. c	10. b
11. c	12. c	13. d	14. d	15. a
16. c	17. c	18. c	19. c	20. a

Chapter 3 Form E

1. c	2. b	3. d	4. a	5. b
6. d	7. b	8. c	9. b	10. d
11. b	12. c	13. b	14. d	15. a
16. b	17. d	18. b	19. a	20. c

Chapter 3 Form F

1. b	2. c	3. a	4. d	5. a
6. b	7. d	8. a	9. c	10 d.
11. b	12. a	13. d	14. b	15. d
16. a	17. b	18. c	19. b	20. c

College Algebra and Trigonometry Test Answers

Chapter 4 Form A

1. $4x^2 - 4x - 4 + \frac{-15}{x-2}$

2. 107

3. $P(-2) = 0$, $x + 2$ is a factor

4. down to the far left, down to the far right

5. 0, 2, 3

6. $P(1) = -2$, $P(2) = 17$, $P(x)$ has a zero between 1 and 2

7. -3 multiplicity 2, 0 multiplicity 3, $\frac{4}{3}$ multiplicity 1, 3 multiplicity 2

8. $\pm 1, \pm 3, \pm 7, \pm 21, \pm\frac{1}{5}, \pm\frac{3}{5}, \pm\frac{7}{5}, \pm\frac{21}{5}$

9. upper bound: 4
   lower bound: -3

10. 3 or 1 positive real roots, 2 or 0 real negative roots

11. $-1, \frac{1}{2}, 4$

12. $-1 - 4i, -i, i$

13. $0, 2, 3, -i, i$

14. $x^4 - 2x^3 - 2x^2 + 8x - 8$

15. $x = -3$, $x = 6$

16. $y = \frac{4}{3}$

17.

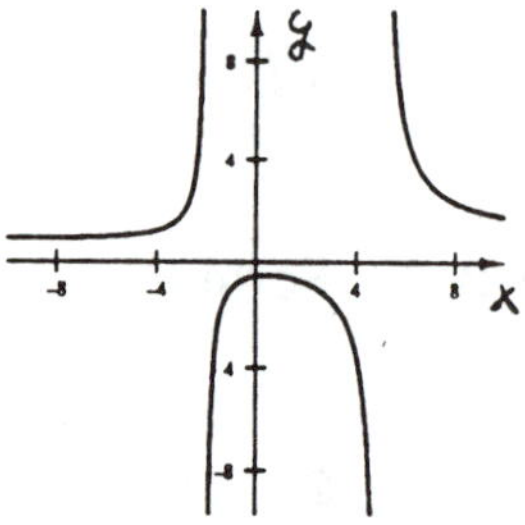

18.

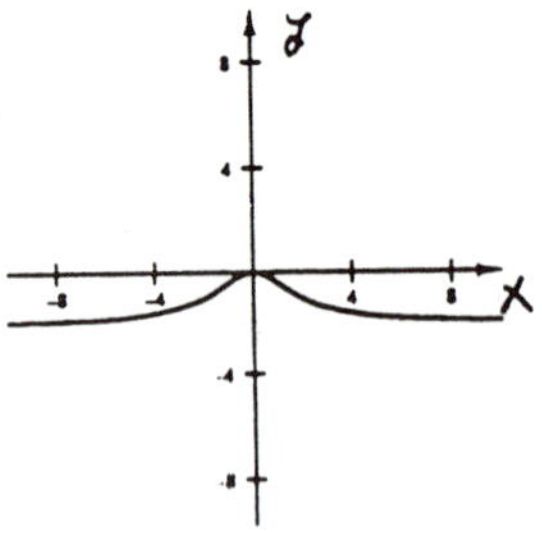

19.

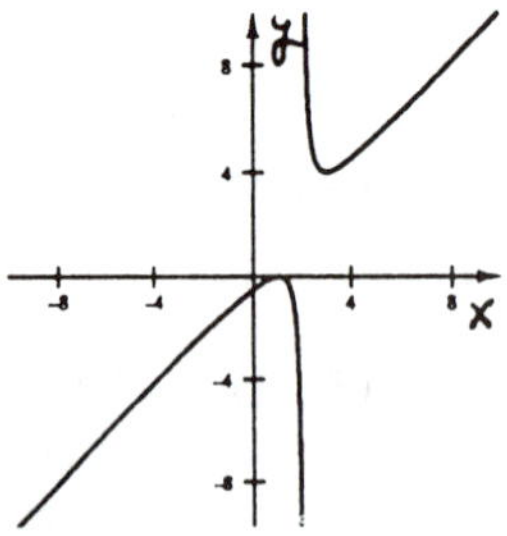

20. ~~0.6875~~ 0.66 0.65662

College Algebra and Trigonometry Test Answers

Chapter 4 Form B

1. $2x^2 - 8x + 18 + \frac{-35}{x+2}$

2. -65

3. $P(-4) = 0$, $x + 4$ is a factor

4. down to the far left
down to the far right

5. $-1, -2, -3$

6. $P(1) = -1$, $P(2) = 7$
$P(x)$ has a zero between 1and 2

7. -4 multiplicity 3, -3 multiplicity 2, $-\frac{4}{3}$ multiplicity 1, 3 multiplicity 2

8. $\pm 1, \pm 5, \pm\frac{1}{2}, \pm\frac{1}{3}, \pm\frac{1}{6}, \pm\frac{5}{2}, \pm\frac{5}{3}, \pm\frac{5}{6}$

9. upper bound: 2
lower bound: -2

10. 2 or 0 positive real roots, 3 or 1 negative real roots

11. $3, \frac{-1 \pm \sqrt{97}}{12}$

12. $-2 - i, -3i, 3i$

13. $\frac{3}{5}, -3 \pm \sqrt{10}$

14. $2x^4 + 3x^3 + 3x - 2$

15. $x = 0, x = -2, x = \frac{4}{3}$

16. $y = \frac{1}{3}$

17. 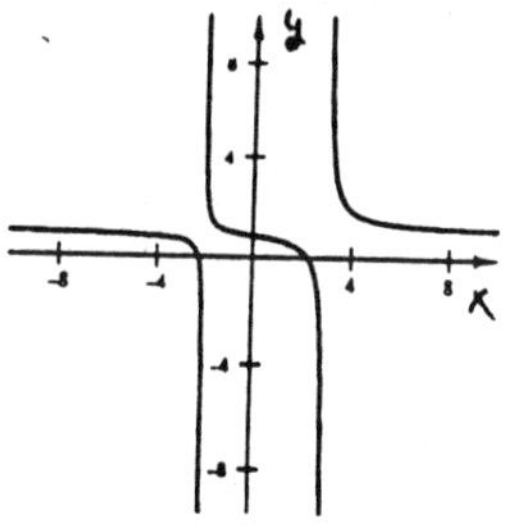

18. 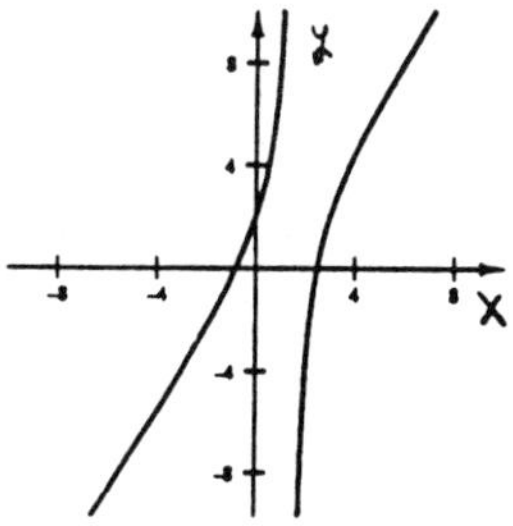

19. 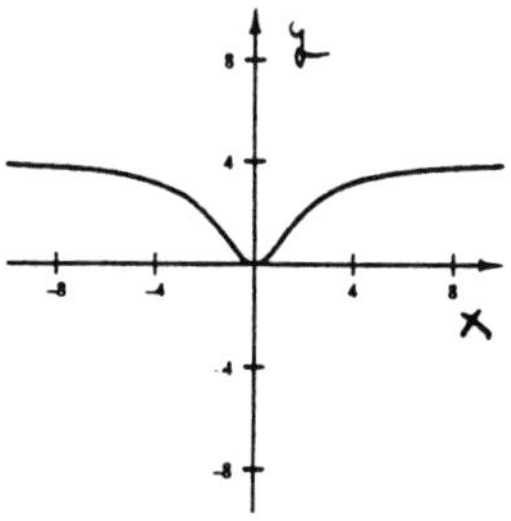

20. 2.8125

College Algebra and Trigonometry Test Answers

Chapter 4 Form C

1. $2x^2 + 11x + 55 + \frac{266}{x-5}$

2. 61

3. $P(-1) = 0$, $x + 1$ is a factor

4. down to far left
up to far right

5. $-2, -1, 1, 3$

6. $P(1) = -6$, $P(2) = 6$,
$P(x)$ has a zero between 1 and 2

7. -9 multiplicity 4, -1 multiplicity 2, 0 multiplicity 1, 1 multiplicity 2

8. $\pm 1, \pm 3, \pm 5, \pm 15, \pm\frac{1}{7}, \pm\frac{3}{7}, \pm\frac{5}{7}, \pm\frac{15}{7}$

9. upper bound: 2
lower bound: -2

10. 2 or 0 positive real zeros, 3 or 1 negative real zeros

11. $-\frac{1}{7}, -3 \pm \sqrt{6}$

12. $1 - i, -i\sqrt{7}, i\sqrt{7}$

13. $-\frac{1}{5}, -4 \pm \sqrt{15}$

14. $x^5 - 3x^4 - 5x^3 + 29x^2 - 30x$

15. $x = -\frac{4}{3}, x = 0, x = 1$

16. $y = \frac{1}{2}$

17.

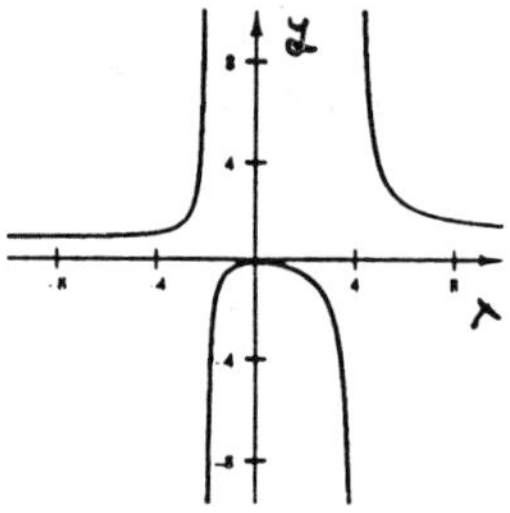

18.

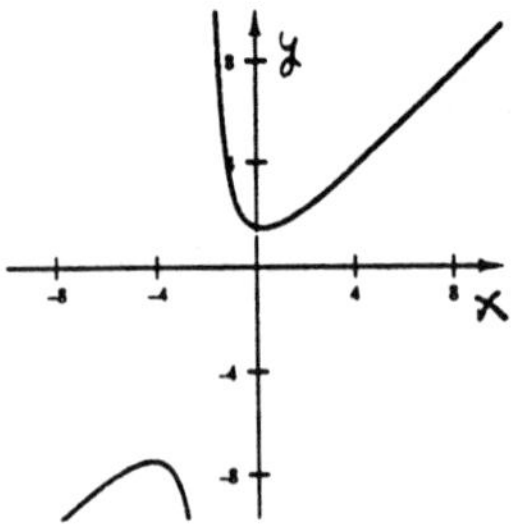

19.

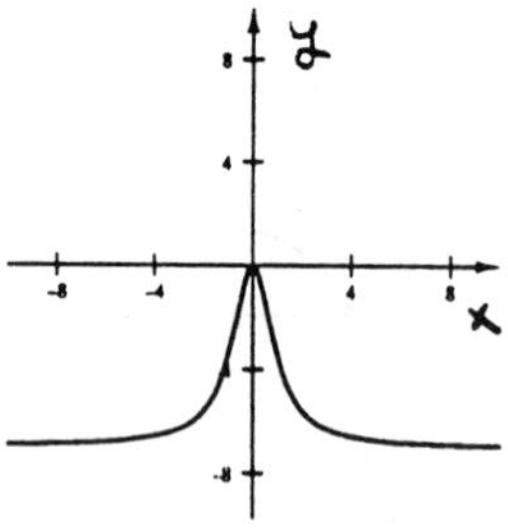

20. 3.1

Chapter 4 Form D

1. b	2. c	3. d	4. c	5. b
6. d	7. a	8. b	9. c	10. b
11. d	12. a	13. c	14. d	15. b
16. a	17. a	18. d	19. b	20. b

Chapter 4 Form E

1. b	2. d	3. a	4. b	5. d
6. a	7. b	8. c	9. b	10. b
11. d	12. a	13. c	14. b	15. d
16. a	17. b	18. d	19. b	20. b

Chapter 4 Form F

1. c	2. d	3. a	4. b	5. d
6. c	7. d	8. b	9. a	10. b
11. c	12. b	13. b	14. b	15. d
16. a	17. b	18. d	19. b	20. d

College Algebra and Trigonometry Test Answers

Chapter 5 Form A

1. 70.6807

2. 4.19265

3. 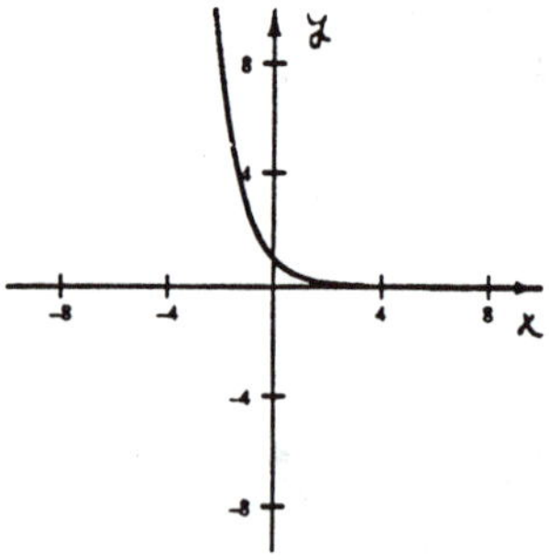

4. 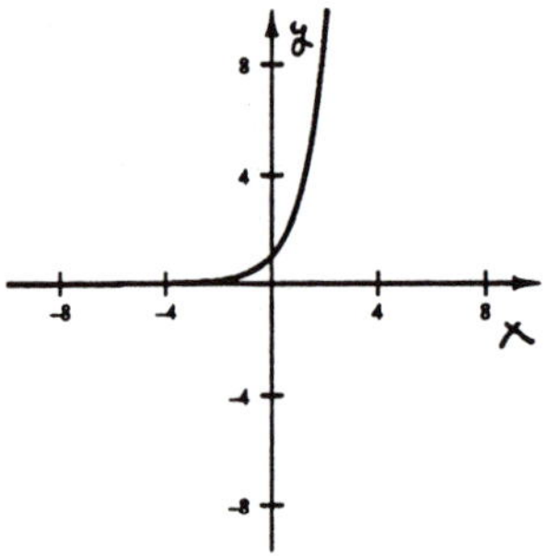

5. 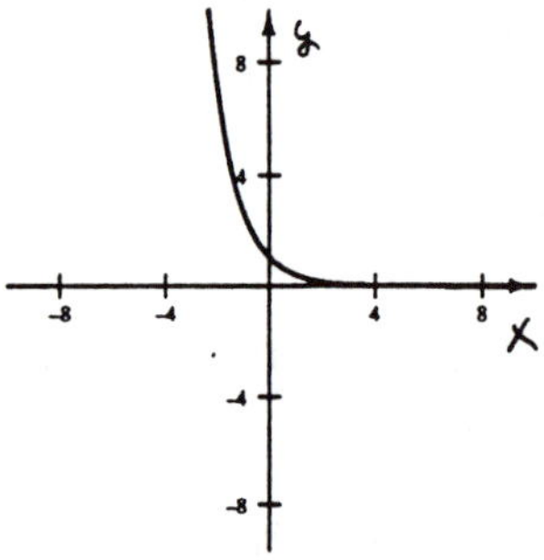

6. $b^c = 2x - 5$

7. $\log_5 b = a$

8. $5 \log_b x + 2 \log_b y - 2 \log_b z$

9. $\log_b y + 4 \log_b z - \frac{2}{3} \log_b x$

10. $\log_b \frac{3x - 4}{\sqrt{2x + 5}}$

11. 0.9603

12. 1.6826

13. 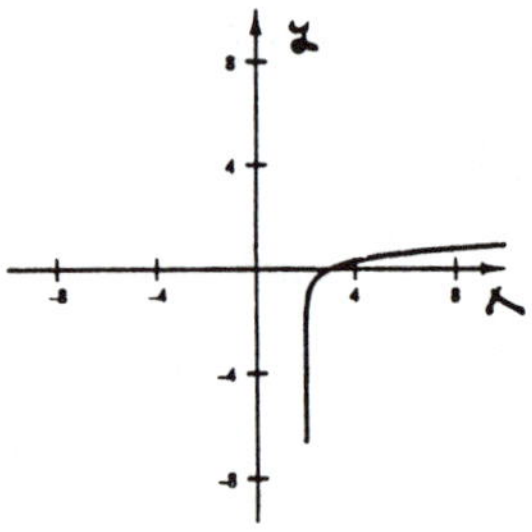

14. 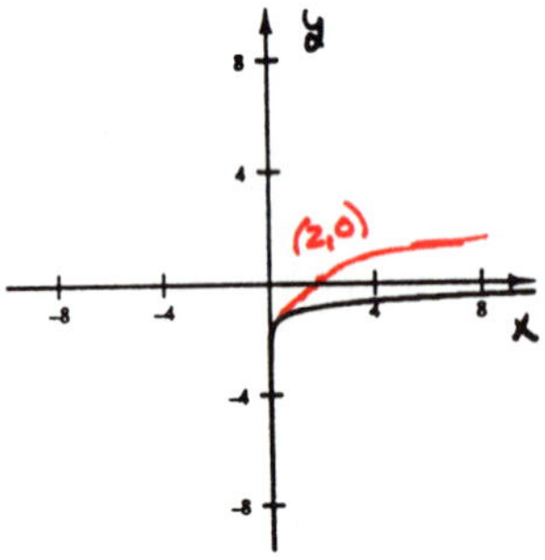

15. 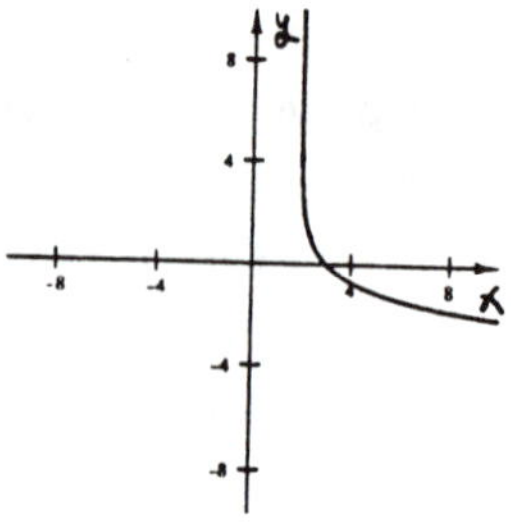

16. $-\frac{1}{4}$

17. 1.7969

18. 70

19. \$32,423.25

20. 1277 hours

Chapter 5 Form B

1. 6.43944

2. 3.96857

3. 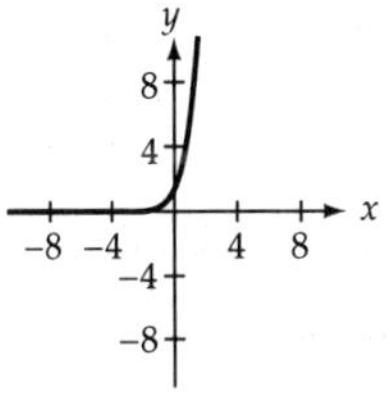

4. 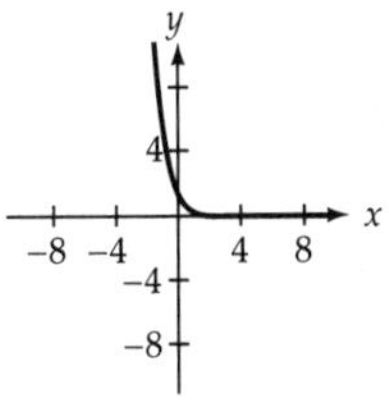

5. 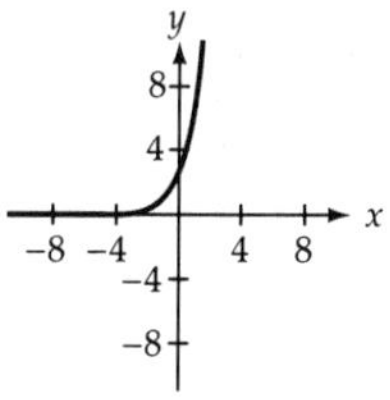

6. $b^s = 2x - 7$

7. $\log_4 y = \sqrt{3x}$

8. $5 \log_b x + 2\log_b z - \log_b y$

9. $\frac{2}{3} \log_b x + \log_b y - \frac{1}{2} \log_b z$

10. $\log_{10} \frac{3x + 1}{x^9}$

11. 0.6770

12. 1.8982

13. 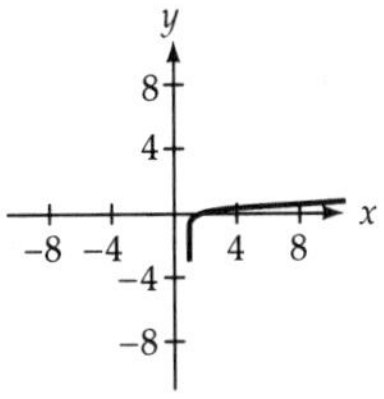

14. 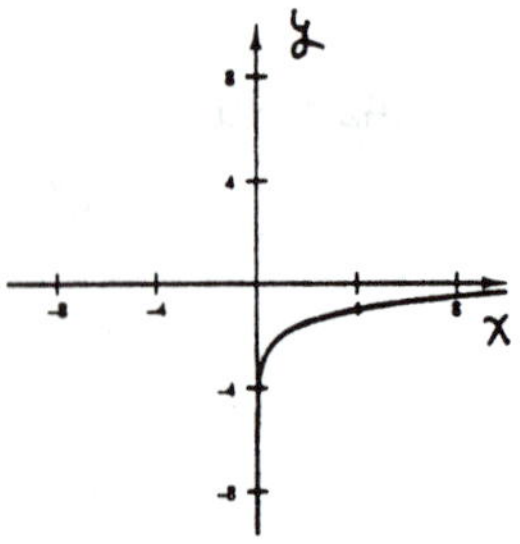

15. 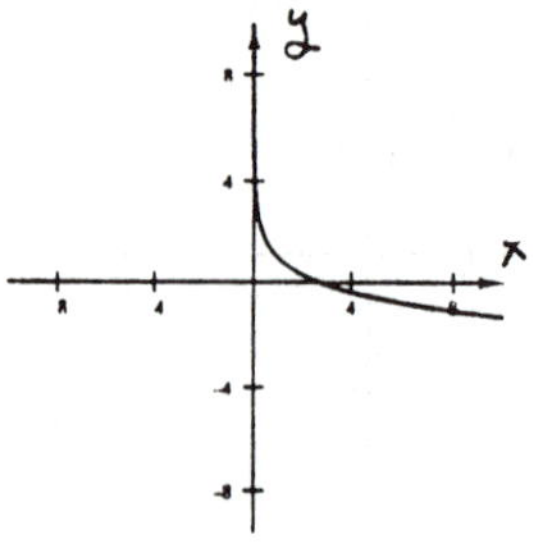

16. $\frac{2}{3}$

17. 3.4022

18. 4

19. ~~\$42,245.35~~ \$32,423.25

20. 12.4%

Chapter 5 Form C

1. 38.6345

2. 2.50929

3. 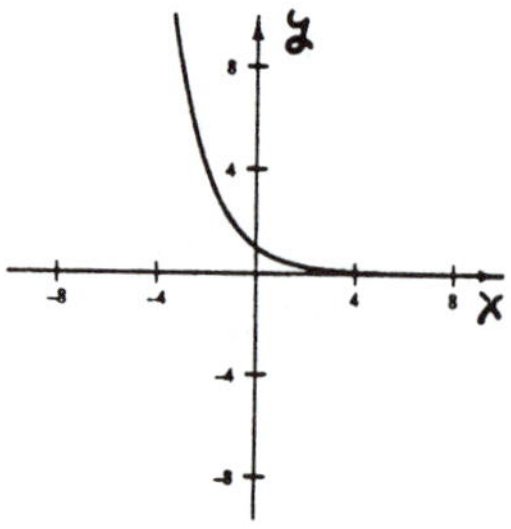

4. 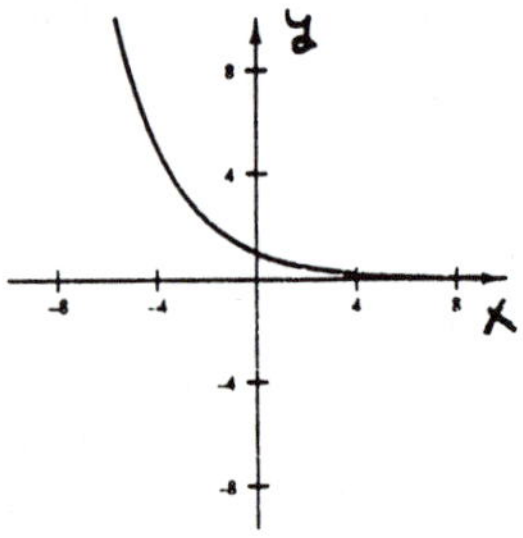

5. 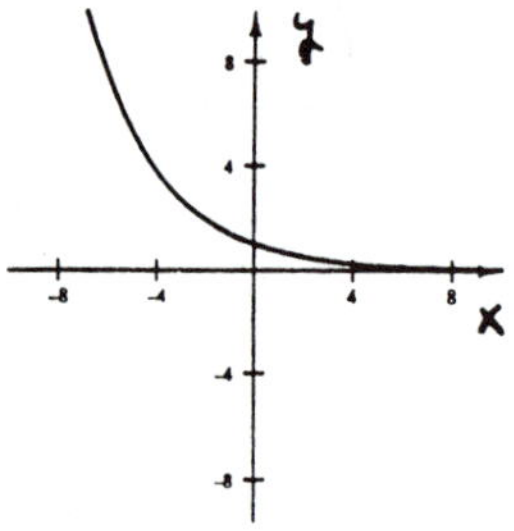

6. $b^r = 4 - 3x$

7. $\log_7 y = x - 1$

8. $2 \log_b x + \log_b y - 7 \log_b z$

9. $\frac{3}{4} \log_b x + \frac{1}{4} \log_b z - 2 \log_b y$

10. $\log_4 \frac{x(x+2)}{2x-3}$

11. 0.7686

12. 1.8614

13. 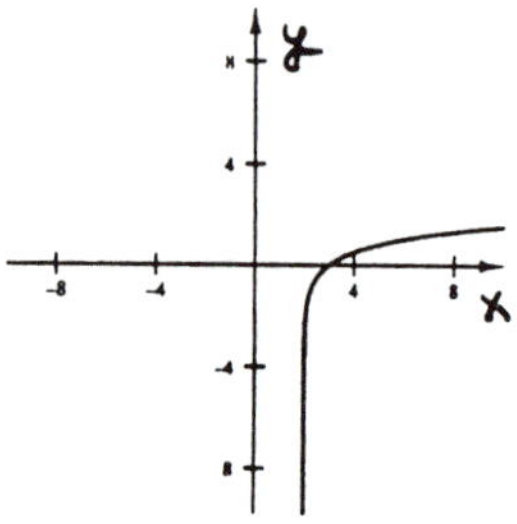

14.

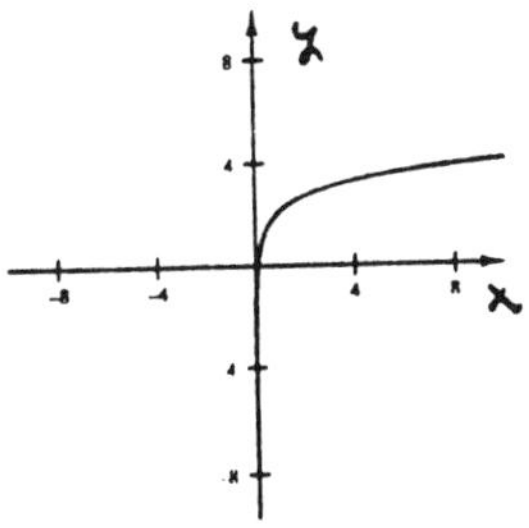

15.

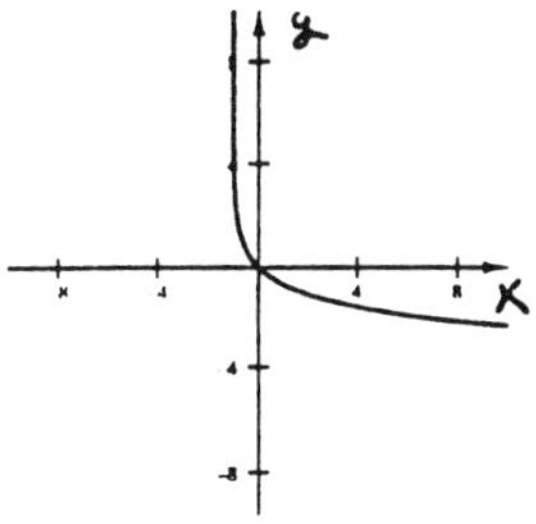

16. 4

17. 3.1972

18. 1453.04

19. $32,461.87

20. 5 min

Chapter 5 Form D

1. b	2. d	3. a	4. d	5. c
6. a	7. c	8. b	9. d	10. c
11. d	12. b	13. b	14. d	15. c
16. a	17. b	18. d	19. b	20. b

Chapter 5 Form E

1. a	2. c	3. b	4. d	5. b
6. b	7. d	8. a	9. c	10. b
11. b	12. a	13. c	14. a	15. d
16. c	17. b	18. c	19. b	20. b

Chapter 5 Form F

1. b	2. c	3. b	4. d	5. a

6. a 7. b 8. a 9. b 10. d

11. a 12. b 13. d 14. a 15. b

16. c 17. b 18. d 19. b 20. d

Chapter 6 Form A

1. $\frac{7\pi}{6}$

2. $\frac{5\pi}{12}$

3. 7π in.

4. 16π radians/s

5. 16.25 ft/s

6. $\sec\theta = \frac{\sqrt{58}}{7}$

7. $\sec 36° \approx 1.2361$

8. $\frac{5}{4}$

9. $\left(-\frac{\sqrt{2}}{2}, -\frac{\sqrt{2}}{2}\right)$

10. $\cos^2\theta$

11. 3π

12. 2, $\frac{2\pi}{3}$, $\frac{\pi}{18}$

13.

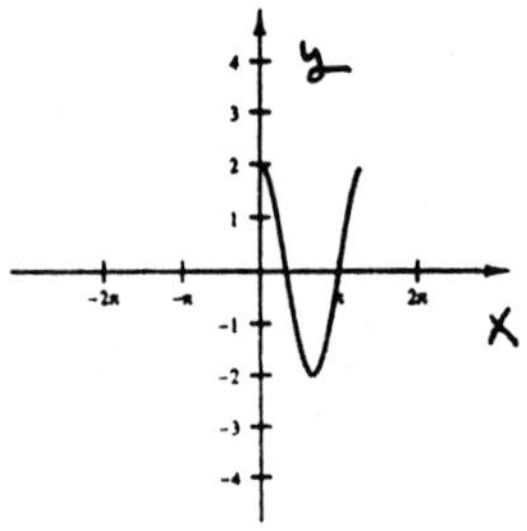

14.

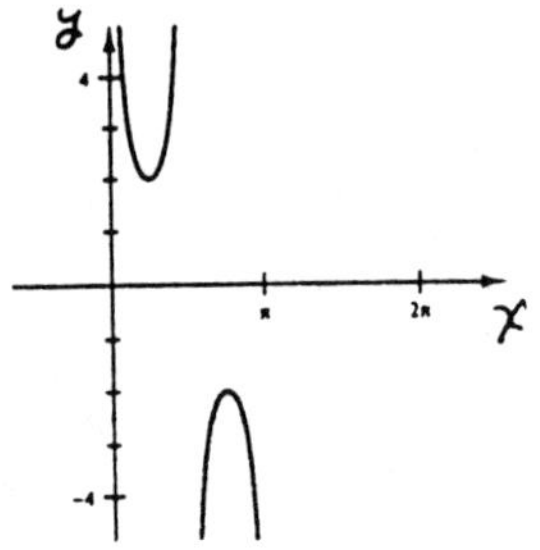

15.

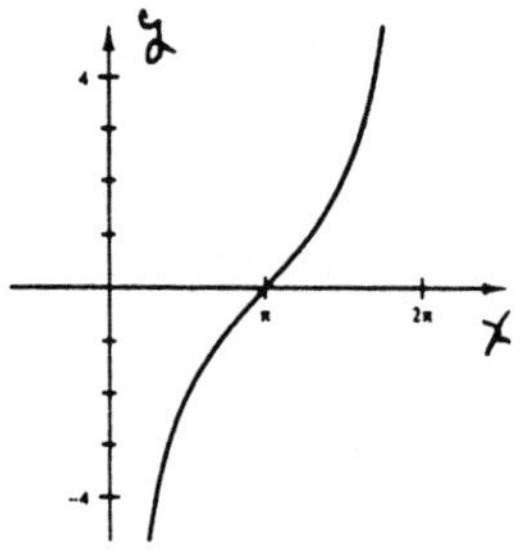

16.

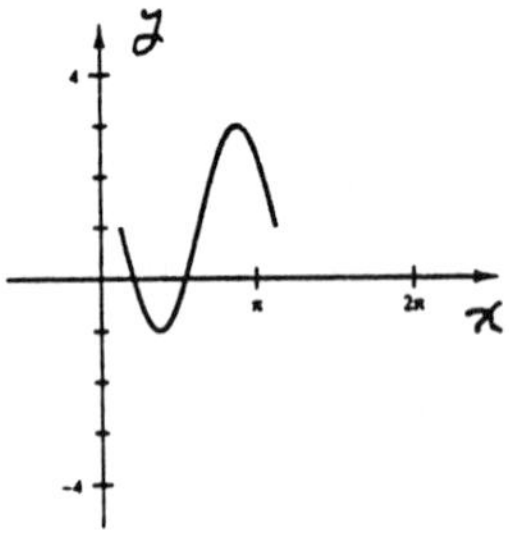

17.

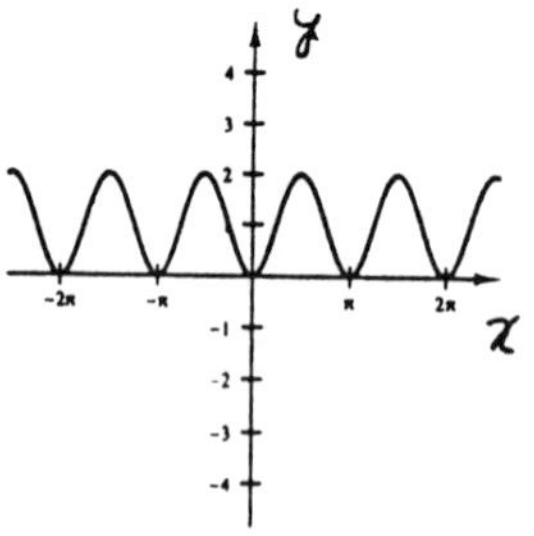

18.

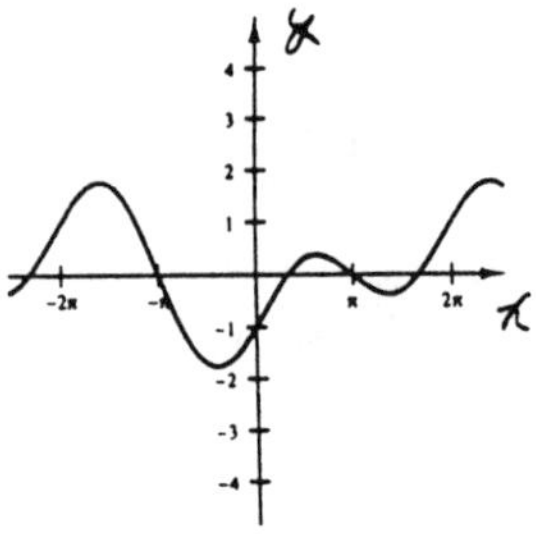

19. 45.5 ft

$y = 4\cos\pi t$

Chapter 6 Form B

1. $\frac{4\pi}{3}$

2. $97.44°$

3. $\frac{75\pi}{4}$ cm

4. 80π radians/s

5. 69 mi/h

6. $\csc\theta = \frac{\sqrt{34}}{3}$

7. $\cot 37 \approx 1.3270$

8. $\frac{-\sqrt{3} + 3\sqrt{2}}{6}$

9. $\left(\frac{\sqrt{3}}{2}, \frac{1}{2}\right)$

10. $-2\cot^2\theta$

11. 4

12. $1, \frac{2\pi}{3}, -\frac{\pi}{3}$

13. 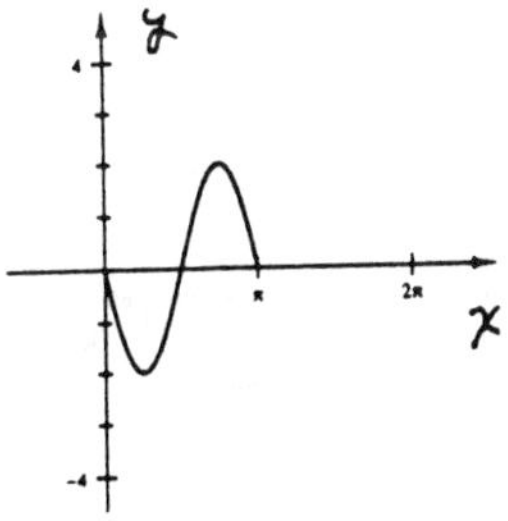

14. 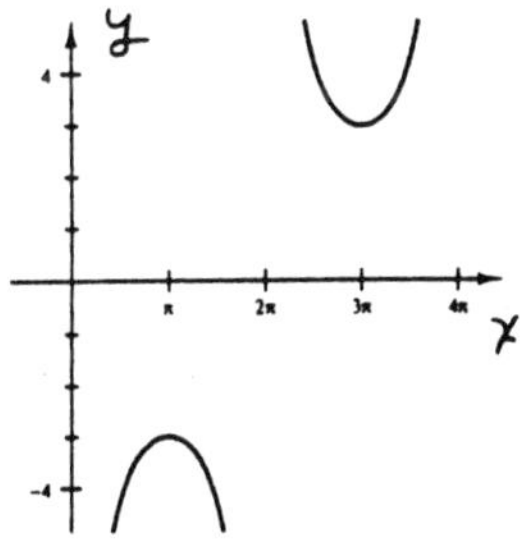

15. 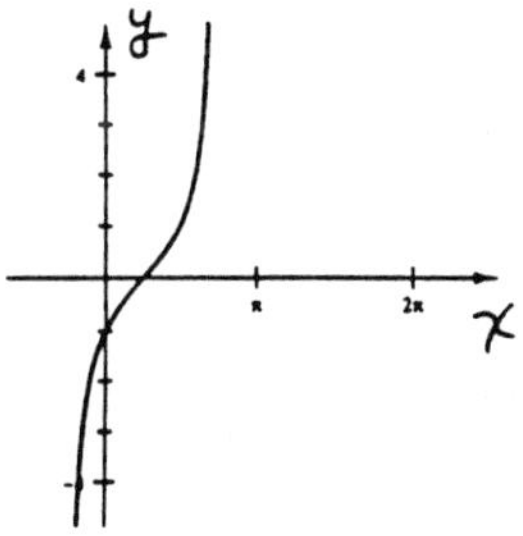

16. 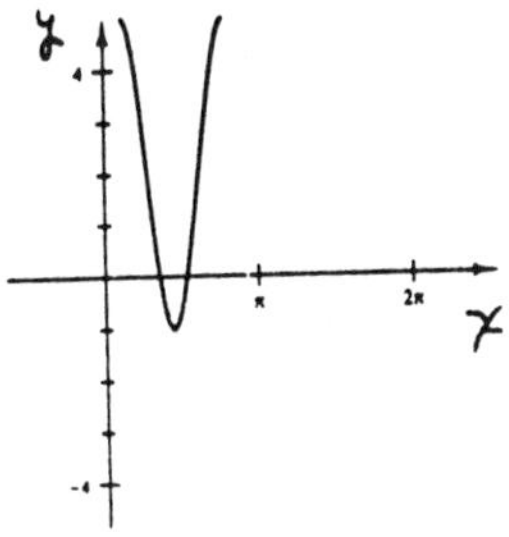

17. 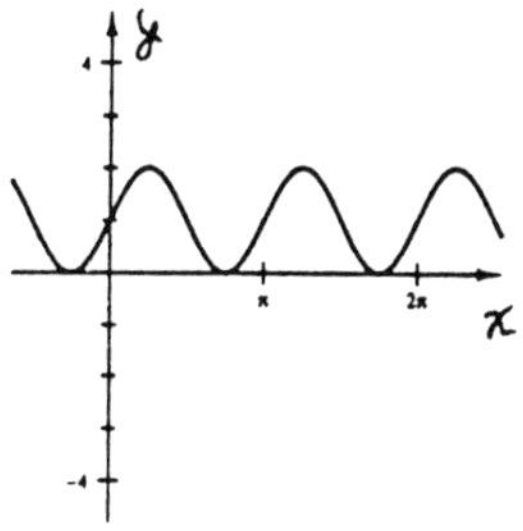

18. 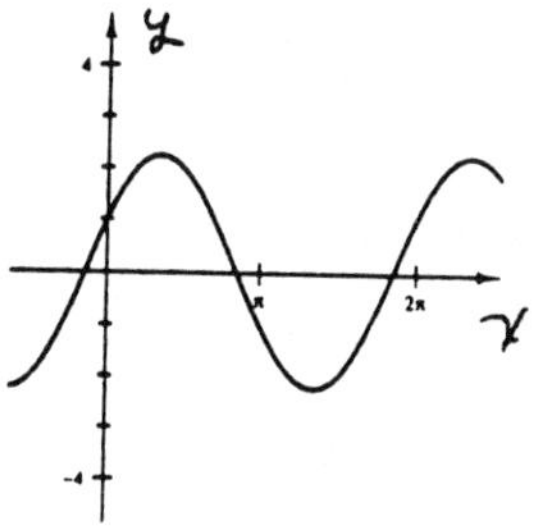

19. 111 ft

20. $y = 12\cos\frac{4\pi}{3}t$

Chapter 6 Form C

1. $\frac{7\pi}{4}$

2. $\frac{7\pi}{15}$

3. $\frac{4}{9}$ radians

4. $\frac{4\pi}{3}$ radians/s

5. 62.5 ft/s

6. $\sqrt{10}$

7. −1.8361

8. $\frac{\sqrt{2}-1}{2}$

9. $\left(\frac{\sqrt{3}}{2}, \frac{1}{2}\right)$

10. $\frac{[1-\sin^2 t]^2}{\sin^2 t}$

11. 3

12. $-\frac{\pi}{10}$

13.

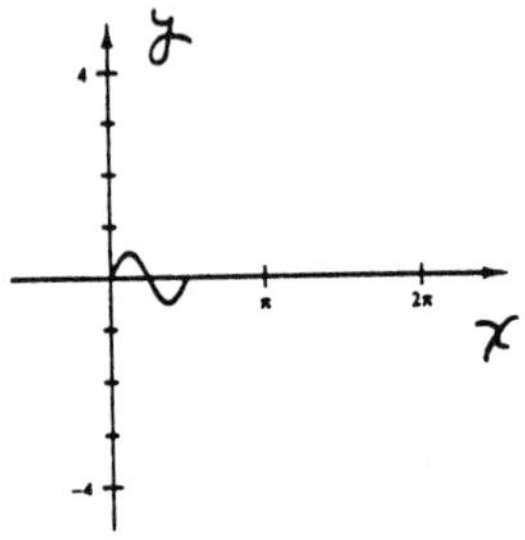

14.

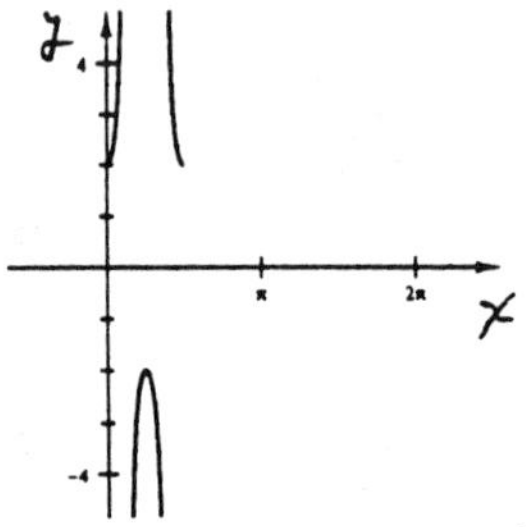

15.

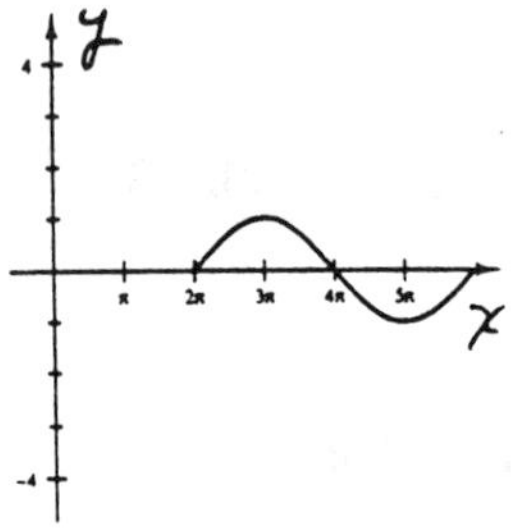

16.

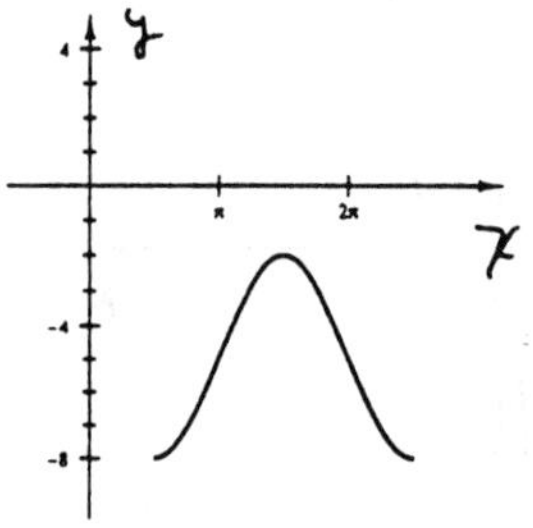

17.

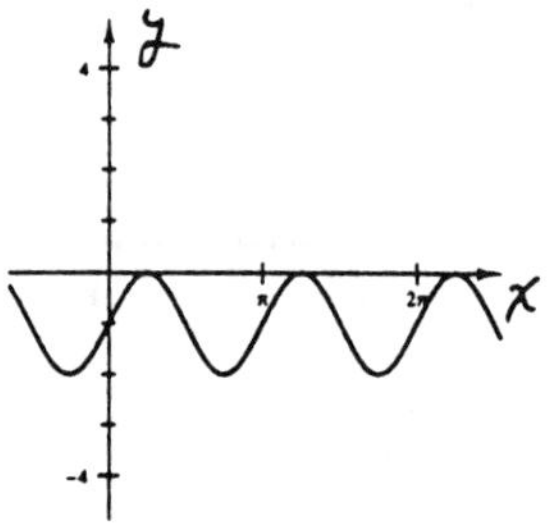

18.

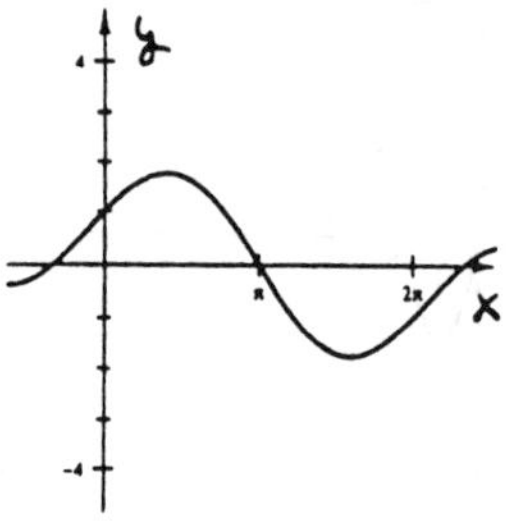

19. 15 m

20. $y = 8 \cos \frac{4}{5}t$

Chapter 6 Form D

1. c	2. d	3. d	4. b	5. c
6. a	7. b	8. c	9. d	10. a
11. a	12. c	13. c	14. d	15. b
16. b	17. c	18. d	19. c	20. b

Chapter 6 Form E.

1. b	2. a	3. d	4. b	5. a

6. d	7. b	8. c	9. c	10. a
11. a	12. c	13. d	14. c	15. c
16. c	17. d	18. b	19. c	20. a

Chapter 6 Form F

1. a	2. c	3. b	4. d	5. b
6. c	7. c	8. a	9. c.	10. a
11. a	12. a	13. b	14. c	15. b
16. a	17. d	18. b	19. b	20. c

Chapter 7 Form A

1. $\cos x \cot x + \sin x = \cos x \cdot \frac{\cos x}{\sin x} + \frac{\sin x}{1} = \frac{\cos^2 x + \sin^2 x}{\sin x} = \frac{1}{\sin x} = \csc x$

2. $\frac{\cos x \tan x + \sin x}{\cos x} = \frac{\cos x \cdot \frac{\sin x}{\cos x} + \sin x}{\cos x} = \frac{\sin x + \sin x}{\cos x} = \frac{2 \sin x}{\cos x} = 2 \tan x$

3. $\frac{\cos^2 x}{1 - \sin x} = \frac{1 - \sin^2 x}{1 - \sin x} = \frac{(1 - \sin x)(1 + \sin x)}{1 - \sin x} = 1 + \sin x$

4. $\frac{\csc x \tan x - \sec x}{\sec x} = \frac{\frac{1}{\sin x} \cdot \frac{\sin x}{\cos x} - \frac{1}{\cos x}}{\sec x} = \frac{\frac{1}{\cos x} - \frac{1}{\cos x}}{\sec x} = 0$

5. $\frac{\sqrt{6} - \sqrt{2}}{4}$ 6. $-\frac{323}{325}$ 7. $-\tan \theta$

8. $\sin x$ 9. $\frac{119}{169}$

10. $\csc x + \cot x = \dfrac{1}{\sin x} + \dfrac{\cos x}{\sin x} = \dfrac{1 + \cos x}{\sin x} = \dfrac{1}{\tan \frac{x}{2}} = \cot \dfrac{x}{2}$

11. $\dfrac{2 \tan \frac{1}{2}x}{1 + \tan^2 \frac{1}{2}x} = \dfrac{2 \tan \frac{1}{2}x}{\sec^2 \frac{1}{2}x} = \dfrac{2\dfrac{\sin \frac{1}{2}x}{\cos \frac{1}{2}x}}{\dfrac{1}{\cos \frac{1}{2}x}} = 2\dfrac{\sin \frac{1}{2}x}{\cos \frac{1}{2}x} \cdot \dfrac{\cos^2 \frac{1}{2}x}{1} = 2 \sin \dfrac{1}{2}x \cos \dfrac{1}{2}x = \sin x$

12. $\dfrac{1}{4}$

13. $y = \sin\left(x - \dfrac{\pi}{3}\right)$

14. $-\dfrac{\pi}{6}$

15. 1.8077

16. $\dfrac{15}{17}$

17.

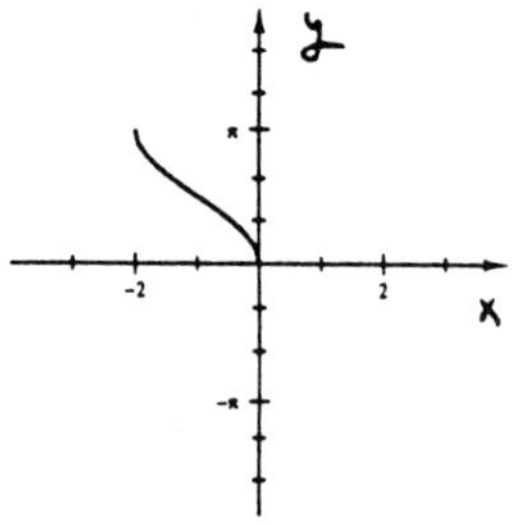

18. 30°, 150°

19. 0°

20. $\dfrac{\pi}{6}, \dfrac{5\pi}{6}, \dfrac{11\pi}{6}$

Chapter 7 Form B

1. $\dfrac{\sin x}{1 + \sin x} - \dfrac{\sin x}{1 - \sin x} = \dfrac{\sin x(1 - \sin x) - \sin x(1 - \sin x)}{1 - \sin^2 x} = \dfrac{\sin x - \sin^2 x - \sin x - \sin^2 x}{\cos^2 x} = \dfrac{-2 \sin^2 x}{\cos^2 x}$

$= -2 \tan^2 x$

2. $\tan x + \cot x = \dfrac{\sin x}{\cos x} + \dfrac{\cos x}{\sin x} = \dfrac{\sin^2 x + \text{cox}^2 x}{\cos x \sin x} = \dfrac{1}{\cos x \sin x} = \sec x \csc x$

3. $\sec^2 x \sin^2 x - \sec^2 x = \sec^2 x(\sin^2 x - 1) = \sec^2 x(-\cos^2 x) = \dfrac{1}{\cos^2 x} \cdot \dfrac{-\cos^2 x}{1} = -1$

4. $\csc x(\cos x + \sin x) = \dfrac{1}{\sin x}(\cos x + \sin x) = \dfrac{\cos x}{\sin x} + 1 = \cot x + 1$

5. $-\frac{\sqrt{2}+\sqrt{6}}{4}$

6. $+\frac{13}{85}$

7. $\sin\left(\frac{3\pi}{2}+\theta\right) = \sin\frac{3\pi}{2}\cos\theta + \cos\frac{3\pi}{2}\sin\theta = -1\cos\theta + 0\cdot\sin\theta = -\cos\theta$

8. $\cos 8x$

9. $-\frac{120}{169}$

10. $\frac{1-\tan^2\frac{x}{2}}{1+\tan^2\frac{x}{2}} = \frac{1-\frac{\sin^2\frac{x}{2}}{\cos^2\frac{x}{2}}}{\sec^2\frac{x}{2}} = \frac{1-\frac{\sin^2\frac{x}{2}}{\cos^2\frac{x}{2}}}{\frac{1}{\cos^2\frac{x}{2}}} = \frac{\cos^2\frac{x}{2}-\sin^2\frac{x}{2}}{1} = \cos 2\left(\frac{x}{2}\right) = \cos x$

11. $\frac{2\tan x}{1+\tan^2 x} = \frac{\frac{2\sin x}{\cos x}}{\sec^2 x} = \frac{\frac{2\sin x}{\cos x}}{\frac{1}{\cos^2 x}} = \frac{2\sin x}{\cos x}\cdot\frac{\cos^2 x}{1} = 2\sin x\cos x = \sin 2x$

12. $\frac{\sqrt{3}+2}{4}$

13. $y = 2\sin\left(x+\frac{\pi}{6}\right)$

14. $-45°$

15. 0.958

16. $\frac{24}{25}$

17.

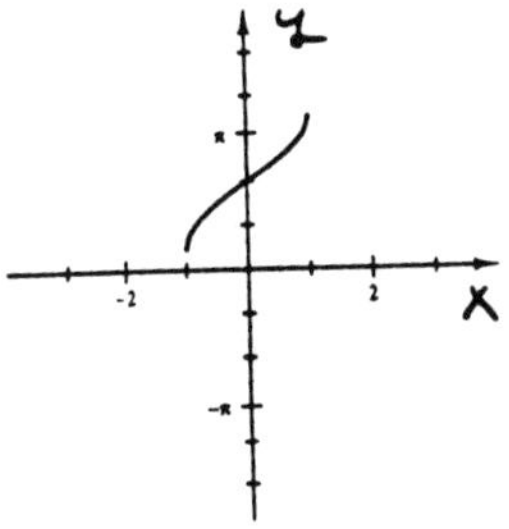

18. 30°, 210°

19. π

20. 45°, 210°, 315°, 330°

Chapter 7 Form C

1. $(1+\tan x)^2 = 1 + 2\tan x + \tan^2 x = (1+\tan^2 x) + 2\tan x = \sec^2 x + 2\tan x$

2. $\tan^2 x\sin^2 x - \tan^2 x = \tan^2 x(\sin^2 x - 1) = \frac{\sin^2 x}{\cos^2 x}(-\cos^2 x) = -\sin^2 x$

3. $\frac{1 + \csc x}{\sec x} = \frac{1}{\sec x} + \frac{\csc x}{\sec x} = \cos x + \cot x$

4. $\tan^2 x - \sin^2 x = \frac{\sin^2 x}{\cos^2 x} - \sin^2 x = \sin^2 x\left(\frac{1}{\cos^2 x} - 1\right) = \sin^2 x(\sec^2 x - 1) = \sin^2 x \tan^2 x$

5. $\frac{-\sqrt{2} + \sqrt{6}}{4}$

6. $\frac{33}{65}$

7. $\sin\left(\frac{\pi}{3} + x\right) = \sin\frac{\pi}{3}\cos x + \cos\frac{\pi}{3}\sin x = \frac{\sqrt{3}}{2}\cos x + \frac{1}{2}\sin x$

8. $\sin 3\theta$

9. $-\frac{24}{25}$

10. $2 \sec x \sin^2 \frac{x}{2} = 2 \sec x\left(\frac{1 - \cos x}{2}\right) = \frac{2 \sec x}{2} - \frac{2 \sec x \cos x}{2} = \sec x - 1$

11. $\cos 2x + 2\sin^2 x = 1 - 2\sin^2 x + 2\sin^2 x = 1$

12. $\frac{\sqrt{3} - 2}{4}$

13. $y = 2\sin\left(x + \frac{5\pi}{6}\right)$

14. $45°$

15. 0.498 radians

16. 0.955 radians

17.

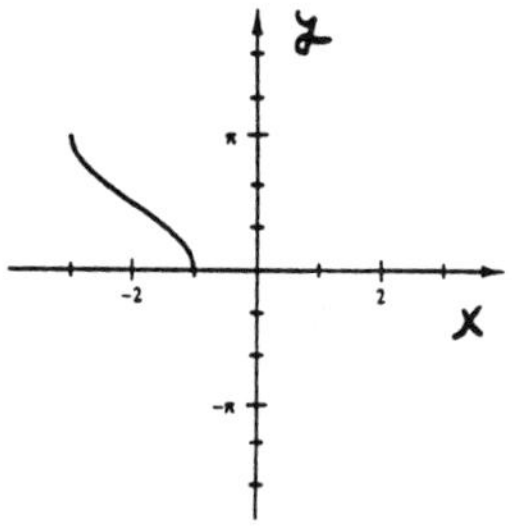

18. 150°, 210°

19. 0°

20. $\frac{7\pi}{6}, \frac{11\pi}{6}$

Chapter 7 Form D

1. b
2. c
3. d
4. a
5. a

6. b	7. c	8. d	9. a	10. b
11. b	12. b	13. a	14. b	15. a
16. d	17. b	18. c	19. b	20. a

Chapter 7 Form E

1. d	2. c	3. b	4. a	5. c
6. a	7. a	8. a	9. b	10. b
11. a	12. c	13. d	14. b	15. b
16. b	17. d	18. d	19. b	20. a

Chapter 7 Form F

1. b	2. c	3. d	4. c	5. a
6. c	7. b	8. b	9. b	10. c
11. a	12. b	13. d	14. c	15. a
16. c	17. c	18. a	19. a	20. c

Chapter 8 Form A

1. $C = 67°$, $b = 16$, $a = 18$

2. 21°

3. 6.8

4. 109°

5. 698

6. 175 square units

7. 108 square units

8. $-8\mathbf{i} - 8\sqrt{3}\,\mathbf{j}$

9. $3\mathbf{i} - 113\mathbf{j}$

10. -5

11. $157°12$

12. $6\sqrt{2}$ cis $135°$

13. $-\frac{3\sqrt{2}}{2} - \frac{3\sqrt{2}}{2}i$

14. $-20\sqrt{3} - 20i$

15. 5 cis $290°$

16. 16

17. $-2i$, $-\sqrt{3} + i$, $\sqrt{3} + i$

18. 113 miles

19. 80 miles

20. $66,888.87

Chapter 8 Form B

1. $A = 77°$. $a = 26$, $c = 16$

2. No triangle is formed

3. 9.8 units

4. $55°$

5. 475 square units

6. 104 square units

7. 1585 square units

8. $2.74\mathbf{i} + 7.52\mathbf{j}$

9. $-18\mathbf{i} - 40\mathbf{j}$

10. -17

11. $8°$

12. 10 cis $150°$

13. $-2\sqrt{3} - 2i$

14. $43.13 + 32.5i$

15. 0.5 cis $213°$

16. $-128 - 128\sqrt{3}\,i$

17. $-4\sqrt{2} + 4\sqrt{2}\,i$, $4\sqrt{2} - 4\sqrt{2}\,i$

18. 260 mph, heading $108°$

19. $4.7°$

20. 156 miles

Chapter 8 Form C

1. $C = 53°$ $a = 22$, $b = 8.4$

2. $28°$

3. 35

4. $95°$

5. 83 square units

6. 266 square units

7. 26 square units

8. $57.4\mathbf{i} - 48.2\mathbf{j}$

9. $\langle 68, -49 \rangle$	10. -36	11. $103°12.$
12. $2\sqrt{3}$ cis $325°$	13. $-5 + 5\sqrt{3}\,i$	14. $-14.49 - 3.88i$
15. $-0.254 - 0.544i$	16. $-4096 + 0i$	17. $-3\sqrt{2} + 3\sqrt{2}\,i,\ 3\sqrt{2} - 3\sqrt{2}\,i$
18. 65 miles	19. 36.8 acres	20. 25.5 miles

Chapter 8 Form D

1. b	2. c	3. d	4. d	5. b
6. c	7. b	8. c	9. c	10. a
11. b	12. d	13. d	14. c	15. a
16. b	17. b	18. d	19. b	20. a

Chapter 8 Form E

1. a	2. c	3. b	4. b	5. a
6. b	7. b	8. c	9. c	10. c
11. a	12. b	13. c	14. b	15. c
16. a	17. d	18. b	19. b	20. b

Chapter 8 Form F

1. b	2. a	3. c	4. a	5. c
6. d	7. b	8. c	9. d	10. a

11. d	12. a	13. a	14. a	15. c
16. b	17. c	18. d	19. d	20. c

Chapter 9 Form A

1. vertex: (0,0), focus: $\left(0, -\frac{1}{32}\right)$, directrix: $y = \frac{1}{32}$

2. vertex: (5, 1), focus: (5, 3), directrix: y = –1

3. $(y + 7)^2 = 12(x - 8)$

4.

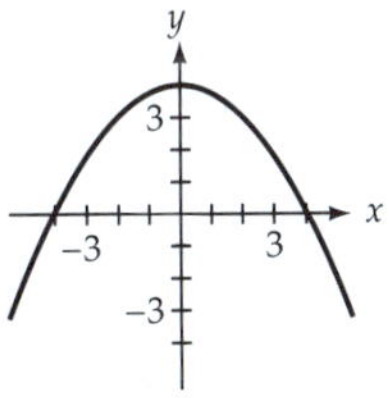

5. vertices: (0, 7), (0, –7), foci: $(0, 2\sqrt{6})$, $(0, -2\sqrt{6})$

6.

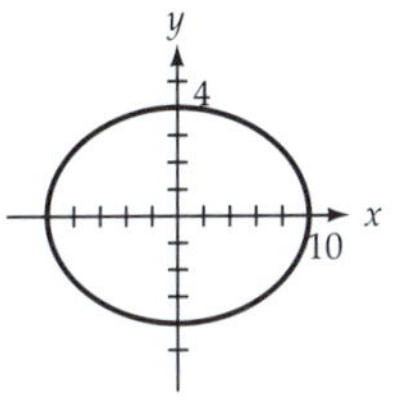

7. vertices: (24, –1), (0, –1),
foci: $(12 - 3\sqrt{15}, -1)$, $(12 + 3\sqrt{15}, -1)$
$8\sqrt{2}$ $8\sqrt{2}$

8. $\frac{(x-3)^2}{36} + \frac{(y-8)^2}{25} = 1$

9. $\frac{3}{5}$

10.

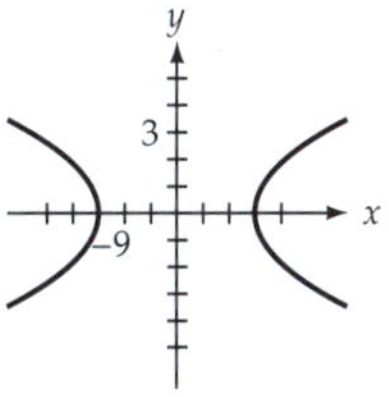

11. vertices: (5, 0), (–5, 0), foci: $(5\sqrt{5}, 0)$, $(-5\sqrt{5}, 0)$, asymptotes: $y = \pm\frac{1}{2}x$

12. 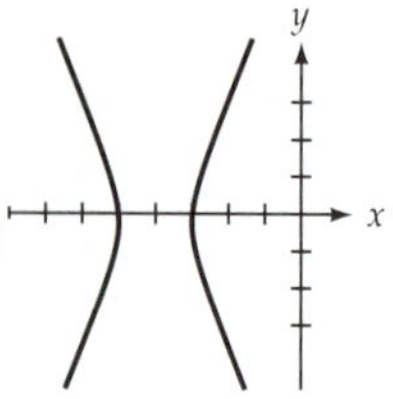

c(−3, −1)
v(−2, −1)(−4, −1)

13. vertices: (3, 0), (3, −10),
foci: $(3, -5 + \sqrt{29})$, $(3, -5 - \sqrt{29})$

14. $\dfrac{(x+2)^2}{4} - \dfrac{(y-3)^2}{11} = 1$

15. 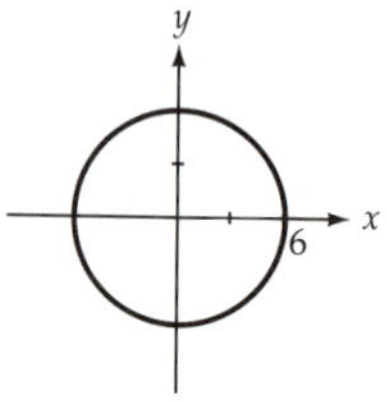

16. 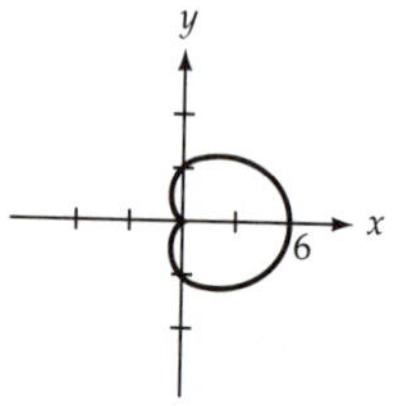

17. 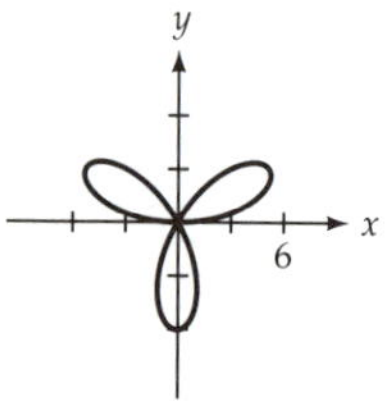

18. $\left(-\dfrac{3\sqrt{2}}{2}, -\dfrac{3\sqrt{2}}{2}\right)$

19. $18y = 81 - x^2$

$x^2 - 18y = 81$

20. $r^2 = \dfrac{25}{3}$

Chapter 9 Form B

1. vertex: (0, 0), focus: $\left(-\dfrac{1}{2}, 0\right)$, directrix: $x = \dfrac{1}{2}$

2. vertex: (−3, 7), focus: (−2, 7), directrix: x = −4

3. $(x+3)^2 = 8(y+5)$

4. 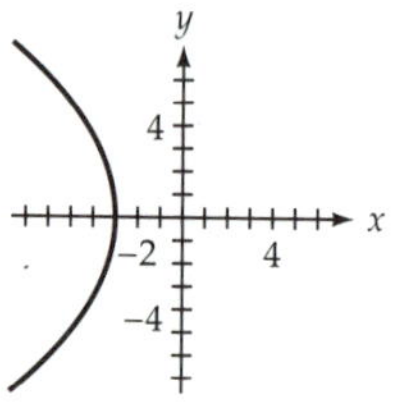

5. vertices: (9, 0), (−9, 0)
foci: $(\sqrt{65}, 0)$, $(-\sqrt{65}, 0)$

6.

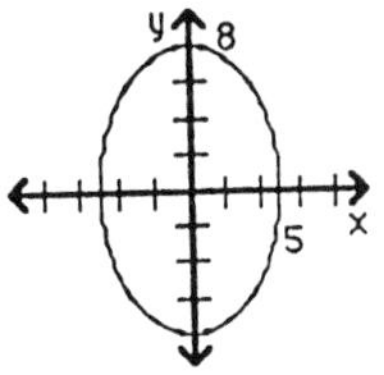

7. vertices: (-4, 5), (-4, 9), foci: $\left(-4,\ 7 + \sqrt{2}\right), \left(-4,\ 7 - \sqrt{2}\right)$

8.

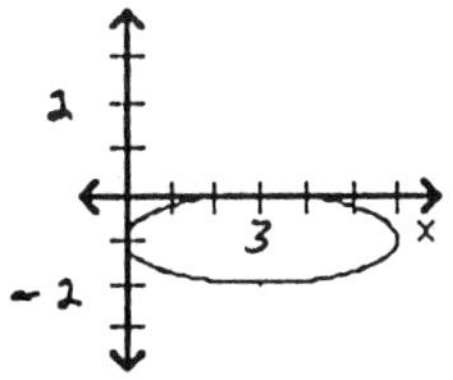

9. $\frac{2}{7}$

10.

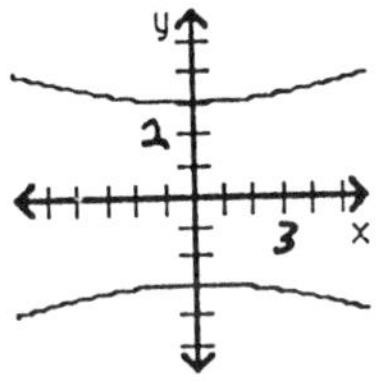

11. vertices: (6, 0), (-6, 0), foci: $\left(3\sqrt{13},\ 0\right), \left(-3\sqrt{13},\ 0\right)$, asymptotes: $y = \pm\frac{3}{2}x$

12.

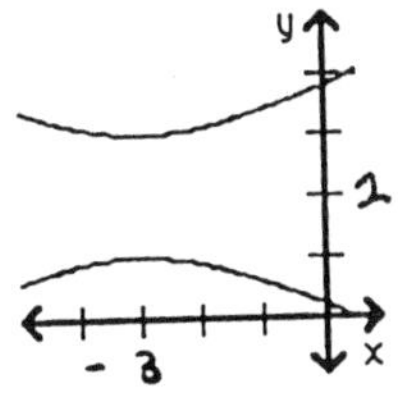

13. vertices: (-24, 3), (6, 3), foci: $\left(-9 + 5\sqrt{13},\ 3\right), \left(-9 - 5\sqrt{13},\ 3\right)$

14. $\frac{(y + 4)^2}{16} - \frac{(x - 7)^2}{15} = 1$

15.

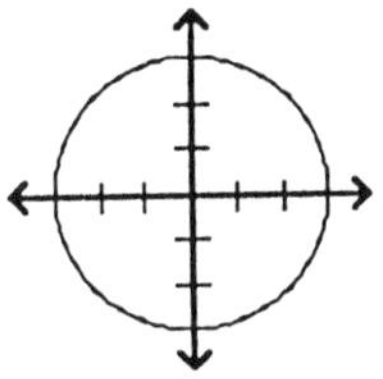

16.

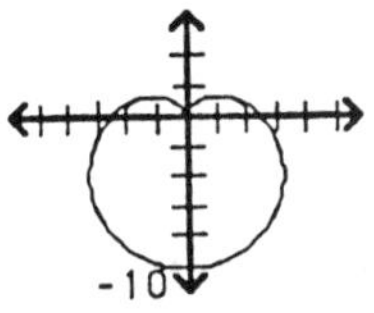

17.

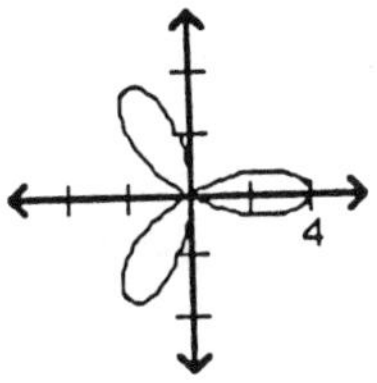

18. $(7,\ -7\sqrt{3}$

19. $r \sin \theta = 8$

20. $y^2 + 4x - 4 = 0$

College Algebra and Trigonometry   Test Answers

Chapter 9   Form C

1. vertex: (0, 0), focus: $\left(\frac{1}{6}, 0\right)$, directrix: $x = -\frac{1}{6}$

2. vertex: (4, 0), focus: $\left(4, -\frac{1}{2}\right)$, directrix: $y = \frac{1}{2}$

3. $(x - 2)^2 = -4(y - 1)$

4.

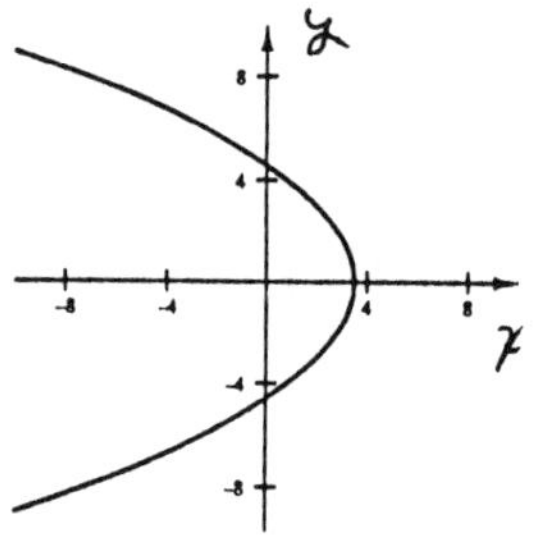

5. vertices: (0, 6), (0, −6), foci: $\left(0, 2\sqrt{6}\right)$, $\left(0, -2\sqrt{6}\right)$

6.

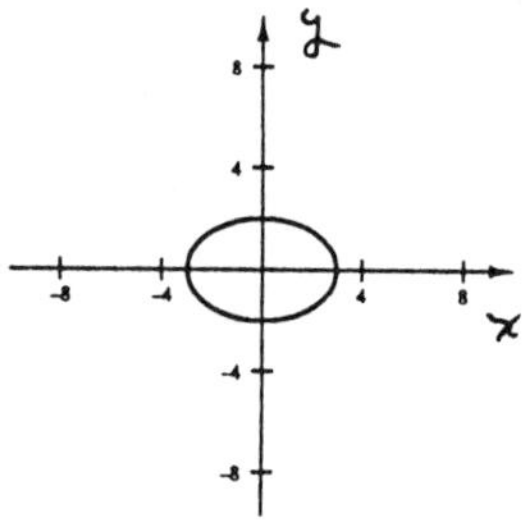

7. vertices: (−1, −2), (5, −2), foci: $\left(2 + \sqrt{5}, -2\right)$, $\left(2 - \sqrt{5}, -2\right)$

8. $\frac{(x - 2)^2}{20} + \frac{y^2}{4} = 1$

9. $\frac{\sqrt{3}}{2}$

10.

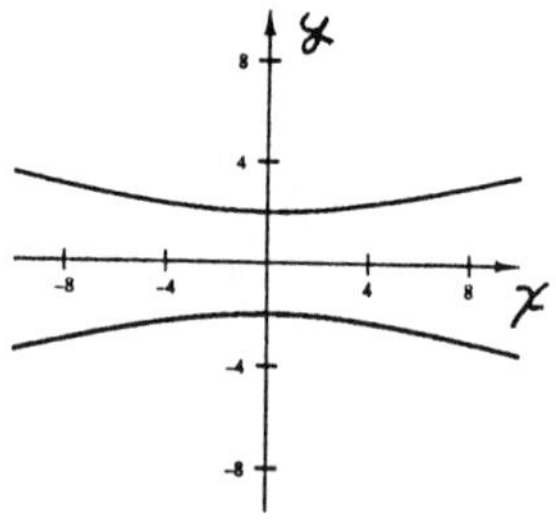

11. vertices: (6, 0), (−6, 0), foci: $\left(\sqrt{61}, 0\right)$, $\left(-\sqrt{61}, 0\right)$, asymptotes: $y = \pm\frac{5}{6}x$

12.

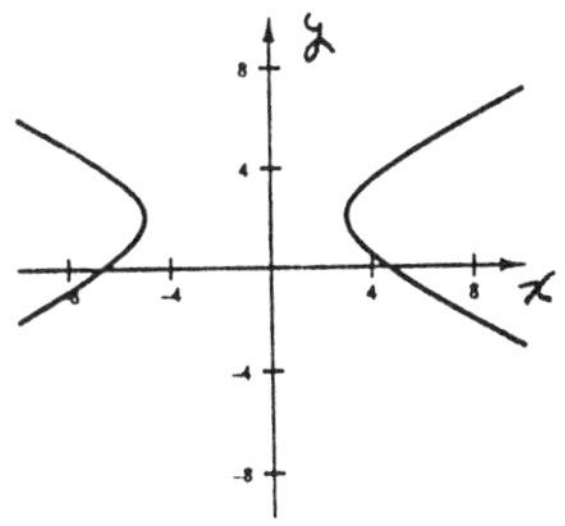

13. vertices: (−2, 8), (−2, −2), foci: $\left(-2, 3 + \sqrt{29}\right)$, $\left(-2, 3 - \sqrt{29}\right)$

14. $\dfrac{(y-4)^2}{20} - \dfrac{(x+5)^2}{16} = 1$

15.

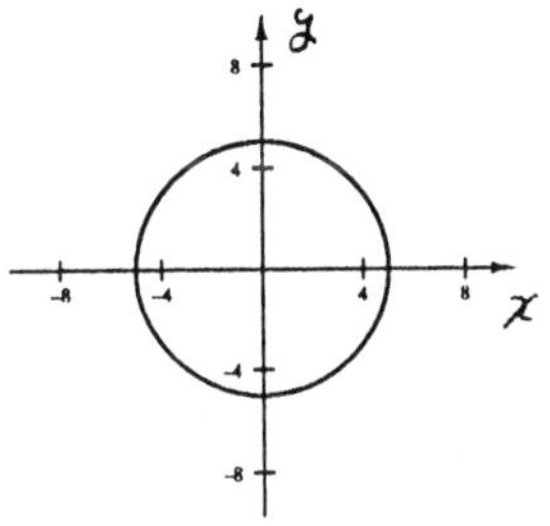

16.

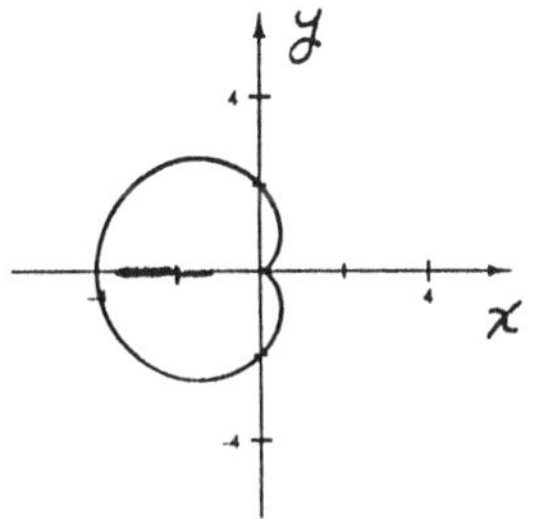

17.

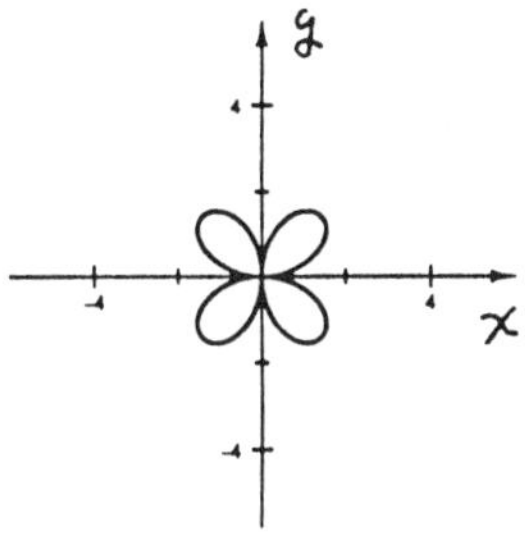

18. $\left(\sqrt{13}, 124°\right)$

19. $x^2 - 2x + y^2 = 0$

20. $r(3 \cos \theta + 2 \sin \theta) = 4$

Chapter 9 Form D

1. b	2. d	3. b	4. c	5. b
6. d	7. c.	8. a	9. a	10. d
11. d	12. b	13. a	14. c	15. a
16. b	17. d	18. c	19. d	20. c

Chapter 9 Form E

1. b	2. a	3. c	4. d	5. d
6. a.	7. b	8. a	9. b	10. d
11. d	12. b	13. c	14. a	15. c
16. a	17. d	18. d	19. a	20. c

Chapter 9 Form F

1. b	2. d	3. a	4. d	5. d
6. b	7. a	8. b	9. a	10. d
11. d	12. d	13. d	14. a	15. d
16. b	17. c	18. b	19. b	20. a

Chapter 10 Form A

1. (−1. −8)

2. Inconsistent

3. $\left(\frac{-15c+28}{11}, \frac{24c-14}{11}, c\right)$

4. (3, −1, 2)

5. $\left(\frac{c}{4}, -\frac{3c}{4}, c\right)$

6. (−6, 3), (1, −4)

7. (−5, 0), (5, 0)

8. (c + 3, −5c, c)

9. 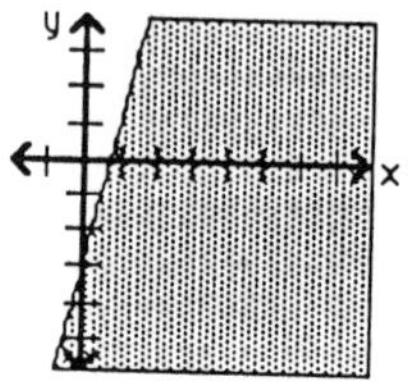

10. 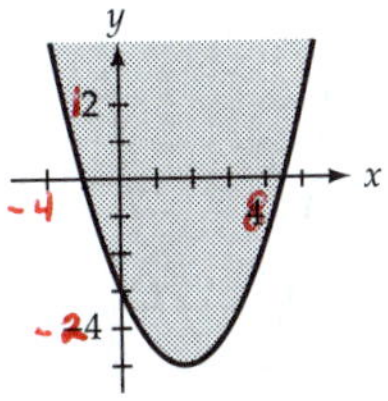

11. 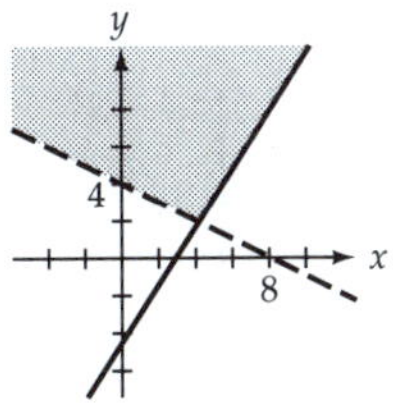

12. 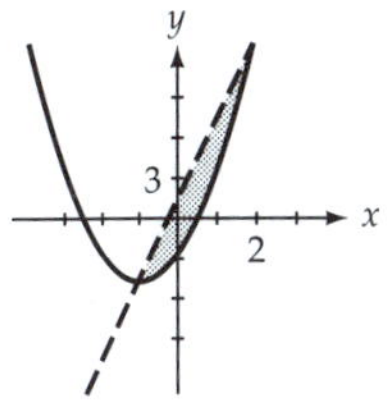

13. 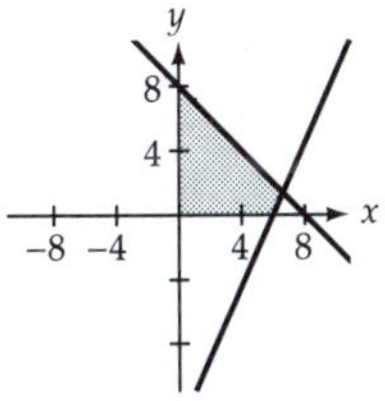

14. $\dfrac{2}{x-3} + \dfrac{4}{x+5}$

15. $\dfrac{2}{x} + \dfrac{5}{x+1} + \dfrac{7}{(x+1)^2}$

16. $\dfrac{-4}{3x+1} + \dfrac{2x+1}{x^2+9}$

17. $\left(x - \dfrac{3}{2}\right)^2 + \left(y - \dfrac{13}{2}\right) = \dfrac{13}{2}$

$x^2+y^2-3x-13y+38=0$

18. Rate of the motorboat: 20 mph

Rate of the current: 4 mph

19. 20% solution: 30 liters, 45% solution: 70 liters

20. 4

Chapter 10 Form B

1. (3, –1)

2. (2, 3)

3. (5, 4, –9)

4. Inconsistent

5. (c + 5, c – 4, c)

6. (c, 5c, –2c)

7. $(1, \sqrt{5}), (1, -\sqrt{5})$

8. (–7, 0), (7, 0)

9. 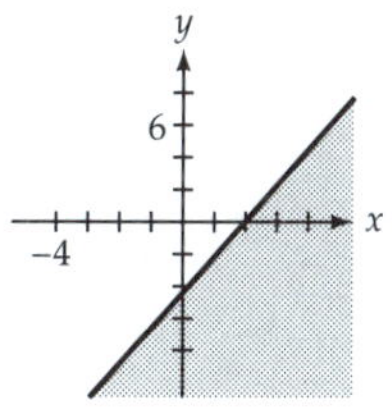

10.

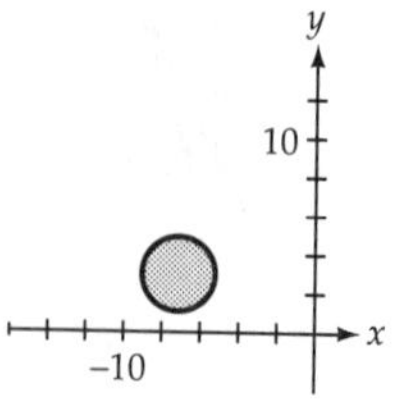

11.

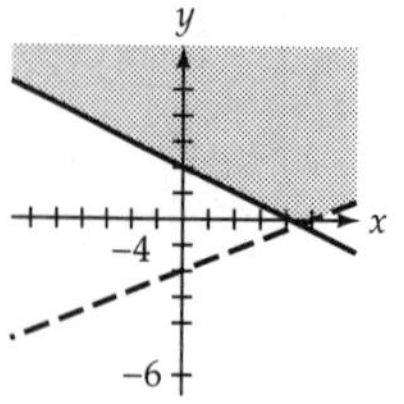

12.

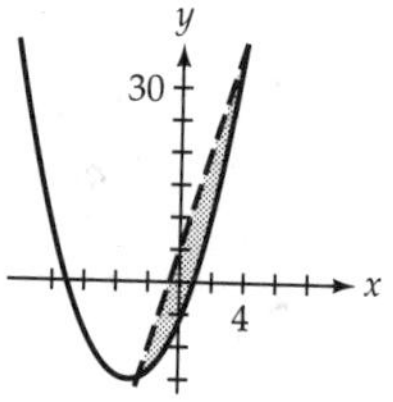

13.

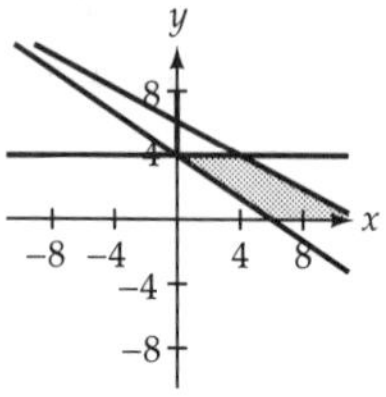

14. $\dfrac{3}{x} + \dfrac{2}{2x+5}$

15. $\dfrac{3}{x} + \dfrac{4}{x-3} + \dfrac{1}{(x-3)^2}$

16. $\dfrac{-2}{5x+2} + \dfrac{4x+3}{x^2+1}$

17. $(x-1)^2 + (y+9)^2 = 169$

18. rate in calm water: 3.5 mph

rate of the current: 1.5 mph

19. 35% solution: 40 g, 70% solution: 60 g

20. 36

Chapter 10 Form C

1. (4, –10)

2. (–1, –2)

3. (2, –3, 1)

4. Inconsistent

5. (16c – 71, 11c – 49, c)

6. $\left(\dfrac{8c}{7}, \dfrac{11c}{14}, c\right)$

7. $(1, \sqrt{3}), (1, -\sqrt{3})$

8. (4, 2), (–1, –3)

9.

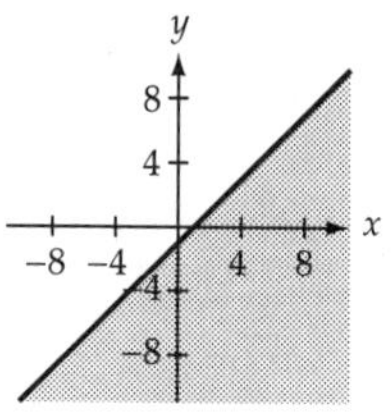

10. 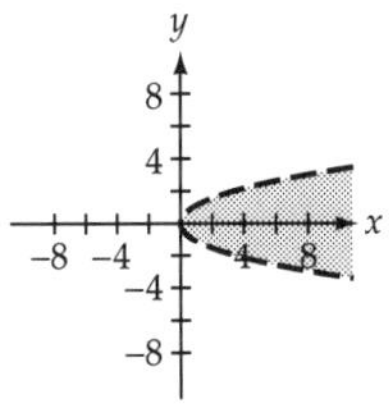

11. 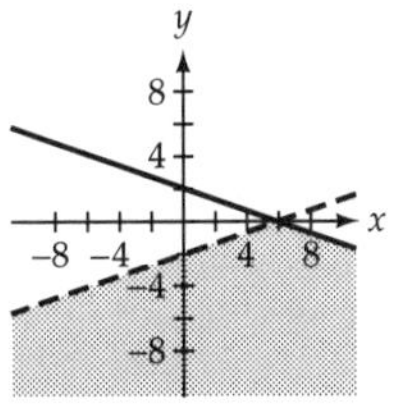

12. 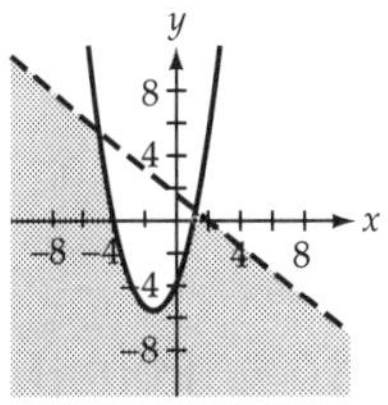

13. 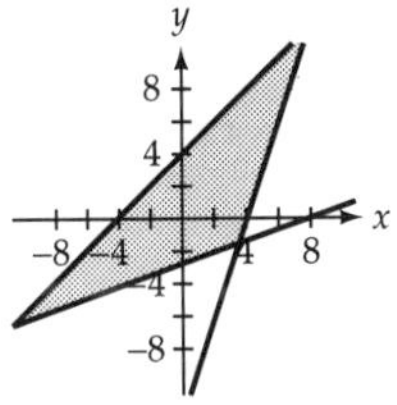

14. $\frac{5}{6(x-4)} + \frac{7}{6(x+2)}$

15. $\frac{-1}{x} + \frac{1}{x^2} + \frac{3}{x-2}$

16. $\frac{2}{3x+1} + \frac{2x+1}{x^2-3}$

17. $(x-1)^2 + (y+9)^2 = 169$

18. rate of the plane: 275 mph

rate of the wind: 25 mph

19. 12 dimes, 17 quarters

20. 20

Chapter 10 Form D

1. a
2. c
3. c
4. b
5. c
6. d
7. c
8. d
9. b
10. b
11. a
12. a
13. b
14. c
15. b
16. d
17. b
18. a
19. b
20. c

Chapter 10. Form E

1. a
2. a
3. c
4. c
5. c
6. a
7. b
8. c
9. d
10. b

11. c	12. d	13. d	14. a	15. c
16. a	17. a	18. d	19. c	20. d

Chapter 10 Form F

1. b	2. c.	3. b	4. d	5. a
6. d	7. c	8. b	9. c	10. b
11. a	12. d	13. c	14. b	15. b
16. c	17. d	18. b	19. a	20. c

Chapter 11 Form A

1. $\begin{vmatrix} 3 & -4 & 1 & -1 \\ 2 & -3 & -4 & 17 \\ 1 & 0 & -5 & 4 \end{vmatrix}, \begin{vmatrix} 3 & -4 & 1 \\ 2 & -3 & -4 \\ 1 & 0 & -5 \end{vmatrix}, \begin{vmatrix} -1 \\ 17 \\ 4 \end{vmatrix}$

2. $$\begin{aligned} -2x + 5y - 2z + 3w &= 1 \\ 4x - 2y + 3z + w &= -4 \\ -x - 4y + 2z + 3w &= -1 \end{aligned}$$

3. (1, 4, −3)

4. (−c + 4, c − 1, c)

5. $\left(\frac{-1-4c}{9}, \frac{-2c+13}{3}, \frac{23-7c}{9}, c \right)$ (printed ~~$\frac{10c-47}{9}$~~ crossed out by hand, with $\frac{-1-4c}{9}$ written above)

6. $\begin{bmatrix} 5 & -10 & -20 \\ -20 & 10 & -10 \end{bmatrix}$

7. Not possible

8. $\begin{vmatrix} -18 & 0 & -6 \\ -22 & -8 & 8 \\ 2 & 6 & 8 \end{vmatrix}$

9. $\begin{vmatrix} 5 & 8 & 19 \\ 16 & -2 & 8 \end{vmatrix}$

10. $\begin{vmatrix} 11 & 2 & 13 \\ -8 & 5 & 6 \\ 4 & -2 & -1 \end{vmatrix}$

11. $\begin{vmatrix} -5 & -3 & -11 \\ 3 & -1 & -14 \\ 19 & 1 & 6 \end{vmatrix}$

12. $\begin{vmatrix} 27 & -2 & 10 \\ 31 & 10 & 5 \\ 1 & -3 & 3 \end{vmatrix}$

13. Not possible

14. $\begin{vmatrix} -2 & -\frac{1}{3} & \frac{2}{3} \\ -6 & -\frac{1}{3} & \frac{5}{3} \\ 3 & \frac{1}{3} & -\frac{2}{3} \end{vmatrix}$

15. 5, −5
−18, 18

16. −3

17. −19

18. ~~(1, 1, −1)~~ (−5, 2)

19. ~~(1, 1, −1)~~
y=1

20. 1 serving food A, 3 servings food B, 4 servings food C

Chapter 11 Form B

1. $\begin{vmatrix} 2 & -3 & 1 & 3 \\ -1 & 4 & 0 & -2 \\ 3 & -5 & 2 & 17 \end{vmatrix}, \begin{vmatrix} 2 & -3 & 1 \\ -1 & 4 & 0 \\ 3 & -5 & 2 \end{vmatrix}, \begin{vmatrix} 3 \\ -2 \\ 17 \end{vmatrix}$

2. $x + 2y + 4z + 5w = 1$
$-2x + 3y - z + 4w = -7$
$2x - 3y + 4z - w = 3$

3. (1, −3, 2)

4. $(-c + 7, c - 3, c)$

5. $\left(\frac{7c - 30}{4}, \frac{30 - c}{8}, 7 - c, c\right)$

6. $\begin{vmatrix} -16 & 0 & -20 \\ 12 & -8 & 12 \\ -24 & -8 & -36 \end{vmatrix}$

7. Not possible

8. $\begin{vmatrix} 10 & 0 & 13 \\ -17 & 2 & -15 \\ 26 & 0 & 37 \end{vmatrix}$

9. $\begin{vmatrix} 18 & -9 & 21 \\ -27 & 31 & -38 \end{vmatrix}$

10. $\begin{vmatrix} 10 & 2 & 14 \\ 28 & 10 & 41 \\ -55 & -4 & -74 \end{vmatrix}$

11. not possible

12. not possible

13. $\begin{vmatrix} 46 & 10 & 65 \\ -36 & -2 & -48 \\ 72 & 22 & 105 \end{vmatrix}$

14. $\begin{vmatrix} 4 & \frac{5}{3} & -\frac{5}{3} \\ \frac{3}{2} & 1 & -\frac{1}{2} \\ -3 & -\frac{4}{3} & \frac{4}{3} \end{vmatrix}$

15. −9, 9

16. 6

17. 1

18. (−1, 2, 1)

19. $\left(\frac{35}{6}, \frac{2}{3}, \frac{5}{2}\right)$

20. 5 servings food A
2 servings food B
1 serving food C

Chapter 11 Form C

1. $\begin{vmatrix} 1 & -3 & 1 & -4 \\ 2 & -3 & -5 & 2 \\ 1 & 7 & -8 & -1 \end{vmatrix}, \begin{vmatrix} 1 & -3 & 1 \\ 2 & -3 & -5 \\ 1 & 7 & -8 \end{vmatrix}, \begin{vmatrix} -4 \\ 2 \\ -1 \end{vmatrix}$

2. $2x + y - 2z + 3w = 4$
$-x + 2y + 3z + w = 5$
$5x + 4y - z = 2w = 3$

3. $\left(\frac{9}{7}, \frac{6}{7}, \frac{45}{7}\right)$

4. (0, 4, 2)

5. (2, c − 4, c, 1)

6. $\begin{vmatrix} -3 & -3 & 3 \\ -3 & 0 & 6 \\ 0 & -6 & -9 \end{vmatrix}$

7. Not possible

8. $\begin{vmatrix} -1 & -7 & 9 \\ 1 & -2 & 6 \\ 0 & -2 & -7 \end{vmatrix}$

9. $\begin{vmatrix} -3 & 8 & 7 \\ 5 & 3 & -1 \end{vmatrix}$

10. $\begin{vmatrix} -1 & 7 & 12 \\ 1 & 2 & 0 \\ 2 & 2 & -1 \end{vmatrix}$

11. $\begin{vmatrix} -4 & 12 & 10 \\ 0 & 8 & -1 \\ -2 & -2 & -4 \end{vmatrix}$

12. Not possible

13. $\begin{vmatrix} 2 & -1 & -6 \\ 1 & -3 & -7 \\ 2 & 6 & 5 \end{vmatrix}$

14. $\begin{vmatrix} -\frac{1}{15} & \frac{8}{15} & \frac{1}{5} \\ -\frac{2}{15} & \frac{1}{15} & \frac{2}{5} \\ \frac{4}{15} & -\frac{2}{15} & \frac{1}{5} \end{vmatrix}$

15. 2, −2

16. 15

17. −1

18. $\left(-\frac{2}{3}, -3, -\frac{1}{3}\right)$

19. $\left(-\frac{5}{7}, \frac{3}{7}, \frac{4}{7}\right)$

20. $\left(\begin{vmatrix} 1 & 0 & 0 \\ 0 & 1 & 0 \\ 0 & 0 & 1 \end{vmatrix} - \begin{vmatrix} .12 & .18 & .08 \\ .06 & .10 & .14 \\ .20 & .15 & ....10 \end{vmatrix}\right)^{-1} \cdot \begin{vmatrix} 60 \\ 60 \\ 45 \end{vmatrix}$

Chapter 11 Form D

1. d	2. c	3. b	4. a	5. a
6. d	7. b	8. c	9. c	10. d
11. b	12. c	13. c	14. a	15. a
16. a	17. b	18. a	19. a	20. a

Chapter 11 Form E

1. d	2. d	3. b	4. b	5. c
6. d	7. b	8. b	9. b	10. d
11. b	12. a	13. b	14. a	15. c

16. c	17. c	18. b	19. b	20. d

Chapter 11 Form F

1. a	2. b	3. b	4. b	5. c
6. d	7. c	8. a	9. b	10. d
11. a	12. d	13. c	14. a	15. c
16. a	17. b	18. a	19. d	20. a

Chapter 12 Form A

1. −12, 0

2. $-\frac{1}{17}$

3. 1, 6, 54

4. geometric

5. arithmetic

6. geometric

7. 160

8. ~~$\frac{153}{140}$~~ $\frac{197}{70}$

9. 363

10. 74

11. $a_n = (-1)^{n-1}\left(\frac{2}{3}\right)^n$

12. 5

13. $\frac{17}{99}$

14. Let n = 1 $\quad 5(1) - 4 = \frac{1[5(1) - 3]}{2}$

$$1 = \frac{1(2)}{2} = 1$$

The statement is true for n = 1

Assume the statement is true for some positive integer k.

$$1 + 6 + 11 + 16 + \ldots + 5k - 4 = \frac{k(5k - 3)}{2}$$

Verify that the statement is true for n = k + 1.

$$1 + 6 + 11 + 16 + \ldots + 5k - 4 + 5(k + 1) - 4 = \frac{(k + 1)(5(k + 1) - 3)}{2}$$

$$\frac{k(5k - 3)}{2} + 5(k + 1) - 4 =$$

$$\frac{5k^2 - 3k}{2} + \frac{2(5k + 5 - 4)}{2} =$$

$$\frac{5k^2 - 3k + 10k + 2}{2} =$$

$$\frac{(5k^2 + 7k + 2}{2} =$$

$$\frac{(k + 1)(5k + 2)}{2} = \frac{(k + 1)[5(k + 2) - 3}{2}$$

Thus the statement is true for all positive integers.

15. Let n = 1 $4^n - 1 = 3$ and $3 = 3(1)$

Assume the statement is true for some positive integer k.

Thus $4^k - 1 = 3p$ for some integer p

Verify the statement is true for n = k + 1.

$$\begin{aligned} 4^{k+1} - 1 &= 4(4^k) - 1 \\ &= 4(3p + 1) - 1 \quad \text{since } 4^k = 3p + 1 \\ &= 12p + + 3 \\ &= 3(4p + 1) \quad \text{and is divisible by 3.} \end{aligned}$$

Thus the statement is true for all positive integers.

16. $x^{15} - 20x^{12}y + 160x^9y^2 - 640x^6y^3 + 1280x^3y^4 - 1024y^5$

17. $2{,}627{,}665{,}920x^8y^8$

18. 292 logs

19. 1,123,200 serial numbers

20. $\frac{7}{11}$

Chapter 12 Form B

1. $\frac{1}{3}, \frac{11}{21}$

2. $-\frac{6}{7}$

3. −2, −8, 72

4. geometric

5. neither

6. arithmetic

7. $\frac{85}{1024}$

8. 441

9. $-\frac{47}{120}$

10. $a_n = 11n - 14$

11. $-20{,}480$

12. 2

13. $\frac{35}{99}$

14. Let $n = 1$. $2 = \frac{1}{2}(1)[3(1) + 1] = \frac{1}{2}(4) = 2$

The statement is true for $n = 1$.

Assume the statement is true for some positive integer k.

$$2 + 5 + 8 + 11 + \ldots + 3k - 1 = \frac{1}{2}k(3k + 1)$$

Verify that the statement is true for $n = k + 1$.

$$2 + 5 + 8 + 11 + \ldots + 3k - 1 + 3(k + 1) - 1 = \frac{1}{2}(k + 1)[3(k) + 1]$$

$$\frac{1}{2}k(3k + 1) + 3(k + 1) - 1 =$$

$$\frac{3k^2 + k}{2} + \frac{2(3k + 2)}{2} =$$

$$\frac{3k^2 + 7k + 4}{2} =$$

$$\frac{1}{2}(k + 1)(3k + 4) = \frac{1}{2}(k + 1)(3[k + 1) + 1]$$

Thus the statement is true for all positive integers.

15. Let $n = 1$ $3^1 < 3^{1+1}$

$3 < 3^2 = 9$

The statement is true for $n = 1$.

Assume the statement is true for some positive integers k.

$$3^k < 3^{k+1}$$

Verify that the statement is true for $n = k + 1$.

that is $3^{k+1} < 3^{(k+1)+1} = 3^{k+2}$ but $3^{k+2} = 3 \cdot 3^{k+1} > 3 \cdot 3^k = 3^{k+1}$

Thus $3^{k+1} < 3^{k+2}$

Thus the statement is true for all positive integers.

16. $x^{10} - 15x^8y + 90x^6y^2 - 270x^4y^3 + 405x^2y^4 - 243y^5$

17. $672{,}000y^{18}$

18. $143,000

19. 40,320 ways

20. $\frac{550}{1463}$

Chapter 12 Form C

1. $\frac{4}{3}, \frac{6}{7}$

2. $-\frac{24}{11}$

3. 2, 8, 72

4. arithmetic

5. neither

6. geometric

7. 40

8. $\frac{437}{60}$

9. 3795

10. 387

11. $a_n = (-1)^{n-1}\left(\frac{3}{5}\right)^n$

12. $\frac{4}{3}$

13. $\frac{47}{99}$

14. Let n =1. $1 = 1[3(1) - 2] = 1$
The statement is true for $n = 1$.

Assume the statement is true for some positive integer k.

$$1 + 7 + 13 + 19 + \ldots + 6k - 5 = k(3k - 2)$$

Verify that the statement is true for $n = k + 1$.

$$1 + 7 + 13 + 19 + \ldots + 6k - 5 + 6(k + 1) - 5 = (k + 1)[3(k + 1) - 2]$$
$$k(3k - 2) + 6(k + 1) - 5 =$$
$$3k^2 - 2k + 6k + 6 - 5 =$$
$$3k^2 + 4k + 1 =$$
$$(3k + 1)(k + 1) = (k + 1)[3(k + 1) - 2]$$

Thus the statement is true for all positive integers.

15. Let $n = 1$ $\quad 1 + 4(1) \leq 5^1$

$1 + 4 \leq 5$

$5 = 5$

The statement is true for $n = 1$

Assume the statement is true for some positive integer k.

$1 + 4k \leq 5^k$

Verify that the statement is true for $n = k + 1$.

That is $\quad 1 + 4(k + 1) \leq 5^{k+1} \qquad 5^{k+1} = 5^k \cdot 5$

$$\begin{aligned} 5^k \cdot 5 &\geq (1 + 4k)5 \\ &= 5 + 20k \\ &= 1 + 4 + 4k + 16k \\ &= 1 + 4(1 + k) + 16k \geq 1 + 4(1 + k) \end{aligned}$$

Thus $5^{k+1}\kappa \geq 1 + 4(k + 1)$.

Thus the statement is true foir all positive integers.

16. $x^5 + 10x^4y^2 + 40x^3y^4 + 80x^2y^6 + 80xy^8 + 32y^{10}$

17. $326{,}592x^4y^{10}$

18. \$258,000

19. 80,640

20. $\frac{49}{400}$

Chapter 12 Form D

1. b	2. c	3. b	4. d	5. d
6. c	7. b	8. d	9. c	10. a
11. c	12. a	13. b	14. b	15. b
16. a	17. a	18. a	19. b	20. a

Chapter 12 Form E

1. c	2. c	3. a	4. c	5. b
6. b	7. c	8. a	9. c	10. d

11. a	12. d	13. c	14. d	15. b
16. b	17. b	18. a	19. b	20. b

Chapter 12 Form F

1. c	2. b	3. a	4. b	5. d
6. c	7. d	8. b	9. d	10. c
11. d	12. a	13. d	14. a	15. c
16. d	17. c	18. c	19. a	20. d

Precalculus
Tests and Answers

1. Identify the property that is illustrated by $(a + b) + c = (b + a) + c$.

2. Given $A = \{2, 3, 4, 8, 9\}$ and $B = \{2, 4, 5, 7, 9\}$, find $A \cap B$.

3. Simplify: $(-3x^{-3}y^{2})^{-3}(-2x^{2}y^{0})^{-2}$

4. Simplify: $\dfrac{(4x^{-2}y^{3})^{-2}(-2x^{4}y^{-1})^{3}}{(-6x^{-2}y^{2})}$

5. Simplify: $(2x-5)(x^{3}+3x-4)$

6. Evaluate the polynomial $x^{4}-3x^{2}+4x-7$ for $x=-4$.

7. Factor: $3x^{4}-3xy^{3}$

8, Factor: $x^{5}+x^{2}y^{3}-y^{2}x^{3}-y^{5}$

9. Simplify: $\dfrac{3x}{x^{2}+2x-8}-\dfrac{2}{x^{2}-4}$

10. Simplify: $\dfrac{x^{2}+2x-8}{x^{3}-1} \div \dfrac{x^{2}+6x+8}{x^{3}+x^{2}+x}$

11. Simplify: $\dfrac{\dfrac{y}{y-2}}{\dfrac{1}{y}+\dfrac{2}{y-2}}$

12. Simplify: $\dfrac{a^{2/3}b^{-3/2}}{a^{-1/3}b^{3/4}}$

13. Simplify: $\dfrac{4a}{\sqrt[4]{16a^5}}$

14. Simplify: $\dfrac{y}{\sqrt{y}-\sqrt{2}}$

15. Simplify (2 - 3i)(4 - 7i) and write the complex number in standard form.

16. Write $\dfrac{3i}{2-3i}$ as a complex number in standard form.

17. Solve: $\dfrac{3}{x+2}+\dfrac{7}{x+2}=\dfrac{5}{x-2}$.

18. Solve: $\sqrt{x+1}-3=\sqrt{x+4}$.

19. Solve: $2x^{1/2}-x^{1/4}-3=0$

20. Solve: $\dfrac{(x-1)(x+4)}{x+1}\leq 0$

21. Solve: $|3x-11|=8$

22. Solve: $|5x+7|\geq 17$

23. The length of a rectangle is 3 feet more than twice the width of the rectangle. If the perimeter of the rectangle is 90 feet, find the length of the rectangle.

24. An accountant invests a bonus of $8000 into two simple interest accounts. On one account the annual simple interest rate is 7.5%, and on the second account the annual simple interest rate is 5.2%. The amount of interest earned for 1 year was $542.50. How much was invested in each account?

25. The first leg of a trip to the country was 150 miles of highway. The second leg, 180 miles of backroads, took 1.5 hours longer than the first part. If the driver was able to travel 10 mph faster on the highway than on the backroads, find the average rate on the backroads.

1. Which of the numbers -8, 0.17, 0, $0.7\overline{4}$, 0.141141114..., $\sqrt{11}$ are rational numbers?

2. Identify the property that is illustrated by (a + b) + c = c + (a + b).

3. Write the inequality $-4 \le x < -1$ using interval notation.

4. Simplify: $\left(-6a^{-2}b^{3}\right)\left(2ab^{-4}\right)^{2}$

5. Simplify: $\dfrac{\left(3x^{-2}y^{4}\right)^{-2}}{\left(-2x^{0}y^{-3}\right)\left(3x^{-4}y^{2}\right)^{3}}$

6. Simplify: $(x-y+2)(2x+y-4)$

7. Factor: $7x^2-28xy+21y^2$

8. Factor: $2ax^2-3bx^2-2ay^2+3by^2$

9. Simplify: $\dfrac{x}{x^2+5x+6}-\dfrac{2}{x^2+2x-3}$

10. Simplify: $\dfrac{x^4+2x^3-x-2}{3x^4+3x^3+3x^2}\cdot\dfrac{4x^2-12x+9}{2x^2+x-6}$

11. Simplify: $a-\dfrac{a}{1+\dfrac{1}{a}}$

12. Simplify: $\dfrac{a^{3/4}b^{-1/3}}{a^{-1/2}b^{-3/2}}$

13. Simplify: $3\sqrt[3]{16a^5b^7} - ab\sqrt[3]{54a^2b^4}$

14. Simplify: $\dfrac{\sqrt{2}}{\sqrt{x}-\sqrt{2}}$

15. Simplify (3 - 2i)(4 + 8i) and write the complex number in standard form.

16. Write $\dfrac{3}{-3-i}$ as a complex number in standard form.

17. Solve: $\dfrac{3}{x-3}+\dfrac{5}{x+3}=\dfrac{1}{x-3}$

18. Solve: $x^4 - 3x^3 + 8x - 24$

19. Solve: $\sqrt{2x+3}-1=\sqrt{3x-5}$

20. Solve: $\dfrac{x^2-7x-8}{x+6}\geq 0$

21. Solve: $|2-5x|=3$

22. Solve: $|5x-1|\leq 9$

23. A biathlon included 12 miles of running and 30 miles of cycling. One contender cycled for 30 minutes longer than he ran, even though his rate of cycling was 7 mph faster than his rate of running. Find his rate of cycling.

24. An experienced word processor can input a report in half the time that a trainee can do the job. Together they can complete the task in 20 minutes. How long would it take the trainee to input a report working alone?

25. You can rent a car for the day from Company A for $20 plus $0.09 a mile. Company B charges $13 plus $0.15 a mile. At what point, in terms of miles driven per day, is it cheaper to rent from Company A?

1. Which of the numbers -5, $3.\overline{4}$, $\sqrt{19}$, $\frac{7}{11}$, π, and 0.565565556... are irrational numbers?

2. Identify the property that is illustrated by the following: 6(a + b) = (a + b)6.

3. Graph the inequality $x < -3$ or $x \geq 0$.

4. Simplify: $\dfrac{\left(-2a^{-1}b\right)^{-3}\left(-3a^{-2}b^{2}\right)^{4}}{\left(6a^{2}b^{-3}\right)^{2}}$

5. Simplify: $\left(x^2+2x-3\right)\left(x^2-2x+3\right)$

6. Factor: $2ax^2-2ax-bx^2+bx$

7. Factor: $x^2y^3+x^2-y^3-1$

8. Simplify: $\dfrac{6}{x^2-16}-\dfrac{5}{x^2+x-12}$

9. Simplify: $\dfrac{6x^2+42x}{6x^2-54}\div\dfrac{x^2+2x-35}{x^2-2x-15}$

10. Simplify: $\dfrac{\dfrac{x}{x+5}}{\dfrac{1}{x}+\dfrac{x-6}{x+5}}$

11. Simplify: $\dfrac{a^{-3/2}b^{5/3}}{a^{1/2}b^{-3/2}}$

12. Simplify: $\dfrac{y}{\sqrt[4]{3^3 y}}$

13. Simplify: $xy\sqrt{18x^2y} - \sqrt{32x^4y^3}$

14. Simplify: $\dfrac{2x}{\sqrt{2} - x}$

15. Simplify $(2 - \sqrt{-1})(3 + \sqrt{-9})$ and write the complex number in standard form.

16. Write $\dfrac{-3i}{-1 + 4i}$ as a complex number in standard form.

17. Solve: $\dfrac{3}{x} + \dfrac{2}{2x - 3} = \dfrac{5}{2x - 3}$

18. Solve: $(2x - 3)(x + 2) - x(x + 4) = 0$

19. Solve: $\sqrt{2x - 1} - 1 = \sqrt{x + 2}$

20. Solve: $\dfrac{x - 2}{x + 1} \geq 2$

21. Solve: $|4 - x| = 5$

22. Solve: $|2x-5| \leq 2$

23. After traveling 110 miles on a 200-mile trip, a motorist ran over a rock, which caused the car to vibrate at higher speeds. The motorist decreased his speed by 20 mph for the final 90 miles of the trip. The last leg took 4/5 hours longer than the first leg. Find the motorist's rate after he ran over the rock.

24. The length of a rectangle is 3 feet more than three times the width of the rectangle. If the area of the rectangle is 90 square feet, find the width of the rectangle.

25. A concert performance brought in $130,800 on the sale of 12,000 tickets. If the tickets sold for $9 and $15, how many of each were sold?

1. Given A = {2, 4, 6, 7, 8, 9} and B = {1, 2, 3, 5, 6, 9}, find $A \cup B$.

   a. {2, 6, 9} b. {1, 2, 3, 4, 5, 6, 7, 8, 9}
   c. {4, 7, 8} d. {1, 3, 5}

2. Find the distance between the points –4 and –21 on the number line.

   a. 17 b. –25 c. 25 d. –17

3. Simplify: $\dfrac{(2xy^{-3})^{-2}(-2x^{-2}y)^2}{4x^2y^2}$

   a. $\dfrac{16y^6}{4x^6}$ b. $\dfrac{16y^4}{x^8}$ c. $\dfrac{y^8}{x^8}$ d. $\dfrac{y^6}{4x^8}$

4. Simplify: $(x^2 - 2x + 3)(x^2 - 3)$

   a. $x^2 - 2x^3 - 6x^2 + 6x - 9$ b. $x^4 - 2x^3 + 6x - 9$
   c. $x^4 - 5x^3 + 6x + 9$ d. $x^4 - 2x^3 + 6x^2 + 6x + 9$

5. Factor: $2ax - 4bx - ay + 2by$

   a. $(a - 2b)(2x - y)$ b. $(a - 2x)(2b - y)$
   c. $(a + 2b)(2x - y)$ d. $(2x - b)(y - a)$

6. Factor: $x^4yz + 27xy^4z^4$

   a. $xyz(x + 3yz)(x^2 + 6xyz + 9y^2z^2)$ b. $xyz(x + 3yz)(x^2 - 3xyz + 9y^2z^2)$
   c. $x^4y^4z^4(yz + 27x)$ d. $xyz(x - 9yz)(x^2 + 9xyz + 9y^2z^2)$

7. Simplify: $\dfrac{5}{x^2 - 25} - \dfrac{4}{x^2 - x - 20}$

   a. $\dfrac{9x - 40}{(x - 5)(x + 5)(x + 4)}$ b. $\dfrac{x + 40}{(x - 5)(x + 5)(x - 4)}$

   c. $\dfrac{1}{(x - 5)(x + 5)(x + 4)}$ d. $\dfrac{x}{(x - 5)(x + 5)(x + 4)}$

8. Simplify: $\dfrac{3x^2 - 12x}{3x^2 + 12x + 12} \bullet \dfrac{x^2 - 3x - 10}{x^2 - 9x + 20}$

a. $\dfrac{x}{(x+2)}$

b. $\dfrac{3(x-5)}{(x+2)(x-4)}$

c. $\dfrac{x(x-4)}{(x+2)(x+4)}$

d. $\dfrac{3(x+4)}{(x+2)(x-4)}$

9. Simplify: $\dfrac{\dfrac{5}{x+5}}{\dfrac{x-7}{x+5} + \dfrac{1}{x}}$

a. $\dfrac{5x}{(x+5)(x+1)}$

b. $\dfrac{5x}{2x-2}$

c. $\dfrac{5x}{(x-5)(x-1)}$

d. $\dfrac{5}{x(x+5)}$

10. Simplify: $\dfrac{x^{3/5}y^{-2/3}}{x^{1/2}y^{1/3}}$

a. $\dfrac{x^{1/10}}{y}$

b. $x^{1/10}y$

c. $\dfrac{x^{11/10}}{y}$

d. $\dfrac{x^{3/10}}{y^{2/9}}$

11. Simplify: $\sqrt[5]{-128x^{17}y^{12}}$

a. $-8x^8y^6\sqrt[5]{2x}$

b. $-2x^3y^2\sqrt[5]{4x^2y^2}$

c. $-4x^2y^2\sqrt[5]{4x^7y^2}$

d. $-2x^3y^2\sqrt[5]{2x^2y^2}$

12. Simplify: $-a\sqrt{18a^3b^3} + 5\sqrt{50a^5b^3}$

a. $2a^2b\sqrt{4ab}$

b. $-3ab\sqrt{2ab}$

c. $28a^2b\sqrt{4ab}$

d. $22a^2b\sqrt{2ab}$

13. Simplify: $(2\sqrt{x} - 3)(2\sqrt{x} + 3)$

a. $4x - 9$ b. $4x - 6\sqrt{x} + 9$ c. $4x - 12\sqrt{x} + 9$ d. $2\sqrt{x} - 9$

14. Simplify: $\dfrac{x}{\sqrt{2} - \sqrt{x}}$

a. $\dfrac{\sqrt{2x} - x}{2 - x}$ b. $\dfrac{x\sqrt{2} + x\sqrt{x}}{2 - x}$

c. $\dfrac{x\sqrt{2} + x}{4 - x^2}$ d. $\dfrac{x\sqrt{2} + x\sqrt{x}}{4 - x}$

15. Simplify $(3 - 2i)(-2 + 4i)$ and write the complex number in standard form.

a. $-14 + 16i$ b. $-14 + 8i$ c. $2 + 16i$ d. $-2 + 16i$

16. Write $\dfrac{-2}{3 + 2i}$ as a complex number in standard form.

a. $-\dfrac{6}{13} + \dfrac{4}{13}i$ b. $-\dfrac{3}{5} + \dfrac{4}{5}i$ c. $\dfrac{3}{5} - \dfrac{2}{5}i$ d. $-\dfrac{6}{13} - \dfrac{2}{13}i$

17. Solve: $\dfrac{4}{x + 5} + \dfrac{2}{x + 5} = \dfrac{9}{x - 5}$

a. $-\dfrac{5}{2}$ b. 5 c. $\dfrac{5}{2}$ d. -25

18. Solve: $3x^2 - x - 1 = 0$

a. $\dfrac{1 \pm \sqrt{13}}{6}$ b. $\dfrac{-1 \pm \sqrt{13}}{2}$ c. $\dfrac{-1 \pm \sqrt{11}}{2}$ d. $-\dfrac{5}{2}$ or 2

19. Solve: $\sqrt{4x + 1} = x - 1$

a. $-1, 1$ b. $\dfrac{3 \pm \sqrt{2}}{2}$ c. 6 d. $\dfrac{3 \pm 2\sqrt{11}}{2}$

20. Solve: $x^4 - 37x^2 + 36 = 0$

a. $\pm i, \pm 6$ b. $\pm 1, \pm 6$ c. $\pm 1, \pm 6i$ d. $\pm i, \pm 6i$

21. Solve: $\frac{(x+1)(x-8)}{x+6} \geq 0$

a. $(-6, -1] \cup [8, \infty)$
b. $(-\infty, 1) \cup [-1, 8]$
c. $[-6, -4] \cup [8, \infty)$
d. $[-8, 1] \cup [6, \infty)$

22. Solve: $|7 - 3x| \geq 5$

a. $\left(-\infty, \frac{2}{3}\right] \cup [4, \infty)$
b. $\left(-\infty, \frac{2}{3}\right] \cup [4, \infty)$
c. $\left[-\infty, \frac{2}{3}\right] \cup [4, 8]$
d. $\frac{2}{3} \leq x \leq 4$

23. A jogger started a course at an average speed of 6 mph. One-half hour later, a cyclist started the same course at an average speed of 10 mph. How long after the jogger started did the cyclist overtake the jogger?

a. 0.75 h b. 1.25 h c. 2 h d. 1.5 h

24. A mason can build a wall in one-fourth the time it takes an apprentice. Together they build the wall in 8 hours. How long would it take the apprentice working alone to build the wall?

a. 10 h b. 20 h c. 40 h d. 30 h

25. One side of a triangle is 2 feet longer than the second side. The third side of the triangle is 1 foot more than twice the second side. The perimeter of the triangle is 15 feet. Find the measures of the three sides of the triangle.

a. 4 ft, 5 ft, 6 ft b. 3 ft, 5 ft, 7 ft c. 2 ft, 4 ft, 9 ft d. 2 ft, 5 ft, 8 ft

1. Identify the property that is illustrated by the following: If p + 2s = t and t = 7 - r, then p + 2s = 7 - r

a. Distributive Property
b. Reflexive
c. Symmetric
d. Transitive

2. Write $|1.5-\pi|$ without absolute value symbols.

a. $\pi-1.5$
b. $-\pi-1.5$
c. $1.5-\pi$
d. $1.5+\pi$

3. Given that $A=\{1,2,5,9,14,17\}$ and $P=\{x|x \text{ is a prime number less than } 20\}$, find $A\cap P$.

a. $\{1,2,5,7,9,11,13,14,17,19\}$
b. $\{1,2,5,17\}$
c. $\{2,5,17\}$
d. $\{1,2,5,9,17\}$

4. Simplify: $\left(-a^3b^{-4}\right)^{-1}\left(3a^{-2}b^3\right)^2$

a. $\dfrac{b^{10}}{9a^7}$
b. $\dfrac{-9b^{10}}{a^7}$
c. $\dfrac{-9a}{b^2}$
d. $-9a^3b^{10}$

5. Simplify: $\dfrac{\left(-5x^{-4}y^3\right)^{-4}}{\left(-2x^0y\right)^5\left(4x^3y^{-1}\right)^{-3}}$

a. $-\dfrac{2x^{25}}{625y^{20}}$
b. $-1250x^{25}y^{20}$
c. $\dfrac{2048x^7}{625y^{10}}$
d. $-\dfrac{64x^7}{625y^{10}}$

6. Simplify: $\left(2x-y^2\right)\left(4x^2+2xy^2+y^4\right)$

a. $8x^3-8x^2y^2+4xy^4-y^6$
b. $8x^3+4xy^4+y^6$
c. $8x^3-y^6$
d. $8x^3-8x^2y^2-y^6$

7. Factor: $81x^3-3y^3$

a. $(9x-y)\left(81x^2+27xy+y^2\right)$
b. $3(3x-7)\left(3x^2+xy+y^2\right)$
c. $3(3x-y)\left(9x^2+3xy+y^2\right)$
d. $3(3x+y)\left(8x^2-3xy+y^2\right)$

8. Factor: $x^4 + 2x^3 - x - 2$

a. $(x-2)(x+1)(x^2-x+1)$

b. $(x+2)(x-1)(x^2+x+1)$

c. $(x+2)(x+1)(x^2-x+1)$

d. $x(x+2)(x+1)(x-2)$

9. Simplify: $\dfrac{4x}{x^2+6x+8} - \dfrac{2}{3x+6}$

a. $\dfrac{2(5x-4)}{3(x+4)(x+2)}$

b. $\dfrac{2}{3(x+2)}$

c. $\dfrac{12x^2-2x-8}{3(x+4)(x+2)}$

d. $\dfrac{8x-8}{3(x+4)(x+2)}$

10. Simplify: $\dfrac{x^3+125}{2x^2+4x-30} \div \dfrac{6x^2+16x-70}{4x^2-100}$

a. $\dfrac{(x-5)(x^2-5x+25)}{(x-3)(3x-7)}$

b. $\dfrac{x^2-5x+25}{3x-7}$

c. $\dfrac{(x-5)(x^2-10x+25)}{(x-3)(3x-7)}$

d. $\dfrac{4(x-5)(x-5)}{(x-3)(3x-7)}$

11. Simplify: $\dfrac{\dfrac{x-2}{x+5}}{\dfrac{2}{x}+\dfrac{x-2}{x+5}}$

a. $\dfrac{x}{x^2+4x+10}$

b. $\dfrac{x(x-2)}{x^2+10}$

c. $\dfrac{x(x-2)}{x+5}$

d. $\dfrac{x-2}{x^2+10}$

12. Simplify: $\dfrac{a^{-3/2}b^{2/3}}{a^{2/3}b^{-5/3}}$

a. $a^{5/6}b^{7/3}$

b. $\dfrac{b}{a^{13/6}}$

c. $a^{5/6}b^{2/3}$

d. $\dfrac{b^{7/3}}{a^{13/6}}$

13. Simplify: $3a\sqrt[3]{-27ab^5} + b\sqrt[3]{125a^4b^2}$

a. $-4ab\sqrt[3]{ab^2}$

b. $-13ab\sqrt[3]{ab}$

c. $4ab\sqrt[3]{ab}$

d. $-4a^2b\sqrt[3]{ab}$

14. Simplify: $\dfrac{2-\sqrt{3}}{\sqrt{x}+\sqrt{3}}$

a. $\dfrac{2\sqrt{x}-2\sqrt{3}+\sqrt{3x}+9}{x-9}$

b. $\dfrac{2\sqrt{x}-2\sqrt{3}-\sqrt{3x}+3}{x-3}$

c. $\dfrac{2\sqrt{x}-2\sqrt{3}+\sqrt{3x}+9}{x-9}$

d. $\dfrac{\sqrt{2x}-2\sqrt{3}+\sqrt{3x}+3}{x-9}$

15. Simplify (3 - 4i)(6 + 2i) and write the complex number in standard form.

a. 10 - 18i b. 1- + 30i c. 26 - 18i d. 18 + 30i

16. Simplify $\dfrac{3i}{1-3i}$ and write the complex number in standard form.

a. $-\dfrac{9}{8}+\dfrac{3}{8}i$ b. $\dfrac{3}{10}-\dfrac{9}{10}i$ c. $-\dfrac{9}{10}+\dfrac{3}{10}i$ d. $-\dfrac{9}{4}+\dfrac{3}{4}i$

17. Solve: $\dfrac{7}{x-6}+\dfrac{3}{x-6}=\dfrac{11}{x+6}$

a. $\dfrac{2}{7}$ b. -88 c. $-\dfrac{2}{7}$ d. 126

18. Solve: $\sqrt{6x+19}=x-6$

a. 17 b. $3\pm 2i\sqrt{2}$ c. $3\pm 2\sqrt{2}$ d. 4, 9

19. Solve: $2(x+3)^{1/2}+3(x+3)^{1/4}-2=0$

a. $1, \dfrac{49}{16}$ b. $\dfrac{23}{8}$ c. -2, $\dfrac{1}{2}$ d. $-\dfrac{47}{16}$

20. Solve: $\dfrac{x^2+7x-78}{x+2}<0$

a. $(-13,-2)\cup(6,\infty)$

b. $(-\infty,-2)\cup(6,\infty)$

c. $(-\infty,-13)\cup(-2,6)$

d. $(-\infty,-13)\cup(-2,6)$

21.Solve: $|2x-1|=9$

a. 4, -5

b. -5, -4

c. -4, 5

d. -4

22. Solve: $|3x-2|\geq 7$

a. $\left(-\infty,-\dfrac{5}{3}\right]\cup[3,\infty)$

b. $-3\leq x\leq\dfrac{5}{3}$

c. $\left(-\infty,\dfrac{5}{3}\right]\cup(3,\infty)$

d. $-\dfrac{5}{4}\leq x\leq 3$

23. A jogger started a course at an average speed of 4.5 mph. A cyclist started the same course 1 hour later at an average speed of 14 mph. How long after the jogger started did the cyclist overtake the jogger? Round to the nearest tenth of an hour.

a. 1.5 h

b. 1.2 h

c. 0.8 h

d. 1.8 h

24. A total of $25,000 is deposited into two simple interest accounts. On one account the annual simple interest rate is 7.4%, and on the second account the annual simple interest rate is 5.5%. The amount of interest earned for one year was $1527. How much is invested in the 5.5% account?

a. $17,000

b. $8000

c. $11,800

d. $15,000

25. A coffee shop decides to blend a coffee that sells for $10 per pound with a coffee that sells for $6 per pound to produce a blend that will sell for $7.50 per pound. How much of the $6 coffee should be used to yield 50 pounds of the new blend?

a. 22 lb

b. 35 lb

c. 18.75 lb

d. 31.25 lb

Precalculus **Chapter 1** **Test** **Form F**

1. Identify the property that identifies the following: 25(a + b) = 25(b + a).

a. Distributive Property
b. Commutative Property of Addition
c. The Symmetric Property
d. Associative Property of Addition

2. Write $-|\pi - 2|$ without absolute value symbols.

a. $\pi - 2$
b. $-\pi - 2$
c. $2 - \pi$
d. $\pi + 2$

3. Write the interval $-12 \leq x \leq 3$ using interval notation.

a. (12, 3)
b. $(-\infty, -12] \cup [3, \infty)$
c. $(-\infty, -12] \cap [3, \infty)$
d. [–12, 3]

4. Simplify: $\dfrac{(2a^2b^{-3})^{-3}}{(6^{-1}a^4b^{-1})^2}$

a. $\dfrac{9b^{11}}{2a^{14}}$
b. $-\dfrac{8b^7}{9a^2}$
c. $\dfrac{9b^7}{2a^2}$
d. $\dfrac{9b^7}{8b^{14}}$

5. Simplify: $(a + b - 3)(a + b + 3)$

a. $a^2 + 2ab + b^2 - 9$
b. $a^2 - 3a + b^2 + 9$
c. $a^2 + 2ab + 6a + b^2 + 6b - 9$
d. $a^2 + 2ab + b^2 + 6b - 9$

6. Factor: $ax - ay + bx - by$

a. $(x - y)(a - b)$
b. $(x + y)(a - b)$
c. $(x - a)(y + b)$
d. $(x - y)(a + b)$

7. Factor: $2a^4b - 16ab^4$

a. $2ab(a - 2b)(a^2 + 2ab + 4b^2)$
b. $2ab(a - 4b)(a - 2b)$
c. $2ab(a - 2b)(a^2 + 6ab + 9b^2)$
d. $2ab(a + 2b)(a^2 - 2ab + 4b^2)$

8. Simplify: $\dfrac{4x}{x^2 + 7x + 12} - \dfrac{3}{5x + 20}$

a. $\dfrac{x - 9}{(x + 3)(x + 4)}$ b. $\dfrac{17x - 9}{5(x + 3)(x + 4)}$ c. $\dfrac{17x - 9}{5(x + 3)(x + 4)}$ d. $\dfrac{23x + 9}{5(x + 4)}$

9. Simplify: $\dfrac{80x^2 + 20x + 5}{25x^2 - 25} \bullet \dfrac{20x^2 - 25x + 5}{64x^3 - 1}$

a. $\dfrac{16x^2 + 4x + 1}{(x + 1)(4x - 1)}$ b. $\dfrac{1}{x + 1}$

c. $\dfrac{4x - 1}{5(x + 1)}$ d. $\dfrac{5(16x^2 + 4x + 1)}{(x - 1)(x + 1)(4x - 1)}$

10. Simplify: $\dfrac{\dfrac{1}{x} - \dfrac{x - 2}{x + 1}}{\dfrac{x}{x + 1}}$

a. $1 + 3x$ b. $\dfrac{1 - x^2}{x^2}$ c. $\dfrac{1 + 3x - x^2}{x^2}$ d. $\dfrac{x^2 - x - 2}{x(x + 1)}$

11. Simplify: $\dfrac{x^{-2/3}y^{3/4}}{x^{1/3}y^{-1/2}}$

a. $\dfrac{y^{5/4}}{x^{1/3}}$ b. $\dfrac{y^{1/4}}{x^{1/3}}$ c. $xy^{1/4}$ d. $\dfrac{y^{5/4}}{x}$

12. Simplify: $\sqrt[3]{-250a^{10}b^6}$

a. $-5a^3b^2\sqrt[3]{2a}$ b. $-5a^3b^2\sqrt[3]{2a^2b}$ c. $-5a^2b\sqrt[3]{2a^2b}$ d. $-25a^2b\sqrt[3]{2a}$

13. Simplify: $(3\sqrt{2} - 2\sqrt{5})^2$

a. $-2 - 12\sqrt{10}$ b. $38 - 2\sqrt{5} + 3\sqrt{2}$ c. $38 - 12\sqrt{10}$ d. $18 - 6\sqrt{10}$

14. Simplify: $\dfrac{3}{\sqrt{a}-\sqrt{b}}$

a. $\dfrac{\sqrt{3a}+\sqrt{3b}}{a-b}$ b. $\dfrac{3\sqrt{a}+3\sqrt{b}}{a+b}$ c. $\dfrac{3\sqrt{a}+3\sqrt{b}}{a^2-b^2}$ d. $\dfrac{\sqrt{a}+\sqrt{b}}{a+b}$

15. Simplify: $(2-\sqrt{-4})(3+\sqrt{-9})$

a. $12-12i$ b. $-12i$ c. 12 d. -12

16. Write $\dfrac{3i}{2+3i}$ as a complex number in standard form.

a. $-\dfrac{9}{5}+\dfrac{6}{5}i$ b. $\dfrac{9}{13}+\dfrac{6}{13}i$ c. $\dfrac{6}{13}-\dfrac{9}{13}i$ d. $-\dfrac{9}{13}-\dfrac{6}{13}i$

17. Solve: $\dfrac{2x-3}{x+1}-2=\dfrac{3}{x}$

a. $-\dfrac{3}{8}$ b. $\dfrac{4\pm\sqrt{6}}{2}$ c. $-\dfrac{2}{3}$ d. $-\dfrac{1}{8}$

18. Solve: $\sqrt{x-3}+1=\sqrt{x+1}$

a. $\dfrac{3}{4}$ b. $-\dfrac{3}{4}$ c. $\dfrac{21}{4}$ d. $\dfrac{15}{4}$

19. Solve: $4x^{2/3}+19x^{1/3}-5=0$

a. $-5, \dfrac{1}{4}$ b. $-125, \dfrac{1}{64}$ c. $-125, 64$ d. $\dfrac{1}{16}, 25$

20. Solve: $\dfrac{x-3}{2x-3}\le 2$

a. $(-\infty, 1]\cup\left(\dfrac{3}{2},\infty\right)$ b. $(-\infty,-1)\cup\ \dfrac{3}{2},\infty$

c. $(-\infty,-1)\cup\left(\dfrac{3}{2},\infty\right)$ d. $(-\infty, 1)\cup[3,\infty)$

21. Solve: $|4x - 7| = 9$

a. $-\frac{1}{2}, 4$ b. 4 c. $-4, \frac{1}{2}$ d. $\frac{9}{2}, \frac{1}{2}$

22. Solve: $|4x - 3| \geq 7$

a. $\left[-1, \frac{5}{2}\right]$

b. $(-\infty, -1] \cup \left[\frac{5}{2}, \infty\right)$

c. $\left(-\infty, -\frac{5}{2}\right) \cup [1, \infty)$

d. $(-\infty, -1] \cup [1, \infty)$

23. A biathlon consisted of 20 miles of running and 40 miles of cycling. One contestant ran 2/5 hour longer than he cycled. His cycling rate was 5 mph more than twice that of his running rate. Find his running rate.

a. 12.5 mph b. 7.5 mph c. 9 mph d. 10 mph

24. A radiator contains 6 liters of a 25% antifreeze solution. How much should be drained and replaced with pure antifreeze to produce a 75% antifreeze solution?

a. 2.5 L b. 4 L c. 3 L d. 4.5 L

25. You can rent a car for the day from Company A for \$20 plus \$0.08 a mile. Company B charges \$8 a day plus \$0.16 a mile. At what point, in terms of miles driven per day, is it cheaper to rent from Company A?

a. Less than 150 miles per day

b. Greater than 117 miles per day

c. Greater than 150 miles per day

d. Greater than 175 miles per day

1. Convert 210° to exact radian measure.

2. Convert $\frac{11\pi}{15}$ to exact degree measure.

3. Find $W\left(\frac{5\pi}{4}\right)$.

4. Find $W\left(-\frac{13\pi}{3}\right)$.

5. Find the length of an arc that subtends a central angle of 105° in a circle with radius of 12 inches.

6. Find the exact value of $\tan \frac{4\pi}{3}$.

7. Determine whether the function defined by $f(x) = \sin x + \tan x$ is even, odd, or neither.

8. Given $\tan \theta = -\sqrt{3}$, where $\frac{\pi}{2} < \theta < \pi$, evaluate $\sec \theta$.

9. Express $\frac{\csc^2 \theta - 1}{\csc^2 \theta}$ in terms of a single trigonometric function.

10. State the period of $y = -3\sin\frac{2}{3}x$.

11. State the amplitude, period, and phase shift for the equation $y = -2\sin\left(3x - \frac{\pi}{6}\right)$.

12. Graph $y = 2\cos\frac{3}{2}x$.

13. Graph $y = 2\csc 2x$.

14. Graph $y = 2\tan\left(\frac{1}{2}x - \frac{\pi}{2}\right)$.

15. Graph $y = -2\sin\left(2x - \frac{\pi}{4}\right) + 1$.

16. Graph $1 - \cos 2x$.

17. Graph $y = \sin x - \cos\frac{x}{2}$.

18. A wheel with a diameter of 26 inches is rotating at 15 radians per second. Find the linear speed of a point on the edge of the wheel.

19. Find the height of a building if the angle of elevation to the top of the building changes from 20° to 35° as the observer moves a distance of 60 feet toward the building.

20. Write the equation for simple harmonic motion, given that the amplitude is 4 inches and the period is 2 seconds. Assume maximum displacement when $t = 0$.

1. Convert 240° to exact radian measure.

2. Convert $-\frac{7\pi}{6}$ to exact degree measurement.

3. Find $W\left(\frac{\pi}{6}\right)$.

4. Find $W\left(-\frac{11\pi}{3}\right)$.

5. Find the length of an arc that subtends a central angle of 225° in a circle with radius of 15 centimeters.

6. Find the exact value of $\sin\left(-\frac{5\pi}{4}\right)$.

7. Determine whether the function defined by $f(x) = x - \sin x$ is even, odd, or neither.

8. Given $\sin\ \theta = -\frac{1}{2}$, when $\pi < \theta < 2\pi$, evaluate $\tan\ \theta$.

9. Express $\frac{1}{1-\sec t} + \frac{1}{1+\sec t}$ in terms of a single trigonometric function.

10. State the period of $y = -2\cot\frac{\pi}{4}x$.

11. State the amplitude, period, and phase shift for the equation $y = -\sin(3x + \pi)$.

12. Graph $y = -2\sin 2x$.

13. Graph $y = -3\csc \frac{x}{2}$.

14. Graph $y = \tan\left(x - \frac{\pi}{4}\right)$.

15. Graph $y = 3\cos\left(3x - \frac{\pi}{3}\right) + 2$.

16. Graph $y = 1 + \sin 2x$.

17. Graph $y = 2\sin x + \cos x$.

18. The 14 inch wheel of a moving automobile is rotating at 14 revolutions per second. Assuming that the wheel is not skidding, find the speed of the automobile.

19. If a car climbs a hill that has a constant angle of 4° for a distance of 0.3 miles, how many vertical feet has it climbed?

20. Write the equation for simple harmonic motion, given that the amplitude is 12 centimeters and the period is 1.5 seconds. Assume maximum displacement when $t = 0$.

Precalculus Chapter 5 Test Form C

1. Convert 315° to exact radian measure.

2. Convert $-\frac{3\pi}{15}$ to exact degree measure.

3. Find $W\left(\frac{13\pi}{6}\right)$.

4. Find $W\left(-\frac{7\pi}{3}\right)$.

5. Find the measure in radians of the central angle of a circle with radius 27 centimeters an d arc length 12 centimeters.

6. Find the exact value of $\cos 675^\circ$.

7. Determine whether the function defined by $f(x) = x^2 - \cos x$ is even, odd, or neither.

8. Given $\cos\theta = -\frac{1}{2}$, where $\frac{\pi}{2} < \theta < \pi$, evaluate $\csc\theta$.

9. Express $\frac{\cos^2 t}{\tan^2 t}$ in terms of the sine function.

10. State the period of $y = -3\cot\frac{\pi}{3}x$.

11. State the amplitude, period, and phase shift for the equation $y = \frac{2}{3}\sin\left(5x + \frac{\pi}{2}\right)$.

12. Graph $y = \frac{1}{2}\sin 4x$.

13. Graph $y = 2\sec 4x$.

14. Graph $y = \sin\left(\frac{1}{2}x - \pi\right)$.

15. Graph $y = -3\cos\left(x - \frac{\pi}{2}\right) - 5$.

16. Graph $y = -1 + \sin 2x$.

17. Graph $y = \sin x + \cos\frac{x}{2}$.

18. A wheel with a diameter of 5 feet is rotating at 25 radians per second. Find the linear speed of a point on the edge of the wheel.

19. The angle of elevation from point A to the top of a tree is 30 degrees. At point B, 6 meters from A and on a line through the base of the tree and A, the angle of elevation is 37 degrees. Find the height of the tree.

20. Write the equation for simple harmonic motion, given that the amplitude is 8 feet and the period is 2.5 seconds. Assume maximum displacement when $t = 0$.

1. Convert 48° to exact radian measure.

a. $\frac{3\pi}{8}$ b. $\frac{2\pi}{15}$ c. $\frac{4\pi}{15}$ d. $\frac{7\pi}{15}$

2. Convert $\frac{7\pi}{15}$ to exact degree measure.

a. 72° b. 84° c. 96° d. 108°

3. Find $W\left(\frac{2\pi}{3}\right)$.

a. $\left(\frac{\sqrt{3}}{2}, -\frac{1}{2}\right)$ b. $\left(-\frac{\sqrt{2}}{2}, \frac{\sqrt{2}}{2}\right)$ c. $\left(-\frac{\sqrt{2}}{2}, \frac{\sqrt{3}}{2}\right)$ d. $\left(-\frac{1}{2}, \frac{\sqrt{3}}{2}\right)$

4. Find $W\left(-\frac{13\pi}{4}\right)$.

a. $\left(\frac{\sqrt{2}}{2}, -\frac{\sqrt{2}}{2}\right)$ b. $\left(\frac{\sqrt{3}}{2}, -\frac{1}{2}\right)$ c. $\left(-\frac{\sqrt{2}}{2}, \frac{\sqrt{2}}{2}\right)$ d. $\left(-\frac{\sqrt{3}}{2}, \frac{1}{2}\right)$

5. Find the length of an arc that subtends a central angle of 105° in a circle of radius 8 inches.

a. 4π in. b. $\frac{14}{3}\pi$ in. c. $\frac{96}{7}\pi$ in. d. $\frac{7}{2}\pi$ in.

6. Find the exact value of $\cos \frac{11\pi}{3}$.

a. $-\sqrt{3}$ b. $-\frac{\sqrt{3}}{3}$ c. $-\frac{\sqrt{2}}{2}$ d. $\frac{1}{2}$

7. Determine whether the function defined by $f(x) = x + \cos x$ is even, odd, or, neither.

a. even b. odd c. neither

8. Given $\csc\theta = -\frac{1}{2}$, where $\frac{3\pi}{2} < \theta < 2\pi$, evaluate $\cos\theta$.

a. $-\frac{1}{2}$ b. $\frac{1}{2}$ c. $-\frac{\sqrt{3}}{2}$ d. $\frac{\sqrt{3}}{2}$

9. Express $\frac{\sec^2 t}{\cot^2 t}$ in terms of the sine function.

a. $\frac{\sin^2 t}{1 - 2\sin^2 t + \sin^4 t}$ b. $\frac{\sin^2 t}{1 - \sin^4 t}$

c. $\frac{1 - \sin^4 t}{\sin^2 t}$ d. $\sin^4 t$

10. State the period of $y = -5\tan\frac{\pi}{2}x$.

a. 2 b. π c. $\frac{2}{\pi}$ d. $\frac{1}{2}$

11. State the amplitude, period, and phase shift for the equation $y = \frac{1}{2}\sin(4x - \pi)$.

a. $\frac{1}{2}, \frac{\pi}{4}, \frac{\pi}{4}$ b. $-\frac{1}{2}, \frac{\pi}{2}, \frac{\pi}{2}$ c. $\frac{1}{2}, \frac{\pi}{2}, \frac{\pi}{4}$ d. $\frac{1}{2}, 4, \pi$

12. Graph $y = \frac{2}{3}\sin 6x$.

a. 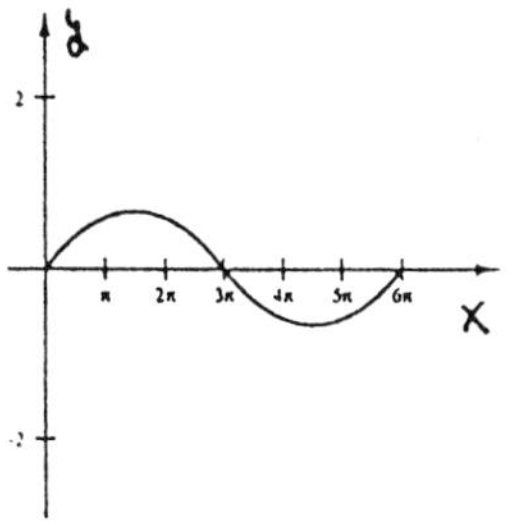

b. 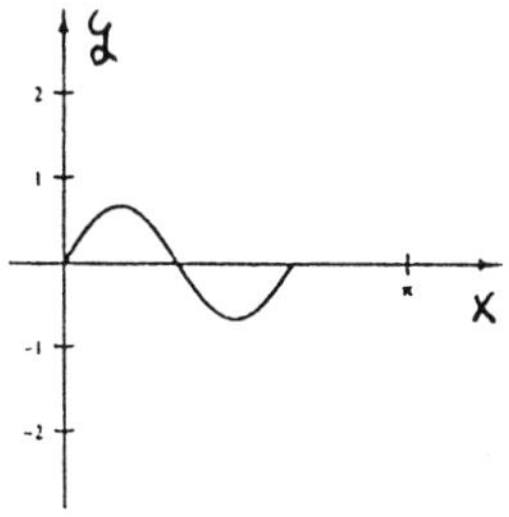

c. 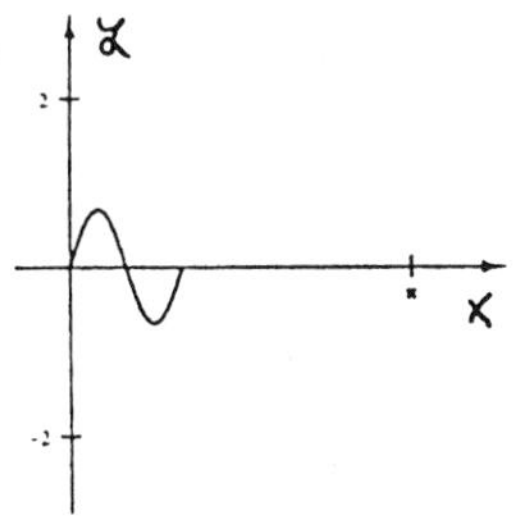

d. 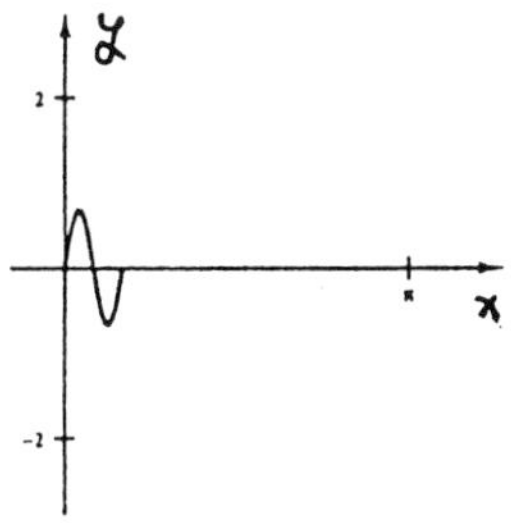

13. Graph $y = \sec 2x$.

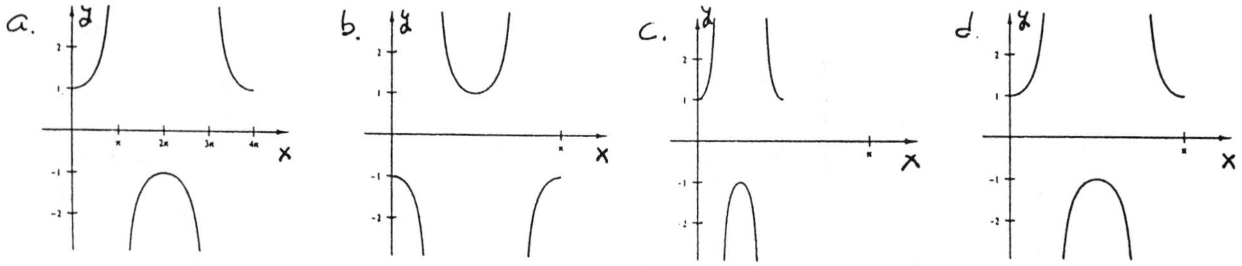

14. Graph $y = -2\tan\dfrac{x}{2}$.

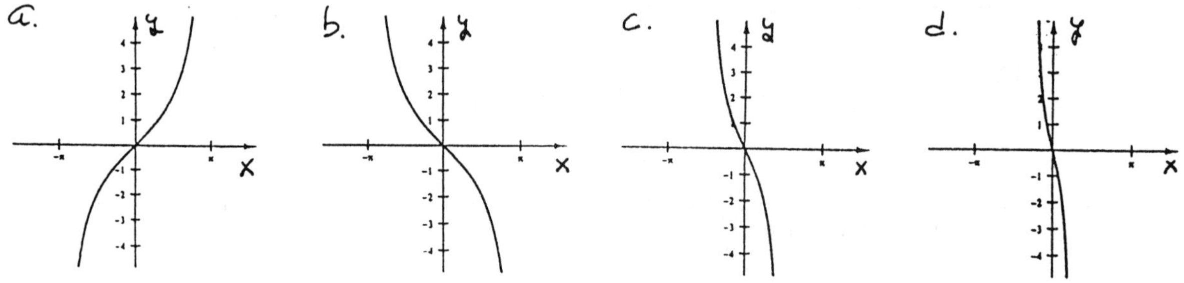

15. Graph $y = \sin\left(\dfrac{1}{4}x + \pi\right)$.

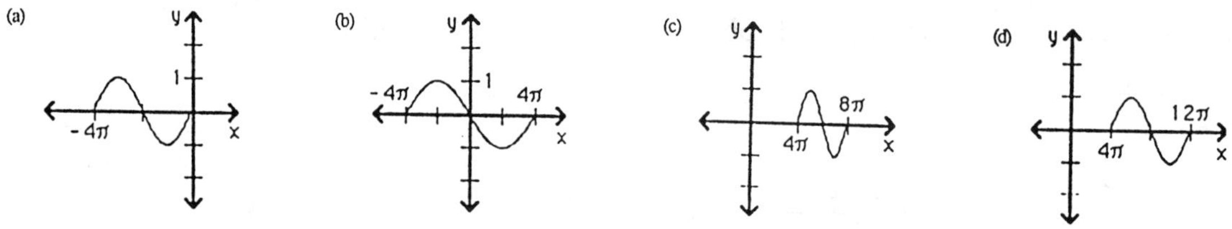

16. Graph $y = 1 + \text{cox}\dfrac{x}{2}$.

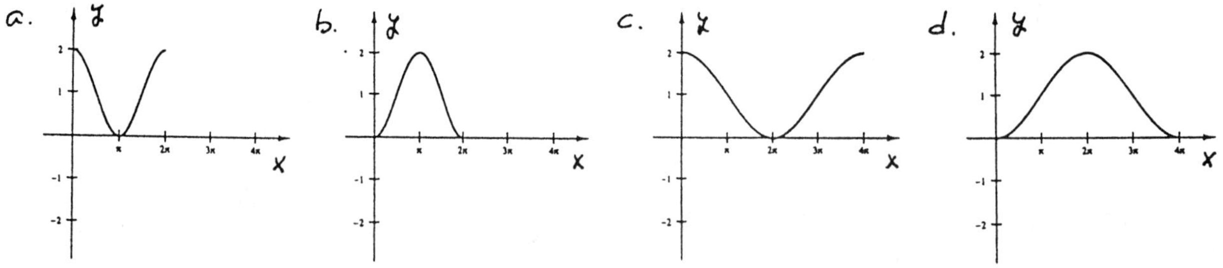

17. Graph $y = 2\sin x + \cos\frac{x}{2}$.

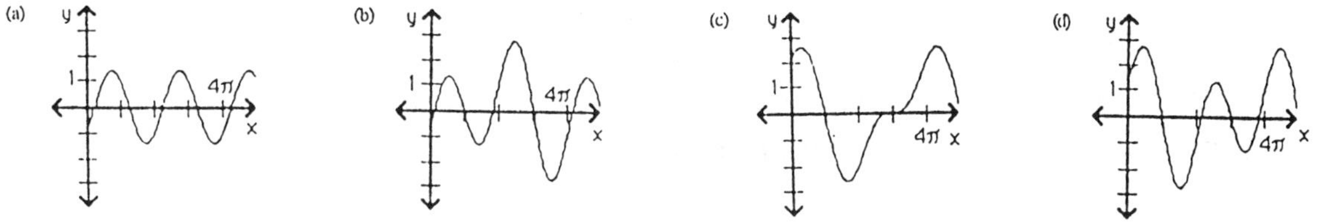

18. A wheel with a diameter of 40 inches is rotating at 22 radians per second. Find the linear speed of a point on the edge of the wheel. Express the answer in ft/s.

a. 880 ft/s b. 73.3 ft/s c. 36.7 ft/s d. 23.5 ft/s

19. A tree casts a shadow of 7.5 feet when the angle of elevation of the sun is $38°$. Find the height of the tree.

a. 14 ft b. 4.0 ft c. 5.9 ft d. 12 ft

20. Write an equation for simple harmonic motion given that the amplitude is 5 centimeters and the period is 4 seconds. Assume maximum displacement at time = 0.

a. $y = 4\cos\frac{2\pi t}{5}$ b. $y = 5\cos\frac{\pi t}{2}$ c. $y = 5\cos\frac{\pi t}{4}$ d. $y = 5\sin\frac{\pi t}{4}$

1. Convert $\frac{11\pi}{15}$ to exact degree measure.

a. 121° b. 132° c. 144° d. 96°

2. Convert 75° to exact radian measurement.

a. $\frac{5\pi}{12}$ b. $\frac{3\pi}{8}$ c. $\frac{7\pi}{12}$ d. $\frac{5\pi}{8}$

3. Find $W\left(\frac{7\pi}{6}\right)$.

a. $\left(-\frac{1}{2},-\frac{\sqrt{3}}{2}\right)$ b. $\left(-\sqrt{3},-1\right)$ c. $\left(-\frac{\sqrt{3}}{2},-\frac{1}{2}\right)$ d. $\left(-1,-\sqrt{3}\right)$

4. Find $W\left(-\frac{11\pi}{6}\right)$.

a. $\left(\frac{1}{2},\frac{\sqrt{3}}{2}\right)$ b. $\left(-\frac{\sqrt{2}}{2},\frac{\sqrt{2}}{2}\right)$ c. $\left(-\frac{1}{2},\frac{\sqrt{3}}{2}\right)$ d. $\left(\frac{\sqrt{3}}{2},\frac{1}{2}\right)$

5. Find the measure in radians of the central angle of a circle with radius 15 3 centimeters and arc length 24 centimeters.

a. $\frac{5}{8}$ b. $\frac{8\pi}{5}$ c. $\frac{4}{5}$ d. $\frac{8}{5}$

6. Find the exact value of $\tan 210^\circ$.

a. $\frac{\sqrt{3}}{3}$ b. $-\frac{\sqrt{3}}{3}$ c. $\sqrt{3}$ d. $-\sqrt{3}$

7. Determine whether the function defined by $f(x) = x^2 - \cos x$ is even, odd, or neither.

a. even b. odd c. neither

8. If $\sin\ \theta = -\frac{\sqrt{3}}{2}$, $\pi \le \theta \le \frac{3\pi}{2}$, find $\sec\ \theta$.

a. $-\frac{2\sqrt{3}}{3}$ b. $\frac{1}{2}$ c. 2 d. -2

9. Express $\frac{1}{1-\sec t} + \frac{1}{1+\sec t}$ in terms of the cotangent function.

a. $-2\cot^2 t$ b. $2\cot^2 t$ c. $\frac{1}{2\cot^2 t}$ d. $1-\cot^2 t$

10. State the period of $y = \frac{3}{2}\cot \pi x$

a. 1 b. π c. $\frac{3}{2}$ d. $\frac{\pi}{2}$

11. State the amplitude, period, and phase shift for the equation $y = -\frac{5}{2}\sin(2\pi x - 2) + 1$.

a. $\frac{5}{2}, 2, \pi$ b. $\frac{5}{2}, 1, \pi$ c. $\frac{5}{2}, 1, \frac{1}{\pi}$ d. $-\frac{5}{2}, 1, \pi$

12. Graph $y = -2\sin\frac{x}{4}$.

(a)
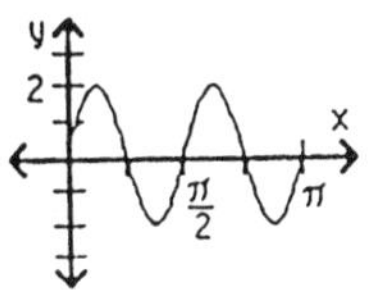

(b)
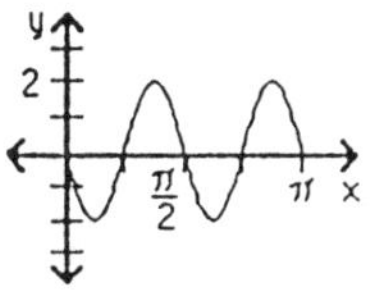

(c)
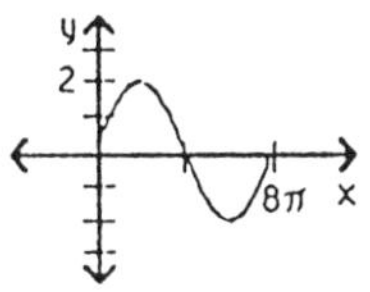

(d)
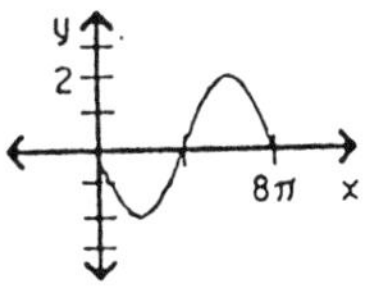

13. Graph $y = -5\csc\frac{x}{3}$.

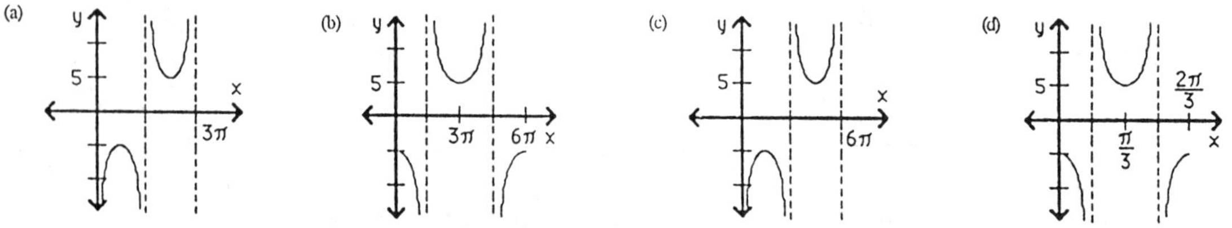

14. Graph $y = -2\tan\frac{x}{3}$.

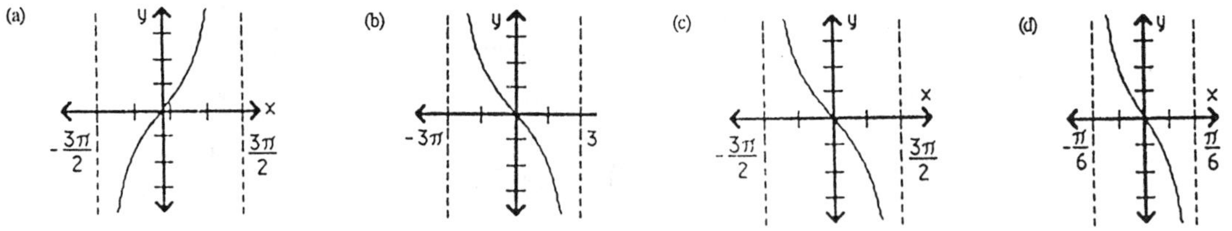

15. Graph $y = -2\cos(x - \pi) - 3$.

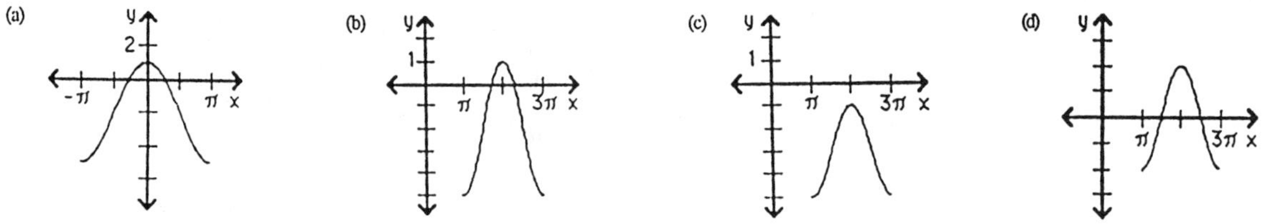

16. Graph $y = 1 + 2\sin\frac{x}{2}$.

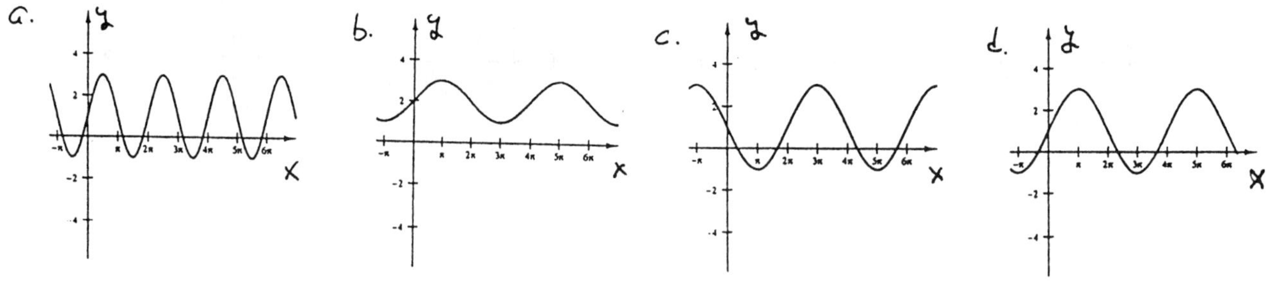

17. Graph $y = \frac{1}{2}x + \sin x$.

a. 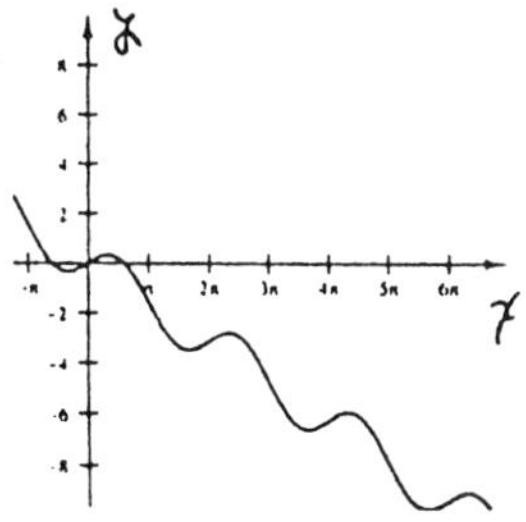

b. 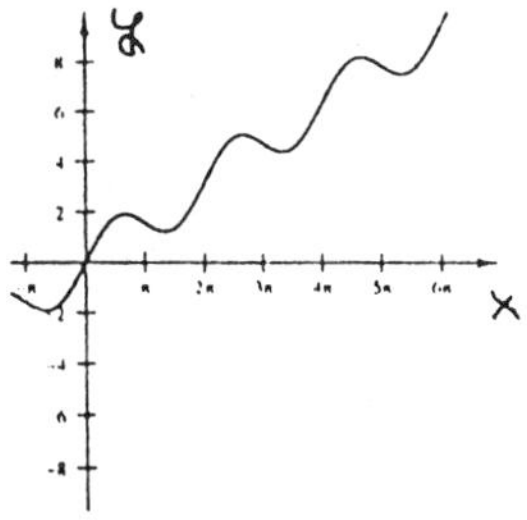

c. 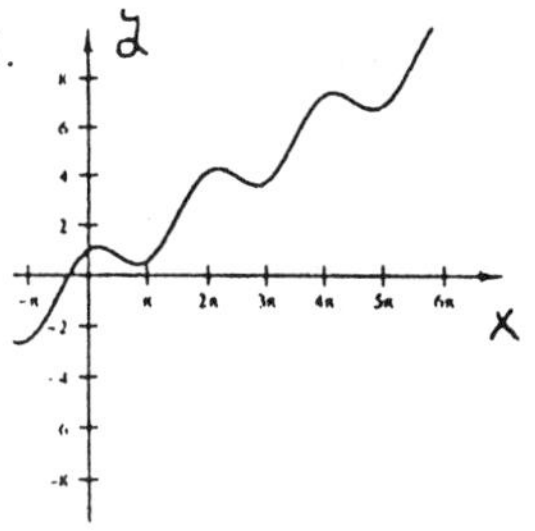

d. 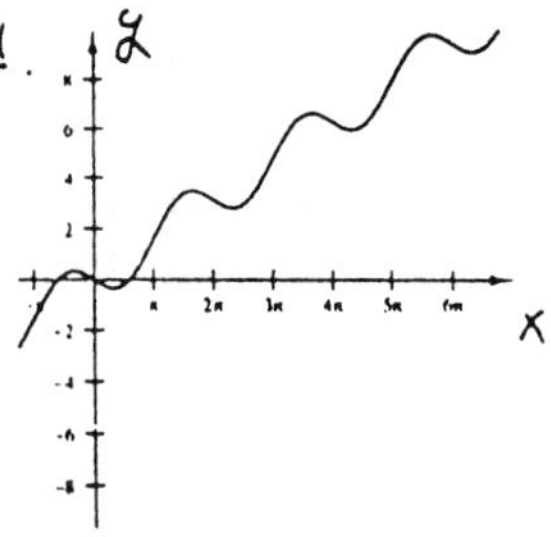

18. A wheel is rotating at 80 rpm. Find the angular speed in radians per second.

a. $\frac{3\pi}{8}$ radians/s b. $\frac{8\pi}{3}$ radians/s

c. $\frac{4\pi}{3}$ radians/s d. 80π radians/s

19. A pulley with a radius of 30 inches uses a belt to drive a pulley with a radius of 18 inches. Find the angle through which the smaller pulley turns as the 30-inch pulley makes one revolution.

a. $\frac{3\pi}{10}$ in. b. $\frac{5\pi}{3}$ in . c. $\frac{10\pi}{3}$ in. d. $\frac{3\pi}{5}$ in.

20. Write the equation for simple harmonic motion given that the amplitude is 11 centimeters and the period is 2 seconds. Assume zero displacement at $t = 0$.

a. $y = 11\sin \pi t$ b. $y = 11\sin\frac{\pi}{2}t$ c. $y = 11\cos 2\pi t$ d. $y = 11\sin\frac{\pi}{4}t$

1. Convert $\frac{13\pi}{4}$ to exact degree measure.

a. $45°$ b. $135°$ c. $225°$ d. $315°$

2. Convert $255°$ to exact radian measure.

a. $-\frac{17\pi}{12}$ b. $-\frac{5}{3}\pi$ c. $-\frac{7\pi}{6}$ d. $-\frac{13\pi}{12}$

3. Find $W\left(\frac{5\pi}{6}\right)$.

a. $\left(\frac{1}{2},-\frac{\sqrt{3}}{2}\right)$ b. $\left(-\frac{1}{2},-\frac{\sqrt{3}}{2}\right)$ c. $\left(-\frac{\sqrt{3}}{2},\frac{1}{2}\right)$ d. $\left(\frac{\sqrt{2}}{2},\frac{1}{2}\right)$

4. Find $W\left(-\frac{2\pi}{3}\right)$.

a. $\left(-\frac{\sqrt{2}}{2},\frac{\sqrt{2}}{2}\right)$ b. $\left(-\frac{1}{2},\frac{\sqrt{3}}{2}\right)$ c. $\left(\frac{\sqrt{3}}{2},-\frac{1}{2}\right)$ d. $\left(-\frac{1}{2},-\frac{\sqrt{3}}{2}\right)$

5. Find the length of an arc that subtends a central angle of $75°$ in a circle of radius 20 inches.

a. $\frac{25}{3}$ in. b. $\frac{\pi}{24}$ in. c. 48π in. d. $\frac{25\pi}{3}$ in.

6. Determine whether the function defined by $f(x) = \sin x + \cos x$ is even, odd, or neither.

a. even b. odd c. neither

7. Find the exact value of $\sec\left(-\frac{7\pi}{4}\right)$.

a. $\sqrt{2}$ b. $\sqrt{3}$ c, $-\sqrt{2}$ d. $\frac{\sqrt{2}}{2}$

8. If $\cos\theta = \frac{1}{2}$, $270^\circ \le \theta \le 360^\circ$, find $\cot\theta$.

a. $\frac{1}{2}$ b. -2 c. $-\frac{\sqrt{3}}{3}$ d. $-\frac{2\sqrt{3}}{3}$

9. Express $\frac{1}{1-\cos t} + \frac{1}{1+\cos t}$ in terms of the cosecant function.

a. $2\csc^2 t$ b. $\frac{2}{\csc^2 t}$ c. $1-\csc^2 t$ d. $1+\csc^2 t$

10. State the period of $y = \frac{1}{2}\tan 8x$.

a. $\frac{\pi}{8}$ b. 8π c. $\frac{\pi}{4}$ d. 4π

11. State the amplitude, period, and phase shift for the equation $y = -\frac{3}{4}\sin\left(3x - \frac{\pi}{2}\right)$.

a. $\frac{3}{4}, \frac{2\pi}{3}, \frac{\pi}{6}$ b. $\frac{3}{4}, \frac{3\pi}{2}, \frac{\pi}{3}$ c. $\frac{4}{3}, \frac{2\pi}{3}, -\frac{\pi}{6}$ d. $\frac{3}{4}, 6\pi, -\frac{\pi}{2}$

12. Graph $y = -5\cos\frac{x}{4}$

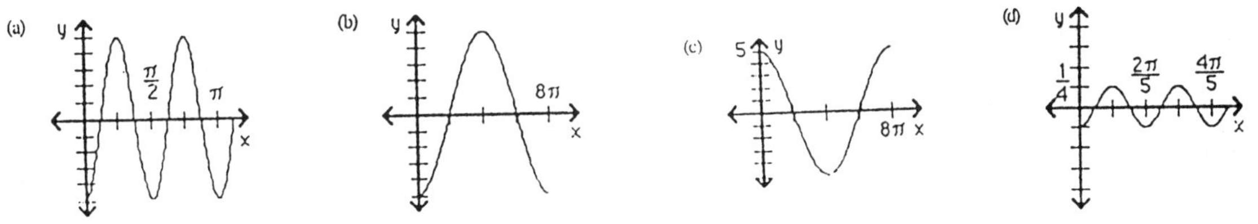

13. Graph $y = -\csc\frac{x}{4}$.

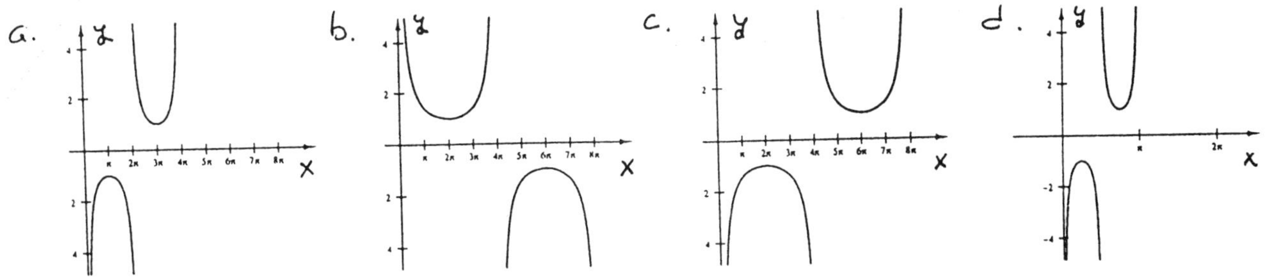

14. Graph $y = 2\tan 4x$.

(a)
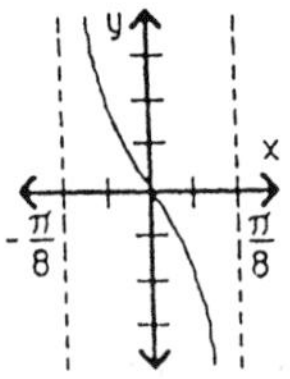

(b)
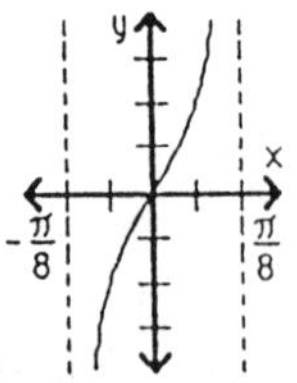

(c)
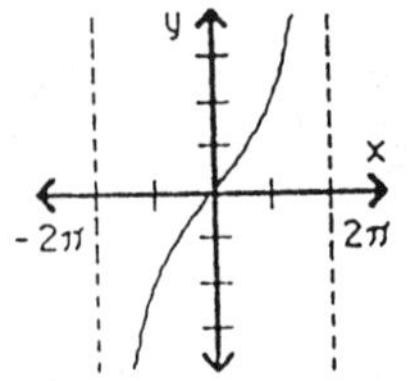

(d)
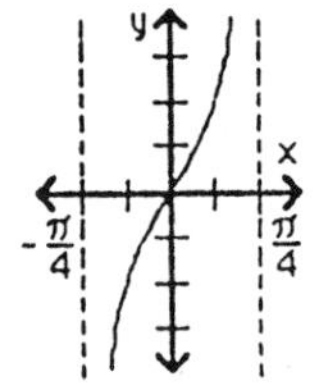

15. Graph $y = \frac{3}{5}\sin\left(x + \frac{\pi}{2}\right) + 4$.

(a)
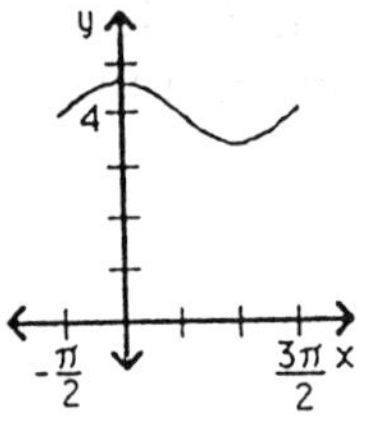

(b)
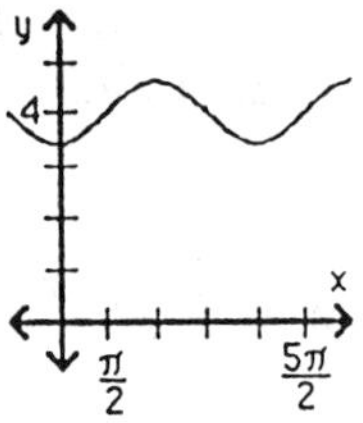

(c)
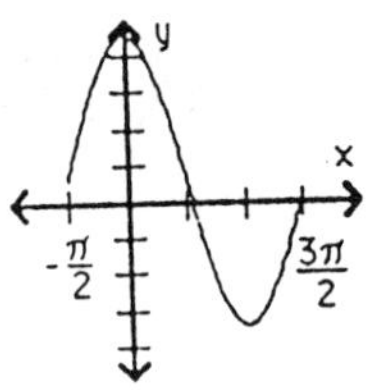

(d)
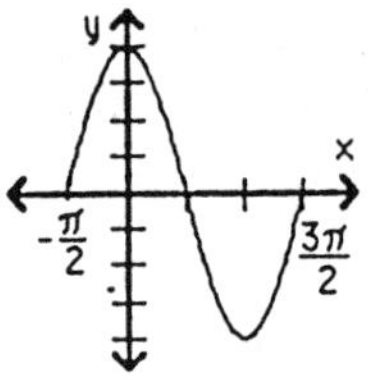

16. Graph $y = 2 - 3\cos\frac{x}{2}$.

a.
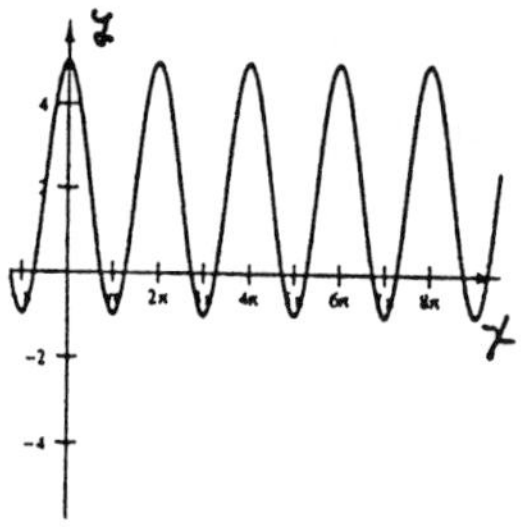

b.
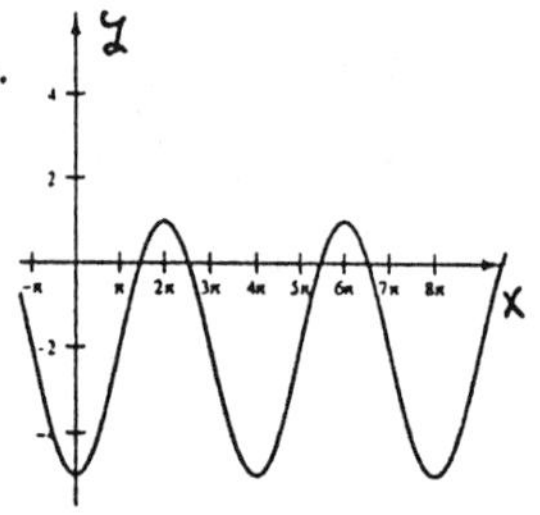

c.
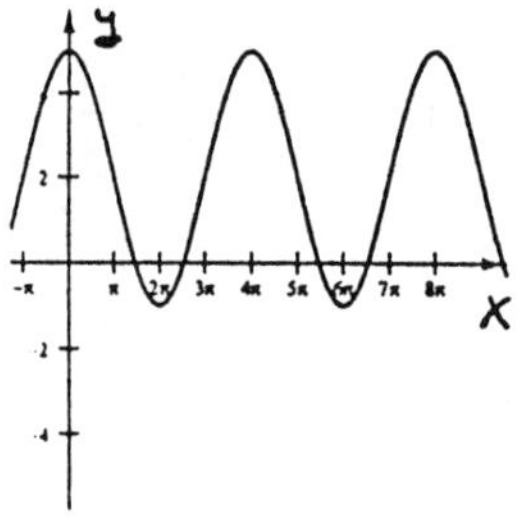

d.
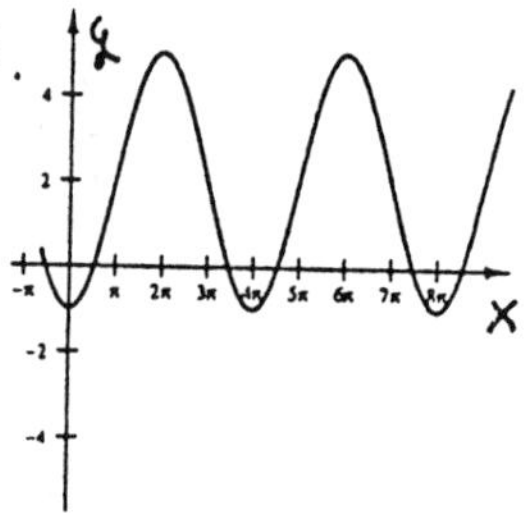

17. Graph $y = x - \sin x$.

(a)
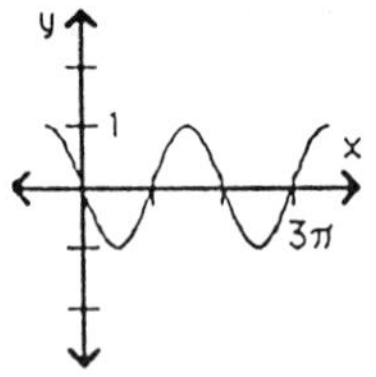

(b)
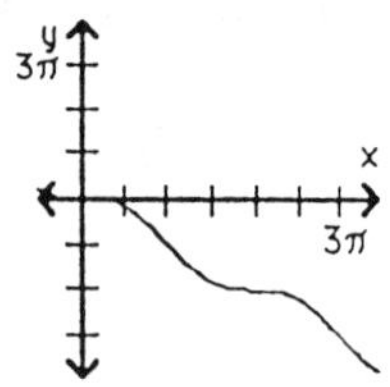

(c)
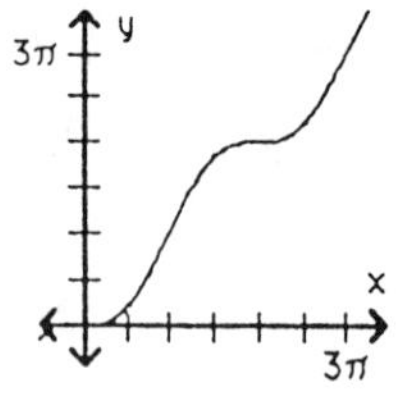

(d)
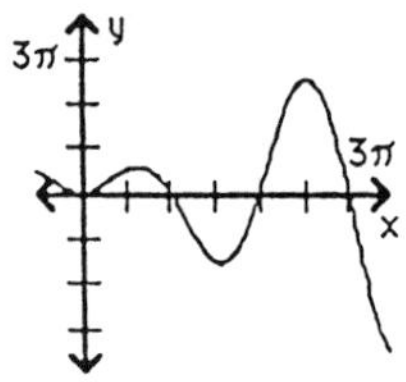

18. A wheel with a diameter of 16 inches uses a belt to drive a second wheel with a diameter of 6 inches. Determine the angle through which the smaller wheel turns as the larger wheel turns makes one revolution.

a. $\frac{16\pi}{3}$ in. b. $\frac{8\pi}{3}$ in. c. $\frac{4\pi}{3}$ in. d. $\frac{11\pi}{3}$ in.

19. The angle of elevation from point A to the top of a building is 27°. At point B, 50 feet from A and on a line through the base of the building and A, the angle of elevation is 40°. Find the height of the building.

a. 25.5 ft b. 64.9 ft c. 55.6 ft d. 73.2 ft

20. Write the equation for simple harmonic motion, given that the amplitude is 4 feet and the period is 3 seconds. Assume maximum displacement at time $t = 0$.

a. $y = 4\cos 4\pi t$ b. $y = 4\cos\frac{3}{2}\pi t$

c. $y = 4\cos\frac{2}{3}\pi t$ d. $y = 4\sin 4\pi t$

Precalculus Test Answers

Chapter 1 Form A

1. Commutative Property of Addition

2. $\{2, 4, 9\}$

3. $-\frac{x^5}{108y^6}$

4. $\frac{x^{18}}{12y^{11}}$

5. $2x^4 - 5x^3 + 6x^2 - 23x - 20$

6. 185

7. $3x(x - y)(x^2 + xy + y^2)$

8. $(x + y)^2(x - y)(x^2 - xy + y^2)$

9. $\frac{3x^2 + 4x - 8}{(x - 2)(x + 2)(x + 4)}$

10. $\frac{x(x - 2)}{(x - 1)(x + 2)}$

11. $\frac{y^2}{3y - 2}$

12. $\frac{a}{b^{9/4}}$

13. $\frac{2\sqrt[4]{a^3}}{a}$

14. $\frac{y\sqrt{y} + y\sqrt{2}}{y - 2}$

15. $-13 - 26i$

16. $-\frac{9}{13} + \frac{6}{13}i$

17. 6

18. no solution

19. $\frac{81}{16}$

20. $(-\infty, -4] \cup (-1, 1]$

21. $1, \frac{19}{3}$

22. $\left\{x \,\middle|\, x \leq -\frac{24}{25} \text{ or } x \geq 2\right\}$

23. 31 ft

24. \$5500 at 7.5%, \$2500 at 5.2%

25. 40 mph

Chapter 1 Form B

1. $-8, 0.17, 0, 0.\overline{74}$

2. Commutative Property of Addition

3. $[-4, 1]$

4. $-\frac{24}{b^5}$

5. $-\frac{x^{16}}{486y^{11}}$

6. $2x^2 - xy - y^2 + 6y - 8$

7. $7(x - y)(x - 3y)$

8. $(2a - 3b)(x - y)(x + y)$

Precalculus Test Answers

9. $\frac{(x-4)(x+1)}{(x+2)(x+3)(x-1)}$

10. $\frac{(x-1)(2x-3)}{3x^2}$

11. $\frac{a}{a+1}$

12. $a^{5/4}b^{7/6}$

13. $3ab^2\sqrt[3]{2a^2b}$

14. $\frac{\sqrt{2x}+2}{x-2}$

15. $28 + 16i$

16. $-\frac{9}{10} + \frac{3}{10}i$

17. $\frac{9}{7}$

18. $-2, 3, -1 + i\sqrt{3}, -1 - i\sqrt{3}$

19. 3

20. $(-6, -1] \cup [8, \infty)$

21. $-\frac{1}{5}, 1$

22. $\left\{x \mid -\frac{8}{5} \leq x \leq 2\right\}$

23. 15 mph

24. 60 min

25. at least 117 miles per day

Chapter 1 Form C

1. $\sqrt{19}, \pi,$ 0.565565556...

2. Commutative Property of Multiplication

3.

4. $-\frac{9b^{11}}{32a^9}$

5. $x^4 - 4x^2 + 12x - 9$

6. $x(x-1)(2a-b)$

7. $(y+1)(y^2-y+1)(x-1)(x+1)$

8. $\frac{X+2}{(X-4)(X+4)(X-3)}$

9. $\frac{x}{x-3}$

10. $\frac{x^2}{x^2-5x+5}$

11. $\frac{b^{19/6}}{a^2}$

12. $\frac{\sqrt[4]{3y^3}}{3}$

13. $-x^2y\sqrt{2y}$

14. $\frac{2x\sqrt{2}+2x^2}{2-x^2}$

15. $9 + 3i$

16. $-\frac{12}{17} + \frac{3}{17}i$

17. 3

18. $\frac{3+\sqrt{33}}{2}, \frac{3-\sqrt{33}}{2}$

19. $6 + 2\sqrt{7}$

20. [–4, 1]	21. –1, 9	22. $\left\{x \mid \frac{3}{2} \le x \le \frac{7}{2}\right\}$
23. 30 mph	24. 5 ft	25. 8200 \$9 tickets 3800 \$15 tickets

Chapter 1 Form D

1. b	2. a	3. d	4. b	5. a
6. b	7. d	8. a	9. c	10. a
11. b	12. d	13. a	14. b	15. c
16. a	17. d	18. a	19. c	20. c
21. a	22. a	23. a	24. c	25. b

Chapter 1 Form E

1. d	2. a	3. c	4. b	5. a
6. c	7. c	8. b	9. a	10. a
11. b	12. d	13. a	14. b	15. c
16. c	17. d	18. a	19. d	20. c
21. c	22. a	23. a	24. a	25. d

Chapter 1 Form F

1. b	2. c	3. d	4. a	5. a
6. d	7. a	8. c	9. b	10. c
11. d	12. a	13. c	14. b	15. c

16. b   17. a   18. c   19. b   20. a

21. a   22. b   23. d   24. b   25. c

Chapter 5 Form A

1. $\frac{7\pi}{6}$

2. $132°$

3. $\left(-\frac{\sqrt{2}}{2}, -\frac{\sqrt{2}}{2}\right)$

4. $\left(\frac{1}{2}, -\frac{\sqrt{3}}{3}\right)$

5. 7π in.

6. $\sqrt{3}$

7. odd function

8. -2

9. $\cos^2\theta$

10. 3π

11. $2, \frac{2\pi}{3}, \frac{\pi}{18}$

12.

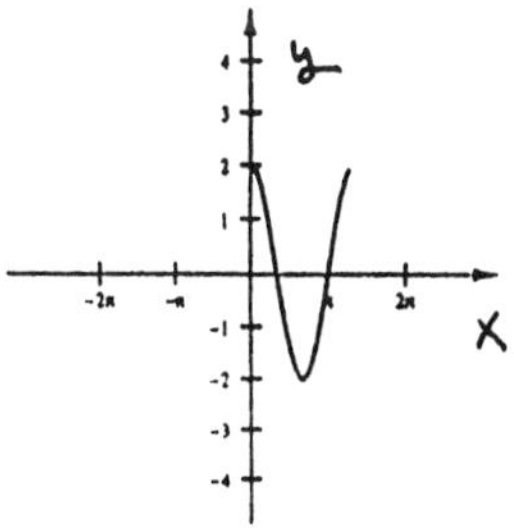

13.

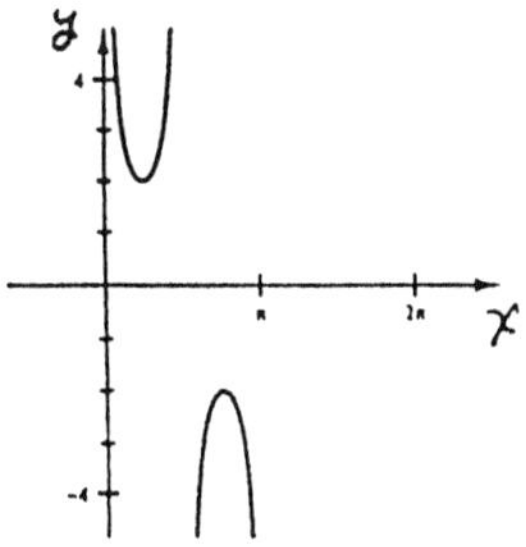

14.

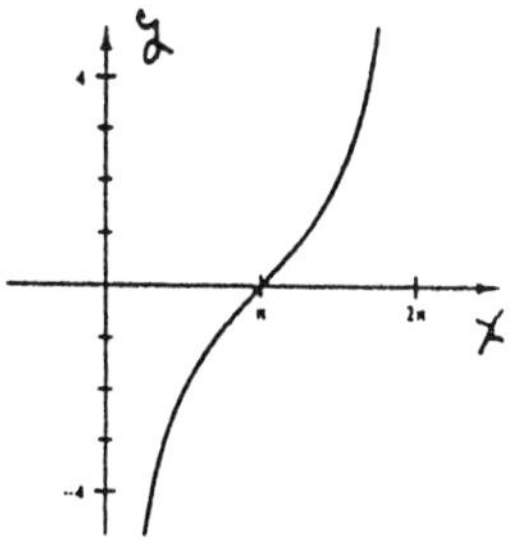

15.

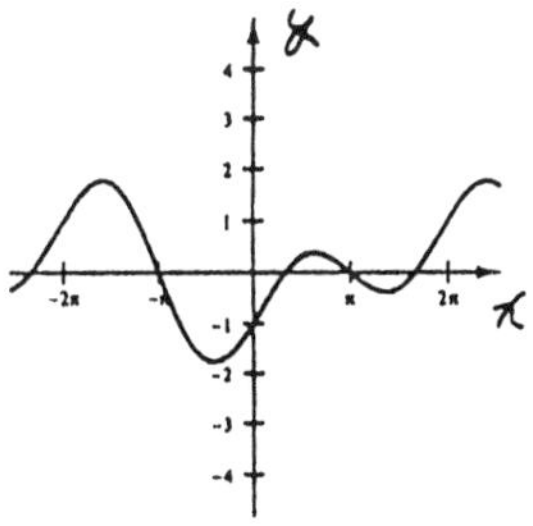

16.

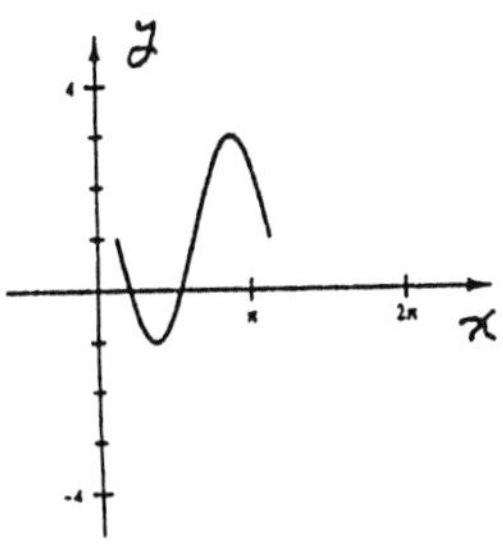

17.

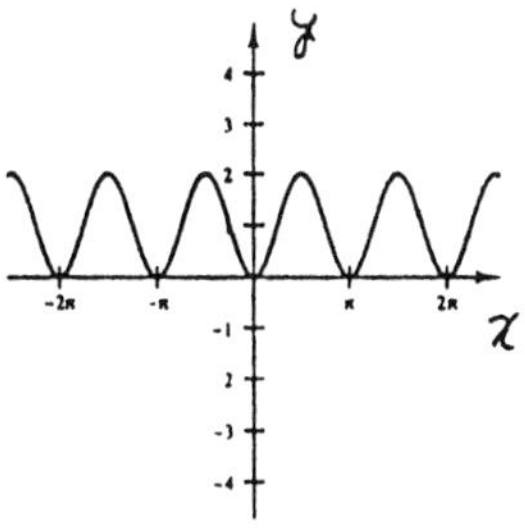

18. 16.25 ft/s

19. 45.5 ft

20. $y = 4\cos\pi t$

Chapter 5 Form B

1. $\frac{2\pi}{4}$

2. $-210°$

3. $\left(\frac{\sqrt{3}}{2}, \frac{1}{2}\right)$

4. $\left(\frac{1}{2}, \frac{\sqrt{3}}{2}\right)$

5. $\frac{75\pi}{4}$ cm

6. $\frac{\sqrt{2}}{2}$

7. odd function

8. $\frac{\sqrt{3}}{3}$

9. $-2\cot^2\theta$

10. 4

11. 1, $\frac{2\pi}{3}$, $-\frac{\pi}{3}$

12.

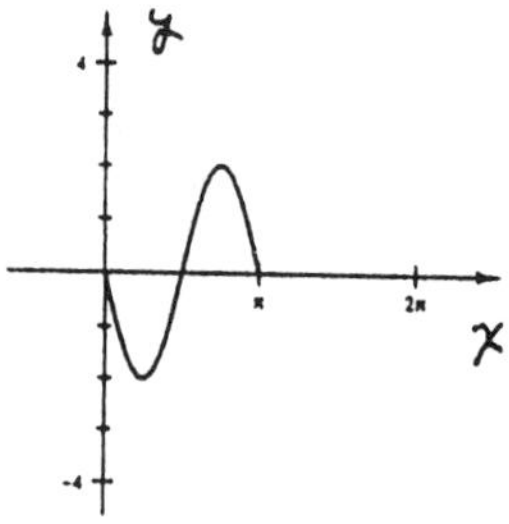

13.

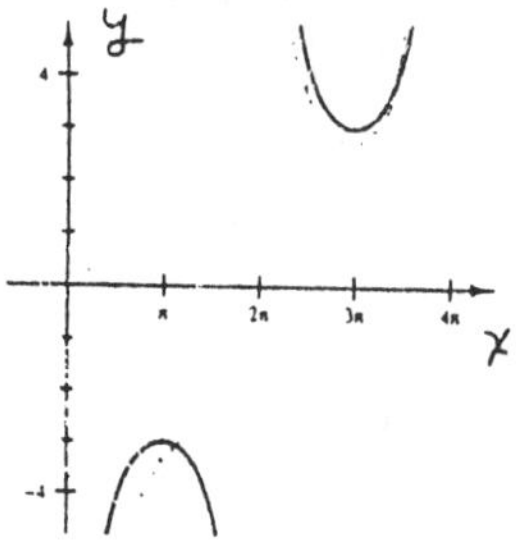

14.

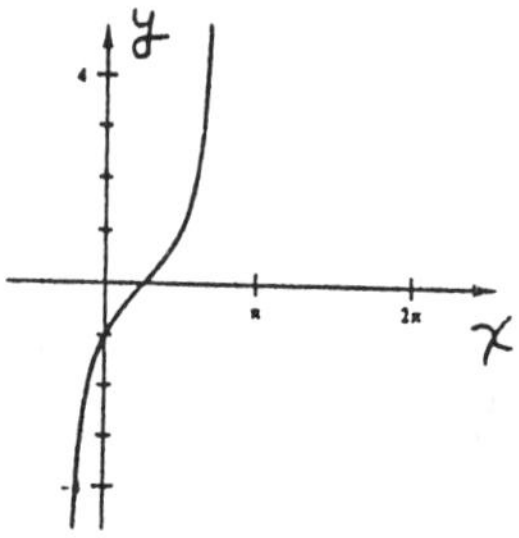

15.

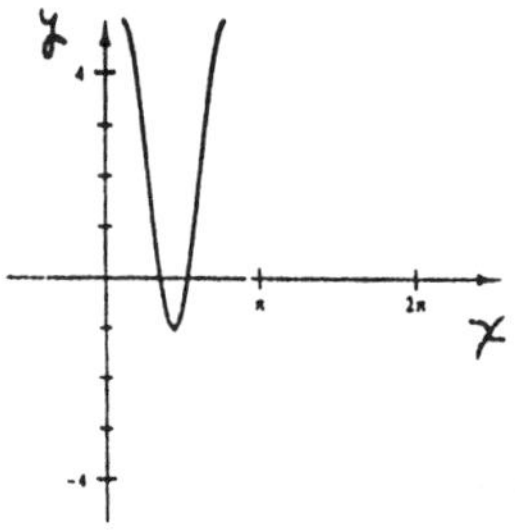

16.

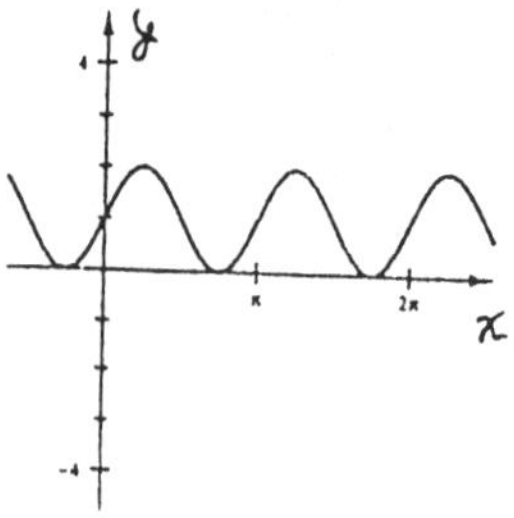

17.

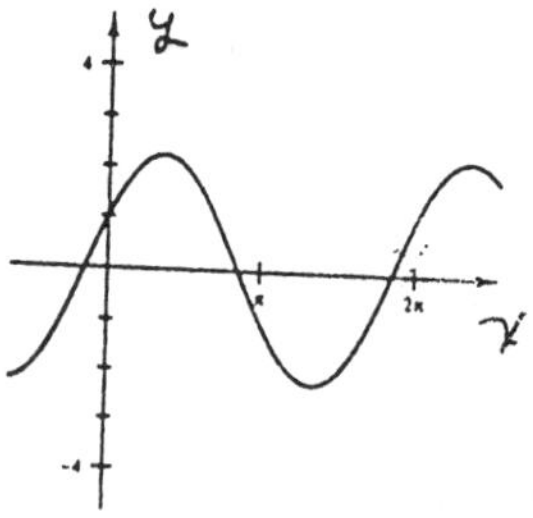

18. 70 mph

19. 110 ft

20. $y = 12\cos\frac{4\pi}{3}t$

Chapter 5 Form C

1. $\frac{7\pi}{4}$

2. $-36°$

3. $\left(\frac{\sqrt{3}}{2}, \frac{1}{2}\right)$

4. $\left(\frac{1}{2}, -\frac{\sqrt{3}}{2}\right)$

5. $\frac{4}{9}$

6. $\frac{\sqrt{2}}{2}$

7. *even*

8. $\frac{2\sqrt{3}}{3}$

9. $\frac{(1 - \sin^2 t)^2}{\sin^2 t}$

10. 3

11. $\frac{2}{3}, \frac{2\pi}{5}, -\frac{\pi}{10}$

12. 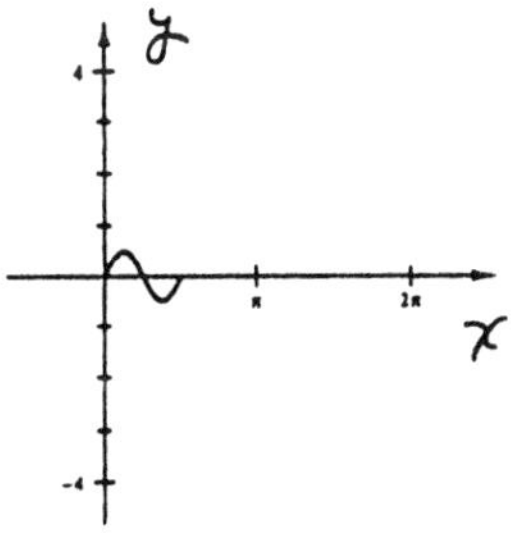

13. 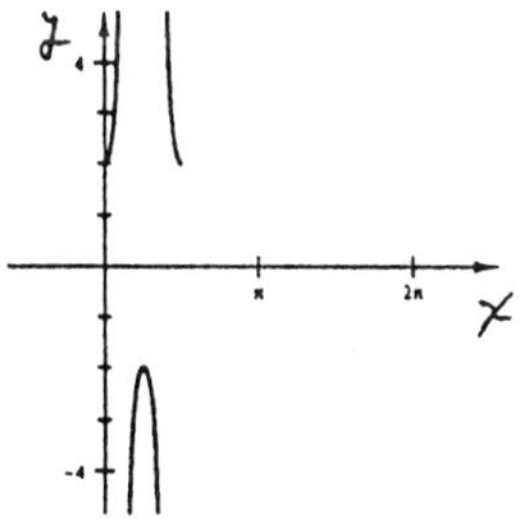

14. 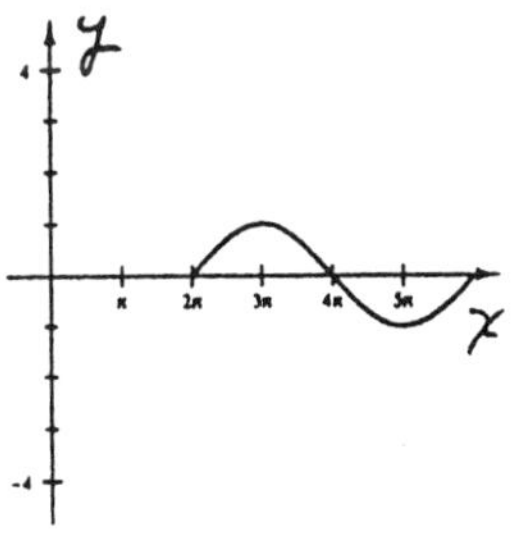

15. 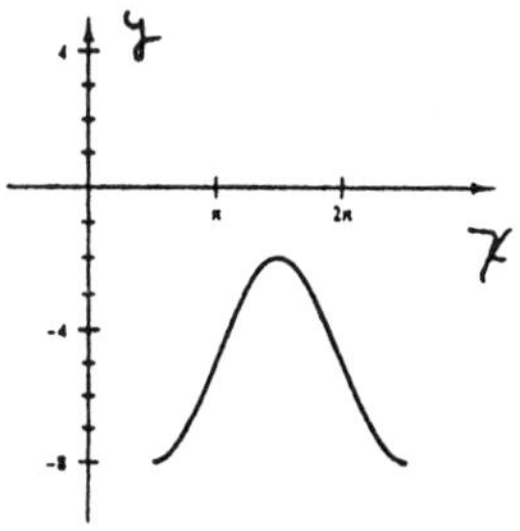

16. 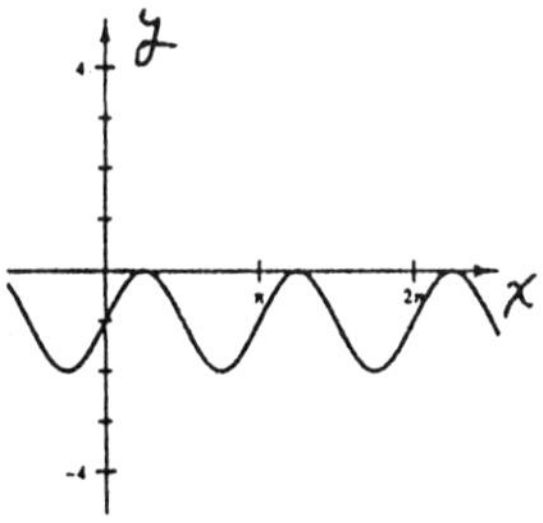

17. 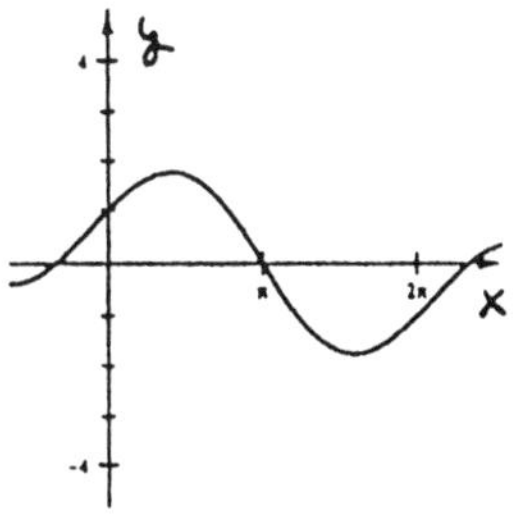

18. 62.5 ft/s

19. 15 m

20. $y = 8 \cos \frac{4\pi}{5} t$

Chapter 5 Form D

1. c
2. b
3. d
4. c
5. b
6. d
7. c
8. d
9. a
10. a

11. c	12. c	13. d	14. b	15. b
16. c	17. d	18. c	19. c	20. b

Chapter 5 Form E

1. b	2. a	3. c	4. d	5. d
6. a	7. a	8. d	9. a	10. a
11. c	12. d	13. c	14. c	15. c
16. d	17. b	18. b	19. c	20. a

Chapter 5 Form F

1. c	2. a	3. c	4. d	5. d
6. c	7. a	8. c	9. a	10. a
11. a	12. b	13. c	14. b	15. a
16. d	17. C	18. a	19. b	20. c

College Trigonometry
Tests and Answers

Trigonometry Chapter 1 Test Form A

1. Solve: $3x - 2(3 - 2x) = 4 - 2(x - 3)$

2. Solve: $7 - 2(3 - 2x) \geq 3x - 4(2x - 3)$

3. Solve: $2x^3 - 3x^2 - 2x + 3 = 0$

4. Solve: $|3 - 4x| < 5$

5. Find the distance between the points (-8, 4) and (-1, -2) and the midpoint of the segment that connects them.

6. Determine the x and y intercepts and then graph the equation $x = -y^2 + 4$.

7. Graph $y = |x + 3| - 2$.

8. Find the center and radius of the circle whose equation is $x^2 - 6x + y^2 + 4y + 9 = 0$.

9. Determine the domain of $f(x) = \frac{3}{x^2 + 3x - 4}$. Evaluate f(-1).

10. Graph $f(x) = -2|x + 1| + 1$. Identify the intervals over which the function is increasing or decreasing.

11. Graph $f(x) = x^2 + 4$. From the graph, determine the domain and range of the function.

12. Use the graph of $f(x) = |x|$ to graph $y = -f(x - 1) + 1$.

13. Which of the following equations defines y as a function of x?

    i. $x^2 + 5y^2 = 10$ ii. $x^2 + 16y = 3$ iii. $2x + 9y = 7$

14. Which of the following defines odd functions?

    i. $f(x) = x^4 - x^2$ ii. $f(x) = x^3 - x$ iii. $f(x) = x^4 + x^2 + 2$

15. Find the difference quotient for $f(x) = x^2 - 1$.

16. Evaluate $(f \circ g)(x)$ for $f(x) = 4x - 9$ and $g(x) = x^2 + 3$.

17. Find the inverse of the function given by the equation $f(x) = \frac{x}{x - 2}$.

18. Find the inverse of $f(x) = x^2 + 2, x \leq 0$. State the domain and range of both f and f^{-1}.

19. The perimeter of a rectangle is 96 feet and the area is 540 square feet. Find the length and width of the rectangle.

20. Hooke's Law states that the distance a spring stretches varies directly as the weight on the spring. A weight of 12 pounds stretches a spring 2.4 in. How far will a weight of 40 pounds stretch the spring?

Trigonometry Chapter 1 Test Form B

1. Solve: $5(x - 2) + 7 = x - 6(x - 2)$

2. Solve: $3x - 4(2 - 5x) \leq 4 - 3(1 - 5x)$

3. Solve: $6x^2 - x - 15 = 0$

4. Solve: $|3 - 2x| > 8$

5. Find the distance between the points (2, 9) and 5, -11) and the midpoint of the segment that connects them.

6. Determine the x and y intercepts and then graph the equation $y = -3x^2 + 3$.

7. Graph $y = x^2 + 4x + 3$.

8. Find the center and radius of the circle whose equation is $x^2 + 10x + y^2 - 6y + 18 = 0$.

9. Determine the domain of $f(x) = \sqrt{5x + 3}$. Evaluate f(3).

10. Graph $f(x) = -2|x + 1| + 1$. Identify the intervals over which the function is increasing or decreasing.

11. Graph $f(x) = x^2 + 4$. From the graph, determine the domain and range of the function.

12. Use the graph of $f(x) = x^2$ to graph $y = -f(x - 3) + 2$.

13. Which of the following equations defines y as a function of x?

    i. $5x - 7y = 9$ ii. $x^2 - 3y^2 = 11$ iii $x^2 + 8y = 2$

14. Which of the following define even functions?

    i. $f(x) = x^8 - x$ ii. $g(x) = x^4 - 3x^3$ iii. $h(x) = x^{10}$

15. Find the difference quotient for $f(x) = x^2 - 3$.

16. Evaluate $(f \circ g)(x)$ for $f(x) = 3x - 4$ and $g(x) = x^2 + 2$.

17. Find the inverse of the function given by the equation $f(x) = \frac{2x}{x + 3}$.

18. Find the inverse of the function given by the equation $f(x) = x^2 - 9$, $x \geq 0$. State the domain and range of both f and f^{-1}.

19. You can rent a car for the day from company A for $18.00 plus $0.10 a mile. Company B charges $12.00 plus $0.18 a mile. Find the number of miles (to the nearest mile) for which it is cheaper to rent from company A.

20. The stopping distance of a car is directly proportional to the square of its speed. A car moving 45 mph can stop in 75 feet. Find the stopping distance of a car that is traveling 55 mph.

Trigonometry Chapter 1 tests Form C

1. Solve: $3 - 2(6 - x) = 4x - 2(4 - 3x)$

2. Solve: $2x - 3(x - 4) \leq 8x - 2(5 - x)$

3. Solve: $8x^3 - 1 = 0$

4. Solve: $|2x - 3| < 2$

5. Find the distance between the points (-4, 3) and (6, 12) and the midpoint of the segment that connects them.

6. Determine the x and y intercepts and then graph the equation $y = -\frac{2}{3}x + 1$.

7. Graph $y = |x - 1| + 2$.

8. Find the center and radius of the circle whose equation is $x^2 - 2x + y^2 + 18y + 33 = 0$.

9. Determine the domain of $f(x) = \sqrt{3 - 7x}$. Evaluate $f(-3)$.

10. Graph $f(x) = -\frac{1}{2}|x + 2| - 2$. Identify the intervals over which the function is increasing or decreasing.

11. Graph $y = 4x^2 + 1$. From the graph, determine the domain and range of the function.

12. Use the graph of $f(x) = |x|$ to graph $y = -f(x + 3) + 2$.

13. Which of the following functions defines y as a function of x?

i. $2x - y^2 = 4$ ii. $\frac{3}{4}x - \frac{5}{8}y = 2$ iii. $x^2 - 2y = 3$

14. Which of the following define even functions?

i. $f(x) = x^2 - 5x$ ii. $g(x) = x^8$ iii. $h(x) = x^6 - x^3$

15. Find the difference quotient for $f(x) = x^2 - 4$.

16. Evaluate $(f \circ g)(x)$ for $f(x) = 8x + 2$ and $g(x) = x^2 - 6$.

17. Find the inverse of the function given by the equation $f(x) = \frac{-x}{x - 1}$.

18. Find the inverse of $f(x) = x^2 + 4$, $x \le 0$. State the domain and range of both f and f^{-1}.

19. A bank offers two checking account plans. Under plan A, you pay $8.00 a month plus $0.05 per check. Under plan B, you pay $3.00 a month plus $0.20 per check. Under what conditions is it less expensive to use plan A?

20. The cost of insulating exterior walls of a house varies jointly with the thickness of the insulation and the area of the walls. It cost $240 to insulate 1500 square feet of walls with insulation 4 inches thick. Find the cost of insulating 2400 square feet of exterior walls with insulation 5 inches thick.

Trigonometry Chapter 1 Test Form D

1. Solve: $3x - 2(4 - 7x) = 15 - 3(x - 3)$

a. $\frac{16}{7}$ b. $\frac{8}{5}$ c. $\frac{6}{5}$ d. $-\frac{6}{5}$

2. Solve: $x^2 - x - 12 \le 0$

a. $[-3, 4]$ b. $(-\infty, -3] \cup [4, \infty)$
c. $[-4, 3]$ d. $(-\infty, -4] \cup [3, \infty)$

3. Solve: $|x - 4| > 8$

a. $[-4, 12]$ b. $(-\infty, -4] \cup [12, \infty)$
c. $(4, 8)$ d. $(-\infty, -4) \cup (12, \infty)$

4. Find the distance between the points (4, –6) and (–2, –5) and the midpoint of the segment that connects them.

a. $\sqrt{5}, \left(3, -\frac{1}{2}\right)$ b. $5\sqrt{5}, \left(2, -\frac{11}{2}\right)$
c. $\sqrt{37}, \left(2, -\frac{11}{2}\right)$ d. $\sqrt{37}, \left(3, -\frac{1}{2}\right)$

5. Solve: $x^3 + x^2 - 4x - 4 = 0$

a. –1, –2, 2 b. 1, –2, 2 c. –1, 2, 4 d. –1, 1, 2

6. Deterine the x and y intercepts and then graph the equation $5x - y = 3$.

a. $\left(\frac{3}{5}, 0\right), (0, -3)$ b. $\left(\frac{5}{3}, 0\right), (0, -3)$ c. $\left(\frac{3}{5}, 0\right), (0, 3)$ d. $(3, 0), (0, 5)$

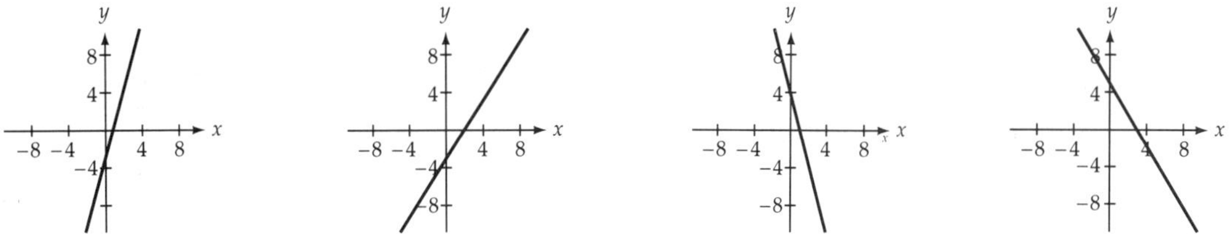

7. Graph $y = |x + 2| - 2$.

a. 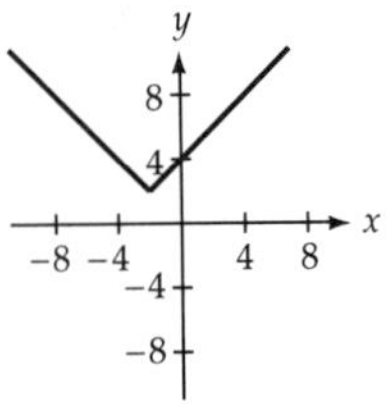

b. 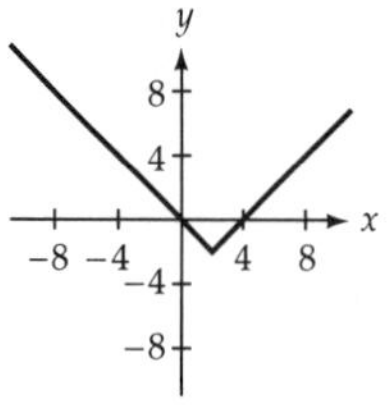

c. 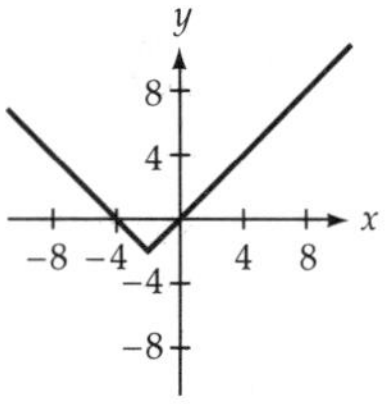

d. 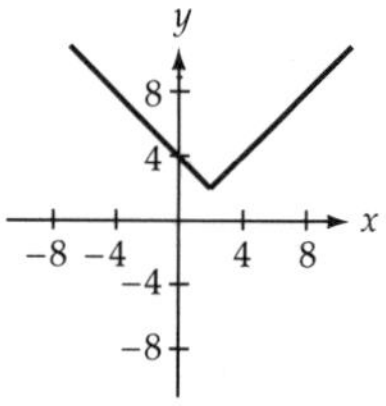

8. Find the center and radius of the circle whose equation is $x^2 + y^2 + 18x - 8y + 93 = 0$.

a. center: (9, −4), radius: 4
b. center: (−9, 4), radius: 4
c. center: (−9, 4), radius: 2
d. center: (9, −4), radius: 2

9. Determine the domain of $f(x) = \sqrt{x^2 - 64}$. Evaluate f(−10).

a. [−8, 8], 6, −6
b. $(-\infty, -8] \cup [8, \infty)$, 6, −6
c. $(-\infty, -64] \cup [64, \infty)$, 6
d. (−8, 8), 6, −6

10. Determine which are graphs of one-to-one functions.

(i) 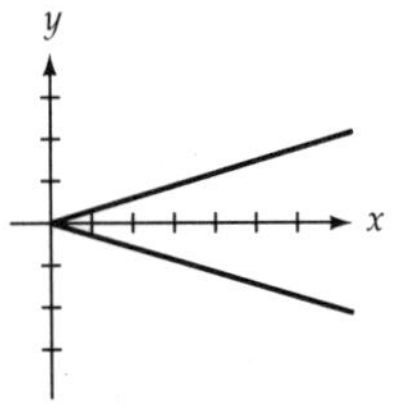

(ii) 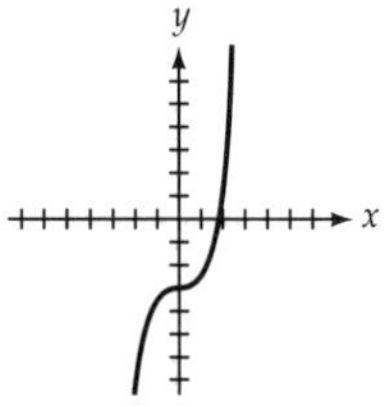

(iii) 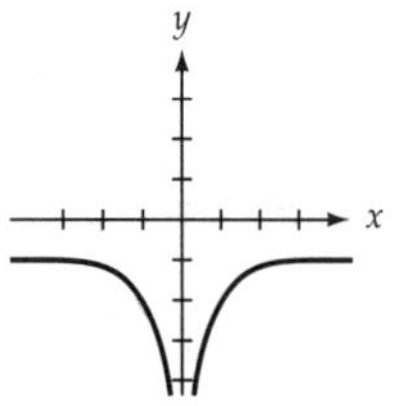

a. i, ii, and iii
b. i and ii
c. ii
d. ii and iii

11. Determine the domain and range of the function $y = 2x^2 - 5$.

a. Domain: all real numbers,
Range: $\{y | y \geq -5\}$
b. Domain: $\left\{x | -\frac{5}{2} \leq x \leq \frac{5}{2}\right\}$
c. Domain: $\{x | x \geq -5\}$,
Range: all real numbers
d. Domain: all real numbers,
Range: $\left\{y | y > -\frac{5}{2}\right\}$

12. Use the graph of $y = x^2$ to graph $y = -f(x + 2) - 3$.

a.
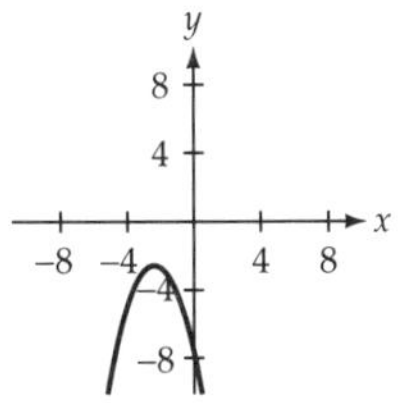

b.
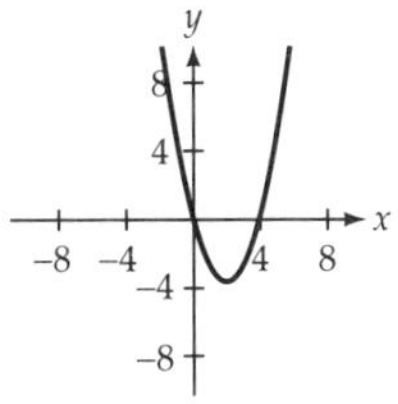

c.
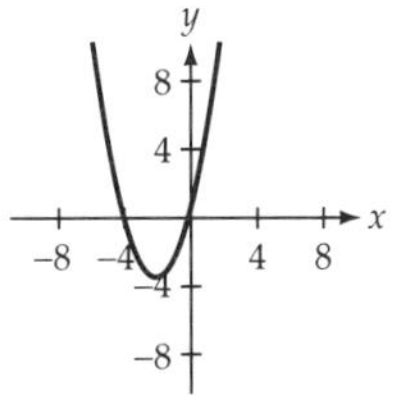

d.
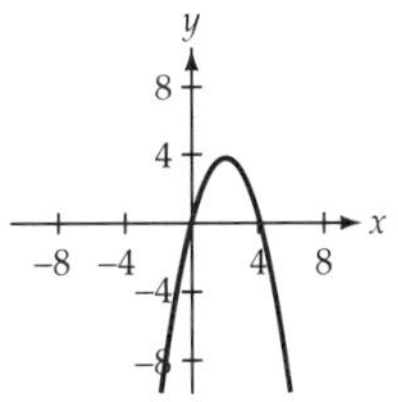

13. Which of the following equations defines y as a function of x.
i. $x^2 - 4y^2 = 12$ ii. $x^2 + 6y = 15$ iii. $8x + 2y = 11$

a. ii and iii b. i c. i, ii, and iii d. iii

14. Which of the following are odd functions?
i. $f(x) = -5|x|$ ii. $g(x) = x^3 + 7$ iii. $h(x) = x^5$

a. ii and iii b. iii c. i, ii, and iii d. i

15. Find the difference quotient for $f(x) = 4x - 1$.

a. $4h + 1$ b. $4h$ c. 4 d. $h + 1$

16. If $f(x) = 8x + 3$ and $g(x) = -3x^2 + 1$, find $f \circ g$.

a. $-24x^2 + 4$ b. $-24x^2 + 11$
c. $-192x^2 - 144x - 26$ d. $-3x^2 - 9x^2 + 8x + 3$

17. Find the inverse of the function given by the equation $f(x) = \dfrac{7x}{x + 5}$, $x \neq -5$.

a. $f^{-1}(x) = \dfrac{x + 5}{7x}$, $x \neq 0$ b. $f^{-1}(x) = -\dfrac{7x}{x + 5}$, $x \neq -5$

c. $f^{-1}(x) = \dfrac{5x}{7 - x}$, $x \neq 7$ d. $f^{-1}(x) = \dfrac{5x}{x - 7}$, $x \neq 7$

18. Find the inverse of $f(x) = x^2 - 3$, $x \leq 0$. State the domain and range of f^{-1}.

a. $f^{-1}(x) = -\sqrt{x+3}$, domain: $\{x | x \geq 3\}$, range: $\{y | y \leq 0\}$

b. $f^{-1}(x) = \sqrt{x+3}$, domain: $\{x | x \geq 3\}$, range: $\{y | y \leq 0\}$

c. $f^{-1}(x) = \frac{1}{x^2 - 3}$, domain: $\{x | -3 \leq x \leq 3\}$, range: $\{y | y \leq 0\}$

d. $f^{-1}(x) = \sqrt{x-3}$, domain: $\{x | x \leq -3 \text{ or } x \geq 3\}$, range: $\{y | y \leq 0\}$

19. The perimeter of a triangle is 48 feet and the area is 135 square feet. Find the length and width of the rectangle.

a. length: 30 ft, width: 18 ft
b. length 22.5 ft, width: 6 ft
c. length: 15 ft, width: 9 ft
d. length: 45 ft, width: 3 ft

20. You can rent a car for the day from company A for \$20.00 plus \$0.08 a mile. Company B charges \$12.00 plus \$0.20 a mile. Find the number of miles (to the nearest mile) for which it is cheaper to rent from company A.

a. 267 mi
b. 67 mi
c. 29 mi
d. 43 mi

Trigonometry Chapter 1 Test Form E

1. Solve: $5 - 7x = -8(2x + 5)$

   a. $\frac{35}{9}$ b. -5 c. $-\frac{43}{23}$ d. 5

2. Solve: $2x^2 - x - 10 \geq 0$

   a. $(-\infty, -2] \cup \left[\frac{5}{2}, \infty\right)$ b. $(-\infty, 2] \cup \left[\frac{5}{2}, \infty\right)$

   c. $(-\infty, 2) \cup \left[\frac{7}{2}, \infty\right)$ d. $(-\infty, -5] \cup [2, \infty)$

3. Solve: $|2x - 3| \leq 7$

   a. $[2, 5]$ b. $[3, 7]$ c. $[-2, 5]$ d. $(-2, 5)$

4. Find the distance between the points $(-4, 15)$ and $(3, -11)$ and the midpoint of the segment that connects them.

   a. $5\sqrt{29}, \left(-\frac{7}{2}, 13\right)$ b. $5\sqrt{29}, \left(-\frac{1}{2}, 2\right)$

   c. $\sqrt{65}, (-1, 4)$ d. $\sqrt{17}, \left(-\frac{1}{2}, 2\right)$

5. Solve: $4x^3 + 4x^2 - 9x + 9 = 0$

   a. $\pm 1, \frac{3}{2}$ b. $-1, \pm\frac{3}{2}$ c. $1, \pm\frac{3}{2}$ d. $-1, -\frac{3}{2}$

6. Determine the x and y intercepts and then graph the equation $4x - 7y + 14 = 0$.

   a. $(2, 0), \left(0, -\frac{7}{2}\right)$ b. $\left(\frac{7}{2}, 0\right), (0, -2)$ c. $\left(-\frac{7}{2}, 0\right), (0, 2)$ d. $(-2, 0), \left(0, \frac{7}{2}\right)$

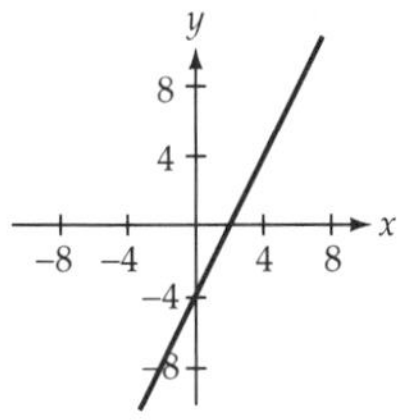
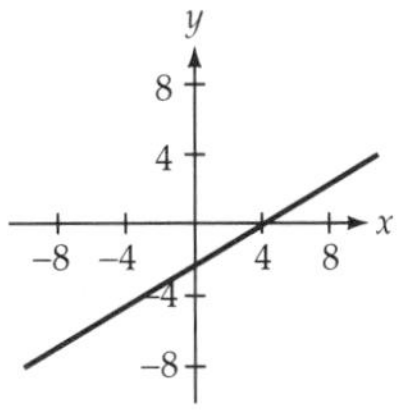
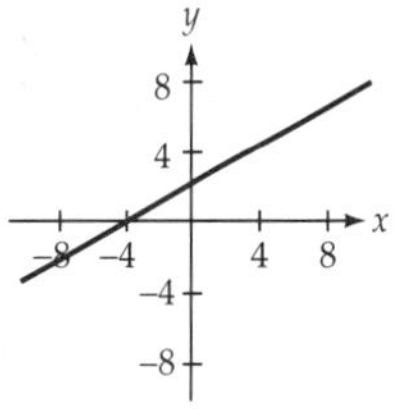
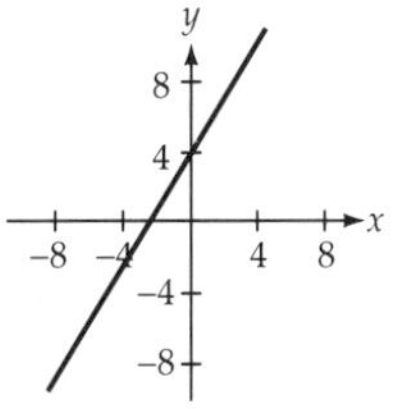

7. Graph $y = 3x^2 - 4$.

a.
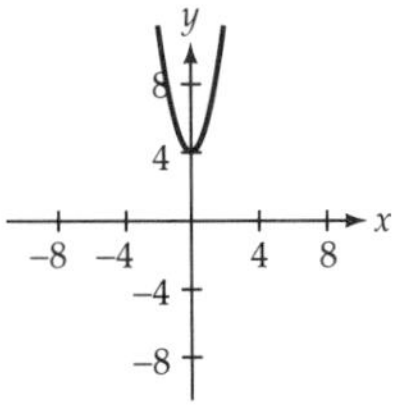

b.
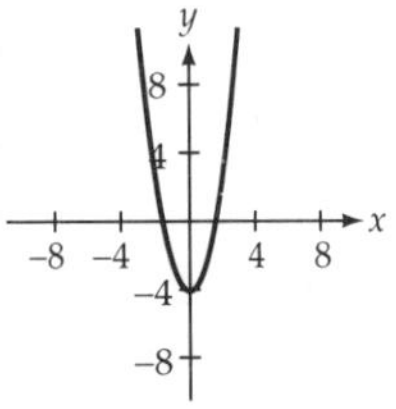

c.
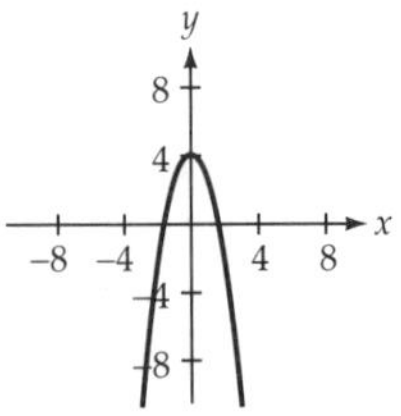

d.
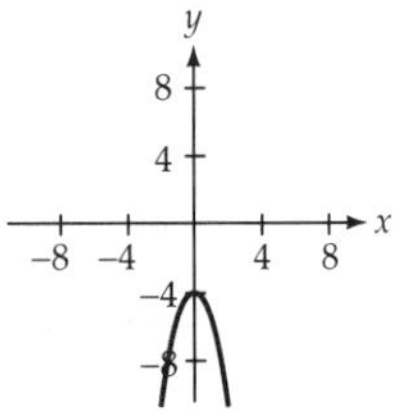

8. Find the center and radius of the circle whose equation is $x^2 + y^2 + 4x - 12y - 24 = 0$.

a. center: (2, –6), radius: 8
b. center: (2, –6), radius: 64
c. center: (–2, 6), radius: 8
d. center: (–2, 6), radius: 64

9. Determine the domain of $f(x) = \sqrt{x^2 + 4}$. Evaluate f(–4).

a. [2, –2], $5\sqrt{2}$
b. $\{y | y \geq -4\}$, $-2\sqrt{5}$
c. $\{x | x \geq -4\}$, 4
d. all the real numbers

10. Determine which are graphs of one-to-one functions.

(i)
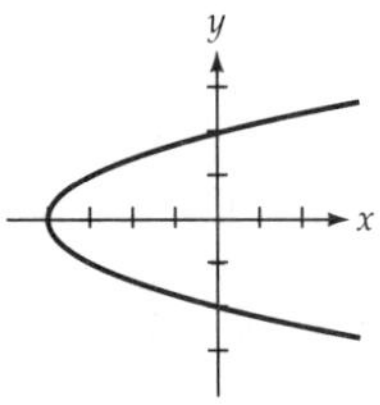

(ii)
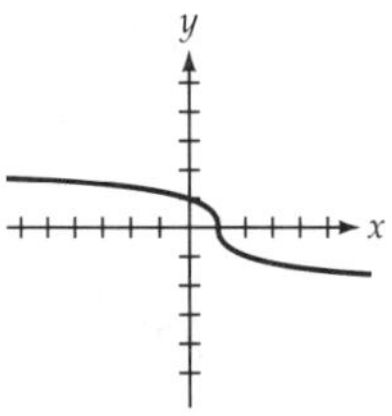

(iii)
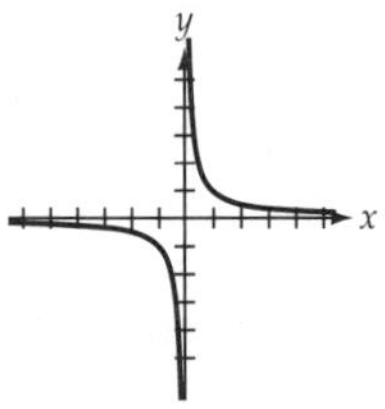

a. i and iii
b. i and ii
c. ii
d. ii and iii

11. Determine the domain and range $y = |x - 2| + 1$.

a. Domain: $(-\infty, \infty)$
Range: $\{y | y \geq -1\}$

b. Domain: $\{x | x \leq 2\}$
Range: $(-\infty, \infty)$

c. Domain: $(-\infty, \infty)$
Range: $\{y | y \geq 1\}$

d. Domain: $\{x | x \leq 2\}$
Range: $\{y | y \geq 1\}$

12. Use the graph of $y = |x|$ to graph $y = f|x + 1| - 2$.

a.
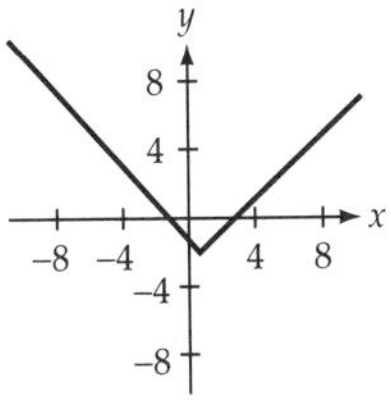

b.
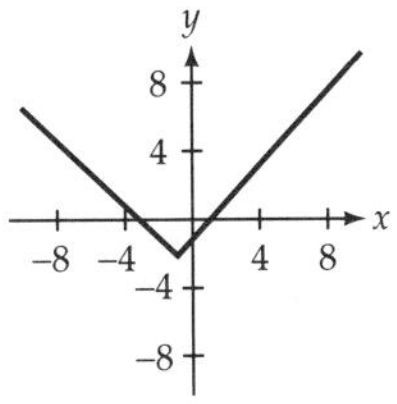

c.
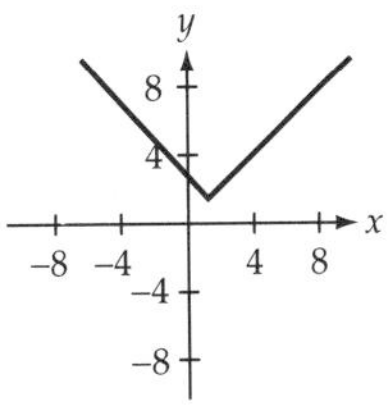

d.
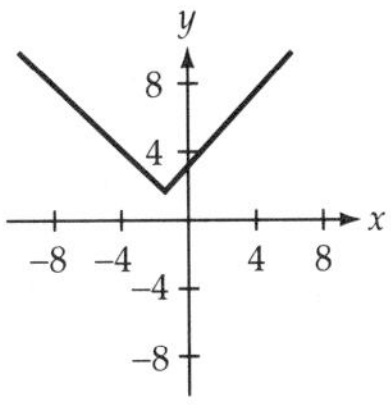

13. Which of the following equations defines y as a function of x.

i. $2x + 15y = 18$ ii. $x^2 + 9y^2 = 12$ iii. $x^2 + 5y = 7$

a. i, ii, and iii b. ii and iii c. i and iii d. i

14. Which of the following are odd functions?

i. $f(x) = x^5 + x^3 + 2$ ii. $g(x) = 4x^6 + 2x$ iii. $h(x) = x^{13}$

a. i and iii b. i c. i, ii, and iii d. iii

15. Find the difference quotient for $y = 2x^2 - 3$.

a. $4x$ b. $4x + h$ c. $4x + 2h$ d. $2x + h$

16. If $f(x) = -3x + 1$ and $g(x) = x^2 + 4$, find $g \circ f$.

a. $-3x^2 - 11$ b. $9x^2 - 6x + 5$ c. $9x^2 + 5$ d. $9x^2 + 6x + 5$

17. Find the inverse of the function given by the equation $f(x) = \frac{8x}{x + 6}$, $x \neq -6$.

a. $f^{-1}(x) = \frac{x + 6}{8x}$, $x \neq 0$ b. $f^{-1}(x) = -\frac{8x}{x - 6}$, $x \neq 6$

c. $f^{-1}(x) = \frac{6x}{8 - x}$, $x \neq 8$ d. $f^{-1}(x) = \frac{6x}{x - 8}$, $x \neq 8$

18. Find the inverse of $f(x) = x^2 + 4$, $x \geq 0$, State the domain and range of f^{-1}.

a. $f^{-1}(x) = -\sqrt{x-4}$, domain: $\{x|x \geq -4\}$, range: $\{y|y \leq 0\}$
b. $f^{-1}(x) = \sqrt{x-4}$, domain: $\{x|x \geq 4\}$, rang: $\{y|y\ |\geq 0\}$
c. $f^{-1}(x) = \sqrt{4-x})$, domain: $\{x|x \leq 4,$ range: $\{y|y \geq 0\}$
d. $f^{-1}(x) = \sqrt{4-x}$, domain: $\{x|x \leq 4\}$, range: $\{y|y \geq 0\}$

19. The stopping distance of a car is directly proportional to the square of its speed. A car moving 40 mph can stop in 65 feet. Find the stopping distance of a car that is traveling 65 mph.

a. 97.5 ft b. 146.25 ft c. 122.5 ft d. 137.5 ft

20. A bank offers two checking account plans. Under plan A you pay \$6.00 a month plus \$0.04 per check. Under plan B, you pay \$2.00 a month plus \$0.10 per check. Under what conditions is it less expensive to use plan A?

a. Writing over 133 checks b. Writing over 33 checks
c. Writing over 66 checks d. Writing over 57 checks

Trigonometry Chapter 1 Test Form F

1. Solve: $2x + 3(4 - 5x) = 2x - 3(2 + x)$

   a. $\frac{9}{7}$ b. $\frac{2}{3}$ c. $-\frac{1}{2}$ d. $\frac{3}{2}$

2. Solve: $-1 - 3(x - 7) \geq 7x - 4(2 - x)$

   a. $\{x|x \leq 2\}$ b. $\{x|x \geq -2\}$ c. $\{x|x \geq -\frac{7}{2}\}$ d. $\{x|x \leq -2\}$

3. Solve: $|2 - x| > 4$

   a. $(-2, 6)$ b. $(-\infty, -2] \cup [6, \infty)$

   c. $(-6, 2)$ d. $(-\infty, -2) \cup (6, \infty)$

4. Find the distance between the points (-5, 11) and (-2, 8) and the midpoint of the segment that connects them.

   a. $9\sqrt{2}, \left(-\frac{7}{2}, \frac{19}{2}\right)$ b. $3\sqrt{2}, \left(-\frac{3}{2}, \frac{3}{2}\right)$

   c. $\sqrt{58}, \left(-\frac{3}{2}, \frac{19}{2}\right)$ d. $3\sqrt{2}, \left(\frac{7}{2}, \frac{19}{2}\right)$

5. Solve: $4x^2 - 5x - 21 = 0$

   a. $-\frac{7}{4}, 3$ b. $-3, \frac{7}{4}$ c. $-\frac{3}{4}, 7$ d. $-\frac{3}{2}, \frac{7}{2}$

6. Determine the x and y intercepts and then graph the equation $4x - 3y = 9$.

   a. $(-3, 0), \left(0, \frac{9}{4}\right)$ b. $\left(\frac{9}{4}, 0\right), (0, -3)$ c. $\left(\frac{4}{3}, 0\right), (0, 3)$ d. $\left(-\frac{4}{3}, 0\right), (0, 3)$

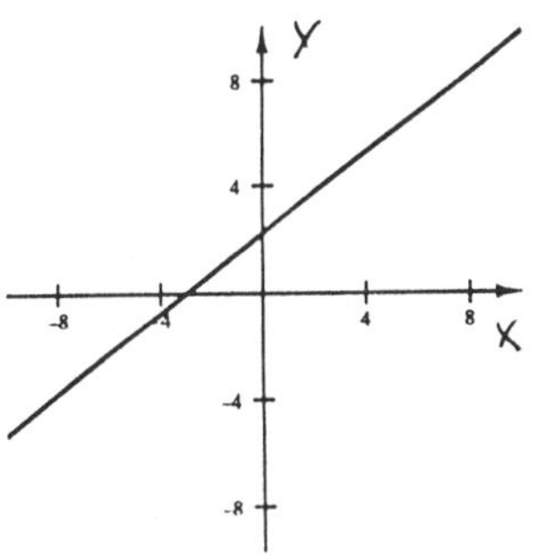

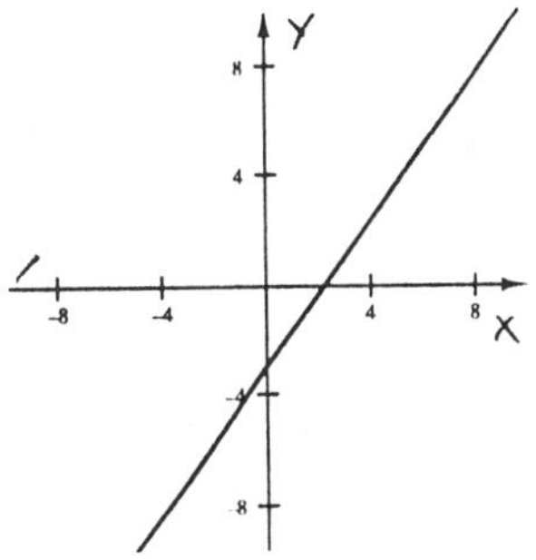

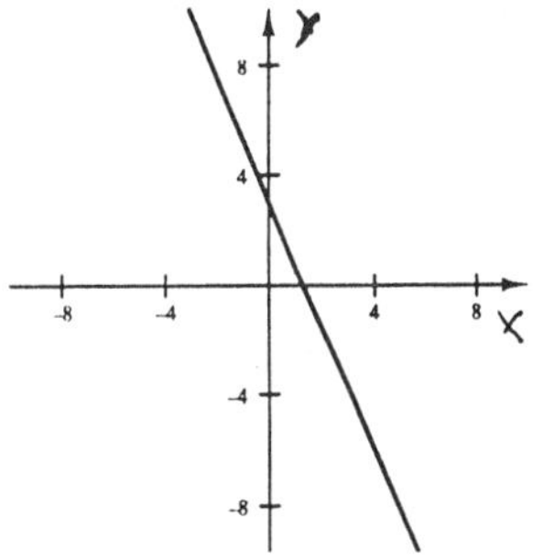

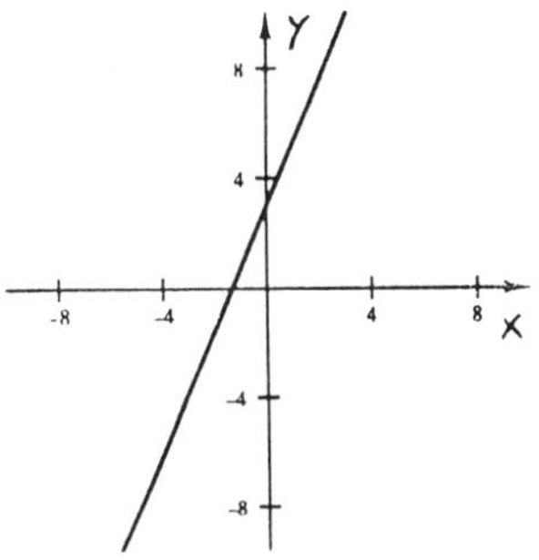

7. Graph $x = y^2 - 4$.

a.

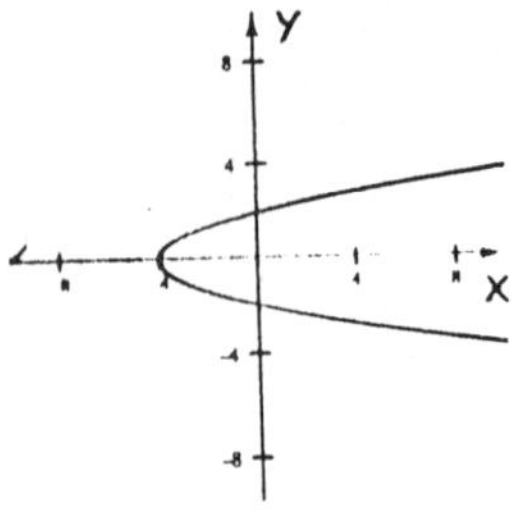

b.

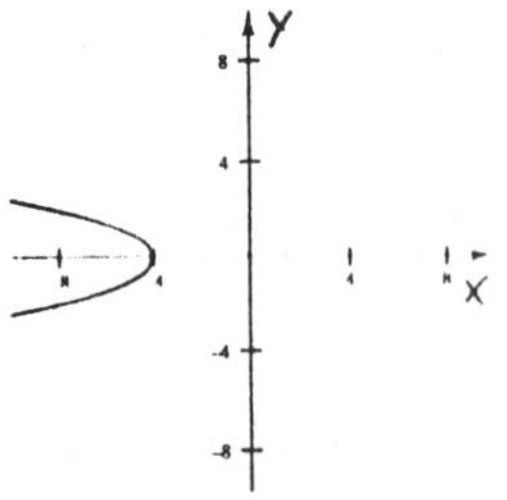

c.

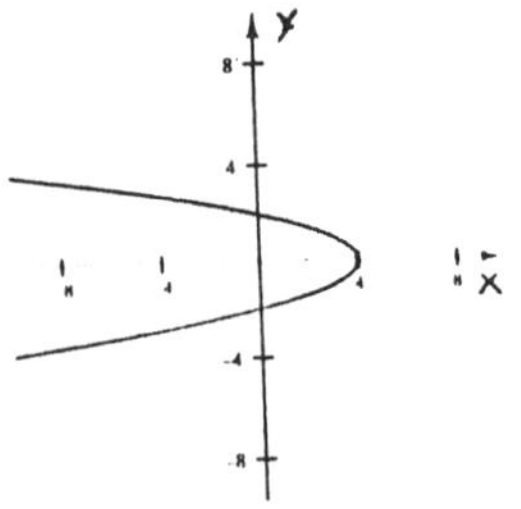

d.

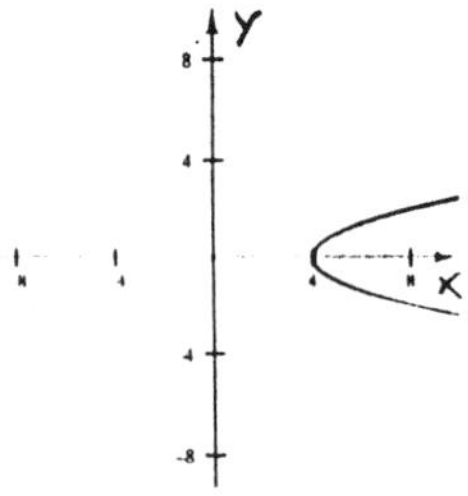

8. Find the center and radius of the circle whose equation is $x^2 - 8x + y^2 + 10y + 25 = 0$.

a. center: $(-4, 5)$, radius: 4

b. center: $(-4, 5)$, radius: 16

c. center: $(8, 10)$, radius: 16

d. center: $(4, -5)$, radius: 4

9. Determine the domain of $f(x) = \sqrt{25 - x^2}$. Evaluate $f(-4)$.

a. $(-\infty, 5] \cup [5, \infty)$, $\sqrt{41}$

b. $[-5, 5]$, 3

c. $(-25, 25)$, ± 3

d. $(-\infty, 5) \cup (5, \infty)$, 9

10. Determine which graphs are one-to-one functions.

i.

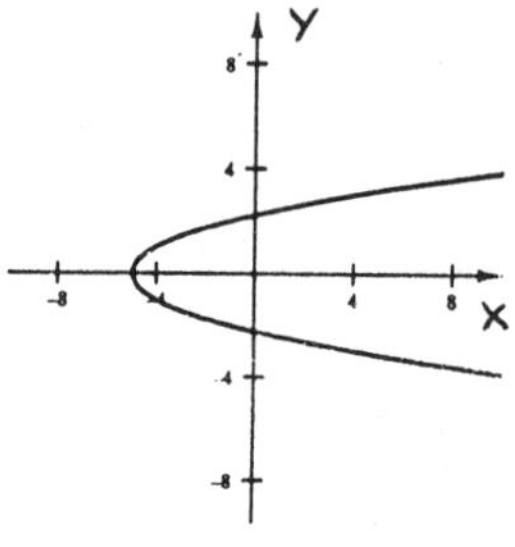

ii.

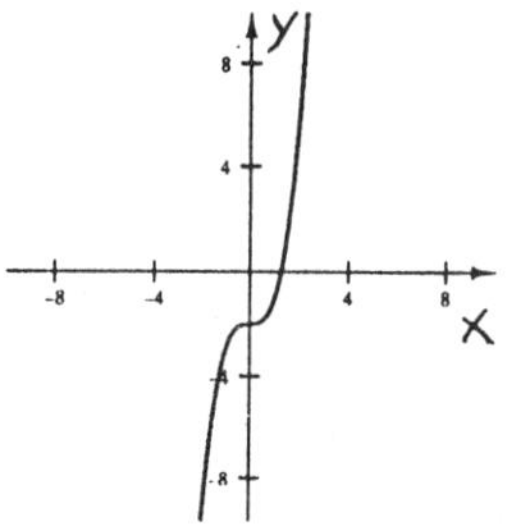

iii.

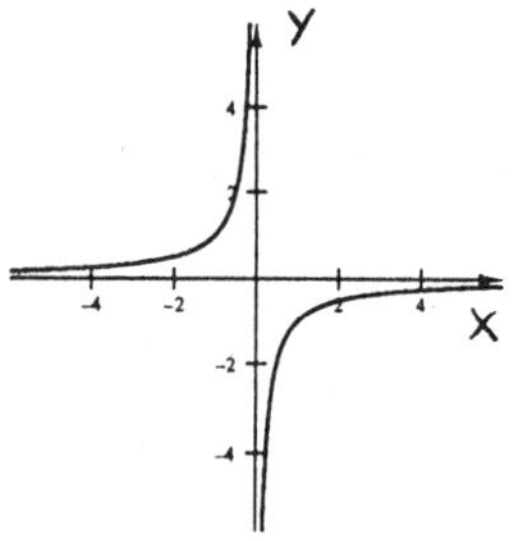

a. ii and iii

b. i and ii

c. ii

d i and iii

11. Determine the domain of $y = -\sqrt{9 - x^2}$.

a. $\{x \mid x \le 9\}$

b. $\{x \mid x \ge 3\}$

c. $\{x \mid x \le 3\}$

d. $\{x \mid x \ge -3\}$

12. Use the graph of $y = x^2$ to graph $-f(x-3)+2$

a. 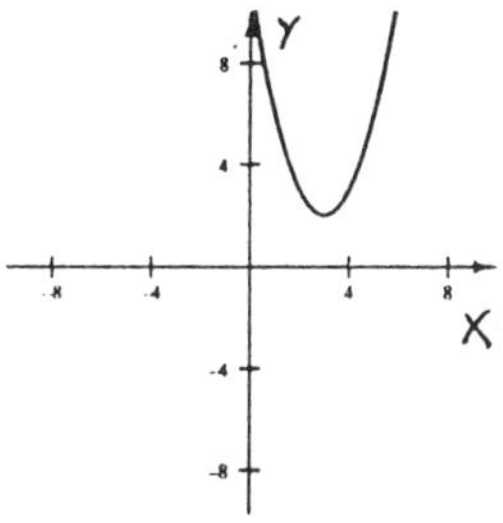

b. 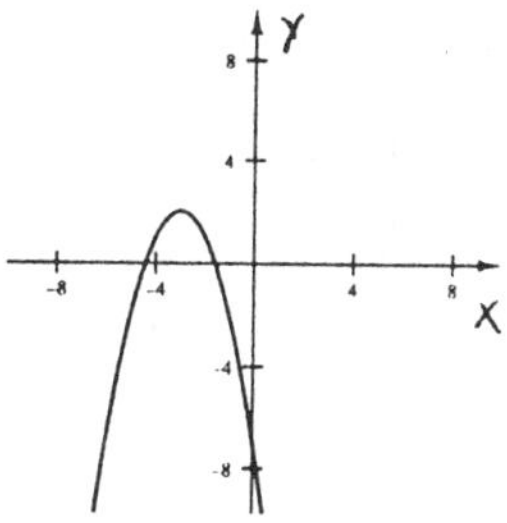

c. 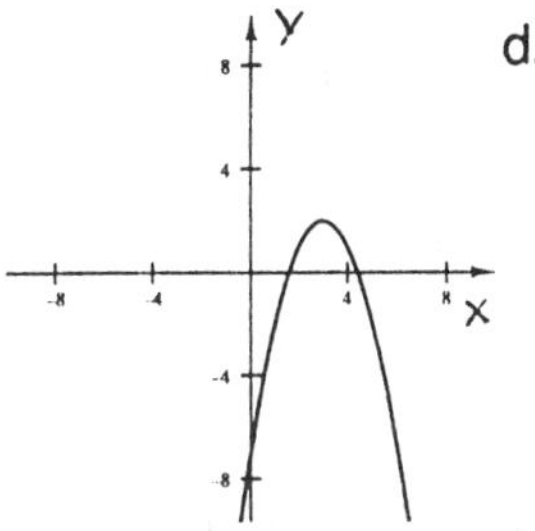

d. 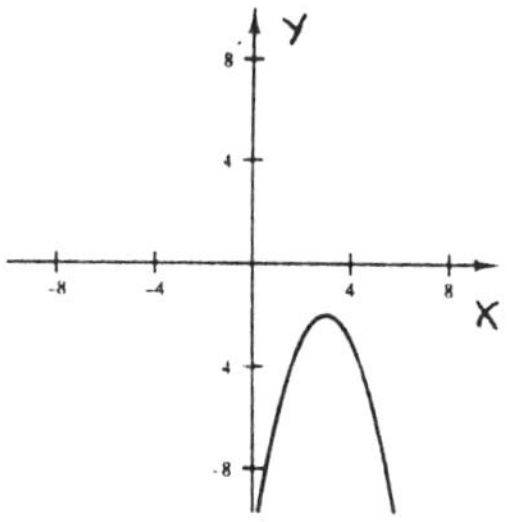

13. Which of the following functions defines y as a function of x?

i. $3x + \frac{1}{2}y = 14$ ii. $4x^2 - 9y^2 = 16$ iii. $x - y^2 = 4$

a. 1 and ii b. i c. ii and iii d. i and iii

14. Which of the following are odd functions?

i. $f(x) = |-8x|$ ii. $g(x) = 2x^2 - 1$ iii. $3x^3 - 5x$

a. i i i b. i and ii c. i i d. ii and iii

15. Find the difference quotient for $f(x) = 3x^2 - 5$.

a. $6x$ b. $x + 3h$ c. $6x + 3h$ d. $3h$

16. If $f(x) = -2x + 1$ and $g(x) = x^2 - 4$, find $f \circ g$.

a. $x^2 - 7$ b. $4x^2 - 4x - 3$ c. $-2x^2 + 9$ d. $2x^2 - 7$

17. Find the inverse of the function given by the equation $f(x) = \frac{2x}{x-2}$, $x \neq 2$.

a. $f^{-1}(x) = \frac{2x}{x-2}$, $x \neq 2$

b. $f^{-1}(x) = \frac{x}{x+2}$, $x \neq -2$

c. $f^{-1}(x) = \frac{2x-1}{x}$, $x \neq 0$

d. $f^{-1}(x) = \frac{x+2}{x}$, $x \neq 0$

18. Find the inverse of $f(x) = x^2 - 8, x \geq 0$. State the domain and range of f^{-1}.

a. $f^{-1}(x) = \sqrt{8 - x}$, domain: $\{x \mid -8 \leq x \leq 8\}$, range: $\{y \mid y \geq 0\}$
b. $f^{-1}(x) = -\sqrt{x + 8}$, domain: $\{x \mid x \geq -8\}$, range: $\{y \mid y \geq 0\}$
c. $f^{-1}(x) = \sqrt{x + 8}$, domain: $\{x \mid x \geq -8\}$, range: $\{y \mid y \geq 0\}$
d. $f^{-1}(x) = \sqrt{x - 8}$, domain: $\{x \mid x \leq -8 \text{ or } x \geq 8\}$, range: $\{y \mid y \geq 0\}$

19. Hooke's Law states that the distance a spring stretches varies directly as the weight on the spring. A weight of 5 kilograms stretches a spring 18 centimeters. How far will a weight of 12 kilometers stretch the spring?

a. 36 cm b. 48 cm c. 54 cm d. 43.2 cm

20. The cost of insulating exterior walls of a house varies jointly with the thickness of the insulation and the area of the walls. It cost \$400 to insulate 1200 square feet of walls with insulation 3 inches thick. Find the cost of insulating 3000 square feet of exterior walls with insulation 5 inches thick.

a. \$600 b. \$900 c. \$1266.67 d. \$1666.67

Trigonometry Chapter 4 Test Form A

1. Solve triangle ABC if A = 62°, B = 51°, and c = 19.

2. Solve triangle ABC if A = 48°, c = 40 and a = 25.

3. Find B in triangle ABC if A = 34°, b = 7, and a = 11.

4. In triangle ABC, B = 71°, a = 4, and c = 7. Find side b.

5. In triangle ABC, a = 26, b = 15, and c = 17. Find Angle A.

6. Given angle C = 61° side a = 55, and side b = 29. Find the area of triangle ABC.

7. Given angle A = 35°, angle C = 80°, and side a = 15, find the area of triangle ABC.

8. Given side a = 12, side b = 22, and side c = 18, find the area of triangle ABC.

9. A vector has a magnitude of 16 and direction of 240°. Write an equivalent vector in the form $\mathbf{v} = a_1\mathbf{i} + a_2\mathbf{j}$.

10. Find the magnitude and direction of the vector $\langle -4, 6 \rangle$.

11. Find a unit vector **u** in the direction of $\mathbf{v} = \langle -1, 3 \rangle$.

12. Find $2\mathbf{u} - 9\mathbf{v}$ given the vectors $\mathbf{u} = 6\mathbf{i} - 7\mathbf{j}$ and $\mathbf{v} = \mathbf{i} + 11\mathbf{j}$.

13. Find the horizontal and vertical components of the vector of magnitude 20 and direction angle 320°. Write the answer in the form $a_1\mathbf{i} + a_2\mathbf{j}$.

14. Find the dot product of $\mathbf{u} = -4\mathbf{i} + \mathbf{j}$ and $\mathbf{v} = 3\mathbf{i} + 7\mathbf{j}$.

15. Given $\mathbf{v} = -2\mathbf{i} + 5\mathbf{j}$ and $\mathbf{w} = 4\mathbf{i} + 6\mathbf{j}$, find $\text{proj}_{\mathbf{w}}\mathbf{v}$.

16. Find the angle to the nearest degree between the vectors $\mathbf{u} = \langle -2, 3 \rangle$ and $\mathbf{v} = \langle 6, -4 \rangle$.

17. Find the area of a parallelogram whose sides are 14 feet and 24 feet and the shortest diagonal is 18 feet.

18. A motorist drove 86 miles at a heading of 146°. The motorist then drove 74 miles at a heading of 236°. Find the distance of the motorist from the starting point.

19. One ship leaves a port at 10:00 A.M. traveling at 14 mph at a heading of 48°. At 12:00 noon another ship leaves the port traveling at 20 mph at a heading of 160°. Find the distance between the ships at 2:00 P.M.

20. A triangular commercial piece of real estate is priced at $12.50 per square foot. Find the cost of the lot measuring 90 ft by 140 ft by 120 ft.

Trigonometry Chapter 4 Test Form B

1. Solve triangle ABC if B = 65°, C = 38°, and b = 24.

2. Solve triangle ABC if A = 35°, c = 55, and a = 30.

3. Find C in triangle ABC if A = 56°, c = 14 and a = 8.0.

4. In triangle ABC, B = 43°, a = 8, and c = 14. Find side b.

5. In triangle ABC, a = 22, b = 18, and c = 14. Find angle B.

6. Given angle B = 22°, side a = 34, and side b = 45, find the area of triangle ABC.

7. Given A = 110°, angle C = 42°, and side a = 25, find the area of triangle ABC.

8. Given side a = 60, side b = 72, and side c = 54, find the area of triangle ABC.

9. A vector has magnitude 8.0 and direction 70°. Write an equivalent vector in the form $\mathbf{v} = a_1\mathbf{i} + a_2\mathbf{j}$.

10. Find the magnitude and direction of the vector $\mathbf{v} = 10\mathbf{i} - 4\mathbf{j}$.

11. Find a unit vector $\mathbf{u}$ in the direction of $\mathbf{v} = \langle 7, 24 \rangle$.

12. Find $4\mathbf{u} - 6\mathbf{v}$ given the vectors $\mathbf{u} = 3\mathbf{i} - 7\mathbf{j}$ and $\mathbf{v} = 5\mathbf{i} + 2\mathbf{j}$.

13. Find the horizontal and vertical components of the vector of magnitude 4p and direction angle 140°. Write the answer in the form $a_1\mathbf{i} + a_2\mathbf{j}$.

14. Find the dot product of $\mathbf{u} = -3\mathbf{i} + 2\mathbf{j}$ and $\mathbf{v} = 5\mathbf{i} - \mathbf{j}$.

15. Given $\mathbf{v} = 6\mathbf{i} + 3\mathbf{j}$ and $\mathbf{w} = 3\mathbf{i} + 7\mathbf{j}$, find $\text{proj}_{\mathbf{w}}\,\mathbf{v}$.

16. Find the angle to the nearest degree between the vectors $\mathbf{u} = \langle -2, 5 \rangle$ and $\mathbf{v} = \langle -1, 4 \rangle$.

17. Find the area of a parallelogram whose sides are 22 m and 40 m and the shortest diagonal is 30 m.

18. A pilot is flying at a heading of 120° at 250 mph. A 55 mph wind is blowing from the southwest at a heading of 36°. Find the ground speed and course of the plane.

19. A 45-pound force keeps a 550-pound object from sliding down an inclined ramp. Find the angle of the ramp.

20. Two ships left a port at the same time. One ship traveled at a speed of 15 mph at a heading of 55°. The other ship traveled at a speed of 22 mph at a heading of 115°. Find the distance between the two ships after 8 hours of travel.

Trigonometry Chapter 4 Test Form C

1. Solve triangle ABC if $A = 105°$, $B = 22°$, and $c = 18$.

2. Solve triangle if $B = 27°$, $a = 15$, and $b = 6$.

3. Find A in triangle ABC if $B = 66°$, $a = 56$, and $b = 108$.

4. In triangle ABC, $B = 39°$, $a = 56$, and $c = 42$. Find side b.

5. In triangle ABC, $a = 42$, $b = 32$, and $c = 55$. Find angle C.

6. Given angle $B = 112°$, side $a = 12$, and side $c = 15$, find the area of triangle ABC.

7. Given angle $A = 44°$, angle $B = 82°$, and side $c = 25$, find the area of triangle ABC.

8. Given side $a = 9$, side $b = 12$, and side $c = 6$, find the area of triangle ABC.

9. A vector has a magnitude of 75 and direction of 320°. Write an equivalent vector in the form $\mathbf{v} = a_1\mathbf{i} + a_2\mathbf{j}$.

10. Find the magnitude and direction of the vector $\langle 12, -7 \rangle$.

11. Find a unit vector $\mathbf{u}$ in the direction of $\mathbf{v} = 6\mathbf{i} - 8\mathbf{j}$.

12. Find $-5\mathbf{u} + 8\mathbf{v}$ given the vectors $\mathbf{u} = \langle -4, 5 \rangle$ and $\mathbf{v} = \langle 6, -3 \rangle$.

13. Find the horizontal and vertical components of the vector of magnitude 32 and direction angle 285°. Write the answer in the form $a_1\mathbf{i} + a_2\mathbf{j}$.

14. Find the dot product of $\mathbf{u} = -3\mathbf{i} - 8\mathbf{j}$ and $\mathbf{v} = -4\mathbf{i} + 6\mathbf{j}$.

15. Given $\mathbf{v} = 4\mathbf{i} - 8\mathbf{j}$, and $\mathbf{w} = \mathbf{i} + \mathbf{j}$, find $\text{proj}_{\mathbf{w}}\,\mathbf{v}$.

16. Find the angle to the nearest degree between the vectors $\mathbf{u} = \langle 3, 5 \rangle$ and $\mathbf{v} = \langle -6, 2 \rangle$.

17. Find the area of a parallelogram whose sides are 32 inches and 60 inches and the shortest diagonal is 45 inches.

18. A motorist drove 55 miles at a heading of 112°. The motorist then drove 35 miles at a heading of 202°. Find the distance of the motorist from the starting point.

19. Find the number of acres in a pasture whose shape is a triangle measuring 560 yards by 800 yards by 640 yards. (An acre is 4840 square yards.)

20. Two fire lookouts are located 15 miles apart. Lookout A is at a bearing of N35°W of lookout B. A fire was sighted at a bearing of N55°E from A and N30°E from B. Find the distance of the fire from lookout A.

1. Solve triangle ABC if $A = 53°$, $B = 79°$, and $c = 15$.

   a. $C = 48°$, $a = 13$, $b = 22$
   b. $C = 48°$, $a = 16$, $b = 20$
   c. $C = 48°$, $a = 20$ $b = 16$
   d. $C = 48°$, $a = 13$, $b = 20$

2. Find B in triangle ABC if $A = 34°$, $b = 16$, and $a = 19$.

   a. 40°, 140°,
   b. 28°, 152°
   c. 28°
   d. No solution

3. In triangle ABC, $B = 44°$, $a = 29$, and $c = 17$. Find side b.

   a. 39
   b. 420
   c. 150
   d. 21

4. In triangle ABC, $b = 30$, $a = 23$, and $c = 18$. Find angle A.

   a. 129°
   b. 29°
   c. 52°
   d. 151°

5. In triangle ABC, $B = 38°$, $a = 15$, and $b = 15$. Find side c.

   a. 20
   b. No solution
   c. 18
   d. 22

6. Given angle $A = 52°$, side $c = 96$, and side $b = 72$, find the area of triangle ABC.

   a. 3560
   b. 2700
   c. 3600
   d. 2840

7. Given angle $A = 64°$, angle $C = 23°$, and side $b = 15$, find the area of triangle ABC.

   a. 35
   b. 52
   c. 40
   d. 46

8. Given side $a = 23$, side $b = 29$, and side $c = 12$, find the area of triangle ABC.

   a. 260
   b. 130
   c. 78
   d. 147

9. A vector has a magnitude of 15 and direction 230°. Write an equivalent vector in the form $\mathbf{v} = a_1\mathbf{i} + a_2\mathbf{j}$.

   a. $\mathbf{v} = 9.6\mathbf{i} - 11\mathbf{j}$
   b. $\mathbf{v} = -9.6\mathbf{i} + 11\mathbf{j}$
   c. $\mathbf{v} = -9.6\mathbf{i} - 11\mathbf{j}$
   c. $\mathbf{v} = 9.6\mathbf{i} + 11\mathbf{j}$

10. Find the magnitude and direction of the vector $\mathbf{v} = \langle -6, 8 \rangle$.

a. 10, 126° b. 10, 53°c. c. $2\sqrt{7}$, 126°d. d. $2\sqrt{7}$, 53°

11. Find the unit vector $\mathbf{u}$ in the direction $\mathbf{v} = 5\mathbf{i} + 2\mathbf{j}$.

a. $\mathbf{u} = \frac{1}{5}\mathbf{i} + \frac{1}{2}\mathbf{j}$

b. $\mathbf{u} = 5\sqrt{2}\,\mathbf{i} + 2\sqrt{2}\,\mathbf{j}$

c. $\mathbf{u} = \frac{5\sqrt{29}}{29}\mathbf{i} + \frac{2\sqrt{29}}{29}\mathbf{j}$

d. $\mathbf{u} = \frac{5}{29}\mathbf{i} + \frac{2}{29}\mathbf{j}$

12. Find $3\mathbf{u} - 7\mathbf{v}$ given the vectors $\mathbf{u} = 5\mathbf{i} + 11\mathbf{j}$ and $\mathbf{v} = 2\mathbf{i} - 3\mathbf{j}$.

a. $\mathbf{i} + 12\mathbf{j}$ b. $\mathbf{i} + 30\mathbf{j}$ c. $\mathbf{i} + 54\mathbf{j}$ d. $13\mathbf{i} + 30\mathbf{j}$

13. Find the horizontal and vertical components of the vector of magnitude 12 and direction angle 145°. Write the answer in the form $a_1\mathbf{i} + a_2\mathbf{j}$.

a.. $-\sqrt{2}\,\mathbf{i} + \sqrt{2}\,\mathbf{j}$ b. $-9.8\mathbf{i} + 6.9\mathbf{j}$ c. $6.9\mathbf{i} - 9.8\mathbf{j}$ d. $-6.9\mathbf{i} + 9.8\mathbf{j}$

14. Find the dot product of $\mathbf{u} = 2\mathbf{i} - 7\mathbf{j}$ and $\mathbf{v} = \mathbf{i} + 8\mathbf{j}$.

a. -54 b. -13 c. 58 d. -12

15. Given $\mathbf{v} = 8\mathbf{i} + 12\mathbf{j}$ and $\mathbf{w} = 2\mathbf{i} + 5\mathbf{j}$, find $\text{proj}_{\mathbf{w}}\,\mathbf{v}$.

a. $\frac{44\sqrt{29}}{29}$ b. $\frac{76\sqrt{29}}{29}$ c. $\frac{76\sqrt{7}}{7}$ d. $\frac{19\sqrt{13}}{13}$

16. Find the angle between the vectors $\mathbf{u} = \langle 3, -1 \rangle$ and $\mathbf{v} = \langle -4, 11 \rangle$.

a. 38° b. 128° c. 142° d. 52°

17. Find the area of a parallelogram whose sides are 16 feet and 24 feet and the longer diagonal is 20 feet.

a. 158.7 ft^2 b. 317.5 ft^2 c. 210.6 ft^2 d. 188.4 ft^2

18. A 180-pound box is dragged 12 feet along a level floor. Find the work done if a force of 80-pounds at an angle If 32° is used.

a. 12,211 ft-lb b. 960 ft-lb c. 509 ft-lb d. 814 ft-lb

19. Two ships left a port at the same time. One ship traveled at a speed of 16 mph at a heading of 55°. The other ship traveled at a speed of 20 mph at a heading of 220°. Find the distance between the two ships after 4 hours of travel.

a. 124 mi b. 143 mi c. 155 mi d. 139 mi

20. Find the area of a parallelogram with sides of 12 ft and 16 ft and with one angle of 66°.

a. 170 ft^2 b. 120 ft^2 c. 88 ft^2 d. 60 ft^2

1. Solve triangle ABC if A = 82°, B = 70°, and c = 23.

   a. C = 28°, a = 49, b = 46
   b. C = 28°, a = 49, b = 22
   c. C = 28°, a = 32, b = 46
   d. C = 28°, a = 32, b = 46

2. Find B in triangle ABC if A = 23°, b = 19, and a = 11.

   a. no solution
   b. 42°
   c. 42°, 138°
   d. 35°

3. In triangle ABC, B = 86°, a = 18, and c = 12. Find side b.

   a. 22
   b. 21
   c. 438
   d. 19

4. In triangle ABC, a = 52, b = 17, and c = 40. Find angle A.

   a. 90°
   b. 127°
   c. 53°
   d. 37°

5. In triangle ABC, C = 42°, b = 27, and c = 18. Find side a.

   a. 18
   b. 22
   c. 25
   d. No solution

6. Given triangle C = 78°, side a = 49, and side b = 56, find the area of triangle ABC.

   a. 1340
   b. 571
   c. 285
   d. 2680

7. Given angle B = 35°, angle C = 72°, and side b = 18, find the area of triangle ABC.

   a. 214
   b. 257
   c. 305
   d. 269

8. Given side a = 22, side b = 30, and side c = 36, find the area of triangle ABC.

   a. 420
   b. 329
   c. 382
   d. 297

9. A vector has a magnitude of 26 and direction 310°. Write an equivalent vector in the form $\mathbf{v} = a_1\mathbf{i} + a_2\mathbf{j}$.

   a. **v** = 279**i** + 196**j**
   b. **v** = 7**i** - 20**j**
   c. **v** = 17**i** - 20**j**
   d. **v** = 20**i** - 17**j**

10. Find the magnitude and direction of the vector $\langle 4, -8 \rangle$.

a. $4\sqrt{5}$, 296.6° b. $4\sqrt{3}$, 296.6° c. $4\sqrt{5}$, 63.4° d. $4\sqrt{3}$, 63.4°

11. Find a unit vector **u** in the direction **v** = $\langle -9, -12 \rangle$.

a. $\mathbf{u} = \langle \frac{-9\sqrt{21}}{21}, \frac{-12\sqrt{21}}{21} \rangle$

b. $\mathbf{u} = \langle \frac{-3\sqrt{63}}{63}, \frac{-4\sqrt{63}}{63} \rangle$

c. $\mathbf{u} = \langle \frac{-3}{5}, \frac{-4}{5} \rangle$

d. $\mathbf{u} = \langle \frac{3}{5}, \frac{4}{5} \rangle$

12. Find 3**u** - 7**v** given the vectors **u** = 5**i** + 11**j** and **v** = 2**i** - 3**j**.

a. **i** + 12**j** b. **i** + 30**j** c. **i** + 54**j** d. 13**i** + 30**j**

13. Find the horizontal and vertical components of the vector of magnitude 40 and direction angle 300°. Write the answer in the form $a_1\mathbf{i} + a_2\mathbf{j}$.

a. $20\mathbf{i} - 20\sqrt{3}\,\mathbf{j}$ b. $-20\mathbf{i} + 20\mathbf{j}$ c. $20\sqrt{3}\,\mathbf{i} - 20\mathbf{j}$ d. $-\sqrt{3}\,\mathbf{i} + \mathbf{j}$

14. Find the dot product of **u** = $\langle 4, 1 \rangle$ and **v** = $\langle -2, 5 \rangle$.

a. 13 b. 3 c. - 3 d. - 13

15. Given **v** = −3**i** + 4**j** and **w** = −2**i** + 3**j**, find $\text{prog}_{\mathbf{w}}\,\mathbf{v}$.

a. $\frac{18\sqrt{5}}{5}$ b. $\frac{18}{5}$ c. $\frac{6\sqrt{13}}{13}$ d. $\frac{18\sqrt{13}}{13}$

16. Find the angle to the nearest degree between the vectors **u** = 5**i** + 8**j** and **v** = -2**i** + 9**j**.

a. 45° b. 135° c. 71° d. 19°

17. Find the area of a parallelogram whose sides are 18 meters and 40 meters and the longer diagonal is 52 meters.

a. 302.6 m^2 b. 605.2 m^2 c. 455.8 m^2 d. 408.4 m^2

18. A motorist drove 52 miles at a heading of 152°. The motorist then drove 90 miles at a heading of 242°. Find the distance of the motorist from the starting point.

a. 142 mi b. 104 mi c. 73 mi d. 68 mi

19. A triangular industrial piece of real estate is priced at $7.40 per square foot. Find the cost of the lot measuring 420 feet by 510 feet by 360 feet.

a. $666,298 b. $552,958 c. $77,724 d. $432,560

20. One ship leaves a port at 12:00 noon traveling at 15 mph at a heading of 310°. At 1:00 P.M. another ship leaves the port traveling at 24 mph at a heading of 70°. Find the distance between the ships at 4:00 P.M..

a. 94 mi b. 114 mi c. 124 mi d. 76 mi

1. Solve triangle ABC if A = 35°, B = 78°, and c = 83.

   a. C = 67°, a = 88, b = 52 b. C = 67°, a = 52, b = 88

   c. C = 67°, a = 64, b = 77 d. C = 67°, a = 77, b = 64

2. Find A in triangle ABC if C = 125°, a = 18, and c = 42.

   a. 21° b. 15° c. 28° d. 42°

3. In triangle ABC A = 32°, b = 66, and c = 42. Find side a.

   a. 26 b. 32 c. 38 d. 44

4. In triangle ABC, a = 37, b = 18, and c = 23. Find angle A.

   a. 129° b. 29° c. 52° d. 151°

5. In triangle ABC, A = 72°, c = 32 and a = 30. Find side b.

   a. 28 b. 34 c. 26 d. No solution

6. Given angle C = 85°, side a = 27, and side b = 19, find the area of triangle ABC.

   a. 22 b. 45 c. 256 d. 511

7. Given angle C = 105°, angle B = 19°, and side c = 54, find the area of triangle ABC.

   a. 315 b. 382 c. 512 d. 407

8. Given side a = 10, side b = 14, and side c = 18, find the area of triangle ABC.

   a. 35 b. 70 c. 46 d. 82

9. A vector has a magnitude of 31 and a direction 130°. Write an equivalent vector in the form $\mathbf{v} = a_1\mathbf{i} + a_2\mathbf{j}$.

   a. $\mathbf{v} = -20\mathbf{i} - 24\mathbf{j}$ b. $\mathbf{v} = 24\mathbf{i} - 20\mathbf{j}$

   c. $\mathbf{v} = -20\mathbf{i} + 24\mathbf{j}$ d. $\mathbf{v} = 67\mathbf{i} - 122\mathbf{j}$

10. Find the magnitude and direction of the vector $\langle 8, -15\rangle$.

a. -17, 298° b. 17, 298°c. c. $\sqrt{161}$, 62° d. 17, 62°

11. Find a unit vector **u** in the direction of **v** = $\langle 1, -1\rangle$.

a. $\mathbf{u} = \langle \sqrt{2}, -\sqrt{2} \rangle$ b. $\mathbf{u} = \langle \frac{1}{2}, -\frac{1}{2} \rangle$

c. $\mathbf{u} = \langle \frac{\sqrt{2}}{2}, -\frac{\sqrt{2}}{2} \rangle$ d. $\mathbf{u} = \langle -\frac{\sqrt{2}}{2}, \frac{\sqrt{2}}{2} \rangle$

12. Find 6**u** - 15**v** given the vectors u = 8**i** - 2**j** and v = **i** - 9**j**.

a. 33**i** - 57**j** b. 53**i** - 21**j** c. 53**i** + 33**j** d. 43**i** + 33**j**

13. Find the horizontal and vertical components of the vector of magnitude 2 and direction angle 110°. Write the answer in the form $a_1\mathbf{i} + a_2\mathbf{j}$.

a. 0.68**i** - 1.9**j** b. -0.68**i** + 1.9**j** c. -0.86**i** + 2.8**j** d. -1.9**i** + 0.68**j**

14. Find the dot product of u = i -5j and v =6i + 4j.

a. -14 b. 14 c. -1 d. 26

15. Given **v** = 5**i** - 2**j** and **w** = 3**i** + **j**, find $\text{proj}_{\mathbf{w}}\,\mathbf{v}$.

a. $\frac{5\sqrt{8}}{8}$ b. $\frac{8\sqrt{5}}{5}$ c. $\frac{-8\sqrt{5}}{5}$ d. $\frac{4\sqrt{5}}{5}$

16. Find the angle to the nearest degree between the vectors **u** = $\langle -1, 4\rangle$ and **v** = $\langle 6, 2\rangle$.

a. 4° b. 42° c. 68° d. 86°

17. Find the area of the parallelogram whose sides are 50 inches and 80 inches and the longer diagonal is 110 inches.

a. 1833 in^2 b. 3433 in^2 c. 3660 in^2 d. 2866 in^2

18. A motorist drove 95 miles at a heading of 139°. The motorist then drove 82 miles at a heading of 229°. Find the distance of the motorist from the starting point.

a. 48 mi b. 177 mi c. 64 mi d. 125 mi

19. A plane is flying at an airspeed of 300 mph at a heading of 140°. A wind of 50 mph is blowing from the west. Find the ground speed of the plane.

a. 312 mph b. 318 mph c. 328 mph d. 334 mph

20. A triangular commercial piece of real estate is priced at $8.60 per square foot. Find the cost of the lot measuring 230 feet by 190 feet by 300 feet.

a. $207,678 b. $167,988 c. $187,898 d. $257,788

1. Write $3 + \sqrt{-16}$ in the form a + bi.

2. Simplify $\left(3 + \sqrt{-9}\right) + \left(12 - \sqrt{-64}\right)$. Write the answer in the form a + bi.

3. Simplify: $\left(\sqrt{-32}\right)\left(\sqrt{-2}\right)$

4. Simplify: i^{462}

5. Simplify: $(3 + 5i) - (-2 + 8i)$

6. Simplify: $(-3 - 5i)(2 + 8i)$

7. Simplify: $\frac{3 + 8i}{2i}$

8. Simplify: $\frac{2 - 5i}{2 + i}$

9. Find the absolute value of $2 - 7i$.

10. Write $4 + 4i$ in trigonometric form.

11. Write $5i$ in trigonometric form.

12. Write $3\left(\cos\frac{2\pi}{3} + i \sin\frac{2\pi}{3}\right)$ in standard form.

13. Write $7(\cos 315° + i \sin 315°)$ in standard form.

14. Simplify $5(\cos 120° + i \sin 120°) \cdot 12(\cos 150° + i \sin 150°)$. Write the answer in standard form

15. Simplify $\dfrac{36\left(\cos\frac{13\pi}{12} + i\sin\frac{12\pi}{3}\right)}{12\left(\cos\frac{\pi}{4} + i\sin\frac{\pi}{4}\right)}$. Write the answer in standard form.

16. Simplify $(2(\cos 30° + i\sin 30°))^8$. Write the answer in trigonometric form.

17. Simplify $\left(\sqrt{3} - i\right)^8$. Write the answer in standard form.

18. Find the four fourth roots of i. Write the answers in trigonometric form.

19. Find the three cube roots of 1 - i. Write the answers in standard form.

20. Find the four solutions of $z^4 + 16 = 0$. Write the answers in trigonometric form.

1. Write $4 - \sqrt{121}$ in the form a + bi.

2. Simplify $(2 - \sqrt{-4}) - (16 + \sqrt{-81})$. Write the answer in the form a + bi.

3. Simplify: $(\sqrt{-8})(\sqrt{-4})$

4. Simplify: i^{1050}

5. Simplify: $(-3 + 8i) - (3 - 12i)$

6. Simplify: $(4 - 8i)(2 + 3i)$

7. Simplify: $\frac{2 - 2i}{i}$

8. Simplify: $\frac{2 - 5i}{1 + i}$

9. Find the absolute value of $8 - i$.

10. Write $\sqrt{3} - i$ in trigonometric form.

11. Write $13i$ in trigonometric form.

12. Write $5(\cos 240° + i \sin 240°)$ in standard form.

13. Write $14(\cos \frac{5\pi}{4} + i \sin \frac{5\pi}{4})$ in standard form.

14, Simplify $2(\cos 135° + i \sin 135°) \cdot 8(\cos 15° + i \sin 15°)$. Write the answer in standard form.

15. Simplify $\frac{75(\cos 60° + i \sin 60°)}{5(\cos 270° + i \sin 270°)}$. Write the answer in standard form.

16. Simplify $[4(\cos 25° + i \sin 25°)]^4$. Write the answer in trigonometric form.

17. Simplify $(\sqrt{3} + i)^6$. Write the answer in standard form.

18. Find the four fourth roots of i. Write the answers in trigonometric form.

19. Find the three cube roots of i - i. Write the answer in standard form.

20. Find the three solutions of $x^3 - 125 = 0$. Leave the answers in trigonometric form.

Trigonometry Chapter 5 Tests Form C

1. Write $8 - \sqrt{-144}$ in the form a + bi.

2. Simplify $(5 - \sqrt{-49}) - (2 + \sqrt{-4})$. Write the answer in the form a + bi.

3. Simplify: $(\sqrt{-125})(\sqrt{-5})$

4. Simplify: i^{677}

5. Simplify: Simplify: $(4 - i) - (-3 + 11i)$

6. Simplify: $(8 + 3i)(1 - 6i)$

7. Simplify: $\frac{4 - 3i}{3i}$

8. Simplify: $\frac{3 - i}{2 - 7i}$

9. Find the absolute value of $-9 + 2i$.

10. Write $3 + 3i$ in trigonometric form.

11. Write $-2i$ in trigonometric form.

12. Write $7(\cos 330° + i \sin 330°)$ in standard form.

13. Write $11(\cos\frac{3\pi}{4} + i \sin\frac{3\pi}{4})$ in standard form.

14. Simplify $8(\cos 42° + i \sin 42°) \cdot 3(\cos 14° + i \sin 14°)$. Write the answer in trigonometric form.

15. Simplify $\frac{33(\cos 210° + i \sin 210°)}{11(\cos 60° + i \sin 60°)}$. Write the answer in standard form.

16. Simplify $[3(\cos 60° + i \sin 60°)]^5$. Write the answer in standard form.

17. Simplify $(1 - i)^8$. Write the answer in trigonometric form.

18. Find the two square roots of $-25i$. Write the answers in standard form.

19. Find the four fourth roots of $\sqrt{3} - i$. Leave the answers in trigonometric form.

20. Find the three solutions of $x^3 + 25 = 0$. Write the answers in standard form.

Trigonometry Chapter 5 Tests Form D

1. Write $-3 - \sqrt{-8}$ as a complex number in standard form.

   a. $-3 - 2i\sqrt{2}$ b. $-3 + 2i\sqrt{2}$ c. $-3 - 8i$ d. $-3 + 8i$

2. Simplify: $\left(\sqrt{-8}\right)\left(\sqrt{-18}\right)$

   a. 12 b. $-12i$ c. -12 d. $12i$

3. Simplify: i^{962}

   a. i b. $-i$ c. 1 d. -1

4. Simplify: $(-5 + 2i) - (4 + 7i)$

   a. $-9 + 9i$ b. $-1 - 5i$ c. $-9 - 5i$ d. $-1 + 5i$

5. Simplify: $-2i(3 + 7i)$

   a. $-14 - 6i$ b. $14 - 6i$ c. $-14 + 6i$ d. -14

6. Simplify: $(2 - 3i)(-2 + 4i)$

   a. $8 - 14i$ b. $-12 + 14i$ c. $-12 - 14i$ d. $8 + 2i$

7. Simplify: $\frac{3 - 5i}{2i}$

   a. $\frac{3}{2} - \frac{5}{2}i$ b. $-\frac{5}{2} + \frac{3}{2}i$ c. $-\frac{3}{2} - \frac{5}{2}i$ d. $-\frac{5}{2} - \frac{3}{2}i$

8. Simplify: $\frac{4 - i}{2 + i}$

   a. $-\frac{7}{5} + \frac{6}{5}i$ b. $\frac{7}{5} - \frac{6}{5}i$ c. $\frac{7}{3} - 2i$ d. $-\frac{7}{3} + 2i$

9. Find the absolute value of $8 - 3i$.

   a. 5 b. $\sqrt{11}$ c. $\sqrt{73}$ d. 73

10. Write −3 in trigonometric form.

a. −3(cos 180° + i sin 180°)
b. 3(cos 90° + i sin 90°)
c. 3(cos 180° + i sin 180°)
d. 3(cos 0° + i sin 0°)

11. Write $1 - i\sqrt{3}$ in trigonometric form.

a. 2(cos 240° + i sin 240°)
b. 2(cos 300° + i sin 300°)
c. $\sqrt{3}$ (cos 300° + i sin 300°)
d. 2(cos 60° + i sin 60°)

12. Write $5(\cos\frac{5\pi}{3} + i\sin\frac{5\pi}{3})$ in standard form.

a. $\frac{5}{2} - \frac{5\sqrt{3}}{2}i$
b. $-\frac{5}{2} + \frac{5\sqrt{3}}{2}i$
c. $\frac{5\sqrt{3}}{2} - \frac{i}{2}$
d. $-5 + 5i\sqrt{3}$

13. Write 12(cos 330° + i sin 330°) in standard form.

a. $-6\sqrt{3} + 6i$
b. $12\sqrt{3} - 12i$
c. $6\sqrt{3} + 6i$
d. $6\sqrt{3} - 6i$

14. Simplify 3(cos 82° + i sin 82°) · 12(cos 112° + i sin 112°). Write the answer in trigonometric form.

a. 36(cos194° + i sin 194°)
b. 15(cos 194° + i sin 194°)
c. 36(cos 30° + i sin 30°)
d. 15(cos 30° + i sin 30°)

15. Simplify $\frac{24(\cos 180° + i\sin 180°)}{48(\cos 300° + i\sin 300°)}$. Write the answer in trigonometric form.

a. $\frac{1}{2}$(cos 120° + i sin 120°)
b. 2(cos 120° − i sin 120°)
c. $\frac{1}{2}$(cos 120° − i sin 120°)
d. $\frac{1}{2}$(cos 240° + i sin 240°)

16. Simplify $\sqrt{2}(\cos 135° + i\sin 135°)^5$. Write the answer in standard form.

a. $4\sqrt{2} - 4i\sqrt{2}$
b. $32\sqrt{2} - 32i\sqrt{2}$
c. 4 − 4i
d. $-4\sqrt{2} - 4i\sqrt{2}$

17. Simplify $\left(2\sqrt{3} - 2i\right)^5$. Write the answer in standard form.

a. $1024\sqrt{3} - 1024\ i$

b. $512\sqrt{3} - 512\ i$

c. $1024\sqrt{3} + 1024\ i$

d. $-512\sqrt{3} - 512\ i$

18. Find the four forth roots of 16. Write the answer in standard form.

a. $-2,\ 2,\ -2i,\ 2i$

b. $-2,\ 2,\ 1 + i,\ 1 - i$

c. $-2,\ 2,\ 2 + 2i,\ 2 - 2i$

d. $-2,\ 2,\ -2,\ + 2i,\ -2,\ -2i$

19. Find the three cube roots of 2 + 2i. Write the answer in trigonometric form.

a. $\sqrt{2}\,(\cos 15° + i \sin 15°)$
$\sqrt{2}\,(\cos 120° + i \sin 120°)$
$\sqrt{2}\,(\cos 255° + i \sin 255°)$

b. $\sqrt[3]{2}\,(\cos 15° + i \sin 15°)$
$\sqrt[3]{2}\,(\cos 120° + i \sin 120°)$
$\sqrt[3]{2}\ \text{cis}\ 255° + i \sin 255°)$

c. $2(\cos 45° + i \sin 45°)$
$2(\cos 175° + i \sin 175°)$
$2(\cos 295° + i \sin 295°)$

d. $\sqrt{2}\,(\cos 30° + i \sin 30°)$
$\sqrt{2}\,(\cos 150° + i \sin 150°)$
$\sqrt{2}\,(\cos 270° + i \sin 270°)$

20. Find the three solutions of $z^3 + 8 = 0$. write the answer in standard form.

a. $-2,\ 2 + 2i\sqrt{3},\ 2 - 2i\sqrt{3}$

b. $-2,\ 1 + i\sqrt{3},\ 1 - i\sqrt{3}$

c. $-2,\ -1 + i\sqrt{3},\ -1 - i\sqrt{3}$

d. $-2,\ 1 + i\sqrt{3},\ -1,\ -i\sqrt{3}$

1. Write $\sqrt{-36} + 11$ in the form a + bi.

a. 11 + 36i b. 11 - 36i c. 11 + 6i d. 11 - 6i

2. Simplify: $(\sqrt{-15})(\sqrt{-15})$

a. 15 b. $\sqrt{15}$ c. -15 d $-\sqrt{15}$

3. Simplify: i^{257}

a. i b. - i c. 1 d. - 1

4. Simplify: (12 - 3i) + (-11 + 5i)

a. 1 - 8i b. 1 + 2i c. -132 - 15i d. 1 - 2i

5. Simplify: 4i(7 - 3i)

a. 12 - 28i b. -12 + 28i c. 40i d. 12 + 28i

6. Simplify: (9 + 2i)(-1 + i)

a. -11 b. -11 - 7i c. -11 + 7i d. 11 + 7i

7. Simplify: $\frac{5 - 8i}{5i}$

a. $-\frac{8}{5} - i$ b. $1 - \frac{8}{3}i$ c. $\frac{8}{5} + i$ d. $-1 + \frac{8}{5}i$

8. Simplify: $\frac{9 - i}{2 + 3i}$

a. $\frac{15}{13} + \frac{23}{13}i$ b. $\frac{21}{13} - \frac{29}{13}i$ c. $-3 + \frac{23}{5}i$ d. $\frac{15}{13} - \frac{29}{13}i$

9. Find the absolute value of 6 - 5i.

a. 11 b. $\sqrt{11}$ c. $\sqrt{61}$ d. 1

10. Write 6 in trigonometric form.

a. $6(\cos 0° = i \sin 0°)$ b. $6(\cos 90° + i \sin 90°)$

c. $6(\cos 45° \ i \sin 45°)$ d. $6(\cos 180° + i \sin 180°)$

11. Write $-\sqrt{3} - i$ in trigonometric form.

a. $4(\cos 120° + i \sin 120°)$ b. $2(\cos 210° i \sin 210°)$

c. $2(\cos 240° + i \sin 240°)$ d. $2(\cos(210° \ i \sin 210°)$

12. Write $21(\cos \frac{\pi}{4} + i \sin \frac{\pi}{4})$ in standard form.

a. $\frac{21\sqrt{2}}{2} + \frac{21\sqrt{2}}{2}i$ b. $21i$

c. -21 d. $-\frac{21\sqrt{2}}{2} + \frac{21\sqrt{2}}{2}i$

13. Write $22(\cos 240° + i \sin 240°)$ in standard form.

a. $-11 + 11i\sqrt{3}$ b. $11\sqrt{3} - 11i$

c. $-11 - 11i\sqrt{3}$ d. $-11\sqrt{3} - 11i$

14. Simplify $6(\cos 32° + i \sin 32°) \cdot (5(\cos 4° + i \sin 4°)$. Write the answer in trigonometric form.

a. $30(\cos 128° \ i \sin 128°)$ b. $11(\cos 28° + i \sin 28°)$

c. $11(\cos 36° + i \sin 36°)$ d. $30(\cos 36° + i \sin 36°)$

15. Simplify $\frac{100(\cos 240° + i \sin 240°)}{20(\cos 60° + i \sin 60°)}$. Write the answer in standard form.

a. -5 b. $-5 + i$ c. -80 d. $4.99 + 0.35i$

16. Simplify $[7(\cos 16° + i \sin 16°)]^3$. Write the answer in trigonometric form .

a. $21(\cos 48° + i \sin 48°)$ b. $21(\cos 4096° + i \sin 4096°)$

c. $343(\cos 48° + i \sin 48°)$ d. $343(\cos 4096° + i \sin 4096°)$

17. Simplify $(\sqrt{3} - i)^4$. Write the answer in standard form.

a. $8 + 8i\sqrt{3}$ b. $-8 - 8i\sqrt{3}$ c. $-16 - 16i\sqrt{3}$ d. $-8 + 8i\sqrt{3}$

18. Find the two square roots of -100i. Write the answer in standard form.

a. $-5\sqrt{2} + 5i\sqrt{2}$, $-5\sqrt{2} - 5i\sqrt{2}$ b. $-5\sqrt{2} + 5i\sqrt{2}$, $5\sqrt{2} + 5i\sqrt{2}$

c. $-5\sqrt{2} + 5i\sqrt{2}$, $5\sqrt{2} - 5i\sqrt{2}$ d. $-5\sqrt{2} - 5i\sqrt{2}$, $5\sqrt{2} + 5i\sqrt{2}$

19. Find the three cube roots of -1 - i. Write the answer in trigonometric form.

a. $\sqrt{2}(\cos 75° + i \sin 75°$
$\sqrt{2}(\cos 195° + i \sin 195°)$
$\sqrt{2}(\cos 315° + i \sin 315°)$

b. $\sqrt[6]{2}(\cos 75° + i \sin 75°)$
$\sqrt[6]{2}(\cos 75° + i \sin 75°)$
$\sqrt[6]{2}(\cos 315° + i \sin 315°$

20. Find the three solutions of $z^3 + 100 = 0$. Write the answers in trigonometric form.

a. $\sqrt[3]{100}(\cos 60° + i \sin 60°)$
$\sqrt[3]{100}(\cos 150° + i \sin 150°)$
$\sqrt[3]{100}(\cos 240° + i \sin 240°)$

b. $\sqrt[3]{100}(\cos 90° + i \sin 90°)$
$\sqrt[3]{100}(\cos 210° + i \sin 210°)$
$\sqrt[3]{100}(\cos 330° + i \sin 330°)$

c. $10(\cos 60° \; i \sin 60°)$
$10(\cos 180° + i \sin 180°)$
$10(\cos 300° + i \sin 300°)$

d. $\sqrt[3]{100}(\cos 60° + i \sin 60°)$
$\sqrt[3]{100}(\cos 180° + i \sin 180°)$
$\sqrt[3]{100}(\cos 300° + i \sin 300°)$

1. Write $\sqrt{-16} + 15$ in the form a + bi.

   a. $15 - 4i$ b. $15 - 16i$ c. $15 + 4i$ d. $15 + 16i$

2. Simplify: $(\sqrt{-5})(\sqrt{-20})$

   a. 10 b. -10 c. 100 d. -100

3. Simplify: i^{236}

   a. -1 b. 1 c. i d. $-i$

4. Simplify $(8 + 7i) + (-2 + 9i)$. Write the answer in the form a + bi.

   a. $6 + 16i$ b. $10 + 16i$ c. $-16 + 63i$ d. $6 - 2i$

5. Simplify: $-5i(3 + 11i)$

   a. $-55 - 15i$ b. $55 + 15i$ c. $55 - 15i$ d. $-26i$

6. Simplify: $(8 + i)(-4 - 6i)$

   a. $-38 - 44i$ b. $-26 - 44i$ c. $-38 + 53i$ d. $-26 - 52i$

7. Simplify: $\frac{8 - 9i}{4i}$

   a. $2 - \frac{9}{4}i$ b. $-\frac{9}{4} - 2i$ c. $\frac{9}{4} + 2i$ d. $\frac{9}{4} - 2i$

8. Simplify: $\frac{5 + i}{1 + 3i}$

   a. $\frac{4}{5} - \frac{7}{5}i$ b. $\frac{1}{5} - \frac{7}{5}i$ c. $\frac{4}{5} - \frac{8}{5}i$ d. $-1 + \frac{7}{4}i$

9. Find the absolute value of $-7 + 2i$.

   a. 53 b. $\sqrt{53}$ c. $\sqrt{5}$ d. $3\sqrt{5}$

10. Write –11 in trigonometric form.

a. 11(cos 180° + i sin 180°) b. 11(cos 0° + i sin 0°)
c. 11(cos 90° + i sin 90°) d. 11(cos 45° + i sin 45°)

11. Write $-2\sqrt{2} - 2i\sqrt{2}$ in trigonometric form.

a. 8(cos 225° + i sin 225°) b. 4(cos 225° + i sin 225°)
c. 4(cos 225° – i sin 225°) d. 4(cos 210° + i sin 210°)

12. Write $17\left(\cos\frac{7\pi}{4} + i\sin\frac{7\pi}{4}\right)$ in standard form.

a. $\frac{17\sqrt{2}}{2} + \frac{17i\sqrt{2}}{2}$ b. $-\frac{17\sqrt{2}}{2} + \frac{17i\sqrt{2}}{2}$
c. $\frac{17\sqrt{2}}{2} - \frac{17i\sqrt{2}}{2}$ d. $-17i$

13. Write 24(cos 150° + i sin 150°) in standard form.

a. $-12 - 12i\sqrt{3}$ b. $-12 + 12i\sqrt{3}$ c. $-12\sqrt{3} - 12i$ d. $-12\sqrt{3} + 12i$

14. Simplify 11(cos 52° + i sin 52°) • 4(cos 4° + i sin 4°). Write the answer in trigonometric form.

a. 15(cos 56° + i sin 56°) b. 44(cos 56° + i sin 56°)
c. 15(cos 208° + i sin 208°) d. 44(cos 208° + i sin 208°)

15. Simplify $\frac{85(\cos 35° + i\sin 35°)}{17(\cos 80° + i\sin 80°)}$. Write the answer in standard form.

a. $-\frac{5\sqrt{23}}{2} - \frac{5i\sqrt{2}}{2}$ b. $\frac{5\sqrt{2}}{2} - \frac{5i\sqrt{2}}{2}$
c. $-34\sqrt{2} - 34i\sqrt{2}$ d. $-34\sqrt{2} + 34i\sqrt{2}$

16. Simplify $[9(\cos 26° + i \sin 26°)]^2$. Write the answer in trigonometric form.

a. $18(\cos 52° + i \sin 52°)$
b. $81(\cos 52° + i \sin 52°)$
c. $81(\cos 676° + i \sin 676°)$
d. $18(\cos 676° + i \sin 676°)$

17. Simplify $(-1 + i)^8$. Write the answer in standard form.

a. 16
b. 16i
c. –16
d. –16i

18. Find the two square roots of –144i. Write the answer in standard form.

a. $-6\sqrt{2} + 6i\sqrt{2}, -6\sqrt{2} - 6i\sqrt{2}$
b. $-6\sqrt{2} - 6i\sqrt{2}, 6\sqrt{2} + 6i\sqrt{2}$
c. $6\sqrt{2} - 6i\sqrt{2}, -6\sqrt{2} - 6i\sqrt{2}$
d. $-6\sqrt{2} + 6i\sqrt{2}, 6\sqrt{2} - 6i\sqrt{2}$

19. Find the three cube roots of $-\sqrt{3} + i$. Write the answers in trigonometric form.

a. $\sqrt[3]{2}(\cos 50° + i \sin 50°)$
$\sqrt[3]{2}(\cos 170° + i \sin 170°)$
$\sqrt[3]{2}(\cos 290° + i \sin 290°)$

b. $\sqrt[3]{2}(\cos 30° + i \sin 30°)$
$\sqrt[3]{2}(\cos 150° + i \sin 150°)$
$\sqrt[3]{2}(\cos 290° + i \sin 290°)$

c. $\sqrt{3}(\cos 50° + i \sin 50°)$
$\sqrt{3}(\cos 170° + i \sin 170°)$
$\sqrt{3}(\cos 290° + i \sin 290°)$

d. $\sqrt[3]{2}(\cos 40° + i \sin 40°)$
$\sqrt[3]{2}(\cos 160° + i \sin 160°)$
$\sqrt[3]{2}(\cos 280° + i \sin 280°)$

20. Find the three solutions of $z^3 + 27 = 0$. Write the answers in standard form.

a. $\frac{3}{2} + \frac{3\sqrt{3}}{2}i$
-3
$\frac{3}{2} - \frac{3\sqrt{3}}{2}i$

b. $3 + 3i\sqrt{3}$
-3
$3 - 3i\sqrt{3}$

c. $\frac{1}{2} + \frac{\sqrt{3}}{2}i$
-3
$\frac{1}{2} - \frac{\sqrt{3}}{2}i$

d. $\frac{\sqrt{3}}{2} + \frac{3}{2}i$
-3
$\frac{\sqrt{3}}{2} - \frac{3}{2}i$

Trigonometry Chapter 6 Form A

1. Find the vertex, focus, and directrix of the parabola given by the equation $x^2 - 10x - 8y + 33 = 0$.

2. Find the vertices and foci of the ellipse given by the equation $4x^2 + 9y^2 - 24x + 18y + 9 = 0$.

3. Find the vertices, foci, and asymptotes of the hyperbola given by the equation $4y^2 + 40y - 25x^2 + 150x - 225 = 0$.

4. Find the vertex and vertices given by the conic $9x^2 + y^2 - 72x + 10y + 160 = 0$.

5. Find the equation of the parabola with directrix $x = 5$ and focus $(11, -7)$.

6. Find the equation of the ellipse with center at $(3, 8)$, foci $(3 + \sqrt{11}, 8)$, $(3 - \sqrt{11}, 8)$, and minor axis of length 10.

7. Find the equation of the hyperbola with vertices at $(0, 3)$ and $(-4, 3)$ and foci $(-2 + \sqrt{15}, 3)$, $(-2 - \sqrt{15}, 3)$.

8. Let $P(-2, 5)$ be a point in an xy-coordinate system. Find the coordinates of P in an x'y'-coordinate system that has been rotated 60°.

9. For the equation $x^2 + 2\sqrt{3}\,xy + 3y^2 + 8x - 8\sqrt{3} = 0$, determine what angle of rotation would eliminate the xy-term.

10. Graph $r = 4 \sin \theta$.

11. Graph $r = 3 + 3 \cos \theta$.

12. Graph $r = 6 \sin 3\theta$.

13. Graph $r = 3 \csc \theta$.

14. Graph $4 = \frac{6}{1 + \cos \theta}$.

15. Graph $r = \frac{4}{4 - \cos\theta}$.

16. Find the rectangular coordinates of the point whose polar coordinates are $\left(3, \frac{5\pi}{4}\right)$.

17. Write $x^2 - \frac{y^2}{4} = 1$ as an equation in polar coordinates.

18. Write $r = \frac{6}{2 - \cos\theta}$ as an equation in rectangular coordinates.

19. Eliminate the parameter and sketch the curve given by the parametric equations $x = t + 2$, $y = t^2$.

20. Eliminate the parameter and sketch the curve given by the parametric equations $x = 2\sin\theta$, $y = \cos\theta - 2$, where $0 \le \theta \le 2\pi$.

1. Find the vertex, focus, and directrix of the parabola given by the equation $x + \frac{1}{2}y = 0$.

2. Find the vertex, focus, and directrix of the parabola given by the equation $4x - y^2 + 14y - 37 = 0$.

3. Find the vertices and foci of the ellipse given by the equation $x^2 + y^2 + 16x - 14y + 77 = 0$.

4. Find the vertices, foci, and asymptotes of the hyperbola given by the equation $-x^2 + 4y^2 - 6x - 16y + 3 = 0$.

5. Find the equation in standard form of the parabola with directrix $y = -7$ and focus $(-3, -3)$.

6. Find the equation in standard form of the ellipse with center $(-5, 4)$, foci $\left(-5, 4 + \sqrt{11}\right)$, $\left(-5, 4 - \sqrt{11}\right)$, and minor axis of length 14.

7. Find the equation in standard form of the hyperbola with vertices at $(7, 0)$ and $(7, -8)$ and foci $\left(7, -4 + \sqrt{31}\right)$, $\left(7, -4 - \sqrt{31}\right)$.

8. Let $(-2, 4)$ be a point in an xy–coordinate system. Find the coordinates of P in an x'y'–coordinate system that has been rotated 30°.

9. For the equation $2\sqrt{3}\,x^2 - 2xy + \sqrt{3}\,y^2 + 2x - 5y + 10 = 0$, determine what angle of rotation would eliminate the xy–term.

10. Graph $r = 3 \cos \theta$.

11. Graph $r = 5 - 5 \sin \theta$.

12. Graph $r = 4 \cos 3\theta$.

13. Graph $r = \csc \theta$.

14. Graph $r = \frac{8}{2 - 4 \sin \theta}$.

15. Graph $r = \frac{2}{2 + \cos \theta}$.

16. Find the rectangular coordinates of the point whose polar coordinates are $\left(14, \frac{5\pi}{3}\right)$.

17. Write $\frac{x^2}{4} - \frac{y^2}{9} = 1$ as an equation in polar coordinates.

18. Write $r + r \cos \theta = 2$ as an equation in rectangular coordinates.

19. Eliminate the parameter and sketch the curve given by the parametric equations $x = t - 3$, $y = t^2 - 1$, $t \geq 0$.

20. Eliminate the parameter and sketch the curve given by the parametric equations $x = 1 + \sin \theta$, $y = \cos \theta$, $0 \leq \theta \leq 2\pi$.

Trigonometry Chapter 6 Test Form C

1. Find the vertex, focus, and directrix of the parabola given by the equation $3y^2 = 2x$.

2. Find the vertex, focus, and directrix of the parabola given by the equation $x^2 - 8x + 2y + 16 = 0$.

3. Find the vertices and foci of the ellipse given by the equation $4x^2 + 9y^2 - 16x + 36y + 16 = 0$.

4. Find the vertices, foci, and asymptotes of the hyperbola given by the equation $4y^2 - 25x^2 - 24y - 100x - 164 = 0$.

5. Find the equation in standard form of the parabola with directrix $y = 2$ and focus $(2, 0)$.

6. Find the equation in standard form of the ellipse with center $(2, 0)$, foci $(6, 0)$ and $(-2, 0)$, and minor axis of length 4.

7. Find the equation in standard form of the hyperbola with vertices at $\left(-5, 4 + 2\sqrt{5}, -5, 4 - 2\sqrt{5}\right)$ and foci $(-5, 10)$ and $(-5, -2)$.

8. Let $P(4, -3)$ be a point in an xy-coordinate system. Find the coordinates of P in an x'y'-coordinate system that has been rotated 120°.

9. For the equation $8x^2 + 5xy + 5y^2 - 2x + 4y - 40 = 0$, determine what angle of rotation would eliminate the xy-term.

10. Graph $r = \frac{2}{\sin \theta}$.

11. Graph $r = 2 - 2 \cos \theta$

12. Graph $r = 2 \sin 2\theta$.

13. Graph $r = 4 \sec \theta$.

14. Graph $r = \dfrac{2}{2 - \cos \theta}$.

15. Graph $r = \dfrac{6}{3 + 2 \sin \theta}$.

16. Find the polar coordinates of the point whose rectangular coordinates are $(-2, 3)$

17. Find the rectangular form of $r = 2 \cos \theta$.

18. Find the polar form of $3x + 2y = 4$.

19. Eliminate the parameter and sketch the curve given by the parametric equations $x = t^2 - 2t$, $y = t - 1$, $t \in R$.

20. Eliminate the parameter and sketch the curve given by the parmetric equations $x = 2 \sin \theta$, $y = 4 \cos \theta$, $0 \le \theta \le 2\pi$.

1. Find the vertex, focus, and directrix of the parabola given by the equation $2y^2 = -3x$.

   a. vertex: (0, 0), focus: $\left(\frac{3}{8}, 0\right)$, directrix: $x = -\frac{3}{8}$
   b. vertex: (0, 0), focus: $\left(-\frac{3}{8}, 0\right)$, directrix: $x = \frac{3}{8}$
   c. vertex: (0, 0), focus: (3, 0), directrix: $x = -3$
   d. vertex: (0, 0), focus: (–3, 0), directrix: $x = 3$

2. Find the vertex, focus, and directrix of the parabola given by the equation $x^2 + 8x - 16y + 80 = 0$.

   a. vertex: (4, –4), focus: (4, 0), directrix: $y = -8$
   b. vertex: (–4, 4), focus: (–4, 20), directrix: $y = -12$
   c. vertex: (4, –4), focus: (8, –4), directrix: $x = 0$
   d. vertex: (–4, 4), focus: (–4, 8), directrix: $y = 0$

3. Find the vertices and foci of the ellipse given by the equation $5x^2 + y^2 + 60x + 4y + 159 = 0$.

   a. vertices: (–11, –2), (–1, –2), foci: $(-6 + 2\sqrt{5}, -2)$, $(-6 - 2\sqrt{5}, -2)$
   b. vertices: (6, 7), (6, –3), foci: $(6, 2 + 2\sqrt{5})$, $(6, 2 - 2\sqrt{5})$
   c. vertices: (6, 7), (6, –3), foci: $(6 + 2\sqrt{5}, 2)$, $(6 - 2\sqrt{5}, 2)$
   d. vertices: (–6, 3), (–6, –7), foci: $(-6, -2 + 2\sqrt{5})$, $(-6, -2 - 2\sqrt{5})$

4. Find the vertices and foci of the hyperbola given by the equation $\frac{(x+7)^2}{9} - \frac{(y+1)^2}{1} = 1$.

   a. vertices: (–4, –1), (–10, –1), foci: $(-7 + \sqrt{10}, -1)$, $(-7 - \sqrt{10}, -1)$
   b. vertices: (–7, 2), (–7, –4), foci: $(-7, -1 + \sqrt{10})$, $(-7, -1 - \sqrt{10})$
   c. vertices: (10, 1), (4, 1), foci: $(7 + \sqrt{10}, 1)$, $(7 - \sqrt{10}, 1)$
   d. vertices: (10, 1), (4, 1), foci: $(7 + 2\sqrt{2}, 1)$, $(7 - 2\sqrt{2}, 1)$

5. Find the equation in standard form of the parabola with directrix $x = 7$ and focus (–1, 3).

   a. $(y - 3)^2 + 12(x - 3)$
   b. $(y - 3)^2 = -16(x - 3)$
   c. $(x + 3)^2 = 3(y - 3)$
   d. $(y - 3)^2 = 3(x - 3)$

6. Find the equation in standard form of the ellipse with center at (–3, 6), foci $(-3 + \sqrt{15}, 6)$, $(-3 - \sqrt{15}, 6)$, and minor axis of length 12.

a. $\frac{(x+3)^2}{159} + \frac{(y-6)^2}{144} = 1$

b. $\frac{(x-3)^2}{39} + \frac{(y+6)^2}{24} = 1$

c. $\frac{(x+3)^2}{51} + \frac{(y-6)^2}{36} = 1$

d. $\frac{(x+3)^2}{39} + \frac{(y-6)^2}{24} = 1$

7. Find the equation in standard form of the hyperbola with vertices at (9, 2), (11, 2) and foci $(10 + \sqrt{6}, 2)$, $(10 - \sqrt{6}, 2)$.

a. $\frac{(y-2)^2}{1} - \frac{(x-2)^2}{5} = 1$

b. $\frac{(x+10)^2}{1} - \frac{(y-2)^2}{5} = 1$

c. $\frac{(x-10)^2}{1} - \frac{(y-2)^2}{5} = 1$

d. $\frac{(x+10)^2}{1} - \frac{(y+1)^2}{6} = 1$

8. Let P(–2, 6) be a point in an xy-coordinate system. Find the coordinates of P in an x'y'-coordinate system that has been rotated 45°.

a. $(-\sqrt{2} + 3\sqrt{2}, 3\sqrt{2} + 2)$

b. $(-3\sqrt{2} + \sqrt{2}, \sqrt{2} - 3\sqrt{3})$

c. $(\sqrt{2}, 3\sqrt{2})$

d. $(-3\sqrt{2}, -\sqrt{2})$

9. For the equation $2x^2 - 3xy + 6y^2 + 3x + 2y + 5 = 0$, determine what angle of rotation would eliminate the xy-term.

a. 42°

b. 28.2°

c. 82.2°

d. 24°

10. Graph r = 10.

a.
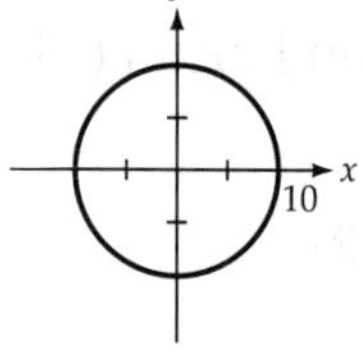

b.
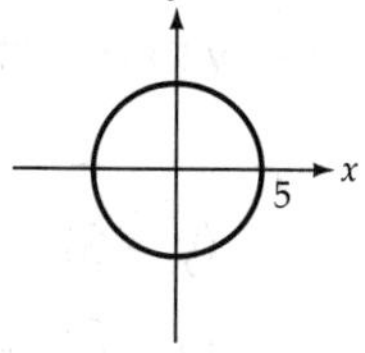

c.
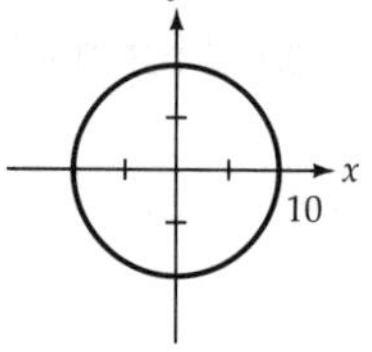

d.
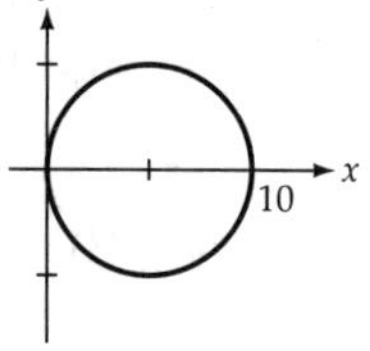

11. Graph $r = 4 + 4\cos\theta$

a. 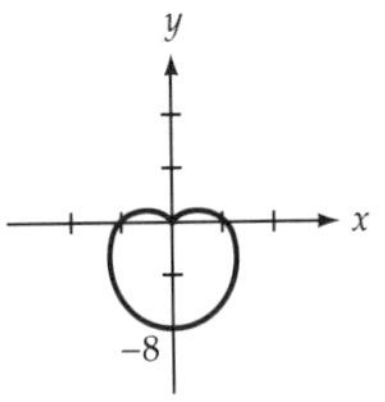

b. 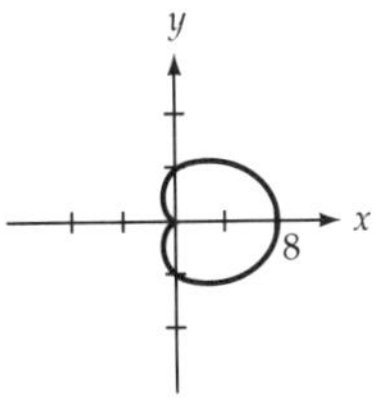

c. 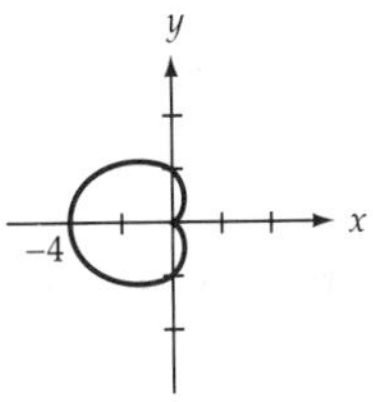

d. 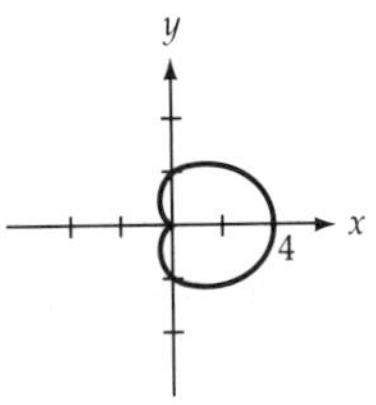

12. Graph $r = -8\sin 3\theta$.

a. 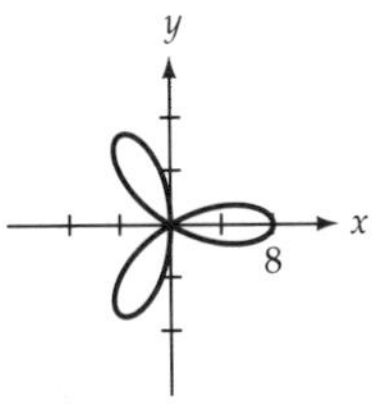

b. 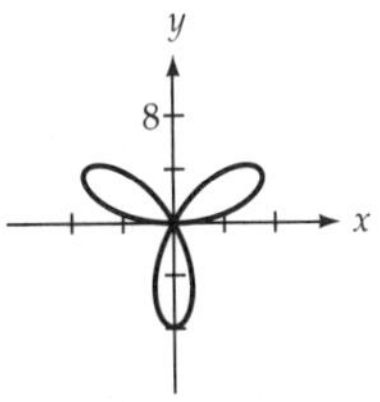

c. 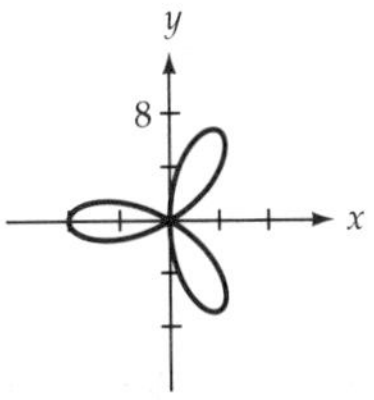

d. 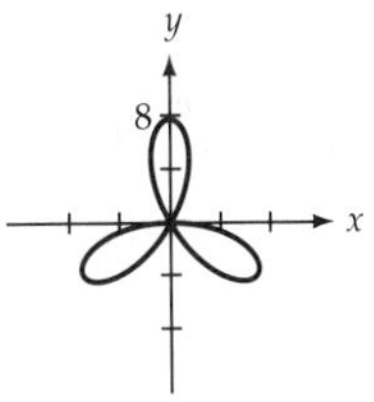

13. Graph $r = -2\csc\theta$.

a. 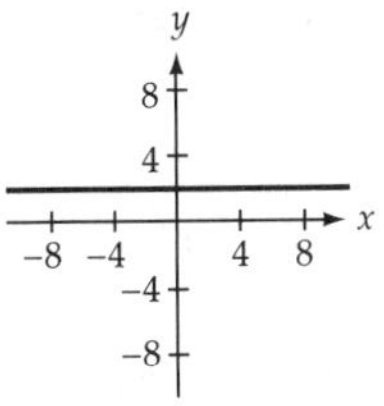

b. 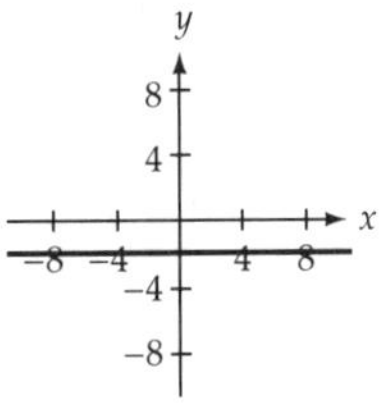

c. 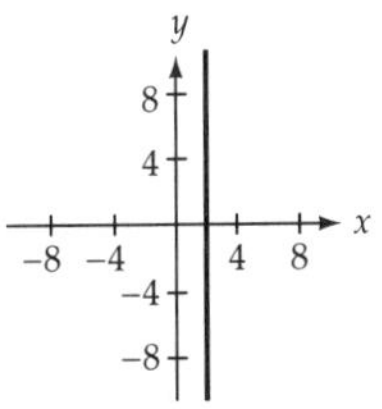

d. 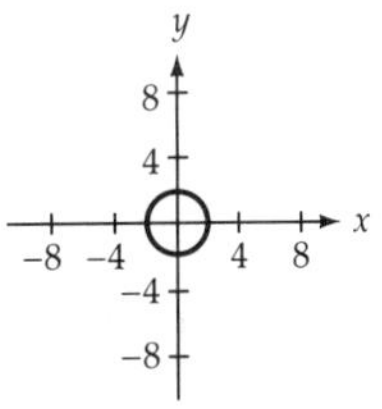

14. Graph $r = \dfrac{2}{2 + \sin\theta}$.

a. 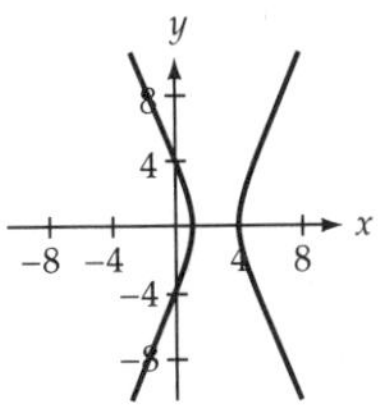

b. 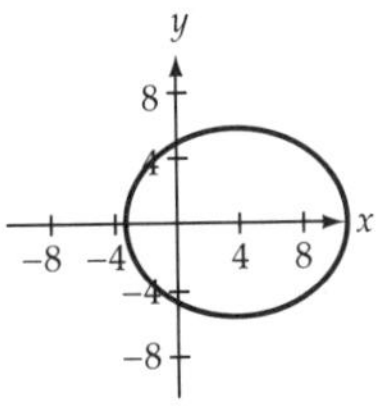

c. 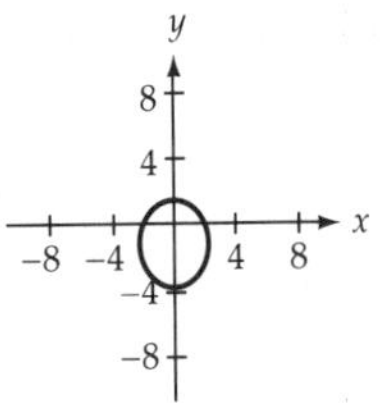

d. 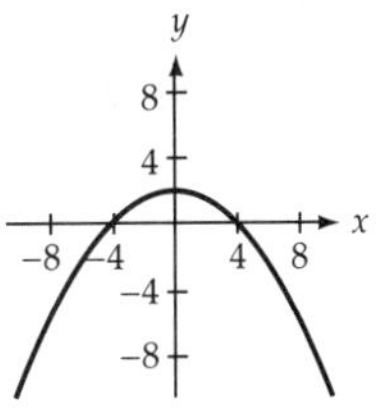

15. Graph $r = \dfrac{6}{4 - \cos\theta}$.

a. 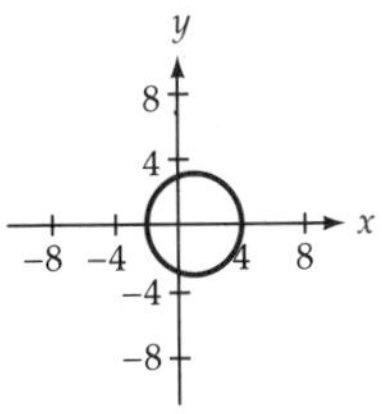

b. 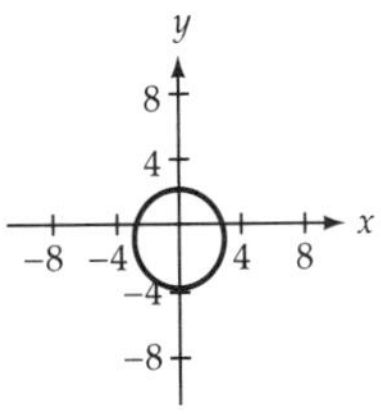

c. 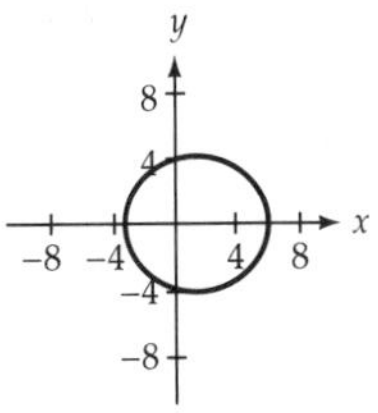

d. 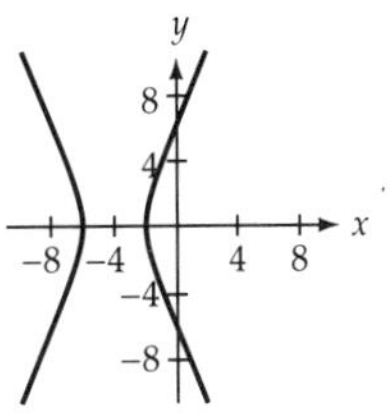

16. Find the polar coordinates of the point whose rectangular coordinates are $(-4\sqrt{3}, 4)$.

a. $\left(8, \frac{11\pi}{6}\right)$ b. $\left(4, \frac{7\pi}{6}\right)$ c. $\left(8, \frac{5\pi}{6}\right)$ d. $\left(8, \frac{7\pi}{6}\right)$

17. Find the rectangular form of the equation $r = 11 \sin \theta$.

a. $x^2 + y^2 = 11$ b. $x^2 = 11y$ c. $y^2 = 11x$ d. $x^2 + y^2 = 11y$

18. Find the polar form of $y^2 = 4x + 4$.

a. $r^2 \sin^2\theta = 4$ b. $r^2 \sin^2\theta - 4r \sin 1 = 4$
c. $r^2 \sin^2\theta - 4r \cos \theta = 4$ d. $r \sin \theta - 4r \cos \theta = 4$

19. Eliminate the parameter and sketch the curve given by the parametric equations $x^2 = t + 1$, $y = 3t - 2$, $t \geq 0$.

a. $3x^2 - 1$ b. $x^2 - 5$ c. $3x^2 - 5$ d. $3x^2 + 2$

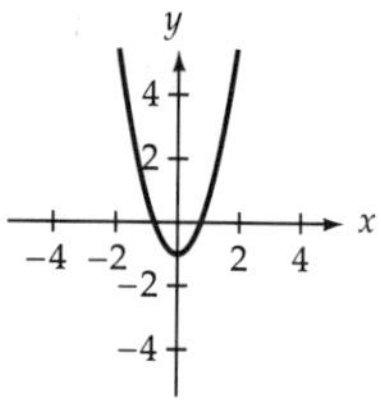 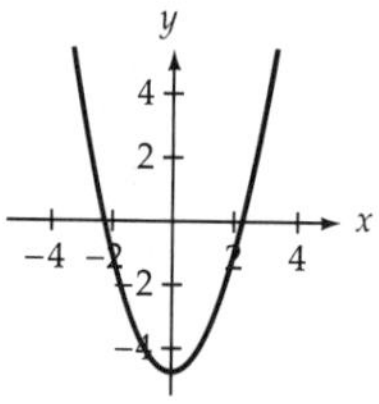 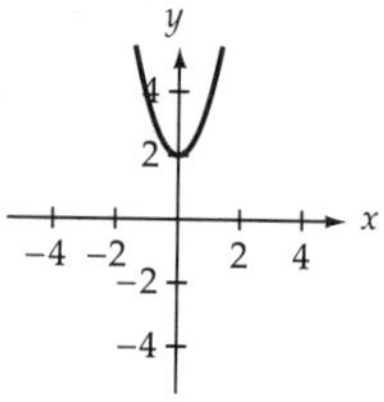 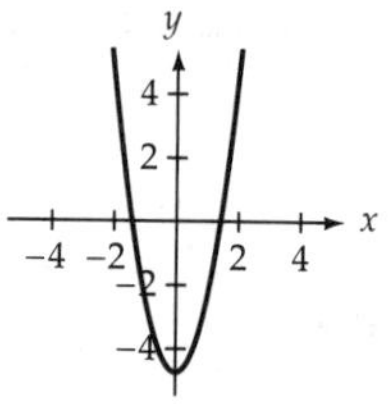

20. Eliminate the parameter and sketch the curve given by the parametric equations $x = 1 + \sin \theta$, $y = \cos \theta$, $0 \leq \theta \leq 2\pi$.

a. $x + (y - 1)^2 = 1$ b. $(x + 1)^2 + y^2 = 1$ c. $\frac{(x-1)^2}{2} + y^2 = 1$ d. $(x - 1)^2 + y^2 = 1$

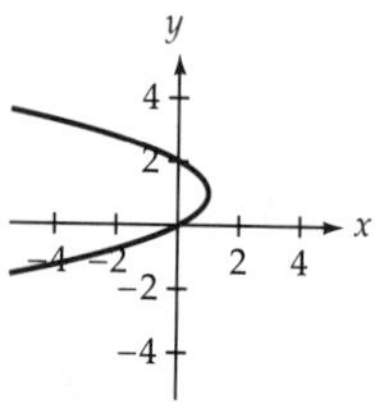 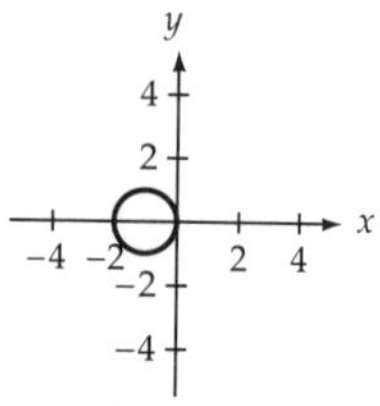 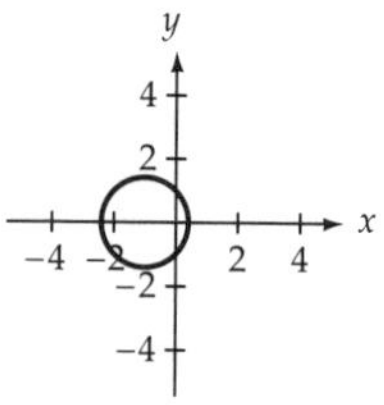 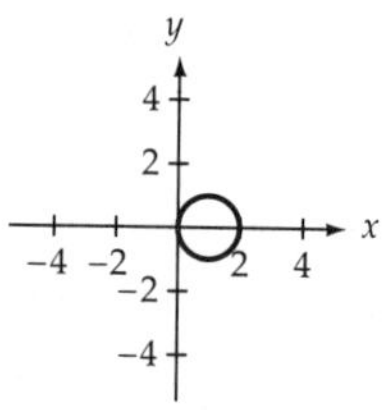

Trigonometry Chapter 6 Test Form E

1. Find the vertex, focus, and directrix of the parabola given by the equation $x^2 = -6y$.

   a. vertex: (0, 0), focus: $\left(0, \frac{3}{2}\right)$, directrix: $y = -\frac{3}{2}$

   b. vertex: (0, 0), focus: $\left(0, -\frac{3}{2}\right)$, directrix $y = \frac{3}{2}$

   c. vertex: (0, 0), focus: $\left(0, \frac{2}{3}\right)$, directrix: $y = -\frac{2}{3}$

   d. vertex: (0, 0), focus: $\left(0, -\frac{3}{2}\right)$, directrix: $y = \frac{2}{3}$

2. Fidn the vertex, focus, and directrix of the parabola given by the equation $24x - y^2 - 2y - 73 = 0$.

   a. vertex: (3, –1), focus: (9, –1), directrix: $x = -3$
   b. vertex: (–1, 3), focus: (5, 3), directrix: $x = -7$
   c. vertex: (1, 3), focus: (1, 9), directrix: $y = -3$
   d. vertex: (3, –1), focus: (3, 5), directrix: $y = -7$

3. Find the vertices and foci of the ellipse given by the equation $x^2 + 2y^2 - 8x - 36y + 162 = 0$.

   a. vertices: (0, 9), (8, 9), foci: $(4 + 2\sqrt{2}, 9)$, $(4 - 2\sqrt{2}, 9)$
   b. vertices: (20, 9), (–12, 9), foci: (12, 9), (–4, 9)
   c. vertices: (–8, –9), (0, –9), foci: $(-4 + 2\sqrt{2}, 9)$, $(-4 - 2\sqrt{2}, 9)$
   d. vertices: (4, 13), (4, 5), foci: $(4, 9 + 2\sqrt{2})$, $(4, 9 - 2\sqrt{2})$

4. Find the vertices and foci of the hyperbola given by the equation $\frac{(y-8)^2}{16} - \frac{(x+3)^2}{1} = 1$.

   a. vertices: (8, 1), (8, –7), foci: $(8, -3 + \sqrt{17})$, $(8, -3 - \sqrt{17})$
   b. vertices: (8, 1), (8, –7), foci: $(8 + \sqrt{17}, -3)$, $(8 - \sqrt{17}, -3)$
   c. vertices: (–3, 12), (–3, 4), foci: $(-3, 8 + \sqrt{17}$, $(-3, 8 - \sqrt{17})$
   d. vertices: (–3, 12), (–3, 4), foci: $(-3, 8 + \sqrt{15})$, $(-3, 8 - \sqrt{15})$

5. Find the equation in standard form of the parabola with directrix $y = 4$ and focus (3, –6).

   a. $(x - 3)^2 = 20(y + 1)$
   b. $(y - 3)^2 = -5(x + 1)$
   c. $(x - 3)^2 = -20(y + 1)$
   d. $(x + 3)^2 = 20(y - 1)$

6. Find the equation in standard form of the ellipse with center at (–2, 6), foci $(-2, 6 + \sqrt{14})$, $(-2, 6 - \sqrt{14})$, and minor axis of length 16.

a. $\frac{(x-2)^2}{64} + \frac{(y+6)^2}{78} = 1$ b. $\frac{(x+2)^2}{64} + \frac{(y-6)^2}{78} = 1$

c. $\frac{(x+2)^2}{32} + \frac{(y-6)^2}{46} = 1$ d. $\frac{(x+2)^2}{78} + \frac{(y-6)^2}{64} = 1$

7. Find the equation in standard form of the hyperbola with vertices at (3, –5), (9, –5) and foci $(3 + \sqrt{21}, -5)$, $(3 - \sqrt{21}, -5)$.

a. $\frac{(x-6)^2}{9} - \frac{(y+5)^2}{15} = 1$ b. $\frac{(x+6)^2}{9} - \frac{(y-5)^2}{15} = 1$

c. $\frac{(x-6)^2}{3} - \frac{(y+5)^2}{18} = 1$ d. $\frac{(x+5)^2}{15} - \frac{(y-3)^2}{9} = 1$

8. Let P(–2, 6) be a point in an xy-coordinate system. Find the coordinates of P in an x'y'-coordinate system that has been rotated 45°.

a. $(-\sqrt{2} + 3\sqrt{2}, 3\sqrt{2} + \sqrt{2})$ b. $(-3\sqrt{2} + \sqrt{2}, \sqrt{2} - 3\sqrt{2})$

c. $(\sqrt{2}, 3\sqrt{2})$ d. $(-3\sqrt{2}, -\sqrt{2})$

9. For the equation $2x^2 - 3xy + 6y^2 + 3x + 2y + 5 = 0$, deterine what angle of rotation would eliminate the xy-term.

a. 42° b. 28° c. 82° d. 24°

10. Graph $\frac{y^2}{64} - \frac{x^2}{4} = 1$.

a.

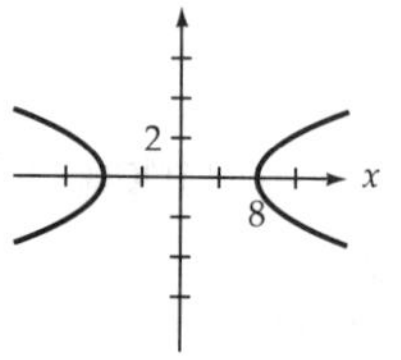

b.

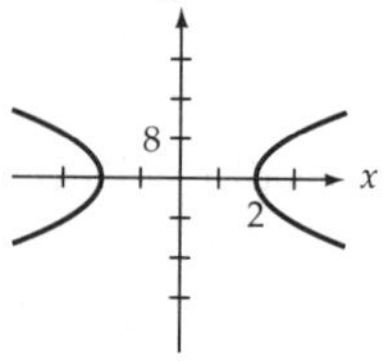

c.

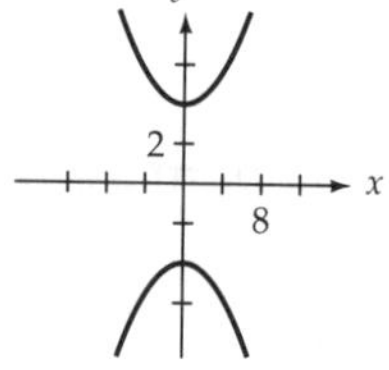

d.

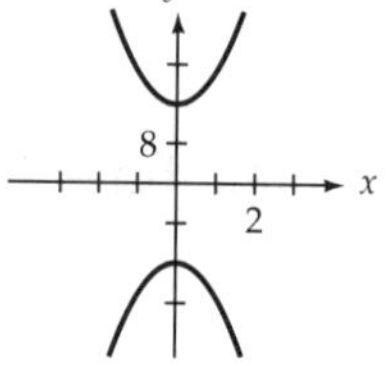

11. Graph r = 7 + 7 sin θ

a.
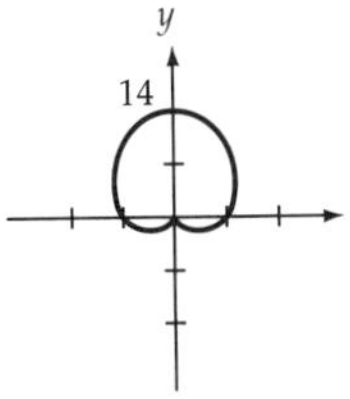

b.
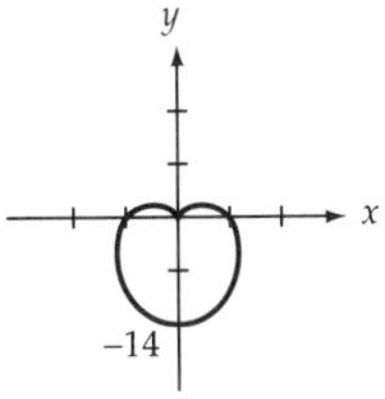

c.
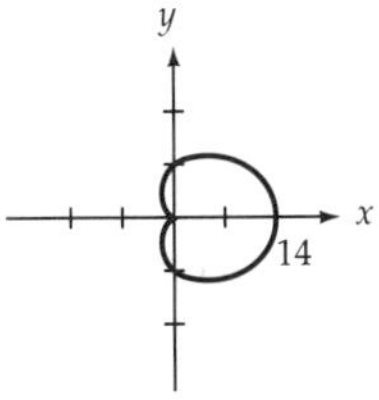

d.
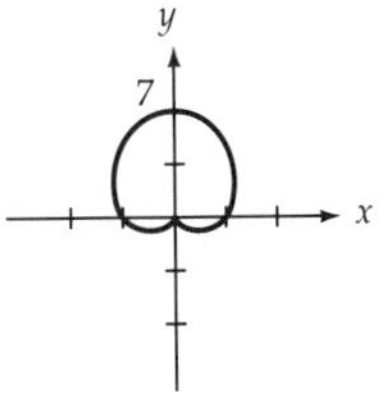

12. Graph r = –2 cos 3θ.

a.
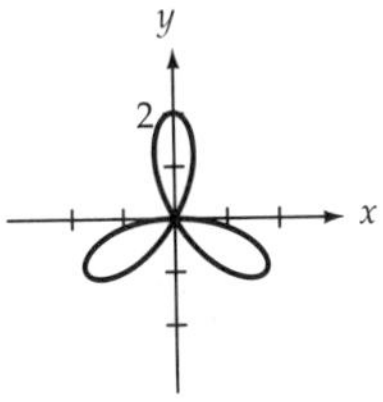

b.
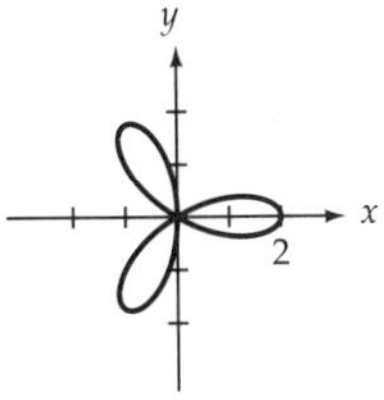

c.
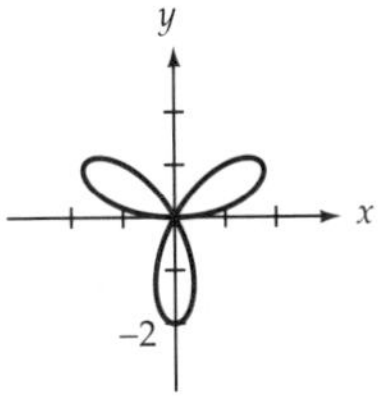

d.
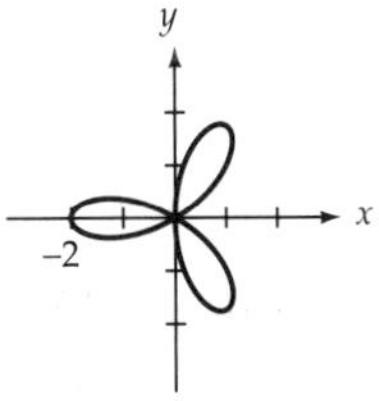

13. Graph r = 5 sin θ.

a.
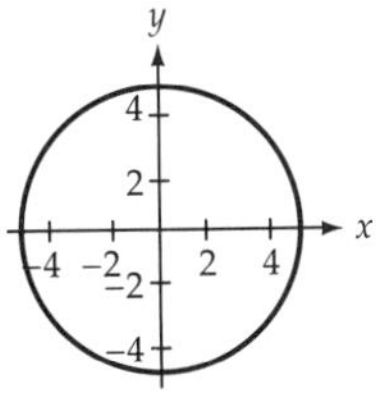

b.
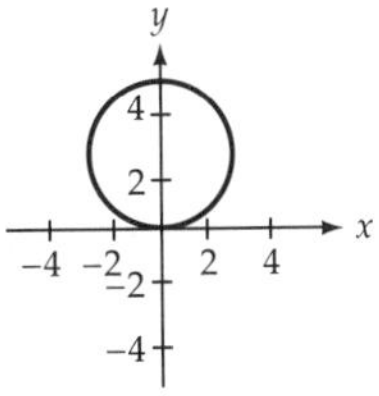

c.
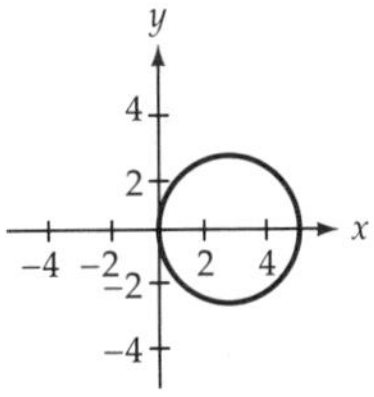

d.
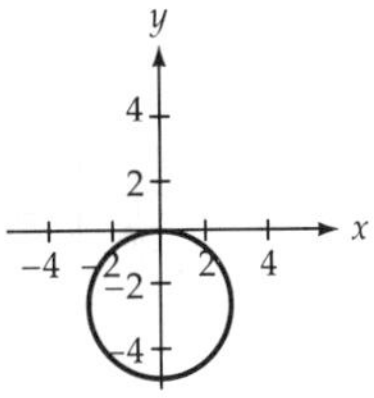

14. Graph $r = \dfrac{2}{1 - 2\sin\theta}$.

a.
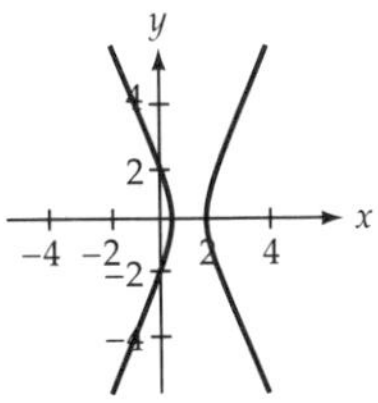

b.
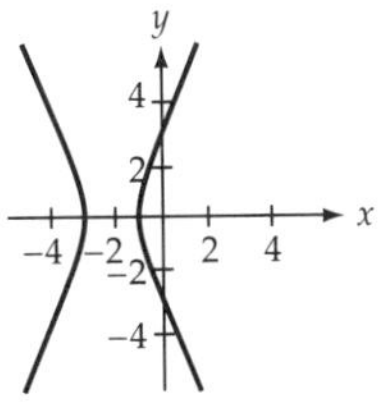

c.
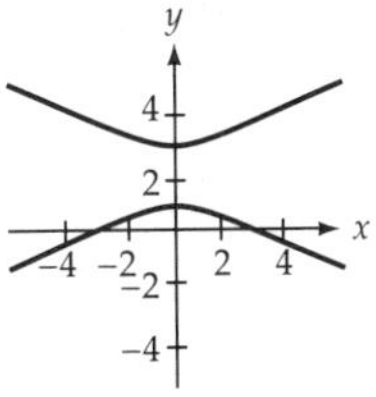

d.
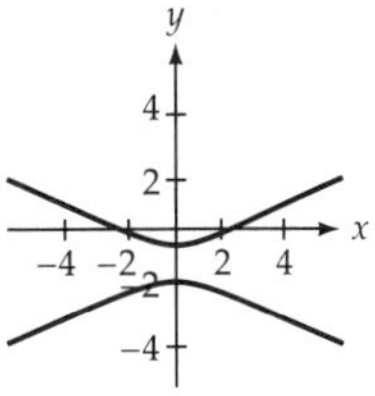

15. Graph $r = \dfrac{4}{2 + \cos\theta}$.

a.
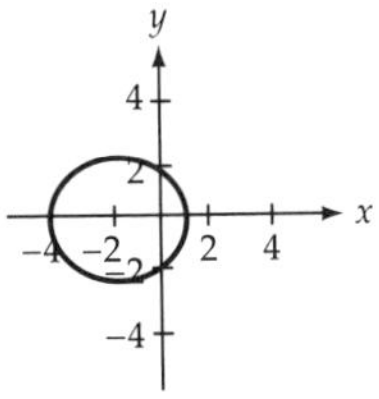

b.
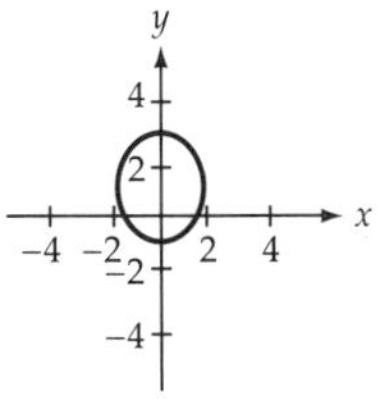

c.
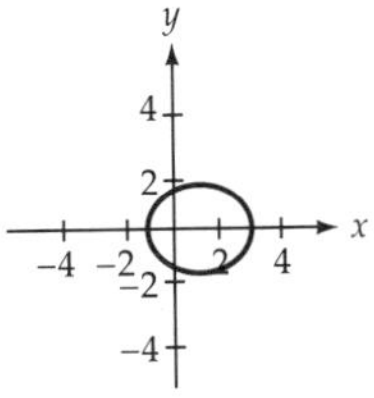

d.
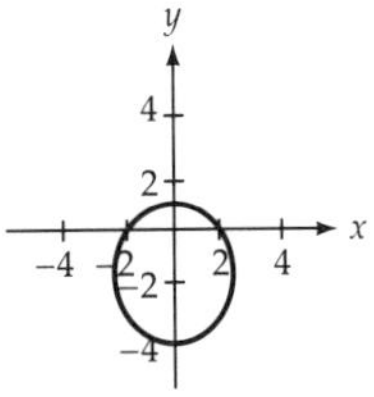

16. Find the rectangular coordinates of the point whose polar coordinates are $\left(-6, \dfrac{7\pi}{6}\right)$

a. $(-3, -3\sqrt{3})$ b. $(-3\sqrt{3}, -3)$ c. $(3, 3\sqrt{3})$ d. $(3\sqrt{3}, 3)$

17. Find the polar form of $x = 5$.

a. $r\cos\theta = 5$ b. $r\sin\theta = 5$ c. $r\cos\theta = 25$ d. $r\sin\theta = 25$

18. Find the rectangular form of $r = 2\cos\theta$.

a. $x + y = 2$ b. $\sqrt{x^2 + y^2} = 2$ c. $x^2 - 2x + y^2 = 0$ d. $x = 2y$

19. Eliminate the parameter and sketch the curve given by the parametric equations $x = t^2 + 1$, $y = 1 - 2t$, $t \ge 0$.

a. $(y - 1)^2 = 4(x + 2)$

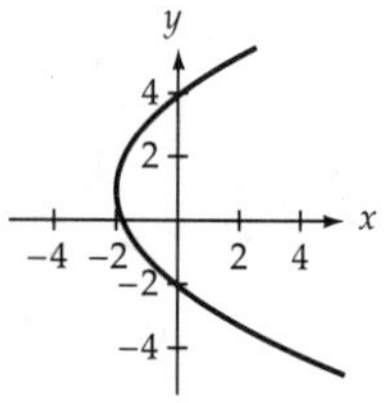

b. $(x - 2)^2 = 4(y + 2)$

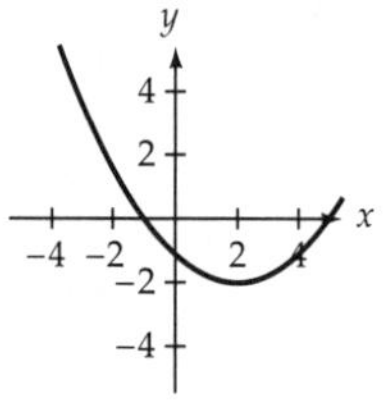

c. $(y + 1)^2 = 4(x - 2)$

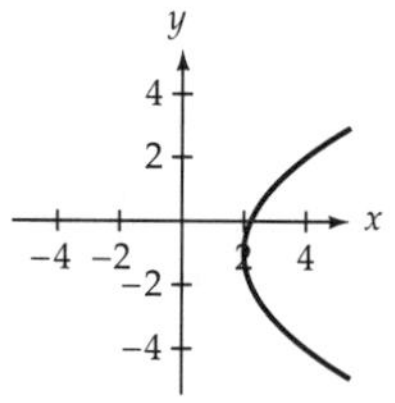

d. $(x - 1)^2 = 4(y - 2)$

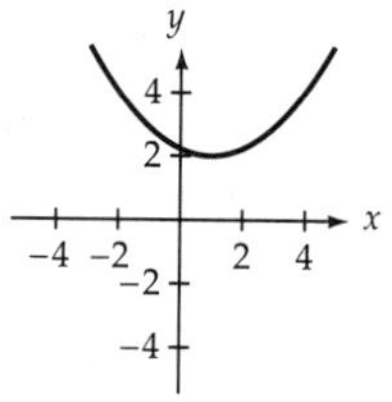

20. Eliminate the parameter and sketch the curve given by the parametric equations $x = 1 - \cos\theta$, $y = 2 + \sin\theta$, $0 \le \theta \le 2\pi$.

a. $(x + 1)^2 + (y + 2)^2 = 1$

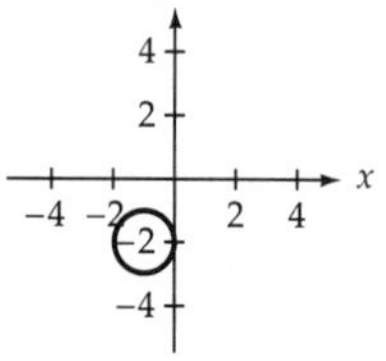

b. $(x - 1)^2 + (y - 2)^2 = 1$

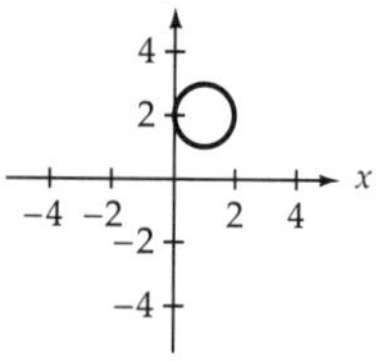

c. $(x - 2)^2$ $(y + 1)^2 = 1$

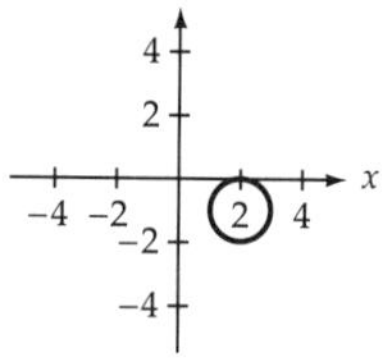

d. $\dfrac{x^2}{1} + \dfrac{y^2}{4} = 1$

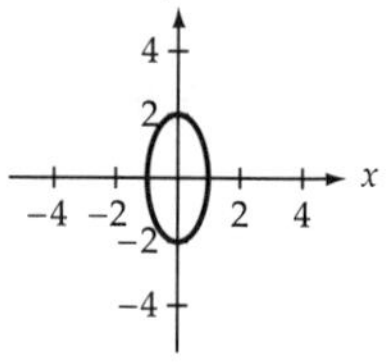

1. Find the vertex, focus, and directrix of the parabola given by the equation $6x = 4y^2$.

a. vertex: (0, 0), focus: $\left(-\frac{3}{8}, 0\right)$, directrix: $y = \frac{3}{8}$

b. vertex: (0, 0), focus: $\left(\frac{3}{8}, 0\right)$, directrix: $x = -\frac{3}{8}$

c. vertex: (0, 0), focus: $\left(-\frac{3}{8}, 0\right)$, directrix: $x = \frac{3}{8}$

d. vertex: (0, 0), focus: $\left(0, \frac{3}{8}\right)$, directrix: $y = -\frac{3}{8}$

2. Find the vertex, focus, and directrix of the parabola given by the equation $x^2 - 4x - 8y - 20 = 0$.

a. vertex: (3, 2), focus: (–6, 2), directrix: $x = -5$

b. vertex: (4, –3), focus: (4, –6), directrix: $y = 1$

c. vertex: (–2, 3), focus: (–2, 1), directrix: $y = 5$

d. vertex: (2, –3), focus: (2, –1), directrix: $y = -5$

3. Find the vertices and foci of the ellipse given by the equation $x^2 + 16y^2 - 6x + 64y + 57 = 0$.

a. vertices: (–1, –2), (7, –2), foci: $(3 + \sqrt{15}, -2)$, $(3 - \sqrt{15}, -2)$

b. vertices: (–2, –1), (–2, 7), foci: $(-2, 3, + \sqrt{15})$, $(-2, 3 - \sqrt{15})$

c. vertices: (1, –2) (7, –2), foci: $(3 + \sqrt{15}, 2)$, $(3 - \sqrt{15}, 2)$

d. vertices: (–1, –2), (7, –2), foci: $(3 + \sqrt{17}, -2)$, $(3 - \sqrt{17}, -2)$

4. Find the vertices and foci of the hyperbola given by the equation $\frac{(x-3)^2}{16} - \frac{(y+4)^2}{36} = 1$.

a. vertices: (–4, 7), (–4, –1), foci: $(-2, 4 + 2\sqrt{13})$, $(-2, 4 - 2\sqrt{13})$

b. vertices: (7, –4), (–4, –1), foci: $(4 + 2\sqrt{5}, -4)$, $(4 - 2\sqrt{5}, -4)$

c. vertices: (10, –6), (–4, –6), foci: $(3 + 2\sqrt{5}, -6)$, $(3 - 2\sqrt{5}, -6)$

d. vertices: (7, –4), (–1, –4), foci: $(3 + 2\sqrt{13}, -4)$, $(3 - 2\sqrt{13}, -4)$

5. Find the equation in standard form of the parabola with directrix $y = -5$ and focus (4, 3).

a. $(x - 4)^2 = 16(y + 1)$

b. $(y - 3)^2 = 4(x - 4)$

c. $(x + 4)^2 = 4(y + 1)$

d. $(x - 4)^2 = 4(y - 1)$

Trigonometry Chapter 6 Test Form F

6. Find the equation in standard form of the ellipse with center (–2, 3), foci (–2, 7) and (–2, –1), and minor axis of length 4.

a. $\frac{(x-2)^2}{20}+\frac{(y+3)^2}{4}=1$

b. $\frac{(x+2)^2}{20}+\frac{(y-3)^2}{4}=1$

c. $\frac{(x+2)^2}{4}+\frac{(y-3)^2}{20}=1$

d. $\frac{(x-2)^2}{4}+\frac{(y+3)^2}{20}=1$

7. Find the equation in standard form of the hyperbola with vertices at (1, 4) and (–5, 4) and foci (3, 4) and (–7, 4).

a. $\frac{(x+2)^2}{9}-\frac{(y-4)^2}{16}=1$

b. $\frac{(x-2)^2}{9}-\frac{(y+4)^2}{16}=1$

c. $\frac{(x-2)^2}{16}-\frac{(y-4)^2}{9}=1$

d. $\frac{(y-4)^2}{16}-\frac{(x+2)^2}{9}=1$

8. Let P(6, –4) be a point in an xy-coordinate system. Find the coordinates of P in an x'y'-coordinate system that has been rotated 60°.

a. $(3+2\sqrt{3},\ 2+3\sqrt{3})$

b. $(3-2\sqrt{2},\ -2-3\sqrt{3})$

c. $(2\sqrt{3}-3),\ -3\sqrt{3}-2)$

d. $(2\sqrt{3}+2,\ 3\sqrt{3}+2)$

9. For the equation $5x^2+\sqrt{3}\,xy+8y^2+6x-8y-55=0$, determine what angle of rotation would eliminate the xy-term.

a. 60°

b. 54°

c. 30°

d. 42°

10. Graph $4y^2-8y-x^2-4x=16$.

a.

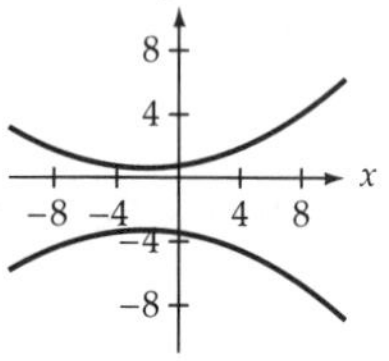

b.

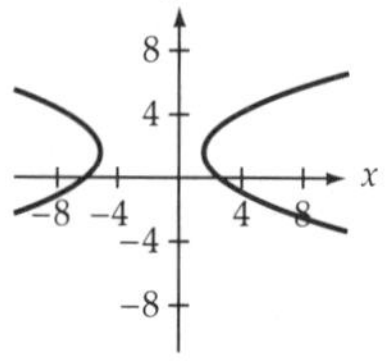

c.

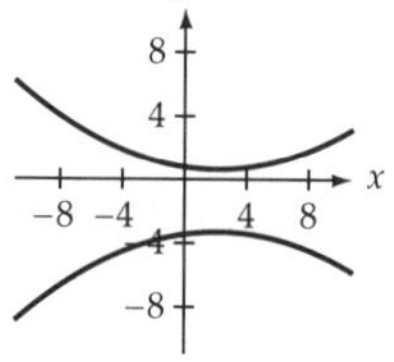

d.

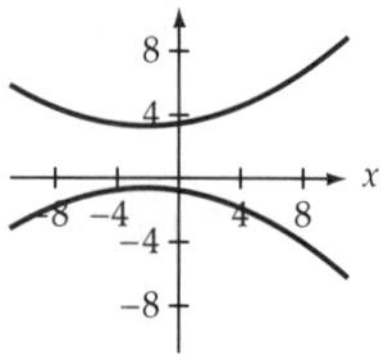

11. Graph $r = -2 \sin \theta$

a. 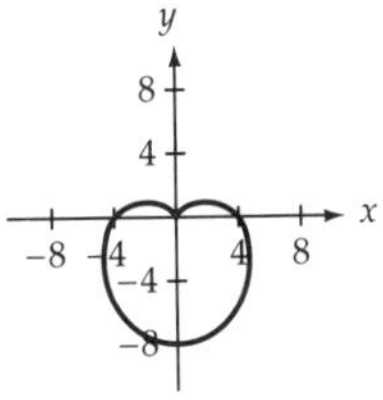

b. 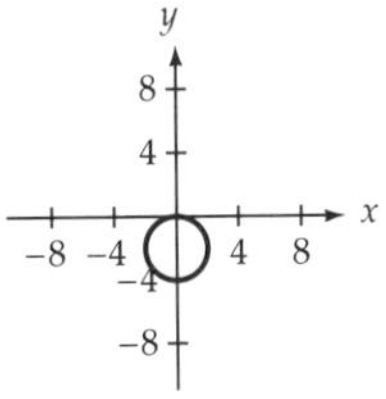

c. 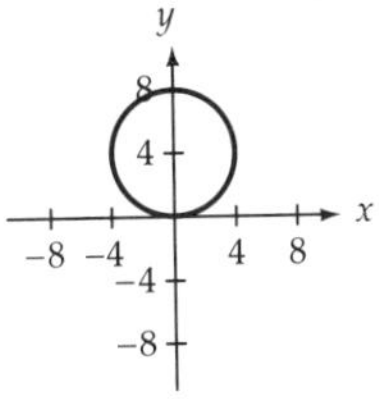

d. 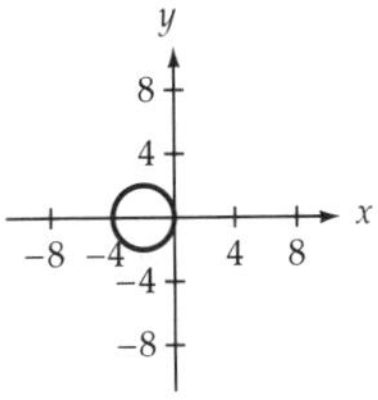

12. Graph $r = 3 - \sin \theta$.

a. 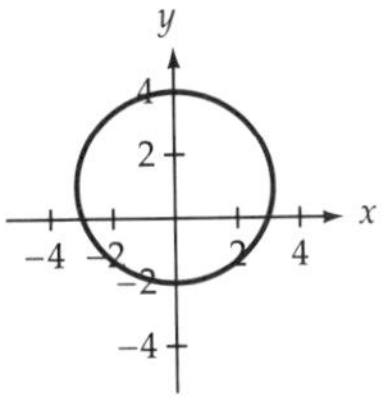

b. 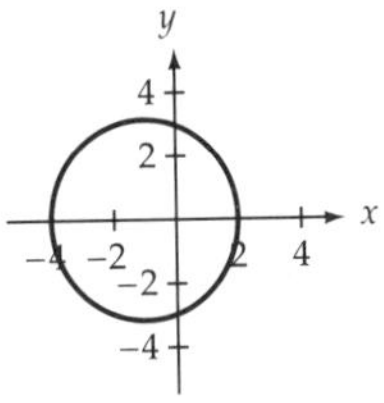

c. 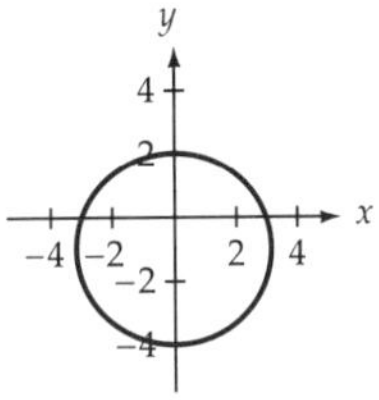

d. 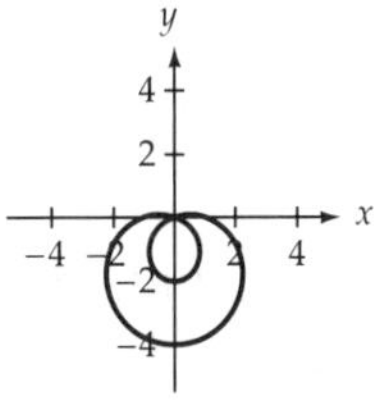

13. Graph $r = -4 \cos \theta$.

a. 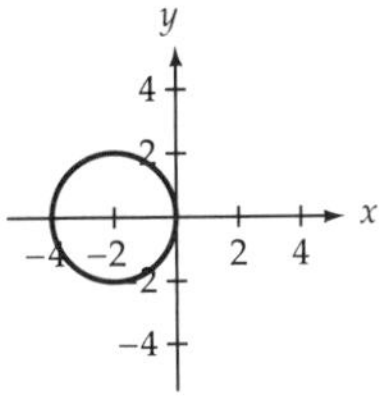

b. 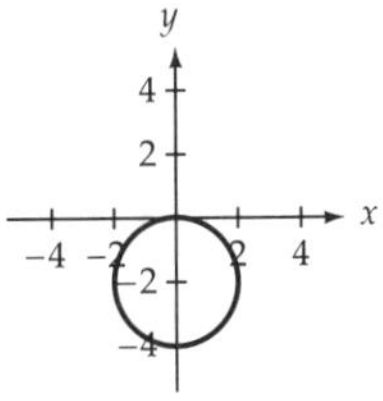

c. 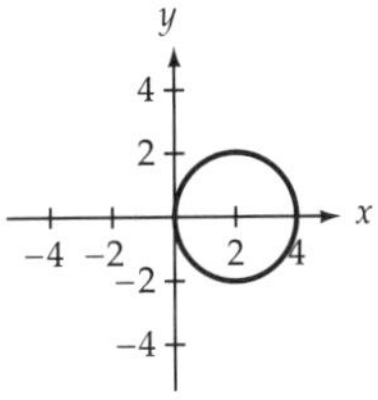

d. 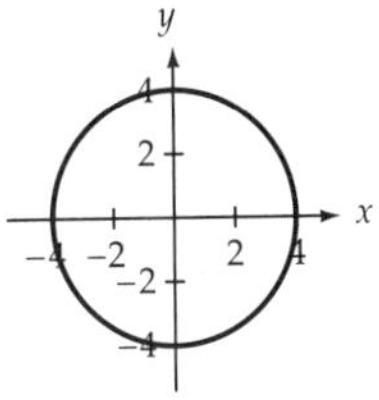

14. Graph $r = \dfrac{6}{4 - 2 \sin \theta}$.

a. 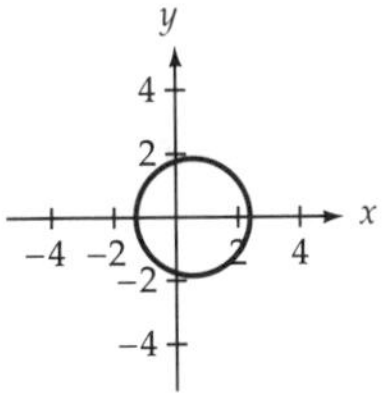

b. 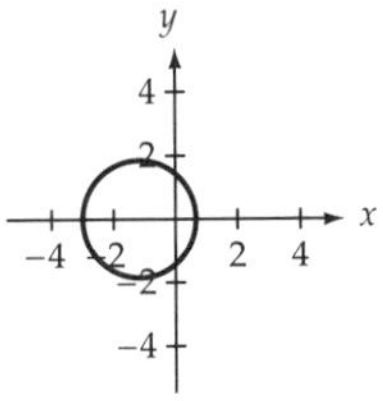

c. 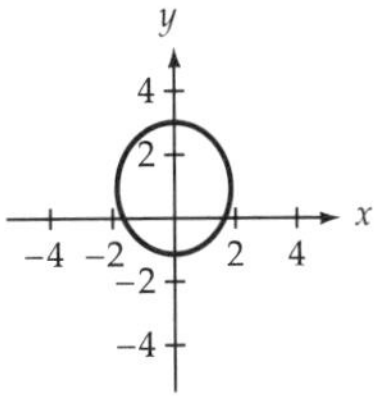

d. 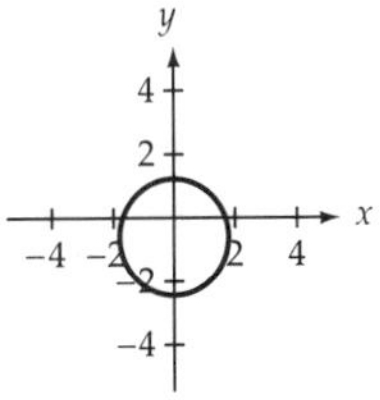

15. Graph $r = \dfrac{4}{1 + \cos \theta}$.

a. 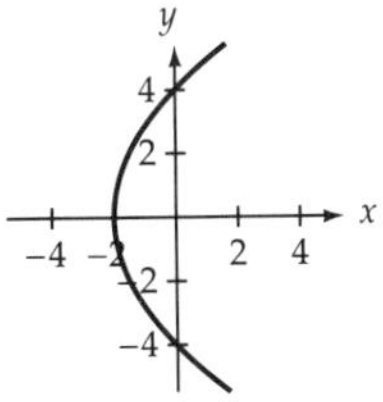

b. 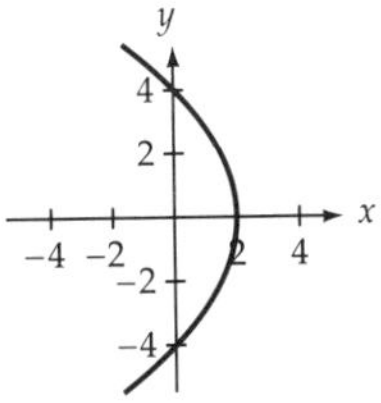

c. 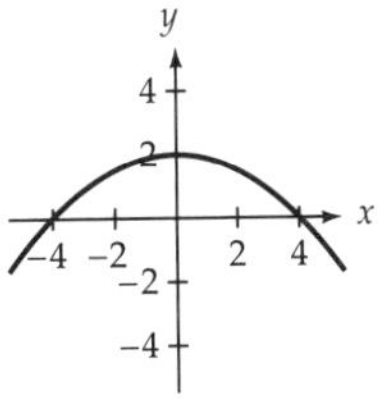

d. 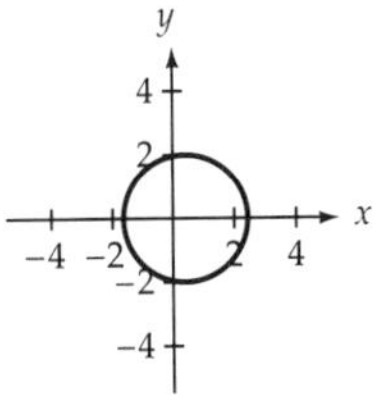

16. Find the polar coordinates of the point whose rectangular coordinates are (–5, 5).

a. $\left(5, \frac{3\pi}{4}\right)$ b. $\left(5\sqrt{2}, \frac{3\pi}{4}\right)$ c. $\left(-5, \frac{3\pi}{4}\right)$ d. $\left(5\sqrt{2}, \frac{5\pi}{4}\right)$

17. Find the polar form of $x^2 - y^2 = 1$.

a. $r \cos \theta = 1$ b. $r^2 \cos 2\theta = 1$ c. $r^2 \sin 2\theta = 1$ d. $r^2 \sin^2 \theta = 1$

18. Find the rectangular form of $r + r \sin \theta = 4$.

a. $x^2 = 16 - 8y$ b. $y^2 = x^2 - 16$ c. $x^2 = y - 4$ d. $\sqrt{x^2 + y^2} = 4$

19. Eliminate the parameter and sketch the curve given by the parametric equations $x = t^2 + t$, $y = t + 1$, $t \geq 0$.

a. $x - 1 = (y + 1)^2$ b. $x = (y + 1)^2$ c. $x = y^2 - 3y + 1$ d. $x = y^2 + 3y$

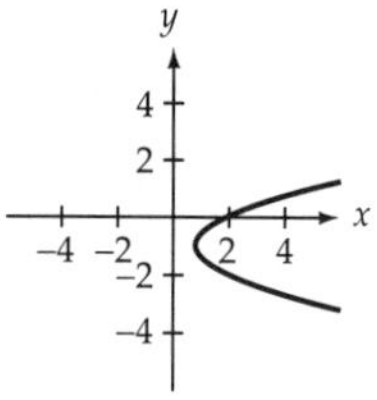

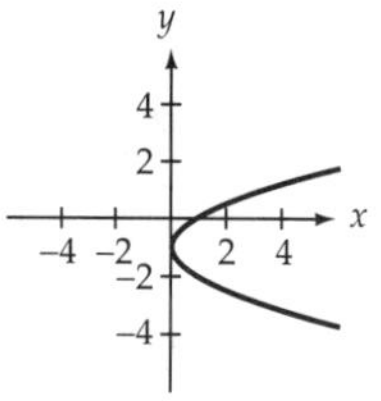

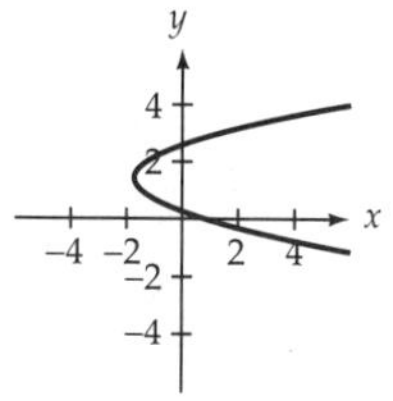

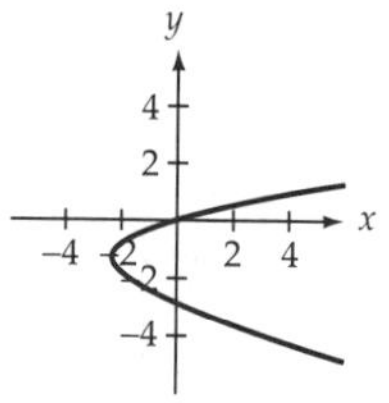

20. Eliminate the parameter and sketch the curve given by the parametric equations $x = \cos \theta + 2$, $y = \sin \theta$, $0 \leq \theta \leq 2\pi$.

a. $x^2 + (y - 2)^2 = 1$ b. $(x - 2)^2 + y^2 = 1$ c. $\frac{x^2}{1} = \frac{(y\ 2)^2}{4} = 1$ d. $(x + 2)^2 - y^2 = 1$

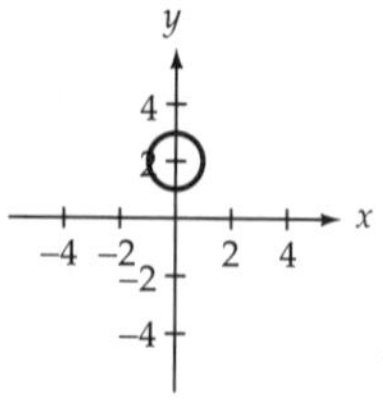

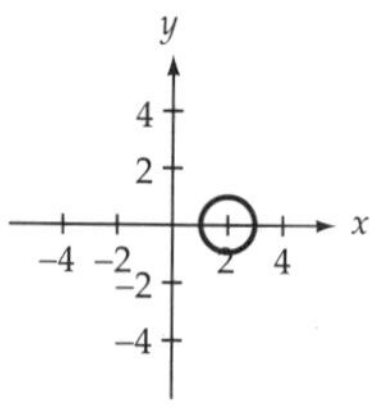

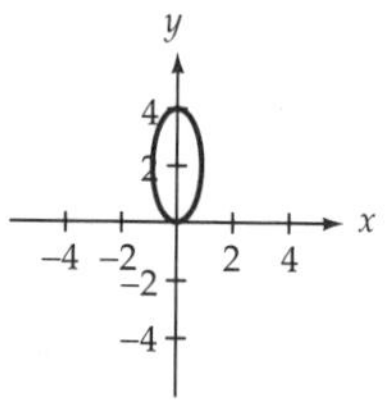

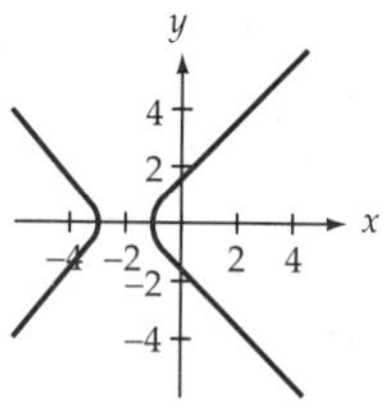

Trigonometry Test Answers

Chapter 1 Form A

1. $\frac{16}{9}$

2. $x \geq \frac{11}{9}$

3. $-1, 1, \frac{3}{2}$

4. $\left\{x \mid -\frac{1}{2} < x < 2\right\}$

5. $\sqrt{85}, \left(-\frac{9}{2}, 1\right)$

6. (4, 0), (0, -2), (0, 2)

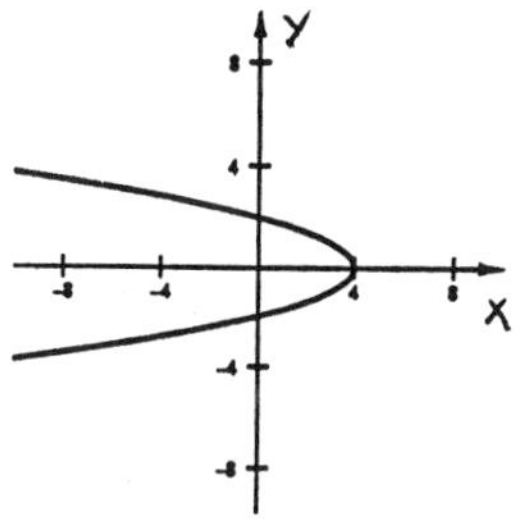

7.

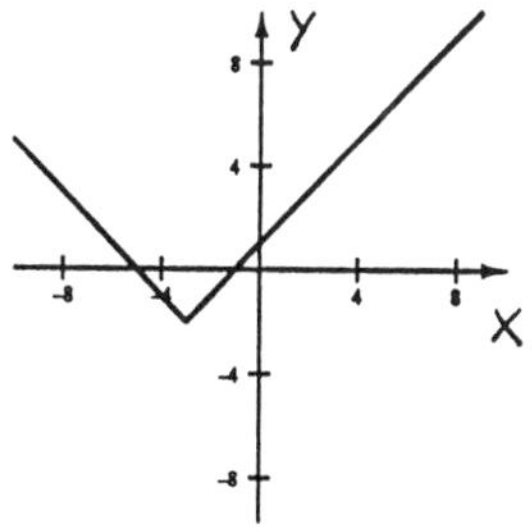

8. center: (3, -2),
radius: 2

9. Domain: $\{x \mid x \neq -4, x \neq 1\}$
$f(-1) = -\frac{1}{2}$

10. in creasing: $(-\infty, -1]$
decreasing: $[-1, \infty)$

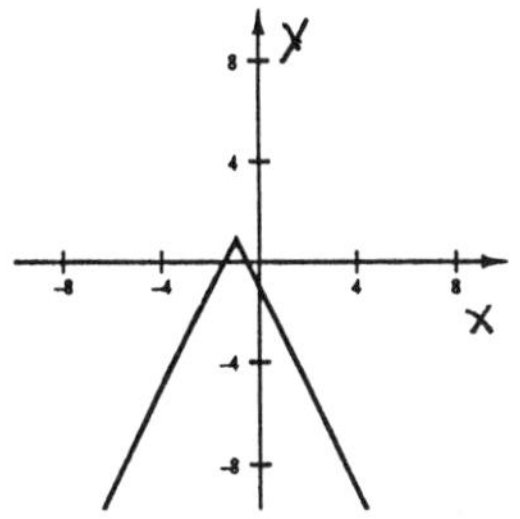

11. Domain: R,
Range: $[4, \infty)$

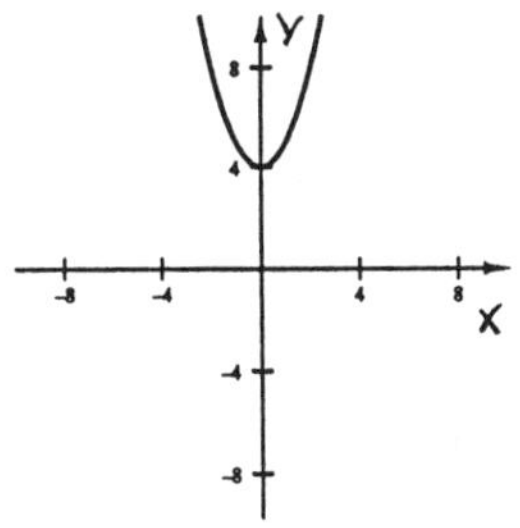

12.

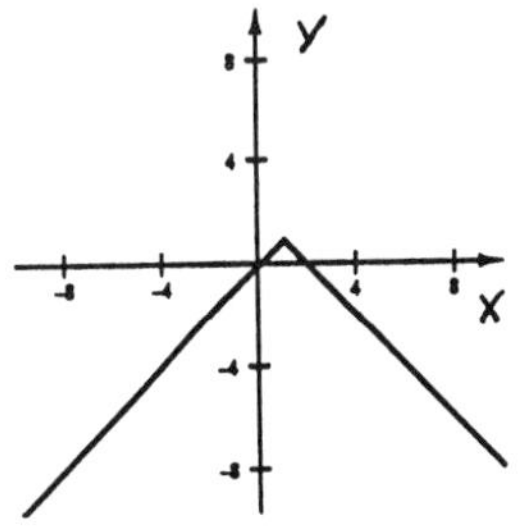

13. ii and iii

14. i i

15. $2x + h$

16. $4x^2 + 3$

17. $f^{-1}(x) = \frac{2x}{x - 1}, x \neq 1$

18. $f^{-1}(x) = -\sqrt{x - 2}$
domain of f: $x \leq 0$,
range of f: $y \geq 2$
domain of f^{-1}: $x \geq 2$
range of f^{-1}: $y \leq 0$

19. L = 30 ft, W = 18 ft

20. 8 in.

Trigonometry Test Answers

Chapter 1 Form B

1. $\frac{3}{2}$

2. $x \le \frac{9}{8}$

3. $-\frac{3}{2}, \frac{5}{3}$

4. $\left\{x \mid x < -\frac{5}{2} \text{ or } x > \frac{11}{2}\right\}$

5. $\sqrt{409}, \left(\frac{7}{2}, -1\right)$

6. (0, 3), (–1, 0), (1, 0)

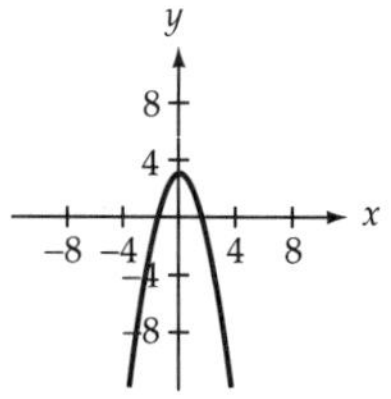

7.

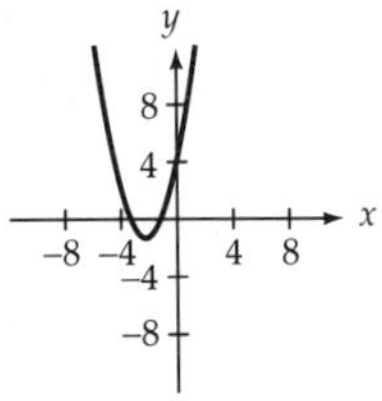

8. center: (–5, 3)
radius: 4

9. Domain: $\left[-\frac{5}{3}, \infty\right)$
$f(3) = 2\sqrt{3}$

10. increasing: $(-\infty, 1]$
decreasing: $[1, \infty)$

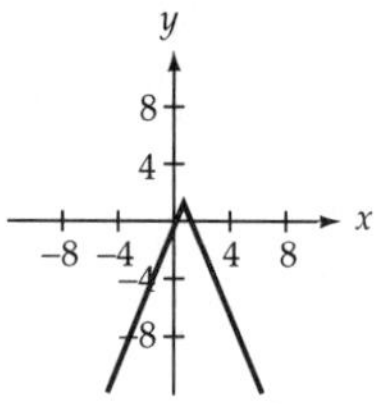

11. Domain: R
Range: $[-2, \infty)$

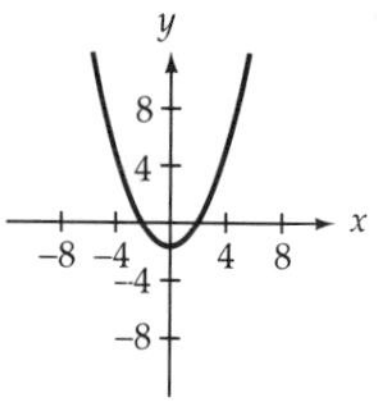

12.

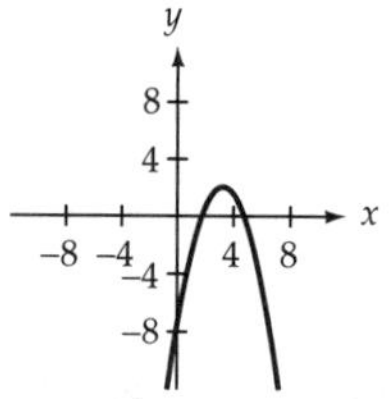

13. i and iii

14. iii

15. $2x + h$

16. $3x^2 + 2$

17. $\frac{3x}{2 - x}, x \ne 2$

18. $f^{-1}(x) = \sqrt{x + 9}$
Domain of f: $x \ge 0$
Range of f: $y \ge -9$
Domain of f^{-1}: $x \ge -9$
Range of f^{-1}: $y \ge 0$

19. Greater than 75 mi

20. 112 ft

Trigonometry Test Answers

Chapter 1 Form C

1. $-\frac{1}{8}$

2. $x \geq 2$

3. $\frac{1}{2}$, $-\frac{1}{4} + \frac{\sqrt{3}}{4}i$, $-\frac{1}{4} - \frac{\sqrt{3}}{4}i$

4. $\left\{x \mid \frac{1}{2} < x < \frac{5}{2}\right\}$

5. $\sqrt{181}$, $\left(1, \frac{15}{2}\right)$

6. $(0, 1)$, $\left(\frac{3}{2}, 0\right)$

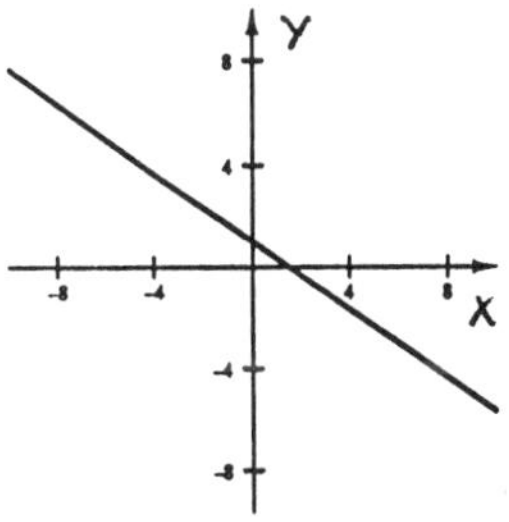

7.

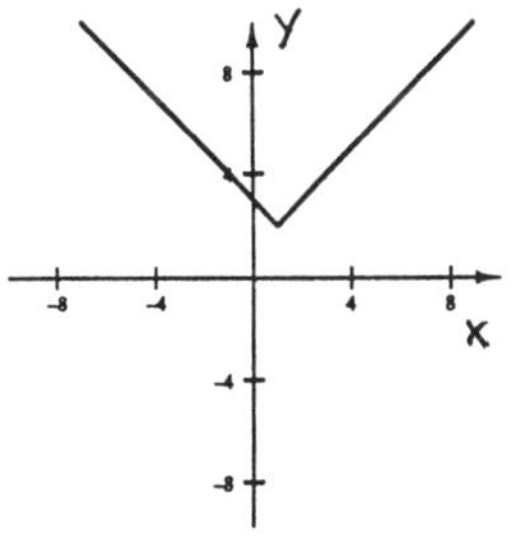

8. center: $(1, -9)$

radius: 7

9. Domain: $\left(-\infty, \frac{3}{7}\right]$, $f(-3) = 2\sqrt{6}$

10. increasing: $(-\infty, -2]$
decreasing: $[-2, \infty)$

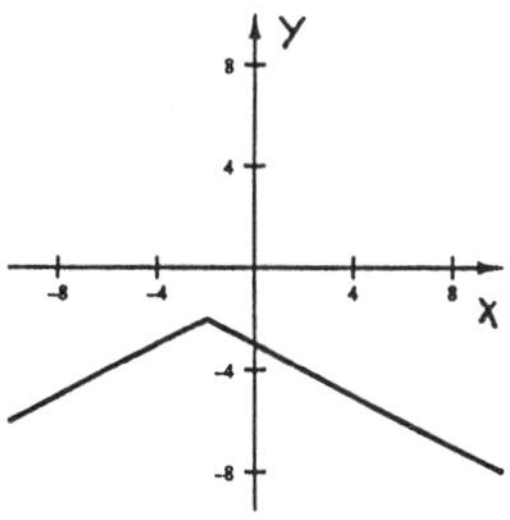

11. Domain: R,
Range: $[1, \infty)$

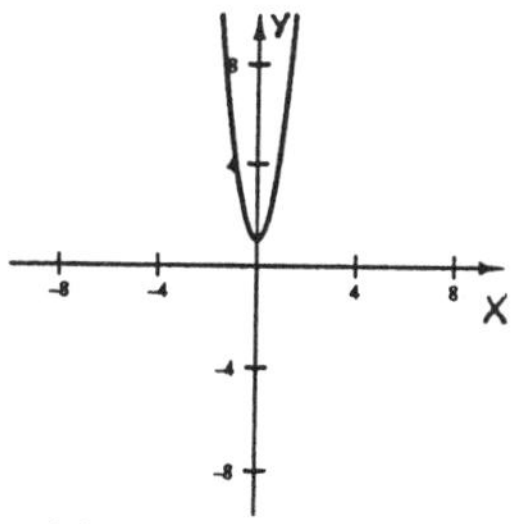

12.

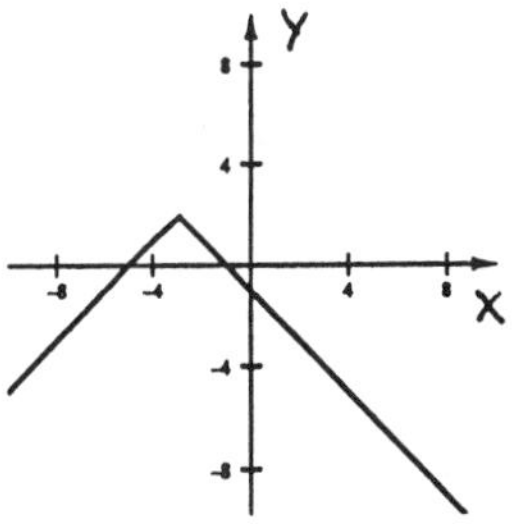

13. ii and iii

14. ii

15. $2x + h$

16. $8x^2 - 46$

17. $f^{-1}(x) = \frac{x}{x+1}$, $x \neq -1$

18. $f^{-1}(x) = -\sqrt{x - 4}$

Domain of f: $x \leq 0$
Range of f: $y \geq 4$
Domain of f^{-1}: $x \geq 4$
Range of f^{-1}: $y \leq 0$

19. Greater than 33 checks

20. $480

Trigonometry Test Answers

Chapter 1 Form D

1. b	2. a	3. d	4. c	5. a
6. a	7. c	8. c	9. b	10. c
11. a	12. a	13. a	14. b	15. c.
16. b	17. c	18. a	19. c	20. b

Chapter 1 Form E

1. b	2. a	3. c	4. b	5. c
6. c	7. b	8. c	9. d	10. d
11. c	12. b	13. c	14. d	15. c
16. b	17. c	18. b	19. d	20. c

Chapter 1 Form F

1. d	2. a	3. d	4. d	5. a
6. b	7. a	8. d	9. b	10. a
11. c	12. c	13. b	14. a	15. c
16. c	17. a	18. c	19. d	20. d

Trigonometry Test Answers

Chapter 4 Form A

1. $C = 67°$, $b = 16$, $a = 18$
2. no triangle
3. 21°
4. 6.8
5. 109°
6. 698 square units
7. 175 square units
8. 108 square units
9. $-8\mathbf{i} - 8\sqrt{3}\,\mathbf{j}$
10. $2\sqrt{13}$, 124°
11. $\mathbf{u} = \langle -\frac{\sqrt{10}}{10}, \frac{3\sqrt{10}}{10} \rangle$
12. $3\mathbf{i} - 113\mathbf{j}$
13. $15.3\mathbf{i} - 12.9\mathbf{j}$
14. -5
15. $\frac{11\sqrt{13}}{13}$
16. 157°
17. 250 ft^2
18. 113 mi
19. 80 mi
20. $66,888.87

Chapter 4 Form B

1. $A = 77°$, $a = 26$, $c = 16$
2. no triangle
3. no triangle
4. 9.8
5. 55°
6. 287 square units
7. 104 square units
8. 1585 square units
9. $2.74\mathbf{i} + 7.52\mathbf{j}$
10. $2\sqrt{29}$, 338°
11. $\mathbf{u} = \left\langle \frac{7}{25}, \frac{24}{25} \right\rangle$
12. $-18\mathbf{i} - 40\mathbf{j}$
13. $-30.6\mathbf{i} + 25.7\mathbf{j}$
14. -17
15. $\frac{39\sqrt{58}}{58}$
16. 8°
17. 651 m^2
18. 260 mph, heading 108°
19. 4.7°
20. 156 mi

Trigonometry Test Answers

Chapter 4 Form C

1. $C = 53°$, $a = 22$, $b = 8.4$		2. no triangle
3. 28°	4. 35	5. 95°
6. 83 square units	7. 266 square units	8. 26 square units
9. $57.4\mathbf{i} - 48.2\mathbf{j}$	10. $\sqrt{193}$, 330°	11. $\mathbf{u} = \frac{3}{5}\mathbf{i} - \frac{4}{5}\mathbf{j}$
12. $\langle 68, -49 \rangle$	13. $8.3\mathbf{i} - 30.9\mathbf{j}$	14. −36
15. $-2\sqrt{2}$	16. 103°	17. 1413 in^2
18. 65 mi	19. 36.8 acres	20. 32 mi

Chapter 4 Form D

1. b	2. c	3. d	4. c	5. b
6. b	7. c	8. b	9. c	10. a
11. c	12. c	13. b	14. a	15. b
16. b	17. b	18. d	19. b	20. a

Chapter 4 Form E

1. a	2. c	3. b	4. b	5. d
6. a	7. b	8. b	9. c	10. a
11. c	12. c	13. a	14. c	15. d
16. a	17. b	18. b	19. b	20. b

Trigonometry Test Answers

Chapter 4 Form F

1. b
2. a
3. c
4. a
5. d
6. c
7. d
8. b
9. c
10. b
11. c
12. d
13. b
14. a
15. d
16. d
17. c
18. d
19. d
20. c

Chapter 5 Form A

1. $3 + 4i$
2. $15 - 5i$
3. -8
4. -1
5. $5 - 3i$
6. $36 - 34i$
7. $4 - \frac{3}{2}i$
8. $-\frac{1}{5} - \frac{12}{5}i$
9. $\sqrt{53}$
10. $4\sqrt{2}(\cos 45° + i \sin 45°)$
11. $5(\cos 90° + i \sin 90°)$
12. $-\frac{3}{2} + \frac{3\sqrt{3}}{2}i$
13. $\frac{7\sqrt{2}}{2} - \frac{7\sqrt{2}}{2}i$
14. $-60i$
15. $-\frac{3\sqrt{3}}{2} + \frac{3}{2}i$
16. $256(\sin 240° + i \sin 240°)$
17. $-128 + 128i\sqrt{3}$
18. $\cos 22.5° + i \sin 22.5°$
    $\cos 112.5° + i \sin 112.5°$
    $\cos 202.5° + i \sin 202.5°$
    $\cos 292.5° + i \sin 292.5°$
19. $-0.290 + 1.082i$
    $-0.792 - 0.792i$
    $1.082 - 0.290i$
20. $2(\cos 45° + i \sin 45°)$
    $2(\cos 135° + i \sin 135°)$
    $2(\cos 225° + i \sin 225°)$
    $2(\cos 315° + i \sin 315°)$

Trigonometry Test Answers

Chapter 5 Form B

1. $4 - 11i$
2. $-14 - 11i$
3. $-4\sqrt{2}$
4. -1
5. $-6 + 20i$
6. $32 - 4i$
7. $-2 - 2i$
8. $-\frac{3}{2} - \frac{7}{2}i$
9. $\sqrt{65}$
10. $2(\cos 330° + i \sin 330°)$
11. $13(\cos 90° + i \sin 90°)$
12. $-\frac{5}{2} - \frac{5\sqrt{3}}{2}i$
13. $-7\sqrt{2} - 7i\sqrt{2}$
14. $-8\sqrt{3} + 8i$
15. $-\frac{15\sqrt{3}}{2} + \frac{15}{2}i$
16. $256(\cos 100° + i \sin 100°$
17. -64
18. $\cos 67.5° + i \sin 67.5°$
    $\cos 157.5° + i \sin 157.5°$
    $\cos 247.5° + i \sin 247.5°$
    $\cos 337.5° + i \sin 337.5°$
19. $1.366 + 0.366i$
    $-1 + i$
    $-0.366 - 1.366i$
20. $5(\cos 0° + i \sin 0°)$
    $5(\cos 120° + i \sin 120°)$
    $5(\cos 240° + i \sin 240°)$

Chapter 5 Form C

1. $8 - 12i$
2. $3 - 9i$
3. -25
4. i
5. $7 - 12i$
6. $26 - 45i$
7. $-1 - \frac{4}{3}i$
8. $\frac{13}{53} + \frac{19}{53}i$
9. $\sqrt{85}$
10. $3\sqrt{2}(\cos 45° + i \sin 45°)$
11. $2(\cos 270° + i \sin 270°)$
12. $\frac{7\sqrt{3}}{2} - \frac{7}{2}i$
13. $-\frac{11\sqrt{2}}{2} + \frac{11\sqrt{2}}{2}i$

14. $24(\cos 56° + i \sin 56°)$

15. $-\frac{3\sqrt{3}}{2} + \frac{3}{2}i$

16. $\frac{243}{2} - \frac{243\sqrt{3}}{2}i$

17. $16(\cos 0 + i \sin 0)$

18. $-\frac{5\sqrt{2}}{2} + \frac{5\sqrt{2}}{2}i, \quad \frac{5\sqrt{2}}{2} - \frac{5\sqrt{2}}{2}i$

19. $1.19(\cos 82.5° + i \sin 82.5°)$
$1.19(\cos 172.5° + i \sin 172.5°)$
$1.19(\cos 262.5° + i \sin 262.5°)$
$1.19(\cos 352.5° + i \sin 352.5°)$

20. $5i$
$\frac{-5\sqrt{3}}{2} - \frac{5}{2}i$
$\frac{5\sqrt{3}}{2} - \frac{5}{2}i$

Chapter 5 Form D

1. a 2. c 3. d 4. c 5. b

6. a 7. d 8. b 9. c 10. c

11. b 12. a 13. d 14. a 15. c

16. c 17. d 18. a 19. a 20. b

Chapter 5 Form E

1. c 2. c 3. a 4. b 5. d

6. c 7. a 8. d 9. c 10. a

11. b 12. a 13. c 14. d 15. a

16. c 17. b 18. c 19. b 20. d

Chapter 5 Form F

1. c 2. b 3. b 4. a 5. c

6. d	7. b	8. a	9. b	10. a
11. b	12. c	13. d	14. b	15. b
16. b	17. a	18. d	19. a	20. a

Chapter 6 Form A

1. Vertex: (5, 1), Focus: (5, 3), directrix: $y = -1$

2. Vertices: (6, −1), (0, −1), Foci: $\left(3 + \sqrt{5}, -1\right), \left(3 - \sqrt{5}, -1\right)$

3. Vertices: (3, 0), (3, −10), Foci: $\left(3, -5 + \sqrt{29}\right), \left(3, -5 - \sqrt{29}\right)$,
asymptotes: $y + 5 = \pm\frac{5}{2}(x - 3)$

4. Vertices: (4, −2), (4, −8), Foci: $\left(4, -5 + 2\sqrt{2}\right), \left(4, -5 - 2\sqrt{2}\right)$

5. $(y + 7)^2 = 12(x - 8)$

6. $\frac{(x-3)^2}{36} + \frac{(y-8)^2}{25} = 1$

7. $\frac{(x+2)^2}{4} - \frac{(y-3)^2}{11} = 1$

8. $\left(\frac{-2 + 5\sqrt{3}}{2}\right), \left(\frac{5 + 2\sqrt{3}}{2}\right)$

9. 60°

10.

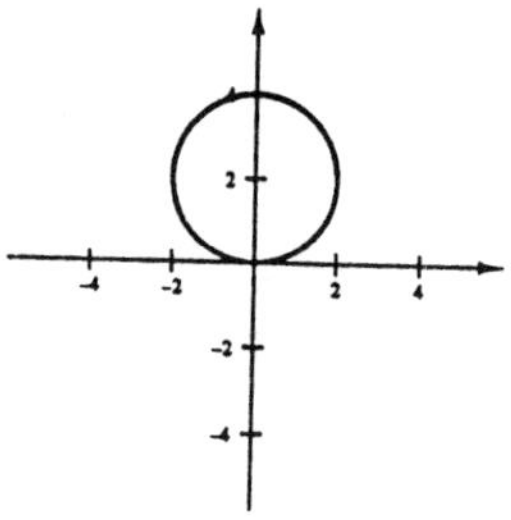

11.

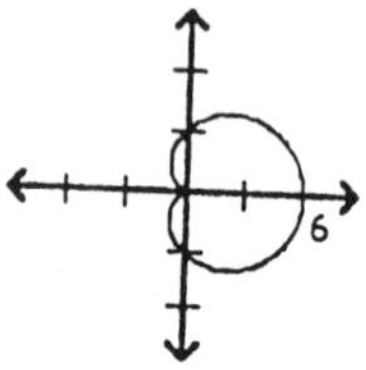

12.

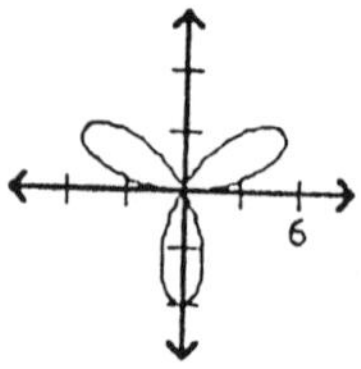

13.

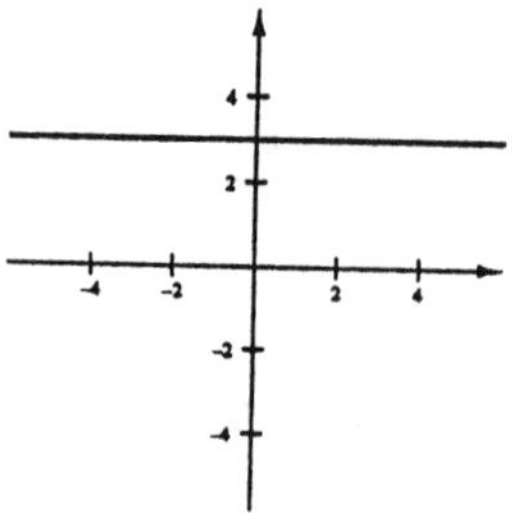

14. 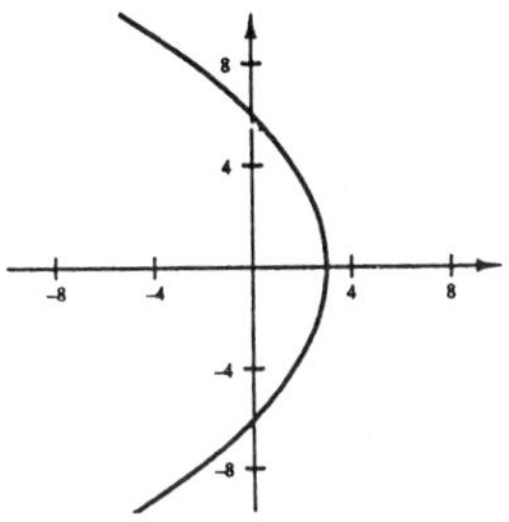

15. 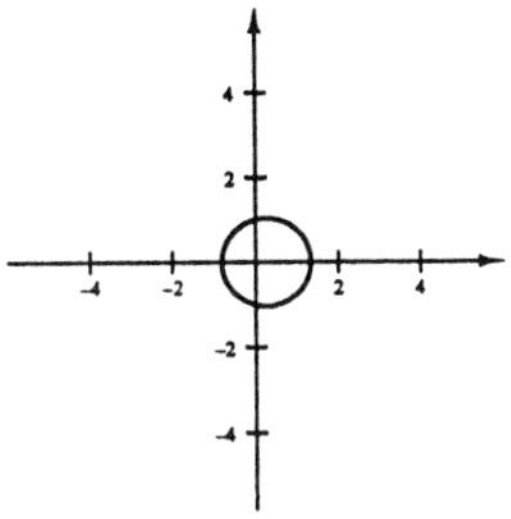

16. $\left(-\frac{3\sqrt{2}}{2}, -\frac{3\sqrt{2}}{2}\right)$

17. $r^2 = \frac{4}{5\cos^2\theta - 1}$

18. $3x^2 + 4y^2 - 12x - 36 = 0$

19. $y = (x - 2)^2$

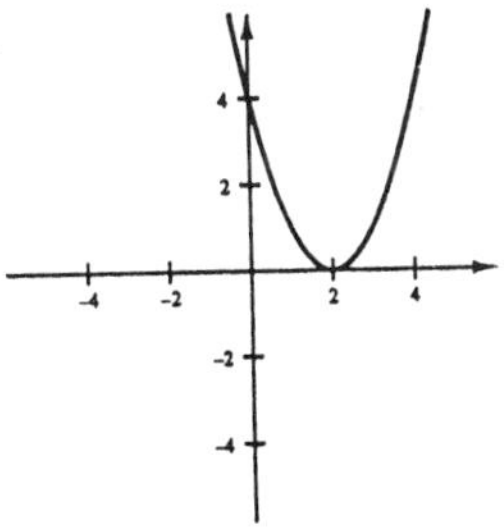

20. $\frac{x^2}{4} + (y + 1)^2 = 1$

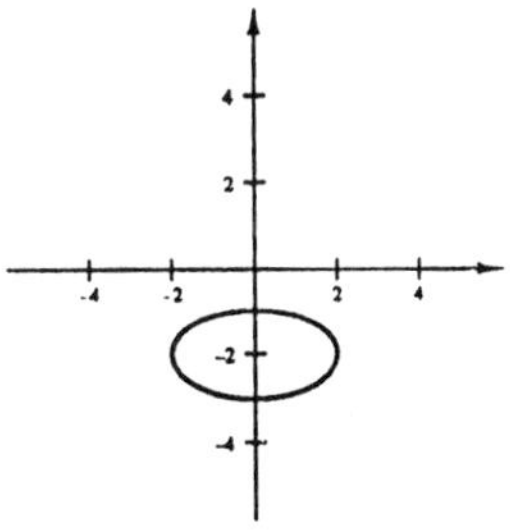

Chapter 6 Form B

1. Vertex: $(0, 0)$, Focus: $\left(-\frac{1}{2}, 0\right)$, Directrix: $x = \frac{1}{2}$

2. Vertex: $(-3, 7)$, Focus: $(-2, 7)$, Directrix: $x = -4$

3. Vertices: $(2, 7)$, $(14, 7)$, Foci: $\left(8 + 6\sqrt{2}, 7\right)$, $\left(8 - 6\sqrt{2}, 7\right)$

4. Vertices: $(-3, 3)$, $(-3, 1)$: Foci: $\left(-3, 2 + \sqrt{5}\right)$, $\left(-3, 2 - \sqrt{5}\right)$,
   Asymptotes: $y - 2 = \pm\frac{1}{2}(x + 3)$

5. $(x + 3)^2 = 8(y + 5)$

6. $\frac{(x + 5)^2}{49} + \frac{(y - 4)^2}{60} = 1$

7. $\frac{(y + 4)^2}{16} - \frac{(x - 7)^2}{15} = 1$

8. $\left(-\sqrt{3} + 2, 2\sqrt{3} + 1\right)$

9. 65.4°

10.

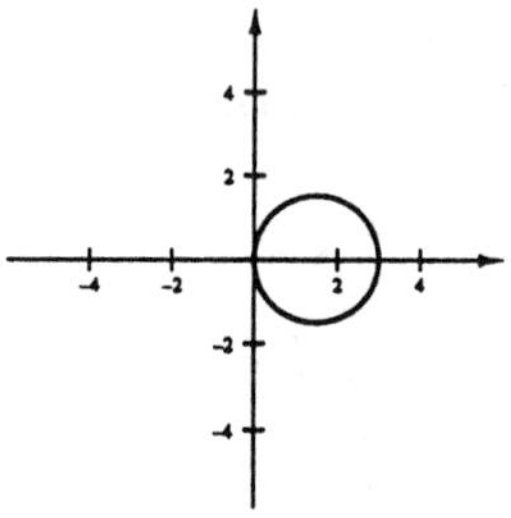

11.

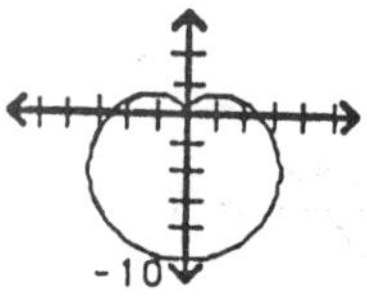

12.

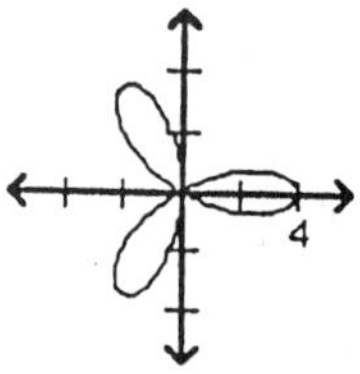

13.

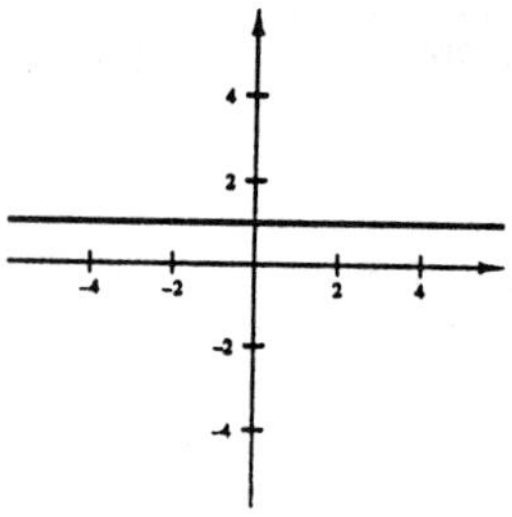

14.

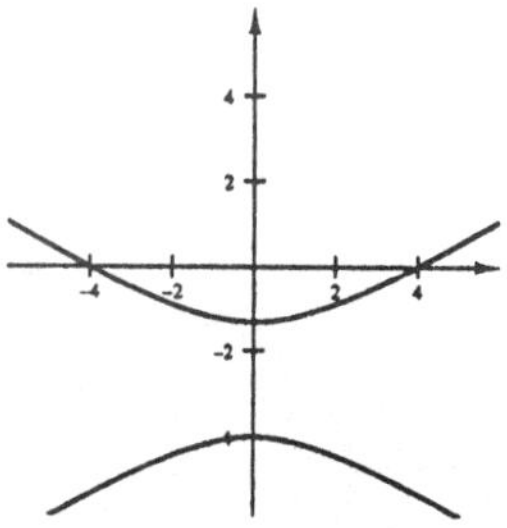

15.

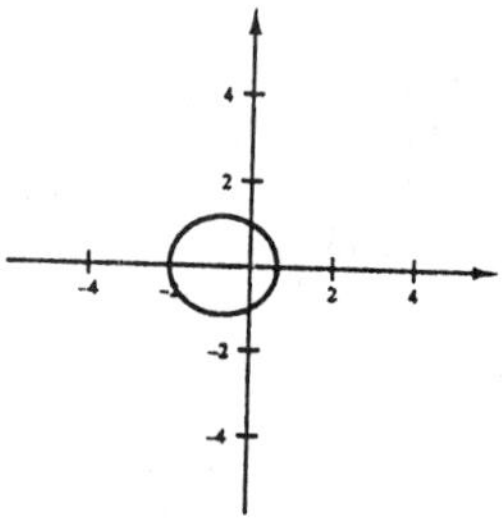

16. $\left(7, -7\sqrt{3}\right)$

17. $r^2 = \dfrac{36}{9 - 13\sin^2\theta}$

18. $y^2 = 4 - 4x$

19. $y + 1 = (x + 3)^2$

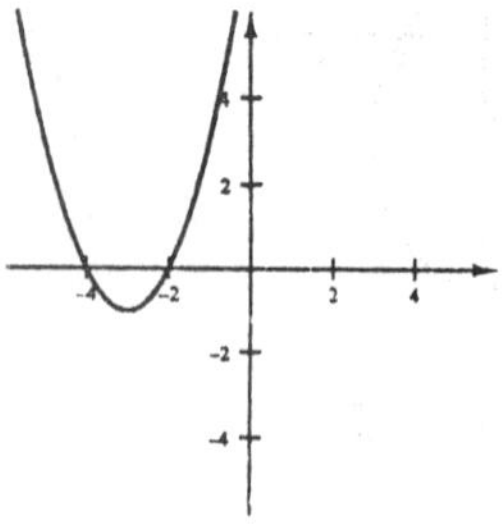

20. $(x - 1)^2 + y^2 = 1$

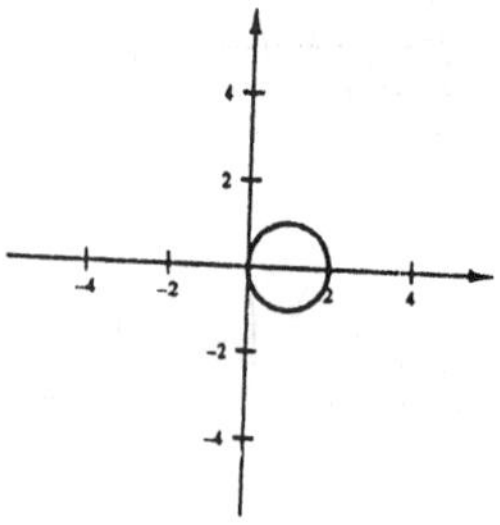

Chapter 6 Form C

1. Vertex: (0, 0), Focus: $\left(\frac{1}{6}, 0\right)$, Directrix: $x = -\frac{1}{6}$

Trigonometry Test Answers

2. Vertex: (4, 0), Focus: $\left(4, -\frac{1}{2}\right)$, Directrix: $y = \frac{1}{2}$

3. Vertices: (−1, −2), (5, −2), Foci: $\left(2 + \sqrt{5}, -2\right)$, $\left(2 - \sqrt{5}, -2\right)$

4. Vertices: (−2, 8), (−2, −2), Foci: $\left(-2, 3 + \sqrt{29}\right)$, $\left(-2, 3 - \sqrt{29}\right)$,
Asymptotes: $y - 3 = \pm\frac{5}{2}(x + 2)$

5. $(x - 2)^2 = -4(y - 1)$

6. $\frac{(x - 2)^2}{20} + \frac{y^2}{4} = 1$

7. $\frac{(y - 4)^2}{20} - \frac{(x + 5)^2}{16} = 1$

8. $\left(\frac{-4 - 3\sqrt{3}}{2}, \frac{3 - 4\sqrt{3}}{2}\right)$

9. 29.5°

10.

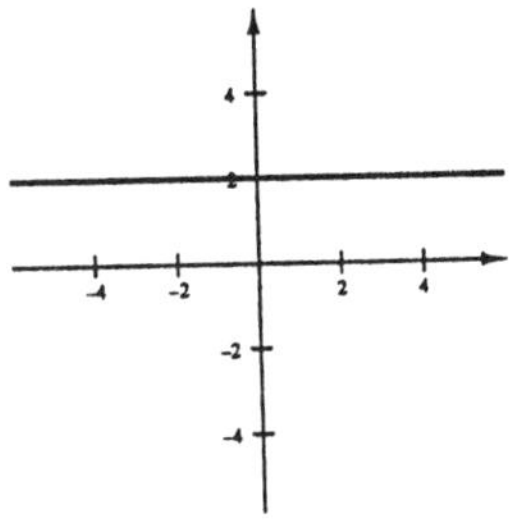

11.

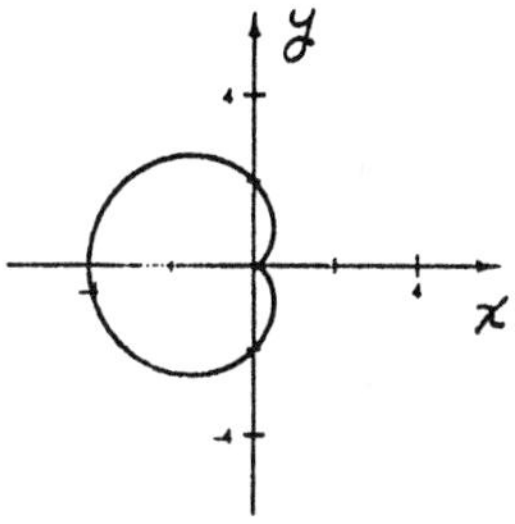

12.

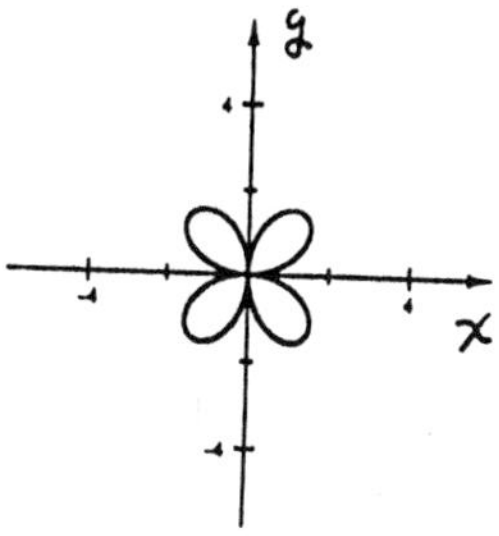

13.

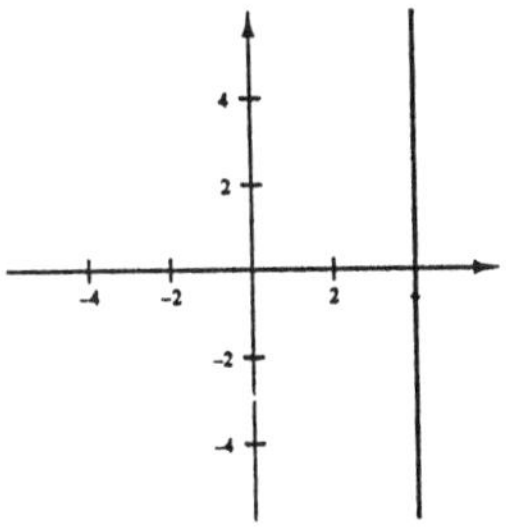

14.

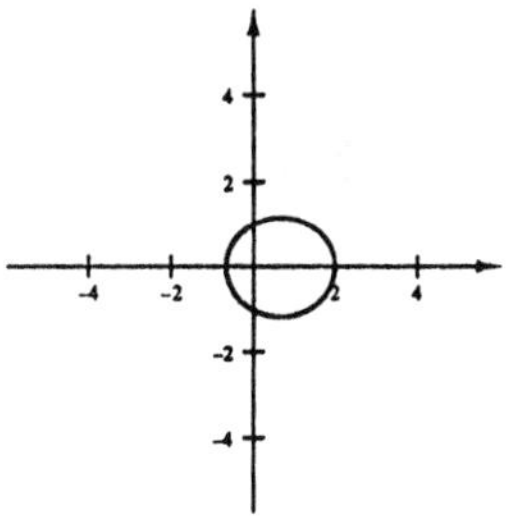

15.

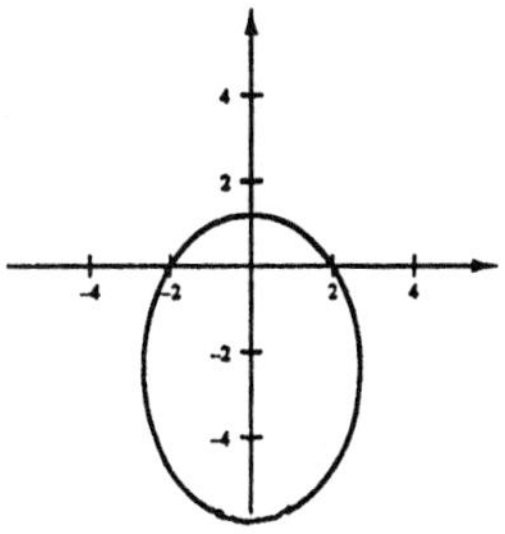

16. $\left(\sqrt{13}, 124°\right)$

17. $x^2 - 2x + y^2 = 0$

18. $r = \frac{4}{3 \cos \theta + 2 \sin \theta}$

19. $x = y^2 - 1$

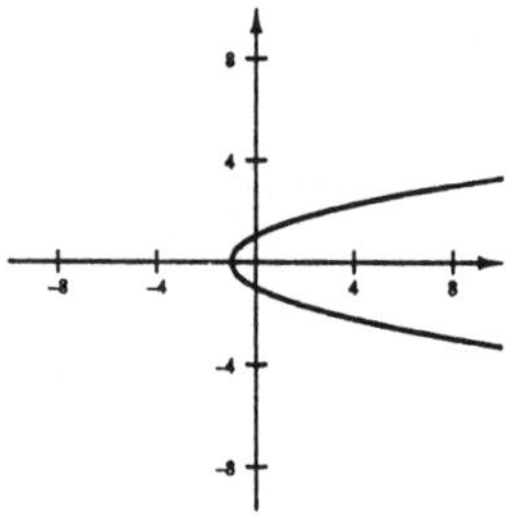

20. $\frac{x^2}{4} + \frac{y^2}{16} = 1$

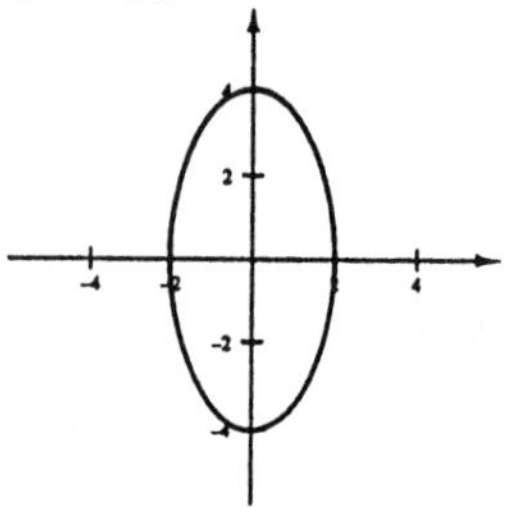

Chapter 6 Form D

1. b	2. d	3. d	4. a	5. b
6. c	7. c	8. a	9. b	10. a
11. b	12. d	13. b	14. c	15. a
16. c	17. d	18. c	19. c	20. d

Chapter 6 Form E

1. b	2. a	3. a	4. c	5. c
6. b	7. a	8. d	9. c	10. d
11. a	12. d	13. b	14. d	15. a
16. d	17. a	18. c	19. a	20. b

Chapter 6 Form F

1. b	2. d	3. a	4. d	5. a
6. c	7. a	8. b	9. c	10. d
11. b	12. c	13. a	14. c	15. b
16. b	17. b	18. a	19. d	20. b